THROMBIN
Structure and Function

THROMBIN
Structure and Function

Edited by
Lawrence J. Berliner
The Ohio State University
Columbus, Ohio

Plenum Press • New York and London

Library of Congress Cataloging-in-Publication Data

Thrombin : structure and function / edited by Lawrence J. Berliner.
       p.   cm.
   Includes bibliographical references and index.
   ISBN 0-306-43991-3
   1. Thrombin.   I. Berliner, Lawrence J.
QP93.5.T472   1992
612.1'15--dc20   92-23213
                                                                CIP

ISBN 0-306-43991-3

© 1992 Plenum Press, New York
A Division of Plenum Publishing Corporation
233 Spring Street, New York, N.Y. 10013

Printed in the United States of America

CONTRIBUTORS

Judy L. Aschner • Departments of Physiology and Pediatrics, The Albany Medical College, Albany, New York 12208

Rachel Bar-Shavit • Department of Oncology, Hadassah University Hospital, Jerusalem 91120, Israel

Miriam Benezra • Department of Oncology, Hadassah University Hospital, Jerusalem 91120, Israel

Lawrence J. Berliner • Department of Chemistry, The Ohio State University, Columbus, Ohio 43210-1173

Ingemar Björk • Department of Veterinary Medical Chemistry, Swedish University of Agricultural Sciences, Uppsala Biomedical Center, Uppsala, Sweden

Wolfram Bode • Max-Planck-Institut für Biochemie, D 8033 Martinsried, Germany

Darrell H. Carney • Department of Human Biological Chemistry and Genetics, University of Texas Medical Branch, Galveston, Texas 77550

Elisabetta Dejana • Instituto di Ricerche Farmacologiche, Milano, Italy

Joe G. N. Garcia • Departments of Medicine, Physiology, and Biophysics, Indiana University School of Medicine, Indianapolis, Indiana 46202

Kenneth D. Gibson • Baker Laboratory of Chemistry, Cornell University, Ithaca, New York 14853-1301

Nicholas J. Greco • Biomedical Research and Development, The Jerome H. Holland Laboratory, American Red Cross, Rockville, Maryland 20855

Robert Huber • Max-Planck-Institut für Biochemie, D 8033 Martinsried, Germany

Chih-Min Kam • School of Chemistry and Biochemistry, Georgia Institute of Technology, Atlanta, Georgia 30332

L. Lorand • Department of Biochemistry, Molecular Biology and Cell Biology, Northwestern University, Evanston, Illinois 60208

Asrar B. Malik • Departments of Physiology and Cell Biology, The Albany Medical College, Albany, New York 12208

John M. Maraganore • Biogen, Inc., Cambridge, Massachusetts 02142

Feng Ni • Biotechnology Research Institute, National Research Council of Canada, Montréal, Québec, Canada H4P 2R2

Steven T. Olson • Division of Biochemical Research, Henry Ford Hospital, Detroit, Michigan 48202

James C. Powers • School of Chemistry and Biochemistry, Georgia Institute of Technology, Atlanta, Georgia 30332

J. T. Radek • Department of Biochemistry, Molecular Biology and Cell Biology, Northwestern University, Evanston, Illinois 60208

Timothy J. Rydel • Department of Chemistry, Michigan State University, East Lansing, Michigan 48824

Valerie Sabbah • Department of Oncology, Hadassah University Hospital, Jerusalem 91120, Israel

Harold A. Scheraga • Baker Laboratory of Chemistry, Cornell University, Ithaca, New York 14853-1301

Stuart R. Stone • MRC Centre, University of Cambridge, Cambridge CBZ 2QH United Kingdom

Alexander Tulinsky • Department of Chemistry, Michigan State University, East Lansing, Michigan 48824

Israel Vlodavsky • Department of Oncology, Hadassah University Hospital, Jerusalem 91120, Israel

George D. Wilner • Department of Medicine, Division of Hematology, The Albany Medical College, Albany, New York 12208

PREFACE

Research on thrombin structure and function has progressed significantly over the past three decades. We are continually discovering new functions for this enzyme in biology. Yet, until quite recently, a full, detailed, three-dimensional picture of its structure was difficult to attain. We believe that this text represents a turning point and, more appropriately, a new starting point for thrombin studies. Our goal for this text is to present a thorough and rounded-out coverage of thrombin chemistry and biochemistry in order to provide the biochemist and physiologist with an excellent desk reference on almost any thrombin-related problem.

This volume is organized into three general thrombin topic areas: Structure, Biochemistry, and Physiology. In Part 1, Structure, we open with the complete three-dimensional x-ray structures of two inhibited human thrombin complexes, one of which is the thrombin–hirudin complex. These complexes are also addressed in the chapter on structural studies in solution, which include NMR, ESR, and fluorescence. Part 2, Biochemistry, includes chapters on synthetic thrombin inhibitors, protein inhibitors (e.g., antithrombin III, hirudin), and thrombin interactions with factor XIII. Part 3, Physiology, covers such topics as chemotactic activities, interactions with cell surfaces, and the vascular endothelium.

Throughout the volume, we recognize the efforts of several scientists who have helped unravel the secrets of this enzyme over the years: Walter Seegers, Ken Brinkhous, Kent Miller, Staffan Magnusson, Earl Davies, Ken Mann, Roger Lundblad, Shosuke Okamoto, Sadaaki Iwanaga, and John Fenton. Were it not for these scientists' laboratories' development of procedures for the isolation of highly pure bovine and human thrombins and other coagulation factors, many of the research results described here would have taken much longer to achieve.

We hope that the value of this text remains for several years, following the lead of earlier contributions by some of the authors acknowledged above. As editor, I am open to your comments, suggestions, and criticisms on the content and quality of this book.

Lawrence J. Berliner

Columbus, Ohio

CONTENTS

1. STRUCTURE

2. BIOCHEMISTRY

3. PHYSIOLOGY

Part 1

STRUCTURE

Chapter 1

X-RAY CRYSTAL STRUCTURES OF HUMAN α-THROMBIN AND OF THE HUMAN THROMBIN–HIRUDIN COMPLEX

Wolfram Bode, Robert Huber, Timothy J. Rydel,
and Alexander Tulinsky

1. THROMBIN STRUCTURE AND FUNCTION

α-Thrombin is a glycosylated trypsinlike serine proteinase (Magnusson, 1971), generated in the penultimate step of the blood coagulation cascade from the circulating plasma protein prothrombin. Upon autocatalytic and factor Xa cleavage, the functional two-chain molecule α-thrombin is generated. In the case of the human species this consists of the 36-residue A chain and the 259-residue B chain (Butkowski *et al.*, 1977; Thompson *et al.*, 1977; Walz *et al.*, 1977; Degen *et al.*, 1983). The two chains are covalently

Wolfram Bode and Robert Huber • Max-Planck-Institut für Biochemie, D 8033 Martinsried, Germany. **Timothy J. Rydel and Alexander Tulinsky** • Department of Chemistry, Michigan State University, East Lansing, Michigan 48824.

Thrombin: Structure and Function, edited by Lawrence J. Berliner. Plenum Press, New York, 1992.

connected by a disulfide bridge. The B chain carries an asparagine-linked sugar chain (Fenton *et al.*, 1977a; Butkowski *et al.*, 1977); it has also been shown to be homologous to the catalytic domains of other pancreatic and coagulation/fibrinolytic trypsinlike proteinases (Jackson and Nemerson, 1980). Upon further autolytic or proteolytic cleavage, more species (in particular β- and γ-thrombin) are generated which retain some activity against small synthetic substrates, but have lost most or all clotting activity (Lundblad *et al.*, 1979; Fenton, 1981, 1986; Berliner, 1984; Elion *et al.*, 1986; Hofsteenge *et al.*, 1988).

Thrombin plays a central role in thrombosis and hemostasis (Fenton, 1981) and is also implicated in wound healing and various disease processes (Fenton, 1988a). α-Thrombin converts fibrinogen into fibrin, which consequently aggregates and forms the interconnecting network of thrombi. Thrombin is furthermore able to activate several coagulation and plasma factors, such as factors V, VIII, XIII, protein S, and protein C. The thrombomodulin complex of thrombin exhibits enhanced reactivity toward protein C (Esmon *et al.*, 1982) but has lost all other procoagulant activity (Esmon *et al.*, 1983; Hofsteenge and Stone, 1987). Thrombin is effectively inhibited by only a very few endogenous protein inhibitors, such as α_2-macroglobulin and the serpins antithrombin III (Rosenberg and Damus, 1973), heparin cofactor II (Tollefsen *et al.*, 1982), and protease nexin I (Cunningham and Farrell, 1986; Gronke *et al.*, 1987). The interaction with these serpins is considerably enhanced by the acidic glycosaminoglycan heparin (Rosenberg, 1977; Björk and Lindahl, 1982; Li *et al.*, 1976; Olson and Shore, 1982; Wallace *et al.*, 1989). Thrombin binds very tightly to and is selectively inhibited by hirudin, a protein isolated from the medicinal leech (Walsmann and Markwardt, 1981).

Thrombin is also a potent cell activator exhibiting marked effects on a variety of cells (Jackson and Nemerson, 1980). Some of these cellular interactions require proteolytically active forms of thrombin, whereas others are nonenzymatic receptor-mediated processes that also function with active-site-blocked thrombin species (Shuman, 1986; Fenton, 1986); in some cellular processes, receptor binding and enzymatic activity of thrombin are coupled (McGowan and Detwiler, 1986; Carney *et al.*, 1986). Thrombin induces platelet aggregation and stimulates platelet secretion. It causes mitogenesis in fibroblasts (Chen and Buchanan, 1975; Tollefsen *et al.*, 1974; Cunningham and Farrell, 1986) and macrophagelike cells (Bar-Shavit *et al.*, 1986) and exhibits chemotactic properties (Bar-Shavit *et al.*, 1983, 1984; Bizios *et al.*, 1986).

As with the other related serine proteinases, thrombin possesses a catalytic triad of a serine, a histidine, and an aspartic acid residue. Thrombin cleaves a variety of proteins for which the amino acid sequence around the scissile peptide bond differs considerably (Blombäck *et al.*, 1977; Scheraga, 1977). Thrombin exhibits primarily a trypsinlike specificity, i.e., cleaves behind arginine and lysine residues, with a clear preference for Arg–X bonds (Lottenberg *et al.*, 1983). It is, however, much more selective than trypsin toward macromolecular substrates in that it cleaves many fewer peptide bonds (Blombäck *et al.*, 1967; Hogg and Blombäck, 1978). The P_2 position (Schechter and Berger, 1967; P_1, P_2, P_3 and P'_1, P'_2 designate substrate/inhibitor residues amino- and carboxy-terminal to the scissile peptide bond, respectively, S_1, S_2, S_3 and S'_1, S'_2 the corresponding subsites of the cognate proteinases) of typical protein substrates (Blombäck *et al.*, 1977) is frequently occupied by a proline residue, and thrombin is particularly active toward peptidic substrates and inhibitors with a proline at this position (Kettner and Shaw, 1981); a D-phenylalanine residue at P_3 makes such peptides even more reactive toward thrombin. Rather reactive and selective chloromethyl and aldehyde inhibitors have been developed with the peptidyl moiety D-PheProArg (Bajusz *et al.*, 1978; Kettner and Shaw, 1979).

In contrast to most other serine proteinases, the thrombin specificity is not determined by subsites surrounding the active residues alone. For efficient cleavage of fibrinogen (as well as of isolated Aα chains) the availability and integrity of a thrombin exosite quite distant from the catalytic residues is important (Liem and Scheraga, 1974; Hageman and Scheraga, 1974; van Nispen *et al.*, 1977; Blombäck *et al.*, 1977; Nagy *et al.*, 1982; Marsh *et al.*, 1982; Berliner, 1984). This "fibrin(ogen)-recognizing exosite" has also been shown to be involved in binding of other proteins such as fibrin monomers, fibrin E domain and its NDSK fragment, thrombomodulin, and hirudin (see below), and of negatively charged cell surfaces, glass, and cation-exchange resins (Liu *et al.*, 1979; Fenton *et al.*, 1981; Berliner *et al.*, 1985; Kaminski and McDonagh, 1987; Fenton *et al.*, 1988; Vali and Scheraga, 1988; Kaczmarek and McDonagh, 1988).

The existence of an "apolar binding site" located close to the catalytic center has been inferred from binding and proflavine displacement studies (Thompson, 1976; Berliner and Shen, 1977). This site has been suggested as accounting for thrombin specificity with tripeptide substrates (Sonder and Fenton, 1984) and for accommodation of large hydrophobic residues amino-terminal of the fibrinogen Aα-cleavage site (Meinwald *et al.*, 1980; Marsh *et al.*, 1983; Ni *et al.*, 1989a).

Several dysfunctional genetic variants of human (pro)thrombin have been found; in three of them the site of mutation has been localized (Miyata *et al.*, 1987; Henriksen and Mann, 1988, 1989).

Three-dimensional models have been proposed for the thrombin B chain based on sequence homology with bovine chymotrypsin and trypsin

(Magnusson et al., 1975; Bing et al., 1981, 1986; Greer, 1981a,b; Furie et al., 1982; Sugawara et al., 1986; Fenton and Bing, 1986; Toma and Suzuki, 1989). These models have provided a general impression of the arrangement of sites involved in the various interactions of thrombin. An adequate understanding of the specific and characteristic functions of thrombin, however, requires knowledge of an experimentally determined structure. Single crystals of bovine (Tsernoglou et al., 1974) and human (McKay et al., 1977) thrombin had been described but did not lead to a structure solution in the latter case.

We have been able to obtain quite good crystals of human α-thrombin inhibited by D-PheProArg chloromethylketone (PPACK; Bode et al., 1989a; Skrzypczak-Jankun et al., 1989). These crystals allowed the elucidation of the high-resolution structure of human α-thrombin and its interaction with PPACK (Bode et al., 1989a). Using the refined coordinates of this PPACK-thrombin structure as a search and initial phasing model, the x-ray crystal structures of two slightly different human α-thrombin–hirudin complexes (Rydel et al., 1990; Grütter et al., 1990), of two different recombinant synthetic peptide (hirugen and hirulog 1) α-thrombin complexes (Skrzypczak-Jankun et al., 1991), and of the homologous noninhibited bovine thrombin (Martin, Kunjummen, Kumar, Bode, Huber, and Edwards, in preparation) have been elucidated. In addition, the crystal structures of bovine trypsin complexed with the powerful arginine, benzamidine, and isocoumarin based thrombin inhibitors have been determined (Matsuzaki et al., 1989; Bode et al., 1990; Chow et al., 1990); based on the experimental PPACK structure, their probable complexes with thrombin have been modeled, providing plausible explanations for the selective and tight binding of these inhibitors toward thrombin (Bode et al., 1990).

2. CRYSTAL STRUCTURE ANALYSIS OF PPACK-HUMAN α-THROMBIN

PPACK-inhibited human α-thrombin (provided by Drs. Jan Hofsteenge and Stuart R. Stone) was crystallized from 16% PEG 6000/0.2 M phosphate buffer pH 7.0 (microdialysis) or from about 20% PEG 6000/0.5 M NaCl/0.2 M phosphate buffer pH 6–7 by vapor diffusion (Bode et al., 1989a). The crystals are of orthorhombic space group $P2_12_12_1$ with cell constants $a = 87.74$ Å, $b = 67.81$ Å, $c = 61.07$ Å. They contain one molecule per asymmetric unit; larger crystals show reflections beyond 1.9 Å resolution. Reflection data were collected with the FAST television area detector and with film cameras (Bode et al., 1989a, 1992). Orientation and position of a starting model derived by homologous modeling were determined by

Patterson search methods. This model was completed and its structure crystallographically refined by a cyclic procedure involving inspection of electron density maps, interactive graphics-aided modeling, and intermittent cycles of least-squares crystallographic refinement with energy restraints.

The PPACK-thrombin model described by Bode *et al.* (1989a) had an R value of 0.171 for 14,466 reflections between 6.0 and 1.9 Å. It comprised 290 amino acid residues; it lacked four carboxy-terminal residues of the A chain and the carboxy-terminal Glu-247 of the B chain, while the first five residues of the A chain were only placed tentatively.

Meanwhile, this model has been slightly improved by further remodeling, localization of some hitherto undefined side chains and insertion of additional water molecules (Bode *et al.*, 1992). The essential properties of the previous model were retained. In the final model, both chain segments and all side chains are defined unequivocally; 424 solvent molecules are located; there is now only one peptide bond (before Pro-37) with a *cis* conformation. The R value of the final model is 0.156 for 8.0 to 1.92 Å data (Bode *et al.*, 1992).

2.1. Overall Structure of Human α-Thrombin

The thrombin molecule can be described as a prolate ellipsoid of approximate dimensions $45 \times 45 \times 50$ Å. Its A and B chains are not organized in separate domains, but form a single contiguous body with a highly furrowed surface. Similar to the other trypsinlike serine proteinases (Bode and Huber, 1986), thrombin consists essentially of two interacting six-stranded barrel-like domains, linked by four transdomain "straps," five helical segments, one helical turn, and various surface-located turn structures (Fig. 1, Table I). Three of these "straps" and three of these helices are also present in other known serine proteinase structures; the A chain has a counterpart in chymotrypsin(ogen) (and probably in some other serine proteinases). The catalytic residues, in particular Ser-195, His-57, and Asp-102 [using the "chymotrypsinogen" nomenclature introduced by Bode *et al.* (1989a); see Table II], are located at the junction between both barrels; a quite prominent active-site cleft stretches perpendicular to this junction (see below).

The secondary structure of the polypeptide strands of the A and B chains of thrombin are illustrated schematically in Fig. 2. Each barrel-like domain appears as six antiparallel running strands. The size and extent of both β-sheets formed are remarkably similar to those observed in other trypsinlike serine proteinases. The six helical segments occurring in thrombin are listed in Table I together with their helix type and length.

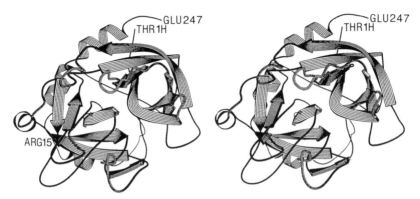

Figure 1. Ribbon plot of the human α-thrombin polypeptide chain; β-strands, helices, and turns are represented by twisted arrows, helical ribbons, and ropes. The view is on the active-site cleft; the A chain runs in the back of the molecule. This figure (from Bode *et al.*, 1992) has been produced with a modified version of a program kindly provided by John Priestle (Priestle, 1988).

The one-turn helix around His-57, the "intermediate helix" around Cys-168, and the long carboxy-terminal helix (of a mixed type) are similar to those found in other serine proteinases; the A-chain helix, the short 3_{10} helix between Tyr-60A and Asp-60E (better described as two interlaced type III turns), and the regular two-turn α-helix around Ala-129 are additional segments or parts of "insertion loops" and seem to be characteristic to thrombin. Many of the β-hairpin loops of thrombin (Fig. 2) are organized as open and/or classical tight 1–4 turns.

Thrombin possesses four disulfide bridges. The first, Cys-1–122, to-

Table I. Helical Segments in Human α-Thrombin

Helix type	Helix extent (number of residues) according to Richardson and Richardson (1990)		
	Hydrogen bonding	Conformation	α-Carbon position
3.6_{13}	Glu-14C to Ser-14I (7)	Glu-14C to Ser-14I (7)	Thr-14B to Tyr-14J (9)
3_{10}	Ala-55 to Leu-58 (5)	Ala-56 to Leu-59 (5)	Ala-56 to Tyr-60A (6)
3_{10}	Tyr-60A to Asp-60E (5)	Pro-60B to Pro-60C (2)	Tyr-60A to Trp-60D (4)
3.6_{13}	Asp-125 to Leu-129C (8)	Arg-126 to Ser-129B (6)	Asp-125 to Glu-131 (10)
3.6_{13}	Glu-164 to Asp-170 (7)	Arg-165 to Lys-169 (5)	Glu-164 to Ser-171 (8) ⎱
3_{10}	Cys-168 to Thr-172 (5)	Lys-169 to Ser-171 (3)	Lys-169 to Thr-172 (4) ⎰
3_{10}	Val-231 to Lys-235 (5)	Val-231 to Trp-237 (7)	His-230 to Leu-234 (5) ⎫
3.6_{13}	Leu-234 to Asp-243 (10)	Ile-238 to Ile-242 (5)	Leu-234 to Ile-242 (9) ⎬
3_{10}	Val-241 to Glu-244 (4)	Val-241 to Glu-244 (4)	Val-241 to Glu-244 (4) ⎭

Table II. Sequence Alignment of Bovine Thrombin (BTHR), Human Thrombin (HTHR), Bovine Trypsin (BTRY), and Bovine Chymotrypsin (BCHY) According to Topological Equivalences; Thrombin "Chymotrypsinogen" Numbering (THRO), Chymotrypsinogen Numbering (CHYG)

THRO	1H	1G	1F	1E	1D	1C	1B	1A	1	2	3	4	5	6	7
BTHR	T	F	G	A	G	E	A	D	C	G	L	R	P	L	F
HTHR	T	F	G	S	G	E	A	D	C	G	L	R	P	L	F
BTRY															
BCHY									C	G	V	P	A	I	Q
CHYG									1	2	3	4	5	6	7

THRO	8	9	10	11	12	13	14	14A	14B	14C	14D	14E	14F	14G	14H
BTHR	E	K	K	Q	V	Q	D	Q	T	E	K	E	L	F	E
HTHR	E	K	K	S	L	E	D	K	T	E	R	E	L	L	E
BTRY															
BCHY	P	V	L	S	G	L	S	—	—	—	—	—	—	—	—
CHYG	8	9	10	11	12	13	14	—	—	—	—	—	—	—	—

THRO	14I	14J	14K	14L	14M	15
BTHR	S	Y	I	E	G	R
HTHR	S	Y	I	D	G	R
BTRY	V	D	D	D	D	K
BCHY	—	—	—	—	—	R
CHYG	—	—	—	—	—	15

THRO	16	17	18	19	20	21	22	23	24	25	26	27	28	29	30
BTHR	I	V	E	G	Q	D	A	E	V	G	L	S	P	W	Q
HTHR	I	V	E	G	S	D	A	E	I	G	M	S	P	W	Q
BTRY	I	V	G	G	Y	T	C	G	A	N	T	V	P	Y	Q
BCHY	I	V	N	G	E	E	A	V	P	G	S	W	P	W	Q
CHYG	16	17	18	19	20	21	22	23	24	25	26	27	28	29	30

THRO	31	32	33	34	35	36	36A	37	38	39	40	41	42	43	44
BTHR	V	M	L	F	R	K	S	CP	Q	E	L	L	C	G	A
HTHR	V	M	L	F	R	K	S	CP	Q	E	L	L	C	G	A
BTRY	V	S	L	N	S	—	—	—	G	Y	H	F	C	G	G
BCHY	V	S	L	Q	D	K	—	T	G	F	H	F	C	G	G
CHYG	31	32	33	34	35	36	—	37	38	39	40	41	42	43	44

THRO	45	46	47	48	49	50	51	52	53	54	55	56	57	58	59
BTHR	S	L	I	S	D	R	W	V	L	T	A	A	H	C	L
HTHR	S	L	I	S	D	R	W	V	L	T	A	A	H	C	L
BTRY	S	L	I	N	S	Q	W	V	V	S	A	A	H	C	Y
BCHY	S	L	I	N	E	N	W	V	V	T	A	A	H	C	G
CHYG	45	46	47	48	49	50	51	52	53	54	55	56	57	58	59

THRO	60	60A	60B	60C	60D	60E	60F	60G	60H	60I	61	62	63	64	65
BTHR	L	Y	P	P	W	D	K	N	H	I	V	D	D	L	L
HTHR	L	Y	P	P	W	D	K	N	H	I	E	N	D	L	L
BTRY	K	—	—	—	—	—	—	—	—	—	—	S	G	I	Q
BCHY	V	—	—	—	—	—	—	—	—	—	T	T	S	D	V
CHYG	60	—	—	—	—	—	—	—	—	—	61	62	63	64	65

(continued)

Table II (continued)

THRO	66	67	68	69	70	71	72	73	74	75	76	77	77A	78	79
BTHR	V	R	I	G	K	H	S	R	T	R	Y	E	R	K	V
HTHR	V	R	I	G	K	H	S	R	T	R	Y	E	R	N	I
BTRY	V	R	L	G	E	D	N	I	N	V	V	E	—	G	N
BCHY	V	V	A	G	E	F	D	Q	G	S	S	S	—	E	K
CHYG	66	67	68	69	70	71	72	73	74	75	76	77	77A	78	79

THRO	80	81	82	83	84	85	86	87	88	89	90	91	92	93	94
BTHR	E	K	I	S	M	L	D	K	I	Y	I	H	P	R	Y
HTHR	E	K	I	S	M	L	E	K	I	Y	I	H	P	R	Y
BTRY	E	Q	F	I	S	A	S	K	S	I	V	H	P	S	Y
BCHY	I	Q	K	L	K	I	A	K	V	F	K	N	S	K	Y
CHYG	80	81	82	83	84	85	86	87	88	89	90	91	92	93	94

THRO	95	96	97	97A	98	99	100	101	102	103	104	105	106	107	108
BTHR	N	W	K	E	N	L	D	R	D	I	A	L	L	K	L
HTHR	N	W	R	E	N	L	D	R	D	I	A	L	M	K	L
BTRY	N	S	N	—	T	L	N	N	D	I	M	L	I	K	L
BCHY	N	S	L	—	T	I	N	N	D	I	T	L	L	K	L
CHYG	95	96	97	—	98	99	100	101	102	103	104	105	106	107	108

THRO	109	110	111	112	113	114	115	116	117	118	119	120	121	122	123
BTHR	K	R	P	I	E	L	S	D	Y	I	H	P	V	C	L
HTHR	K	K	P	V	A	F	S	D	Y	I	H	P	V	C	L
BTRY	K	S	A	A	S	L	N	S	R	V	A	S	I	S	L
BCHY	S	T	A	A	S	F	S	Q	T	V	S	A	V	C	L
CHYG	109	110	111	112	113	114	115	116	117	118	119	120	121	122	123

THRO	124	125	126	127	128	129	129A	129B	129C	130	131	132	133	134	135
BTHR	P	D	K	Q	T	A	A	K	L	L	H	A	G	F	K
HTHR	P	D	R	E	T	A	A	S	L	L	Q	A	G	Y	K
BTRY	P	T	—	S	C	A	—	—	—	—	S	A	G	T	Q
BCHY	P	S	A	S	D	D	—	—	—	F	A	A	G	T	T
CHYG	124	125	126	127	128	129	—	—	—	130	131	132	133	134	135

THRO	136	137	138	139	140	141	142	143	144	145	146	147	148	149	149A
BTHR	G	R	V	T	G	W	G	N	R	R	E	T	W	T	T
HTHR	G	R	V	T	G	W	G	N	L	K	E	T	W	T	A
BTRY	C	L	I	S	G	W	G	N	T	K	S	S	G	T	—
BCHY	C	V	T		G	W	G	L	T	R	Y	T	N	A	—
CHYG	136	137	138	139	140	141	142	143	144	145	146	147	148	149	—

THRO	149B	149C	149D	149E	150	151	152	153	154	155	156	157	158	159	160
BTHR	S	V	A	E	V	Q	P	S	V	L	Q	V	V	N	L
HTHR	N	V	G	K	G	Q	P	S	V	L	Q	V	V	N	L
BTRY	—	—	—	—	S	Y	P	D	V	L	K	C	L	K	A
BCHY	—	—	—	—	N	T	P	D	R	L	Q	Q	A	S	L
CHYG	—	—	—	—	150	151	152	153	154	155	156	157	158	159	160

THRO	161	162	163	164	165	166	167	168	169	170	171	172	173	174	175
BTHR	P	L	V	E	R	P	V	C	K	A	S	T	R	I	R
HTHR	P	I	V	E	R	P	V	C	K	D	S	T	R	I	R
BTRY	P	I	L	S	D	S	S	C	K	S	A	Y	P	G	Q
BCHY	P	L	L	S	N	T	N	C	K	K	Y	W	G	T	K
CHYG	161	162	163	164	165	166	167	168	169	170	171	172	173	174	175

THRO	176	177	178	179	180	181	182	183	184	184A	185	186	186A	186B	186C
BTHR	I	T	D	N	M	F	C	A	G	Y	K	P	G	E	G
HTHR	I	T	D	N	M	F	C	A	G	Y	K	P	D	E	G
BTRY	I	T	S	N	M	F	C	A	G	Y	L	E	—	—	—
BCHY	I	K	D	A	M	I	C	A	G	—	A	S	—	—	—
CHYG	176	177	178	179	180	181	182	183	184	—	185	186	—	—	—

THRO	186D	187	188	189	190	191	192	193	194	195	196	197	198	199	200
BTHR	K	R	G	D	A	C	E	G	D	S	G	G	P	F	V
HTHR	K	R	G	D	A	C	E	G	D	S	G	G	P	F	V
BTRY	G	G	K	D	S	C	Q	G	D	S	G	G	P	V	V
BCHY	—	G	V	S	S	C	M	G	D	S	G	G	P	L	V
CHYG	—	187	188	189	190	191	192	193	194	195	196	197	198	199	200

THRO	201	202	203	204	204A	204B	205	206	207	208	209	210	211	212	213
BTHR	M	K	S	P	Y	N	N	R	W	Y	Q	M	G	I	V
HTHR	M	K	S	P	F	N	N	R	W	Y	Q	M	G	I	V
BTRY	C	S	—	—	—	—	—	—	G	K	L	Q	G	I	V
BCHY	C	K	K	N	—	—	G	A	W	T	L	V	G	I	V
CHYG	201	202	203	204	—	—	205	206	207	208	209	210	211	212	213

THRO	214	215	216	217	—	219	220	221	221A	222	223	224	225	226	227
BTHR	S	W	G	E	—	G	C	D	R	D	G	K	Y	G	F
HTHR	S	W	G	E	—	G	C	D	R	D	G	K	Y	G	F
BTRY	S	W	G	S	—	G	C	A	Q	K	N	K	P	G	V
BCHY	S	W	G	S	S	T	C	S	—	T	S	T	P	G	V
CHYG	214	215	216	217	218	219	220	221	—	222	223	224	225	226	227

THRO	228	229	230	231	232	233	234	235	236	237	238	239	240	241	242
BTHR	Y	T	H	V	F	R	L	K	K	W	I	Q	K	V	I
HTHR	Y	T	H	V	F	R	L	K	K	W	I	Q	K	V	I
BTRY	Y	T	K	V	C	N	Y	V	S	W	I	K	Q	T	I
BCHY	Y	A	R	V	T	A	L	V	N	W	V	Q	Q	T	L
CHYG	228	229	230	231	232	233	234	235	236	237	238	239	240	241	242

THRO	243	244	245	246	247
BTHR	D	R	L	G	S
HTHR	D	Q	F	G	E
BTRY	A	S	N	—	—
BCHY	A	A	N	—	—
CHYG	243	244	245	246	247

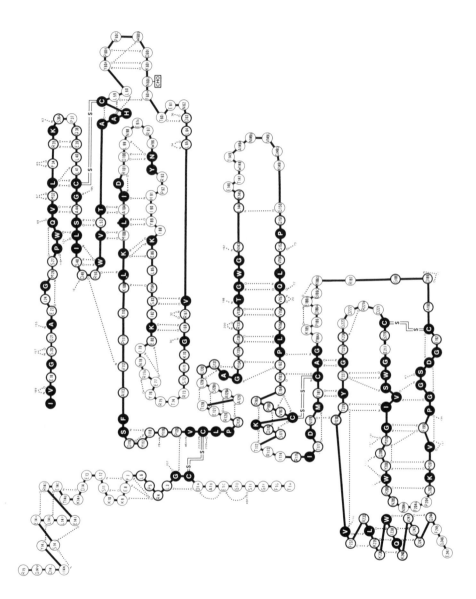

pologically equivalent to that in chymotrypsin(ogen), links the A with the B chain covalently. The other three disulfide bridges, Cys-42–58, Cys-168–182, and Cys-191–220, are also found in most other serine proteinases. The 424 solvent molecules localized in the PPACK-thrombin structure account for almost 30% of all solvent (water) molecules contained in the crystals. Thirty-four of these "water" molecules are located within the protein domain and are in direct contact with at most one bulk water molecule; they are considered to be an integral constituent of the thrombin structure; 16 of them are equivalent to waters found in bovine trypsin (Bode *et al.*, 1992).

2.2. Comparison with Related Serine Proteinase Structures

The α-carbon structure of human α-thrombin optimally superimposed with that of bovine chymotrypsinogen (molecule 1, Wang *et al.*, 1985) obtained by minimizing the r.m.s. deviation of topologically equivalent α-carbon atoms (Rossmann and Argos, 1975) is shown in Fig. 3. One hundred ninety amino acid residues of the thrombin B chain are topologically equivalent (within an r.m.s. deviation of 0.78 Å) with residues in bovine chymotrypsin (Cohen *et al.*, 1981; Tsukuda and Blow, 1985; Blevins and Tulinsky, 1985); in addition, 6 residues of the thrombin A chain are equivalent to the 6 amino-terminal residues of the chymotrypsin propeptide. The 190 equivalent B-chain residues and another 40 residues were designated the sequence numbers of the topologically equivalent chymotrypsinogen residues (Hartley and Kauffman, 1966; Meloun *et al.*, 1966). In addition to secondary structure elements, Fig. 2 defines those human α-thrombin amino acid residues which are topologically equivalent to chymotrypsin residues (Cohen *et al.*, 1981; Tsukuda and Blow, 1985; Blevins and Tulinsky, 1985); solid circles further represent those 84 that possess identical amino acids.

In the thrombin B chain, 195 α-carbon atoms are equivalent to the bovine trypsin structure with an r.m.s. deviation of 0.8 Å (Bode and Schwager, 1975a); at a few sites the thrombin structure is much closer to that of trypsin than to that of chymotrypsin:

◄───

Figure 2. Secondary structure of human α-thrombin. Inter-main-chain hydrogen bonds are selected according to criteria ($E < -0.7$ kcal/mol) given by Kabsch and Sander (1983). Thrombin residues which are topologically equivalent to chymotrypsin residues are shown as thick circles; those which in addition possess identical amino acids are solid circles. (From Bode *et al.*, 1991.)

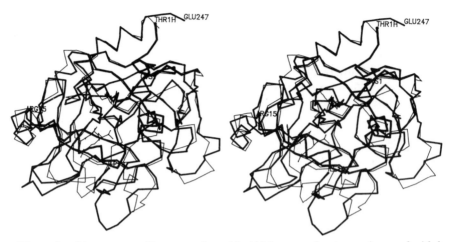

Figure 3. CA structure of human α-thrombin (thick connections) superimposed with bovine chymotrypsinogen (Wang *et al.*, 1985; thin connections). The PPACK molecule is inserted (thin connections), and the thrombin termini are labeled. The view is, as in Fig. 1, toward the active-site cleft (From Bode *et al.*, 1989a.)

1. Thrombin segment 70–80, which resembles the "calcium loop" found in trypsin (Bode and Schwager, 1975b)
2. Gly-184–Tyr-184A
3. Thrombin loop segment 217–224

The generally used chymotrypsinogen numbering of bovine trypsin (Hartley and Shotton, 1971) was used here for thrombin residues. The residual "insertion" residues in the thrombin B chain were marked by letter suffixes; at a few of the junctions between conserved and variable segments, the positioning of the insertions was necessarily somewhat arbitrary.

For A-chain numbering, residue numbers 1 and 15 were assigned to the (chymotrypsin-homologous) cysteine and to the carboxy-terminal arginine residue, respectively. The A-chain residues preceding Cys-1 were designated by a 1 and a letter suffix in alphabetic order in the reverse chain direction; the A-chain residues following Cys-1 were designated by numbers up to 14, followed by 14 and letter suffixes in alphabetic order.

The "chymotrypsinogen" sequence numbering of human thrombin is shown in Table II based on these topological equivalences with chymotrypsin and trypsin (Bode *et al.*, 1989a). The A chain runs from Thr-1H to Arg-15, the B chain starts with Ile-16 and ends up with Glu-247. A nomenclature based on the topological equivalence of the amino acid residues is an enormous advantage, as functional properties already known for the related trypsinlike enzymes can be attributed to distinct sites of the thrombin structure. For comparison purposes, Table II also contains the sequences of bovine chymotrypsinogen (Hartley and Kauffman, 1966; Meloun *et al.*, 1966), bovine trypsinogen (Walsh and Neurath, 1964; Mikes *et al.*, 1966), and bovine thrombin (Magnusson *et al.*, 1975; McGillivray and Davie, 1984).

2.3. The Role of Charged Residues in the Thrombin Structure

Thrombin possesses an exceptionally large number of charged residues compared with the pancreatic trypsinlike enzymes. The A and B chains contain 9 and 31 acidic residues and 6 and 37 basic residues (and 4 histidines), respectively. In addition, thrombin has two terminal amino, two terminal carboxy, and two neuraminic acid groups, the latter terminating the Asn-60G-linked sugar chain (Nilsson *et al.*, 1983). Of the 83 charged thrombin residues, 72 are (at least partially) located on the thrombin surface, representing 42% of all surface residues. Considering "intrinsic" pK values for the ionizable groups of α-thrombin, an overall isoelectric point of 7.8 can be calculated; allowing for the spatial arrangement of the charged groups and their environment, a higher value of 8.4 is obtained (Karshikoff *et al.*, 1989; Bode *et al.*, 1992). These calculated isoelectric points should be compared with the experimental values of 7.3 and 7.6 obtained by isoelectric focusing (Fenton *et al.*, 1977a); however, higher values above (Berg *et al.*, 1979) and below 9 have also been reported for human thrombin (Heuck *et al.*, 1985).

Most of the charged residues are relatively evenly distributed around the thrombin surface. Of the 89 (at pH 7) fully charged groups, 79 are at least partially exposed to bulk water; the remaining 10 are completely buried and inaccessible (Bode *et al.*, 1992). All of these buried charges and many of the surface-located charges are arranged in pairs and clusters of oppositely charged groups. Particularly strong electrostatic interactions are found within the A chain and between the A and B chain (see Fig. 6 and below); 7 salt bridges (2 within the A chain, 3 between A and B, and 2 within the B chain) are completely buried. Extended clusters comprising several hydrogen-bond-connected charged residues stretch around Asp-14 (where they confer stabilization to the A chain and tighten the A–B interaction; see below) and Lys-70 (where they stabilize the structure of loop 70–80 and thus maintain the integrity of the fibrinogen-binding exosite; also see below). Similar to the other activatable trypsinlike serine proteinases, thrombin possesses a buried salt bridge of large electrostatic interaction energy between the ammonium group of Ile-16 (the amino-terminal residue of the B chain) and the side chain carboxyl group of Asp-194; the integrity of this salt bridge has been shown to be important in maintaining the characteristic active enzyme-like conformation (see Bode, 1979).

At a few thrombin surface sites, the positively and negatively charged groups are unevenly distributed, clustering in positively and negatively charged patches. Such clusters give rise to quite high electrostatic field strengths (both positive and negative) outside the thrombin surface (Fig. 4). This charge clustering is the reason for the strong electrostatic interaction of thrombin with anionic structures even at physiological pH values, where its overall charge almost cancels out. Three such surface patches deserve particular attention. The strongest positive electrostatic field is observed for a groove (on "top" of the thrombin molecule in Fig. 4) surrounded by the exposed side chains of Arg-126, Lys-236, Lys-240, and Arg-93, and of some other residues in the more distant neighborhood (see below). It is noteworthy that the electron density for this groove gives no indication of any fixed large solvent anion.

A second surface region of high positive charge density extends along the convex thrombin surface between Lys-149E and Lys-110 (on the "right-hand side" of the active-site cleft, Fig. 5). This positively charged patch carries eight basic amino acid side chains (Lys-149E, Arg-73, Arg-75, Arg-77A, Lys-81, Lys-36, Lys-109, and Lys-110) whose charges are only partially compensated by neighboring carboxylates.

One larger surface patch comprising negatively charged residues extends around the active site of thrombin. It consists of seven aspartic and glutamic acid residues arranged within (Asp-189) and around the

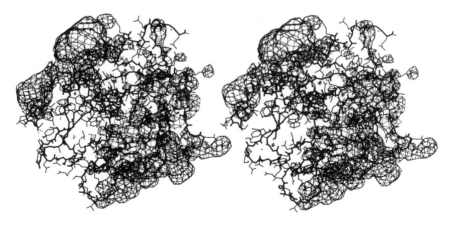

Figure 4. Human α-thrombin overlaid with the electrostatic equipotential surfaces at +3 kcal/mol. The view is, as in Figs. 1 and 2, toward the active-site cleft (Bode *et al.*, 1992).

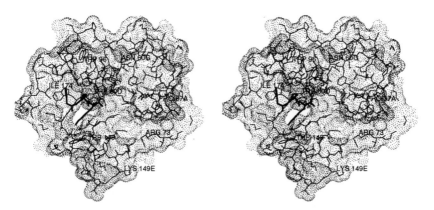

Figure 5. Active-site cleft of human α-thrombin, displayed together with a Connolly dot surface (calculated with a probe of radius 1.4 Å). The view is toward the active-site cleft; the PPACK molecule is emphasized by bold connections. Only parts of the thrombin molecule and of its molecular surface are displayed. A bound substrate polypeptide chain would run from left to right. The surface "hole" close to the center corresponds to the entrance to the specificity pocket. The 60-insertion loop, in particular Trp-60D, is partially obstructing the active site. The fibrinogen-binding exosite is to the right ("Arg-73") (From Bode *et al.*, 1992.)

specificity pocket: Glu-192 (without protein countercharge); Asp-102 (active-site residue, hydrogen bonded with His-57), Asp-194 (involved in a relatively strong internal salt bridge with the ammonium group of Ile-16); and Glu-217, Glu-146, and Glu-97A, each involved in salt bridges with positively charged residues in the neighborhood (Bode *et al.*, 1992).

2.4. Thrombin A Chain

The thrombin A chain is organized mainly in a multiple-turn (in part helical) conformation in a boomeranglike shape and nestles against the B chain opposite to the substrate binding cleft (see Fig. 1). It is topologically similar to the activation peptide of chymotrypsin(ogen) and connected in a similar manner via disulfide bridge Cys-1–Cys-122 to the thrombin B chain. Only three inter-main-chain hydrogen bonds exist between A and B (see Fig. 2). Most of the A–B interactions involve charged side chains; six of them (Fig. 6) represent (mainly) buried salt bridges, and in a further ten, a charged side chain group participates in an interchain hydrogen bond. Stabilization within the A chain also occurs (in particular in its central part; see Fig. 6) preferentially through polar and salt bridge interactions: nine charged side chains of the A chain make intrachain salt bridges; three of them contribute simultaneously to interchain salt bridges mentioned above (Fig. 6).

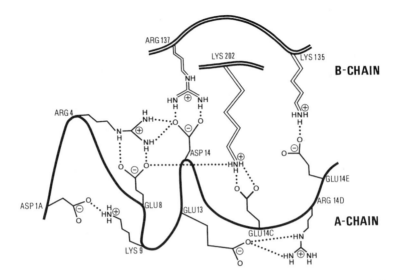

Figure 6. Schematic drawing of the ionic interactions within the thrombin A chain and between the thrombin A chain (bold connections) and B chain (double connections). The arrangement corresponds to a "back" view of the molecule, i.e., Fig. 5 after an approximate 180° rotation around a vertical axis (From Bode *et al.*, 1992.)

The amino-terminal segment of the A chain up to Glu-1C is only defined by relatively weak electron density in the PPACK-α-thrombin crystals. This segment runs along a ridge of the B chain with only a few van der Waals contacts between both chains (Bode *et al.*, 1992). The equivalent A-chain segment seems to be similarly arranged in bovine thrombin; however, in the human thrombin–hirudin crystals (Rydel *et al.*, 1990) and in crystals of hirugen and hirulog complexes of thrombin (Skrzypczak-Jankun *et al.*, 1991) it definitely occupies different positions. It therefore possesses a high degree of flexibility and can adapt differently to various environments.

The central more rigid part of the A chain (between Asp-1A and Tyr-14J) is in a shallow curved groove of the B chain lined by B-chain segments Ser-20–Met-26 and Lys-202–Trp-207. The A chain follows the course observed for the propeptide of chymotrypsin(ogen) up to Leu-6 (Fig. 3); the following residues form a longer, much more exposed loop. Most of the side chains of the charged residues (Asp-14, Arg-4, Glu-8, Glu-14C, Glu-14E) turn inwards and are engaged in an extended salt-bridge/hydrogen bond cluster (Fig. 6) formed with charged side chains extending from the B-chain globule (Arg-137, Lys-202, Lys-135). Most of

these salt bridges are more or less completely buried; electrostatic interactions therefore contribute significantly to the intrachain stability as well
as to the A–B interaction (Bode *et al.*, 1992). The A chain seems, however,
neither to be of great importance for B-chain folding nor for (thermodynamic) stability of the B-chain globule (Hageman *et al.*, 1975).

The carboxy-terminal segment between Thr-14B and Tyr-14J is organized in an amphiphilic α-helix of 1½ turns (Fig. 2 and Table I). The side
chains of Leu-14G and Tyr-14J are involved in hydrophobic contacts
between the A and B chain. The phenolic side chain of Tyr-14J exhibits
enhanced flexibility. The segment from Ile-14K to the carboxy-terminal
Arg-15 is only weakly defined by electron density (Bode *et al.*, 1992). It
deviates considerably from the locations observed in bovine thrombin as
well as in human thrombin–hirudin; it therefore appears to be inherently
flexible.

The corresponding surface of the thrombin B chain has a relatively
similar contour to that of chymotrypsin(ogen); in thrombin, however, most
of the interacting side chains are longer and more polar. As mentioned
above, the B chain alone seems to be thermodynamically stable and to
retain clotting activity (Hageman *et al.*, 1975). Thus, the A chain does not
appear to be a structural element determining specificity of thrombin; its
integration into the B chain might rather be necessary for correct activation cleavage and for other functions of prothrombin (Bode *et al.*, 1992).

2.5. Thrombin B Chain

The tertiary structure of the thrombin B chain is homologous to the
reactive domains of the other trypsinlike serine proteinases of known
spatial structure (Figs. 1–3). A remarkable feature of the thrombin B-chain
globule is, however, a series of more elongated and exposed loops, in
particular around the active-site cleft. Since the unique specificity of
thrombin for macromolecular ligands involves interactions at the surface
of the molecule, it is of interest to inspect those surface segments different
from chymotrypsin(ogen) (see also Fig. 3). The following segments of the
thrombin B chain exhibit new features (Bode *et al.*, 1989a):

1. Loop segment 34–42 (Table II) is similar to the equivalent loop in
   chymotrypsin, but is slightly larger (yet more compact) due to one
   inserted residue (*cis*-proline-37).
2. Due to an insertion of nine residues, segment 59–61 forms a
   completely new β-hairpin loop, which projects considerably into
   the substrate binding cleft (see below).
3. Loop segment 70–80, though quite similar to the "calcium loop"

of bovine trypsin (Bode and Schwager, 1975 a,b), is slightly larger, due to one inserted residue (Arg-77A) (see below).

4. Loop segment 95–100 is also larger due to a single insertion (of Glu-97A), and protrudes more out of the molecule.

5. Segment 125–129C (which contains an insertion of three residues) is organized in a two-turn α-helix (Table I).

6. Loop segment 145–150 is considerably larger due to the insertion of five residues and contains the remarkably exposed indole side chain of Trp-148 (see below).

7. Segment Arg-173–Ile-174 is part of an open, more exposed turn and seems to be important for the hydrophobic lining of the "aryl binding site" (Bode *et al.*, 1990).

8. Loop segment Tyr-184A–Gly-188 is more exposed due to the insertion of four residues (see below).

9. Loop segment Ser-203–Arg-206 (with two inserted residues) makes a larger 1–6 turn.

10. Loop segment Gly-216–Cys-220 is organized almost identically to trypsin, with one residue (218) missing.

11. Loop segment Cys-220–Tyr-225 is organized almost identically to that in trypsin, i.e., Asp-221 or Arg-221A are inserted.

12. The carboxy-terminus is longer by two residues (Gly-246, Glu-247).

All of these deviating segments are located at the surface; most form loops which project out of the molecular surface, but are, nevertheless, remarkably rigid (with the exception of loop segment 145–150; see below). Four of them contain new proline residues, which probably contribute to their rigidity. Three of these loops (in the immediate environment of the active-site cleft) contain tryptophan residues with highly exposed indole side chains (Trp-60D, Trp-96, Trp-148; Fig. 5). A few surface sites which could account for some of the characteristic functions of thrombin are discussed in more detail below.

The positions of some polypeptide segments of the α-thrombin B chain are emphasized in Fig. 7. Some of these cleavage sites (in particular those generated autocatalytically or through trypsin action) may only exist under laboratory conditions and play no important role *in vivo;* they are, nevertheless, extremely valuable assets in investigating the importance of distinct thrombin sites for various thrombin functions. None of the preferred proteolytic cleavage sites shown in Fig. 7 exhibit in PPACK-α-thrombin the "canonical" conformation found in substratelike-binding small protein inhibitors (Huber and Bode, 1978; Laskowski and Kato, 1980). Most of these sites are, however, flexible enough to allow conformational adaptation to the binding sites.

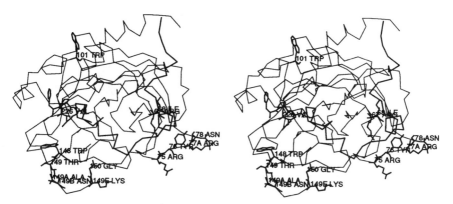

Figure 7. Well-known sites of cleavages and mutations displayed together with the CA tracing of the human α-thrombin. A chain (thick connections) and B chain (thin connections). The view is toward the thrombin active-site cleft with PPACK superimposed. Cleavages at Arg-77A–Asn-78 (and possibly at Arg-75–Tyr-76) and Arg-67–Ile-68 give rise to β-thrombin; cleavages at Trp-148–Thr-149, Ala-149–Thr-149B, Arg-77A, and Lys-149E–Gly-150 yield ξ-, ε-, and γ-thrombin, respectively. In the genetic thrombin variants Quick I, Quick II, and Tokushima, Arg-67, Gly-226, and Arg-101 are replaced by cysteine, valine, and tryptophan, respectively (From Bode *et al.*, 1992.)

2.6. Thrombin Sites of Specific Function

2.6.1. Insertion Loop Leu-59–Asn-62

The large insertion loop Leu-59–Asn-62 plays an important role in the selection of susceptible (macromolecular) substrates and inhibitors, i.e., in recognition and specificity of thrombin (see Fig. 5; Bode *et al.*, 1989a). This loop structure is shown in Fig. 8 together with a few surrounding groups. The whole loop (including all side chains) is well defined by electron density in PPACK-thrombin (Bode *et al.*, 1992), and its internal conformation is generally maintained in each thrombin crystal structure known so far. In particular, segment Leu-59–Phe-60H extends from the main molecular body in the form of a 3_{10} helical turn from Tyr-60A to Trp-60D, terminated by Glu-60E in left-handed helical conformation (Table I). The phenolic side chain of Tyr-60A points toward the S_2 subsite, covering it and packing tightly against Pro-2I of the bound PPACK molecule (see below; PPACK residues are designated by the suffix "I"). The indole moiety of Trp-60D is almost fully exposed to bulk solvent molecules. This part of the exposed loop seems to be particularly stabilized by the two intervening proline residues, which in addition pack against Trp-96 of an adjacent loop (see Fig. 8).

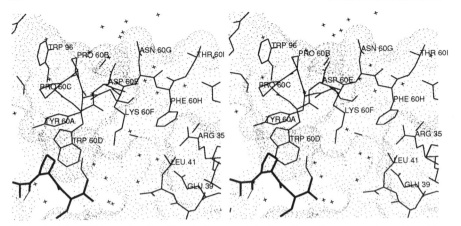

Figure 8. The "60-insertion loop" and the sugar-binding site of human α-thrombin (thin connections) displayed together with part of PPACK (thick connections) and the thrombin Connolly surface. The orientation is similar to Fig. 5. Loop segment Tyr-60A–Pro-60B–Pro-60C–Trp-60D–Asp-60E extends particularly out of the molecule; it probably forms part of the chemotactic domain of thrombin. Asn-60G is pointing away from the active-site cleft; the disordered sugar chain linked to it is not shown (from Bode *et al.*, 1992.)

    The well-defined asparagine carbohydrate-carrying side chain of the succeeding Asn-60G extends from the loop surface in a direction away from the active-site cleft (Fig. 8); in the PPACK-thrombin crystals, there is, however, no continuous density that would indicate the carbohydrate linked to it (Bode *et al.*, 1989a). The bound, presumably disordered, oligosaccharide [consisting of 12 sugar units in a branched chain (Nilsson *et al.*, 1983)] must extend in a direction away from the active-site cleft (Fig. 8). Thus, it should not interfere with macromolecular substrates that bind mainly along the cleft; this is in agreement with the lack of any effect on enzymatic (clotting and esterase) and cellular activities of thrombin on its removal or destruction (Horne and Gralnick, 1983). Concanavalin A, in binding to the α-D-mannose of this sugar chain, might on the other hand come into conflict with an approaching fibrinogen molecule (Skaug and Christenson, 1971; Karpatkin and Karpatkin, 1974; Hageman *et al.*, 1975).

    The quite compact and relatively rigid segment Tyr-60A–Trp-60D is part of a tetradecapeptide with the sequence Tyr-60A–Leu-65 for which extracellular matrix-binding and chemotactic/growth factor activities have been shown (Bar-Shavit *et al.*, 1984, 1986, 1989). It is conceivable that such activities might be attributed to this exposed Tyr-60A–Trp-60D loop of α-thrombin due to its rather unique ridgelike architecture. In thrombin–hirudin, this projecting thrombin loop is smoothed out by the hirudin Pro-46′–Glu-49′ segment packing alongside of it (Rydel *et al.*, 1990); thus, partial masking upon hirudin binding could account for the observed abolishment of chemotactic activity (Bizios *et al.*, 1986; Prescott *et al.*, 1990).

2.6.2. Structure of Segment 67–80 (Fibrinogen-Binding Exosite)

The structure of the loop comprising segment Lys-70–Glu-80 and its surrounding groups is displayed in Fig. 9. This thrombin loop is topologically similar to the so-called "calcium loop" of bovine trypsin (Bode and Schwager, 1975a,b) and the other digestive enzymes (Meyer *et al.*, 1988). The central residue of this loop is Lys-70. Its distal ammonium group occupies an equivalent site to the calcium ion in the pancreatic enzymes. In the thrombin structure, however, only three of the six oxygen ligands of calcium found in bovine trypsin (namely the carbonyl group of residue 72, the carboxylate group of Glu-77, and the carboxylate group of Glu-80) are connected to Lys-70 NZ through short hydrogen bonds, surrounding it with almost tetrahedral geometry (Bode *et al.*, 1992).

The Glu-80 residue, in turn, is involved in another salt bridge/hydrogen bond with the guanidyl group of Arg-67. Thus, the charged groups of four residues (Arg-67, Lys-70, Glu-77, and Glu-80) are cross-connected with one another and form a salt-bridge cluster essentially buried in the molecule; its contribution to the rigidity and integrity of this loop seems therefore to be considerable (Bode *et al.*, 1992). As shown previously for trypsin (Bode, 1979), the integrity of this loop structure confers (thermal) stability to the whole molecule. Thrombin [like porcine kallikrein (Bode *et al.*, 1983) and human leukocyte elastase (Bode *et al.*, 1986a)] seems to carry its endogenous cationic group (70 NZ) to order the 70–80 loop and to maintain a distinct surface contour (Fig. 9).

As discussed above and further shown in Fig. 9b, the slightly notched surface arching over this loop has only positively charged amino acid side chains: in particular, Arg-73, Arg-75, and Arg-77A are surrounded exclusively by other positively charged residues such as Arg-35, Lys-149E, Lys-81, Lys-110, Lys-109, and Lys-36, which are themselves all involved in salt bridges with more peripheral negatively charged residues. The guanidyl group of Arg-67 just touches the surface and is also involved in a salt-bridge cluster beneath the surface (see above). Therefore, Arg-73, Arg-75, and Arg-77A generate a strong positive field around this surface site (Fig. 4; Bode *et al.*, 1992).

The crystal structure analysis of the thrombin–hirudin complex (Rydel *et al.*, 1990; Grütter *et al.*, 1990) revealed that this positively charged thrombin surface patch interacts with the extended hirudin tail segment (see Figs. 16 and 21). Besides these distinct electrostatic interactions of hirudin, the complementary fit of hydrophobic groups seems to be of

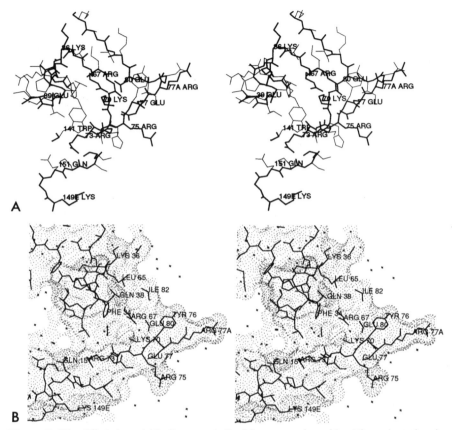

Figure 9. The "fibrin(ogen) binding exosite" of human α-thrombin. The orientation is similar to Fig. 5.

(a) Stick drawing of loop 67–80 and its environment. Lys-70 is located at the center of an extended buried salt bridge network made with Glu-77, Glu-80, and Arg-67. Cleavage of peptide bond Arg-77A–Asn-78 leads to formation of noncoagulant β$_T$-thrombin, in which this loop is presumably unfolded.

(b) Surface residues of the exosite displayed together with the Connolly surface (from Bode *et al.*, 1992.)

particular importance. There is much evidence that this thrombin exosite is also involved in thrombin binding to fibrinogen (Griffith, 1979; Hageman and Scheraga, 1974; Hogg and Blombäck, 1978; Fenton, 1981; Blombäck *et al.*, 1977; van Nispen *et al.*, 1977; Hofsteenge and Stone, 1987; Fenton *et al.*, 1988; Lundblad *et al.*, 1988; Church *et al.*, 1989; Hogg and Jackson, 1989), to fibrin and its E domain (Berliner *et al.*, 1985; Kaczmarek and McDonagh, 1988; Vali and Scheraga, 1988; Kaminski and McDonagh, 1987), to thrombomodulin (Hofsteenge *et al.*, 1986; Hofsteenge and

Stone, 1987; Preissner *et al.*, 1987; Suzuki *et al.*, 1990), and to anionic matrices (see, e.g., Fenton *et al.*, 1988, 1989). It has been shown that the foregoing proteins bind competitively (Hofsteenge *et al.*, 1986), i.e., that they seem to use similar surface patches of this positively charged exosite of thrombin.

The positive field centered on this exosite (Fig. 4) spreads far into the extramolecular space; as a consequence, approaching macromolecules (substrates as well as inhibitors) can be preoriented by the combined influence of this basic exosite and the negatively charged patch around the specificity pocket (see below). Anionic groups of attached proteins need not necessarily interact through direct salt bridges with the cationic thrombin groups, but can still experience the effect of this positive field at some distance from the thrombin surface (see, e.g., the hirudin tail interaction, Fig. 20).

Due to its interaction with various anionic compounds and protein segments, this exosite is usually referred to as "anion-binding exosite" (Fenton, 1986). This could, however, lead to confusion with respect to the putative "heparin-binding site" described below. A more appropriate designation seems to be the term "fibrin(ogen)-recognizing exosite" (Fenton, 1981), due to the importance of this exosite in fibrinogen clotting and fibrin binding.

The exposed peptide bond Arg-77A–Asn-78, in the 70–80 loop (Fig. 7), is particularly susceptible to tryptic and to autocatalytic attack (Boissel *et al.*, 1984; Fenton *et al.*, 1977a; Chang, 1986; Bezeaud and Guillin, 1988; Braun *et al.*, 1988b). This polypeptide segment is relatively mobile and exhibits a main-chain conformation quite different from the canonical conformation of typical serine proteinase protein inhibitor binding loops (Huber and Bode, 1978; Laskowski and Kato, 1980). Upon its cleavage and following excision of peptide Ile-68–Arg-77A, human β-thrombin is formed; thus, this loop is conveniently termed β-loop.

Experimental evidence (Noé *et al.*, 1988) suggests that the whole loop structure unfolds upon cleavage, driven presumably by favorable interactions of charged residues (previously involved in the buried salt cluster) with bulk water. Such a disruption of the exosite surface would explain the tremendous reduction in clotting activity and loss of affinity for fibrin, thrombomodulin, and hirudin, found for $β_T$-thrombin (Braun *et al.*, 1988b; Bezeaud and Guillin, 1988; Hofsteenge *et al.*, 1988) and β-thrombin (Fenton *et al.*, 1977b; Lewis *et al.*, 1987; Stone *et al.*, 1987; Hofsteenge and Stone, 1987; Bezeaud and Guillin, 1988; Hofsteenge *et al.*, 1988). Subsidiary evidence for such an unfolding is that peptide bond Arg-67–Ile-68 (Fig. 7) is quite susceptible to autocatalytic cleavage (Boissel *et al.*, 1984). Being deeply buried, a cleavage of the surface Arg-77A–Asn-78 bond (see Fig. 9) would be necessary prior to a proteolytic attack at this site resulting in β-thrombin.

In the naturally occurring human genetic variant Quick I, Arg-67 (Fig. 7) is replaced by a cysteine residue (Henriksen and Mann, 1988). The surface location of the Arg-67 guanidyl group (Fig. 9) and its integration in an extended internal salt-bridge cluster in normal human α-thrombin suggest that the exosite of this mutant might be disrupted in a similar manner to β-thrombin; this would be in agreement with the tremendous loss in clotting activity observed experimentally.

2.6.3. Insertion Loop Leu-144–Gly-150

The thrombin structure at and around insertion segment Leu-144–Gly-150 as observed in PPACK-thrombin is displayed in Fig. 5. This exposed loop lacks any inter-main-chain hydrogen bonds, but seems to be stabilized in PPACK-thrombin through hydrogen bonds made with its own side chains (Lys-145 and Asn-149B), together with some hydrophobic contacts with a symmetry-related molecule. This loop segment attains a very different conformation and overall shape in thrombin–hirudin (see below, Fig. 23); in the PPACK-thrombin conformation, the Trp-148 side chain would collide with parts of the bound hirudin molecule (Rydel *et al.*, 1990). Although changes in circular dichroism upon hirudin binding seem to indicate a conformational change induced in the thrombin moiety (Mao *et al.*, 1988), it is questionable whether the loop structure observed in PPACK-thrombin is the solution conformation (since it is disordered in the hirugen–thrombin and the hirulog–thrombin complexes, Skrzypczak-Jankun *et al.*, 1991), or if it is enforced by crystal or inhibitor contacts (as in the case of the PPACK-thrombin and hirudin–thrombin complexes where it interacts with neighboring thrombin molecules). Thus, there is considerable evidence that this loop does not exist only in these two very different conformations (stabilized via crystal and/or intermolecular contacts), but that it exhibits a significant degree of overall flexibility.

Three proteolyzed human thrombin variants cleaved in this exposed, flexible loop segment have been found and characterized (displayed in Fig. 7), namely

1. ξ-Thrombin, resulting from cathepsin G cleavage at Trp-148–Thr-149 (Brezniak *et al.*, 1990)
2. ε-Thrombin, generated by elastase action on Ala-149A–Asn-149B (Kawabata *et al.*, 1985; Brower *et al.*, 1987)
3. γ-Thrombin, resulting from tryptic or autocatalytic cleavage at Lys-149E–Gly-150 (Fenton *et al.*, 1977b; Bing *et al.*, 1977; Boissel *et al.*, 1984)

Due to the inherently flexible character of this loop, the first two cleavages should not greatly affect the active-site geometry and thus the catalytic activity of thrombin, in agreement with experimental results (Berliner, 1984; Stone *et al.*, *1987; Hofsteenge et al.*, 1988).

The last scissile peptide bond, at Lys-149E–Gly-150 (Fig. 7), is somewhat obstructed in PPACK-α-thrombin and not readily accessible to attacking protease molecules. A primary autolytic cleavage seems, however, to be possible at this site (Chang, 1986), probably due to the inherent mobility of this loop. This scissile peptide bond is adjacent to segment Gln-151–Pro-152, which (together with Glu-39 of the opposing loop) forms the entrance to the "fibrin(ogen)-recognizing exosite" (see above and Fig. 9b). ε-Thrombin, cleaved four bonds before Lys-149E–Gly-150 (Fig. 7), shows only a minor decrease in clotting activity compared with α-thrombin (Hofsteenge *et al.*, 1988). The nearly complete lack of clotting activity of γ_T- and γ-thrombin (Bing *et al.*, 1977; Fenton *et al.*, 1977b) might therefore mainly be caused by the disruption of the "fibrin(ogen)-recognizing exosite" due to excision of peptide Ile-68–Arg-77A (Hofsteenge *et al.*, 1988).

The peptide segment around Ala-149A–Asn-149B is involved in thrombomodulin binding (Suzuki *et al.*, 1990). Other evidence (Noé *et al.*, 1988) indicates that the accessibility and integrity of the 70–80 surface loop are at least also required for efficient thrombin interaction with thrombomodulin. Inactivation of thrombin by antithrombin III is even stimulated by thrombomodulin (Hofsteenge *et al.*, 1986; Preissner *et al.*, 1987). The thrombin interaction site for this endothelial cell-surface protein might therefore reside near the junction between loop segments 70–80 and 144–151.

2.6.4. *Putative "Heparin-Binding Site"*

A thrombin surface patch of positively charged side chains (Arg-126, Lys-236, Lys-240, and Arg-93) surrounded by other basic residues (Arg-101, Arg-233, Arg-165, Lys-169, Arg-175, Arg-173, Arg-97), which in turn contact negatively charged side chains, is shown in Fig. 10. The first four residues in particular give rise to a very strong positive electrostatic field (Fig. 4) (Bode *et al.*, 1992). Despite this strong field, we see no evidence for fixed solvent anions (Bode *et al.*, 1992). This site seems rather to be more suited to binding polyanions such as heparin; this sulfated glycosaminoglycan binds to thrombin (Griffith *et al.*, 1979) and accelerates its inactivation by antithrombin III (Björk and Lindahl, 1982).

Chemical modification experiments on thrombin revealed that Lys-169 and Lys-240 (at the edges of this positively charged patch, Fig. 10) are blocked by thrombin-bound heparin, and that their modification renders

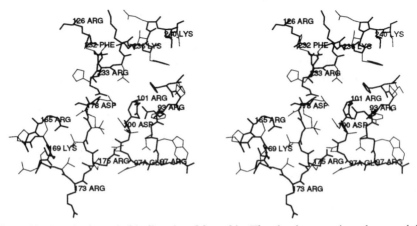

Figure 10. Putative heparin-binding site of thrombin. The view is approximately toward the "top" of the thrombin molecule (compared with Fig. 5). Main-chain segments and all basic side chains in this surface area are emphasized by thick connections. All basic residues and the three acidic residues occurring in that area are assigned by labels. Arg-173, Glu-97A, and Arg-97 (lower edge) are located at the "north rim" of the active-site cleft (Bode *et al.*, 1989a). Segment His-230–Leu-241 (upper edge) is part of the carboxy-terminal B-chain helix. His-91 (right edge) is surrounded exclusively by basic residues. Lys-169 and Lys-240 which are protected from chemical modification upon heparin binding are located at the lower left and at the upper right corner (from Bode *et al.*, 1992.)

thrombin inaccessible to heparin (Church *et al.*, 1989; Griffith, 1979). Cooperative electrostatic interactions of linear heparin molecules (with high linear negative charge density) with this positively charged region of contiguous basic residues could explain the high-affinity binding of heparin and thrombin (Heuck *et al.*, 1985). Furthermore, this putative "heparin-binding site" exhibits (Table 2) a typical heparin binding motif with its helical strand H230–V231–F232–R233–L234–K235–K236–W237–I238–Q239–K240–V241–I242 (Cardin and Weintraub, 1989).

Acceleration of complex formation between antithrombin III and thrombin by heparin may in part be due to a "template mechanism" (Pomerantz and Owen, 1978; Griffith, 1982; Nesheim, 1983) in addition to its effects on the antithrombin III structure . If a docking geometry for antithrombin III bound to thrombin is assumed to be similar to that known for the "smaller" protein inhibitors, the putative thrombin heparin-binding site would be in antithrombin III–thrombin complexes located near the presumed heparin-binding site of antithrombin III (for references, see Huber and Carrell, 1989). It is therefore conceivable that one heparin strand could connect the two components. The Arg-97 and Arg-175 residues, which form a positively charged patch to the "north rim"

(Bode *et al.*, 1989a) of the active-site "canyon" in this putative heparin-binding site (Figs. 5, 10), could further stabilize such an intermediating binding through heparin.

Similar reaction rates of antithrombin III in the presence of heparin with α- and γ-thrombin support the idea that the main heparin recognition sites of antithrombin III-bound thrombin reside outside the "fibrin(ogen)-binding exosite" (Stone and Hofsteenge, 1986). Other evidence (Olson *et al.*, 1986; Hogg and Jackson, 1989) indicates, however, that in the absence of antithrombin III, heparin might bind to both exosites (Naski *et al.*, 1990).

This highly charged putative heparin-binding site might also mediate [in cooperation with the "fibrin(ogen)-binding exosite"] thrombin binding to cation-exchange resins and fibrin (Fenton *et al.*, 1988). Furthermore, the heparin-binding site is close to the amino-terminal Thr-1H of the A chain (Fig. 3); its partial shielding by adjacent pro-parts in prothrombin forms still possessing pro-parts (such as meizothrombin; see Doyle and Mann, 1990) seems therefore also possible without conformational changes in the B-chain globule. Chemical modification experiments further suggest that part of this heparin-binding site (in particular Lys-169 and/or Lys-240) might act simultaneously as a high-affinity thrombin binding site for platelets (White *et al.*, 1981).

In the natural human variant thrombin Tokushima, Arg-101 (close to the positively charged surface patch of the "heparin-binding site," Fig. 7) is replaced by a tryptophan residue (Miyata *et al.*, 1987). As shown in Fig. 11, the Arg-101 side chain points away from the active-site cleft, but is involved with the adjacent Asp-100 in a surface-located salt bridge. According to modeling experiments (Bode *et al.*, 1992), replacement by a tryptophan side chain might disturb the "rim" region and would thus affect primarily binding of more extended substrates and interactions with cells. An effect on the subsequent active-site residue, Asp-102, whose side chain points in the opposite direction (toward the active-site cleft, Fig. 5), cannot be excluded; it would, however, be small, in agreement with the observations (Miyata *et al.*, 1987).

2.6.5. Thrombin RGD Site

Both human and bovine thrombin contain an "R–G–D" segment (Arg-187–Gly-188–Asp-189) "beneath" the specificity pocket. This segment has recently been suggested (Fenton, 1988b) to play a distinct role as a docking site in platelet binding (e.g., through the platelet-specific GPIIb/IIIa integrin receptor, Charo *et al.*, 1987). Glenn *et al.* (1988)

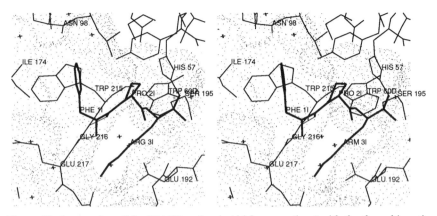

Figure 11. Interaction of the PPACK molecule (thick connections) with the thrombin active site. The two hydrogen bonds between Phe-1I and Gly-216 and the Connolly surface of thrombin are shown in addition. The view is toward the active-site cleft, i.e., similar to Figs. 1, 3–5. The Arg 3I carbonyl group forms a tetrahedral hemiketal group covalently linked with Ser-195 OG and (via the methylene group) with His-57. Pro-2I binds tightly into the S_2 cavity, and the benzyl group of the D-Phe-1I nestles into the hydrophobic "aryl binding site" (from Bode *et al.*, 1992.)

showed that a thrombin-derived 23-amino-acid peptide comprising this Arg–Gly–Asp region binds to the high-affinity receptor on fibroblasts and generates receptor occupancy-dependent mitogenic signals.

In the PPACK-α-thrombin structure, the side chain of Arg-187 runs along the molecular surface and is exposed to solvent; neither Gly-188 nor Asp-189 (except for its carboxylate group which points toward the interior of the specificity pocket, Fig. 12) are, however, accessible to receptor structures of approaching cells or proteins without complete unfolding of this thrombin site. Such an accessibility would seem to be required, however, for proper recognition of the underlying sequence motif.

2.6.6. Active-Site Cleft

The most remarkable feature of the thrombin surface is a deep narrow "canyon" (Fig. 5) containing the catalytic residues and the adjacent substrate binding subsites (Bode *et al.*, 1989a). The catalytic residues (in particular Ser-195, His-57, and Asp-102) divide this cleft into two halves, corresponding to a "fibrinopeptide side" ("west") and a "fibrin side" ("east"). This cleft is bordered and shaped by the four extending polypeptide segments Arg-173–Ile-174, Asp-95–Leu-99, Tyr-60A–Phe-60H, Phe-34–Cys-42 on one side (forming the "north rim," Fig. 5), and the two loop segments Gly-216–Cys-220 and Asn-143–Gln-151 on the opposite side (the "south rim," Fig. 5). The side chains of Ile-174, Tyr-60A, Trp-60D, Lys-60F, Phe-60H, and of Thr-147, Trp-148, respectively, protrude into the active-site cleft, i.e., each rim is lined primarily by hydrophobic groups. The base, in contrast, is covered mainly with polar/charged side chains (Glu-217, Glu-192, Asn-143, Gln-151, Arg-73) and polar main-chain groups.

Around the entrance to the specificity pocket, six negatively charged residues (Asp-189, Glu-192, Asp-102, Glu-217, Glu-146, Glu-97A) are clustered, with Glu-192 (relatively flexible at the base) and Asp-189 (at the bottom of the specificity pocket, Fig. 12) being the only acidic residues not involved in direct salt bridges. The latter contribute substantially to the rather negative field generated at this site (Bode *et al.*, 1992). As described above, the "fibrinogen-recognizing exosite" (at the right-hand side of the active-site cleft, Figs. 5, 9b) gives, in contrast, rise to a strong positive electrostatic field (Fig. 4). This dipolar charge distribution along the active-site cleft would clearly influence the orientation of approaching substrates and inhibitors of large electric moments (such as hirudin, see below).

2.6.6a. Thrombin Active Site and Interaction with PPACK. The PPACK molecule, as it is bound to the active-site cleft of thrombin, is shown in Fig. 11. Its peptidyl moiety juxtaposes the extended thrombin segment Ser-214–Glu-217 in a typically twisted antiparallel manner. The carboxy-terminal Arg-3I carbonyl group is part of a tetrahedral hemiketal structure; i.e., it is linked covalently with Ser-195 OG and, via the methylene group,

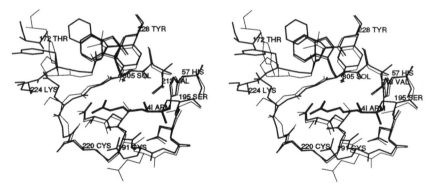

Figure 12. Specificity (S_1) pocket of human α-thrombin (medium connections) superimposed by the equivalent residues of unliganded bovine trypsin (thin connections). Arg-3I of PPACK is shown with thick connections. The view is toward the active-site cleft, i.e., similar to Figs. 1, 3–5, 7–9, 11. The distal guanidinium group of Arg-3I opposes the carboxylate of Asp-189 at the bottom of the pocket. Ser-190 of trypsin is replaced in thrombin by an alanine residue. Sol-305 in thrombin has an equivalent water molecule (Sol-416) in trypsin (from Bode *et al.*, 1992.)

to the imidazole group (ND) of His-57 of thrombin. Compared to its location in thrombin-hirugen (representing an unliganded thrombin), the OG atom of Ser-195 is shifted for 0.6 Å to form this covalent bond.

The thrombin specificity pocket is geometrically quite similar to that of bovine trypsin (Fig. 12) (Bode and Schwager, 1975a). The only different thrombin residue whose side chain points into the pocket is Ala-190, replaced by a serine residue in trypsin. This exchange renders the pocket slightly more apolar, but causes in particular the loss of a hydrogen bond acceptor/donor for the basic group of a P_1-residue (Fig. 12). In consequence, the thrombin S_1 subsite would better accommodate an arginine P_1 side chain; on the other hand, an inserted lysine side chain would fit less tightly in the thrombin pocket, due to the greater space and lack of a hydrogen bond partner (Ser ← Ala-190; Bode *et al.*, 1992). This is in agreement with experimental results concerning the relative affinities for arginine and lysine P_1 side chains (Kettner and Shaw, 1979, 1981; Liem and Scheraga, 1974; Lottenberg *et al.*, 1983).

Another naturally occurring dysfunctional thrombin mutant, Quick II, has the replacent Gly-226 → Val-226 (Henriksen and Mann, 1989). Modeling experiments show that the bulky valine side chain could disturb the Asp-189 conformation at the bottom of the pocket, rather than directly blocking binding of an inserted arginine side chain (Bode *et al.*, 1992). The pocket of Quick II will probably not be nearly as narrowed as in porcine pancreatic elastase (Meyer *et al.*, 1988) or human leukocyte elastase (Bode *et al.*, 1986a). Thus, neither typical trypsin nor elastase substrates should be cleaved efficiently by this mutant, in agreement with the experimentally observed lack of catalytic and clotting activity.

A more extended, quite hydrophobic pocket lies near the entrance to the specificity pocket (Fig. 11), lined by Ile-174, Trp-215, segment 97–99, His-57, Tyr-60A, and Trp-60D. This hydrophobic depression is most certainly the suggested "apolar binding site" (Berliner and Shen, 1977). Its back part is made up of the same structural elements as in trypsin (Trp-215, a shorter segment 97–99, His-57); it is in thrombin, however, further screened off from bulk water by the mainly hydrophobic Tyr-60A–Trp-60D loop on one side and Ile-174 on the other side (to the right and left, respectively, in Fig. 11). The PPACK molecule nestles tightly into this depression. The solvent-accessible surface of the PPACK-thrombin complex is smaller than that of thrombin itself; two thirds of the PPACK surface becomes buried within the complex (Bode *et al.*, 1992).

The pyrrolidine ring of Pro-2I is almost fully shielded from bulk water; it is in a completely hydrophobic environment and seems to be perfectly accommodated through its polyproline II conformation (Fig. 11). This excellent packing in the thrombin S_2 cavity explains the frequent occurrence of proline (and other medium-sized hydrophobic) residues at the P_2 position of naturally occurring thrombin macromolecular substrates. In contrast to trypsin (Liem and Scheraga, 1974; Bajusz et al., 1978), in thrombin it accounts furthermore for the particular reactivity of chloromethyl-ketone inhibitors with P_2 proline residues (Kettner and Shaw, 1981); much larger residues would not pack favorably into the S_2 thrombin cavity.

The (substituted) piperidine rings of the benzamidine (NAPAP) (Stürzebecher et al., 1983) and of the arginine-based (MQPA) (Okamoto et al., 1981) synthetic inhibitors (Fig. 13) probably extend into this pocket where they can be tightly fixed and sandwiched between the flat side of the His-57 imidazole ring and the amino-terminal aryl moiety (Bode et al., 1990). As in the case of PPACK-thrombin, the hydrogen bonds made with Gly-216 N and O determine mainly the docking position of these inhibitors in the thrombin active site.

The side chain of the preceding residue, D-Phe-1I, fits into the notched, mainly hydrophobic cleft made by Ile-174, Trp-215, segment 97–99, and Tyr-60H (Fig. 11). Such an interaction is not only preferred due to favorable hydrophobic contacts; both the perpendicular aryl–aryl ("edge-on") arrangement (Burley and Petsko, 1985) and the aryl–carbonyl contact with the carbonyl group of Glu-97A (see Thomas et al., 1982)

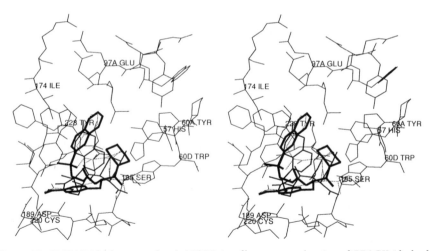

Figure 13. NAPAP (thick connections), MQPA (medium connections), and PPACK (dashed connections) bound to the active site of human α-thrombin (thin connections). The naphthyl, the chinolyl, and the benzyl groups of the inhibitors interact with the "aryl binding site" of thrombin; the side chains both of arginyl residues and of the p-amidino-phenylalanyl residue enter the specificity pocket from different sites; the S_2 subsite of thrombin is occupied to different extents both by pyridine moieties and by the prolyl residue, respectively (from Bode et al., 1992.)

clearly contribute to binding strength (Bode *et al.*, 1990). In natural protein substrates, the variety of amino acids in P_3 is great; generally, L-amino acids in the P_3 position would interact with a different thrombin site. The beneficial effect of large bulky aromatic residues at P_4 connected to a P_3 glycine has been used to construct suitable thrombin substrates (Svendsen *et al.*, 1972). Very recently (Bode *et al.*, 1990; Turk *et al.*, 1991) we have shown that the naphthyl and tosyl moieties of certain benzamidine-derived synthetic inhibitors (see Fig. 13) as well as other aromatic groups in arginine-derived synthetic inhibitors (Matsuzaki *et al.*, 1989) interact in a similar favorable manner with this cleft. As shown below, the phenolic side chain of Tyr-3' of hirudin is also placed in such a manner (Rydel *et al.*, 1990), and there is now good evidence for a similar arrangement of the phenyl moiety of the fibrinogen Aα-chain Phe-8 residue upon binding to thrombin (Meinwald *et al.*, 1980; Ni *et al.*, 1989b; Martin *et al.*, 1992; Stubbs *et al.*, 1992. We have therefore proposed the designation "aryl binding site" (Bode *et al.*, 1990), distinguishing it from the other part of the "apolar binding site" (S_2 cavity).

The extraordinary specificity and reactivity of PPACK for thrombin (Kettner and Shaw, 1979) can thus easily be attributed to the optimal fit of all three residues to their respective thrombin subsites (Fig. 11). In addition, each of these residues seems to exist in a stereochemically and energetically optimal conformation in the thrombin-bound PPACK molecule (Bode *et al.*, 1992). The ammonium and the carbonyl group of D-Phe-1I are in favorable hydrogen-bond contact with Gly-216. In contrast to trypsin, good peptidyl aldehyde inhibitors of thrombin require a free amino terminus (Bajusz *et al.*, 1978), probably to allow the undisturbed formation of this hydrogen bond; it is clear, however, that *N*-alkylation of PPACK (such as in *N*-methyl-PPACK; see Bajusz *et al.*, 1990) should (in solution) not impair hydrogen bond formation.

An L-configurated Phe-1I residue at P_3 would not be able to interact with thrombin in as favorable manner as the D-diastereomer (in particular when linked to a proline) (Fig. 11). Simultaneous hydrogen bond formation through its amino function and favorable aryl interaction are mutually exclusive; the best fit for an L-Phe-1I residue seems to be at a Cα–C dihedral angle of about −60°, with the phenyl group fitting into the "aryl binding site" in a slightly different relative orientation.

2.6.6b. Probable Interaction with Macromolecular Substrates and Inhibitors. The narrow, canyonlike structure of the thrombin active-site cleft is clearly the main source of the remarkable thrombin specificity toward macromolecular substrates and inhibitors (Bode *et al.*, 1989a, 1992). It restricts access of potential scissile peptide bonds to the thrombin catalytic center of most macromolecules simply through steric hindrance, discriminating in this way susceptible substrates by means of shape and conformation. An attempt (Fig. 14) to dock the archetypal inhibitor BPTI (Huber *et al.*, 1970) to the thrombin active site with identical position and orientation as observed for the trypsin(ogen) complexes (Huber *et al.*, 1974; Bode *et al.*, 1978, 1984; Huber and Bode, 1978) results in a collision of residues of the inhibitor binding region with part of the "60-insertion loop" of thrombin, in particular with the side chain of Trp-60D. This steric hindrance cannot be relieved, even upon considerable relative tilting of the inhibitor (Bode *et al.*, 1992). The "60-insertion loop" appears so rigid that strong forces would be required to allow access of the inhibitor. These docking experiments are in agreement with experimental results, which reveal extremely low association constants of about 10^3 M^{-1} for α-thrombin, but almost ten times larger values for the more open and more flexible γ-thrombin (Ascenzi *et al.*, 1988; Berliner *et al.*, 1986).

Similar docking experiments with other "small" serine proteinase protein inhibitors of known spatial structure such as eglin c (Bode *et al.*, 1986b, 1987), the squash seed inhibitor CMTI-I (Bode *et al.*, 1989b), and the turkey ovomucoid third domain (Laskowski *et al.*, 1987; Bode *et al.*, 1986a) show similar restrains (Bode *et al.*, 1992). In each case the binding loop residues on both sides of the scissile peptide bond would essentially fit to the corresponding subsites around the thrombin catalytic site, how-

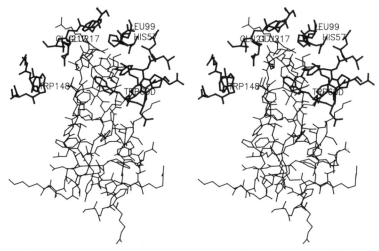

Figure 14. Docking model of the hypothetical complex formed between BPTI (thin connections) and thrombin (thick connections). Only a few residues making up the thrombin substrate binding site are shown. The docking mode of BPTI is as found in the trypsin–BPTI complex (see Bode *et al.*, 1989a). The view is along the active-site cleft of thrombin, which points downwards.

ever; collision with the thrombin "rims" would occur through structural elements extending from the inhibitor core.

In the absence of an experimental structure of an intact serpin inhibitor or of a serpin–serine proteinase complex, one can only speculate on the interaction mode between thrombin and antithrombin III. There are, however, several lines of evidence that serpins bind in an essentially similar canonical manner as the "small" protein inhibitors to their cognate enzymes, namely through an exposed binding loop of canonical conformation (Huber and Bode, 1978; Laskowski and Kato, 1980; Huber and Carrell, 1989; Bode and Huber, 1991); furthermore, the interaction of this loop with thrombin seems to be essentially restricted to the immediate environment of the active-site residues (Chang et al., 1979; Hofsteenge et al., 1988). Thrombin obviously imposes particular restrictions on the shape of the serpin binding region, which must be extremely flat. The core of these serpin inhibitors (Löbermann et al., 1984; Wright et al., 1990; Stein et al., 1990) is much larger than those of the "small" inhibitors; therefore, more intimate molecular interactions with "rim" elements of the thrombin active-site "canyon" are to be expected.

Similar geometric requirements must also be met by the cleavage sites of fibrinogen, as well as of the other coagulation factors susceptible to thrombin action.

3. HIRUDIN

Hirudin is a small 65-residue protein which is the most potent natural inhibitor of thrombin known (Markwardt, 1970). It forms a 1:1 non-covalent complex with α-thrombin with a binding constant in the pico–femtomolar range (Sonder and Fenton, 1984; Stone and Hofsteenge, 1986) but does not require catalytically active thrombin (Stone et al., 1987); however, the dissociation constant of the hirudin PPACK-thrombin complex is 10^6 greater than that with α-thrombin (Stone et al., 1987). Kinetic and equilibrium studies indicate that hirudin interacts simultaneously with the catalytic site and the fibrinogen-binding exosite of thrombin (Fenton, 1981; Chang, 1983). The apolar binding site of thrombin appears to be an important determinant for complex formation since hirudin binding to α-thrombin displaces proflavin (Sonder and Fenton, 1984). Investigations with synthetic peptides have shown that the exosite binds the C-terminal undecapeptide of hirudin (Krstenansky and Mao, 1987; Ni et al., 1990), and the exosite has been shown to contain a number of the lysyl residues of thrombin (Chang, 1989).

The electrostatic and acidic nature of the C-terminal residues appear to be of vital importance in complex formation. Removal of the last seven amino acids of hirudin, which includes two Glu residues and a sulfated tyrosine (in natural hirudin), results in a loss of about 90% of its inhibitory activity, and removal of the last 22 amino acids practically abolishes inhibition (Chang, 1983). Single and multiple mutations of the carboxy-terminal glutamates (57', 58', 61', and 60'; a prime following a residue number designates hirudin) to glutamine cause increases in the dissociation constant of the complex which increases with the number of mutations (Braun *et al.*, 1988a), and each negatively charged Glu residue appears to contribute equally favorably to the binding energy (Stone *et al.*, 1987). Desulfato Tyr-63 hirudin–thrombin complexes possess a higher dissociation constant compared to those with sulfated forms. In addition, the hydrophobic residues in the anionic C-terminal tail are also involved in the interaction. In inhibition studies with hirudin C-terminal peptide mimetics corresponding to residues 55'–64', it was ·shown that Phe-56', Ile-59', Pro-60', and Leu-64' as well as Glu-57' were all sensitive to modification (Krstenansky and Mao, 1987).

Hirudin possesses a two-disulfide, double-loop structure (B, C, D) which is preceded by an ordinary disulfide loop (A) and followed by a 26-residue carboxy-terminal chain (Fig. 15). Furthermore, all hirudin variants share a common disulfide connectivity pattern. They contain six

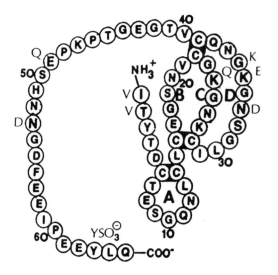

Figure 15. The sequence of recombinant hirudin variant 2-lysine-47 (rHV2-K47). Residues indicated outside circles are those of the most abundant natural iso-inhibitor.

conserved cysteine residues in the N-terminal region which form three characteristic cystine links Cys-6′–Cys-14′, Cys-16′–Cys-28′, and Cys-22′–Cys-39′ (Dodt et al., 1985). Hirudin does not display any topological or sequence homology to any other known serine proteinase inhibitor class. On the other hand, hirudin does share global features with epidermal growth factor domains (EGF), which are structural motifs found in numerous blood proteins (Apella et al., 1988). Alignment of the hirudin and EGF disulfide patterns in a palindromic manner reveals striking similarity in connectivity (Johnson et al., 1989). It is noteworthy, however, that hirudin and EGF share no sequence or functional homology.

The hirudin used in the crystallographic structure determination of the complex was a recombinant form, variant 2 with Lys-47′ (rHV2-K47) (Harvey et al., 1986). It differs from the most abundant natural iso-inhibitor in eight residues and in the absence of a sulfated tyrosine at position 63′ (Fig. 15); natural hirudins are sulfated at this position (Bagdy et al., 1976; Dodt et al., 1984). Although the lack of the sulfated tyrosine leads to a ten-fold reduction in thrombin affinity (Dodt et al., 1988), the binding constants of unsulfated recombinant hirudins remain in the picomolar range (Degryse et al., 1989).

3.1. Hirudin–Thrombin Crystal Structure Analysis

We summarize here the crystallographic structure of the rHV2-K47 human α-thrombin complex at 2.3 Å resolution. A preliminary communication of this work has already appeared elsewhere (Rydel et al., 1990); a lower-resolution structure of a variant 1 hirudin–thrombin complex has also been reported (Grütter et al., 1990). The rHV2-Lys-47 human α-thrombin complex was crystallized as described by Rydel et al. (1990) (rHV2-Lys-47 provided by Dr. Carolyn Roitsch and α-thrombin by Dr. John W. Fenton II). The crystals are tetragonal, space group $P4_32_12$, with cell constants $a = b = 90.39$ Å, $c = 132.97$ Å, one molecule per asymmetric unit; they diffract X rays at 2.3 Å resolution. The orientation and position of a starting model for thrombin in the complex was determined by Patterson search methods (Rydel et al., 1990) utilizing the structure of PPACK-thrombin (Bode et al., 1989a). This structure was refined with the energy restraint–crystallographic refinement program EREF (Jack and Levitt, 1978) as described by Rydel et al., 1990; the R value was 0.193 at this stage, including 207 solvent molecules, for 21,960 reflections between 7.0 and 2.3 Å.

Since the occupancy of solvent molecules was not refined, and since thermal parameters of adjacently bonded atoms were not restrained in EREF refinement, a restrained least-squares refinement program which had these capabilities (PROFFT; Finzel, 1987) was employed to continue the refinement. The final model included the A chain from Gly-1F to Gly-14M, the B chain to Phe-245, the first *N*-acetyl glucosamine moiety of the polysaccharide, and 265 solvent molecules which gave an R value of 0.173. As in the case of PPACK-thrombin, Pro-37 is also in a *cis* conformation.

3.1.1. Hirudin Structure

The structure of hirudin in rHV2-K47–human α-thrombin crystals can be viewed as being comprised of two distinct domains (Fig. 16). The first is a compactly folded globular region corresponding to the N-terminal head (residues Ile-1′–Pro-48′) and the other is the C-terminal tail which exists as two extended stretches of polypeptide chain, residues (Glu-49′– Glu-54′) and (Asp-55′–Pro-60′), respectively, terminating with a near 3_{10} helical type III reverse turn (Glu-61′–Leu-64′). A general helical structure was also deduced for the latter residues by NMR with a synthetic peptide fragment of the hirudin tail (Ni *et al.*, 1990). In addition, the Pro-46′–His-51′ segment, with two prolines, is almost in a perfect polyproline II conformation (Fig. 17).

The folding of the N-terminal domain of hirudin appears to be based on the presence of a three-disulfide core. Residues Cys-6′–Cys-14′ and Cys-16′–Cys-28′ orient nearly perpendicular to one another with a distance of 5 Å between the midpoints of the bridges. Conversely, Cys-16′– Cys-28′ and Cys-22′–Cys-39′ orient nearly parallel with a comparable distance of 5 Å. The loop segments B, C, and D, which form the double loop structure of hirudin (Fig. 15), fold into three unique three-dimensional loops which are stabilized by antiparallel β-structure (Table III). In segment B, residues Cys-14′–Cys-16′ and Asn-20′–Cys-22′ form a short antiparallel β-stretch connected by a type II′ β-turn. In loop D, Lys-27′–Gly-31′ and Gly-36′–Val-40′ form longer β-strands. However, the turn between these strands has little density and is disordered in the rHV2-K47–thrombin complex. It is noteworthy that the turn is hinged to glycine residues on either side which might contribute to the observed flexibility. The result of the disulfide and antiparallel β-interactions is that the N-terminal domain of hirudin is dominated by four spatially distinct loop structures similar to those observed by NMR (Folkers *et al.*, 1989; Haruyama and Wüthrich, 1989). An optimal superposition of 114 CACN atoms of the hirudin structure (4′–30′, 30′–48′) determined by NMR and in the rHV2-K47–thrombin complex gave an r.m.s. difference of 0.86 Å with an r.m.s. difference of 1.95 Å between side chains. These values compare closely with the r.m.s. deviations from the average NMR structure

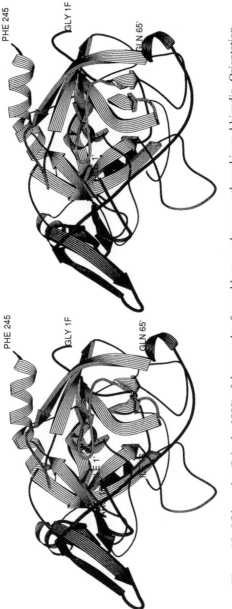
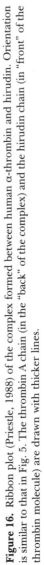

Figure 16. Ribbon plot (Priestle, 1988) of the complex formed between human α-thrombin and hirudin. Orientation is similar to that in Fig. 5. The thrombin A chain (in the "back" of the complex) and the hirudin chain (in "front" of the thrombin molecule) are drawn with thicker lines.

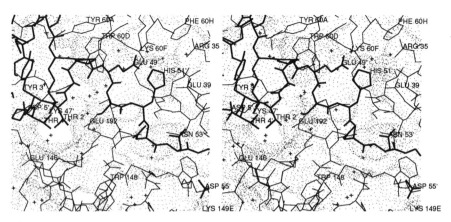

Figure 17. Hirudin segment Pro-46'–Asn-53' and surrounding residues of hirudin (thick connections) and thrombin (thin connections) displayed together with the Connolly surface of the thrombin component and localized solvent molecules (crosses). Segment Pro-46'–His-51' is organized in a polyproline II helix. The surface "hole" (center left) close to Thr-2' marks the entrance to the specificity pocket. Residue Lys-47' whose side chain was presumed to be accommodated in this pocket is located far beyond the thrombin surface (left edge.)

(Folkers *et al.*, 1989). The correspondence suggests that the structure of the N-terminal head remains basically unchanged when complexed with α-thrombin.

The C-terminal tail of hirudin in the rHV2-K47–thrombin complex is an extended polypeptide chain which terminates with a bulky 3_{10} turn (Fig. 16). The first segment of the tail is about 18 Å long, begins with Glu-49' of a polyproline II conformation (Table III, Fig. 17), and ends with three residues of poorly defined electron density (Asn-52'–Gly-54'). The

Table III. Secondary Structural Elements of Hirudin

Element	Type	Residues
β-Structure		
β1	Antiparallel	Cys-14'–Cys-16'; Asn-20'–Cys-22'
β2	Antiparallel	Lys-27'–Gly-31'; Gly-36'–Val-40'
Helix		
H1	Polyproline II	Pro-46'–His-51'
Reverse turns		
T1	Type II'	Glu-17'–Gly-18'–Ser-19'–Asn-20'
T2	Type II	Gly-23'–Lys-24'–Gly-25'–Asn-26'
T3	?	Ser-32'–Asn-33'–Gly-34'–Lys-35'
T4	Type III	Glu-61'–Glu-62'–Tyr-63'–Leu-64'

next extended section begin at a bend, is 16 Å long (Asp-55′–Pro-60′), and terminates with a type III helical turn (Glu-61′–Leu-64′). The extended nature of the carboxy-terminal domain of hirudin in the complex, with no stabilizing interactions back to the N-terminal tail, is in agreement with solution NMR results where it is completely disordered (Folkers *et al.*, 1989; Haruyama and Wüthrich, 1989).

3.1.2. Hirudin–Thrombin Interaction

The ribbon structure of the rHV2-K47–thrombin complex is shown in Fig. 16 from which it can be seen that the amino-terminal head of hirudin binds at the active-site region while the carboxy-terminal tail extends 35 Å across the surface. This structure agrees generally with predictions (Fenton and Bing, 1986; Fenton, 1989; Johnson *et al.*, 1989), with indirect observations (Stone *et al.*, 1987; Noé *et al.*, 1988; Chang, 1989), and with conclusions drawn from the PPACK-thrombin structure (Bode *et al.*, 1989a).

A novel and unexpected interaction occurs between hirudin and α-thrombin in the active site involving the N-terminal three residues of hirudin (Fig. 18), which form a β-*strand* with Ser-214–Gly-219 of thrombin. This segment makes an *antiparallel* interaction in all natural serine proteinase protein inhibitor complexes (Huber and Bode, 1978; Read and

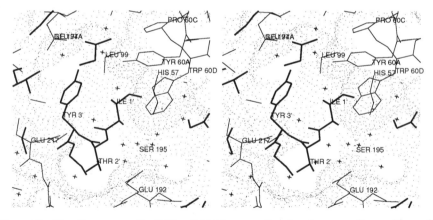

Figure 18. Hirudin segment Ile-1′–Thr-4′ (thick connections) and surrounding residues of thrombin (thin connections) displayed together with the Connolly surface of the thrombin component and localized solvent molecules (crosses). The surface "hole" (lower left) marks the entrance to the specificity pocket. The amino-terminal hirudin segment runs parallel to thrombin segment Ser-214–Gly-219; hirudin residue Ile-1′ occupies the hydrophobic cavity-like S_2 subsite of thrombin; its amino group is in hydrogen bond distance of Ser-195 OG.

James, 1986) and also in PPACK-thrombin (Figs. 11 and 13) (Bode *et al.*, 1989a). Moreover, the amino-terminal nitrogen of Ile-1' is within hydrogen bonding distance of OG of Ser-195 and the carbonyl oxygen of Ser-214. The side chain atoms of Ile-1' and Tyr-3' participate in numerous nonpolar contacts with hydrophobic residues of thrombin (Fig. 18): Ile-1' is in the S_2 subsite of thrombin in contact with His-57, Tyr-60A, Trp-60D, Leu-99, and Trp-215 while the aromatic moiety of Tyr-3' is in the hydrophobic cleft made by Leu-99, Ile-174, and Trp-215 in a manner similar to the D-Phe interaction of PPACK-thrombin (Figs. 11 and 13). This hydrophobic pocket of thrombin is most likely the apolar binding site reported for indole by Berliner and Shen (1977) and elaborated upon by others (Furie *et al.*, 1982; Sonder and Fenton, 1984; Fenton, 1988) and delineated quantitatively by Bode *et al.* (1989a, 1990) in the structure of PPACK-thrombin. Finally, the side chain of the Thr-2' is only at the edge of the S_1 specificity pocket of thrombin and does not interact with Asp-189 like the Arg residue of PPACK (compare Figs. 11 and 18). However, binding through this site is not obligatory, since hirudin binds to inactivated forms of thrombin (Stone *et al.*, 1987).

Most of the N-terminal domain of hirudin is not in contact with the thrombin surface but many dipolar contacts exist at the interface of the two (Fig. 19). Residues Asp-5', Glu-17', Ser-19', and Val-21' form ion pairs or hydrogen bonds with thrombin. The ion pair interactions are: Asp-5'–Arg-221A and Glu-17'–Arg-173 (Fig. 19) while the hydrogen bonds are between Lys-224 NZ and Ser-19' O, and possibly Val-21' O and Glu-217 OE1. In addition, residues Leu-13' and Pro-46' are both involved in van der Waals contacts with Pro-60C of the all-important 60A–60I insertion loop of thrombin (Fig. 17) which narrows the active site cleft (Bode *et al.*, 1989a). The Asp-5', Gly-18', Asn-20', and Lys-24' residues of hirudin also interact with thrombin (Arg-221A, Ser-171–Arg-173, Glu-217, Lys-224, and Pro-60C, respectively), but the interactions are mediated through water molecules.

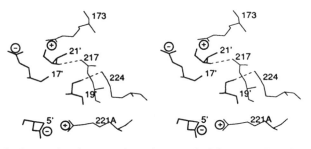

Figure 19. Polar interactions between the amino-terminal domain of hirudin and thrombin. Ion pairs indicated + −; hydrogen bonds are dashed.

The tripeptide segment Pro-46′–Lys-47′–Pro-48′ (Fig. 17) appears to facilitate the hirudin amino-terminal active site interaction but not because of the thrombin substrate-like sequence (Chang, 1985) or because of its similarity with the fragment 1 cleavage site of prothrombin (Dodt et al., 1988). The ε-amino group of Lys-47′ forms hydrogen bonds with the carbonyl oxygen of Asp-5′ and OG of Thr-4′ (Fig. 17). Moreover, the position of the Lys-47′ side chain appears to be maintained by the two proline residues which flank it through a polyproline II conformation (Fig. 17). By forming close contacts with Pro-60C and Trp-60D of the insertion loop, Pro-46′ and Lys-47′ conceivably aid in anchoring the hirudin N-terminal domain in a manner conducive for its terminus to penetrate into the active site (Fig. 18). It is of note here that Lys-47′ has often been viewed as the P_1 residue in the hirudin–thrombin complex (Chang, 1983; Braun et al., 1988a). However, the lysyl side chain of 47′ *does not occupy the specificity pocket of thrombin* and is located about 11 Å away (Fig. 17); nonetheless, it would seem that it exerts some influence on the hirudin active-site interaction.

The carboxy-terminal tail of hirudin, in strong contrast to the amino-terminal head, adopts an unusually long extended conformation in the complex (Figs. 16 and 20). The tail can be viewed as two stretches of polypeptide with a bend at Asp-55′. By means of the extended conformation, the tail can interact with many residues in the fibrinogen-binding exosite (Fenton, 1988a). Of the 17 side chains of the hirudin tail residues, 12 impressively participate in electrostatic or hydrophobic interactions in this general region (Fig. 20, Table IV).

In the first segment of the tail, the side chains of the first three

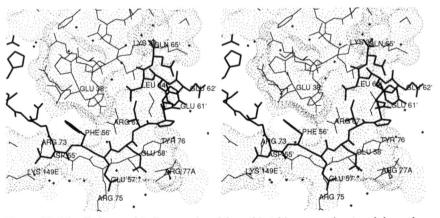

Figure 20. The fibrinogen-binding exosite of thrombin (thin connections) and the carboxy-terminal hirudin segment (thick connections), displayed together with the Connolly surface of the thrombin component. The view is similar to that of Fig. 9b (from Bode et al., 1992.)·

Table IV. Hirudin–α-Thrombin Interactions[a]

Columns are hirudin (H) residues; rows are thrombin (T) residues. Residue one-letter codes are given above their sequence positions.

T \ H	I 1	T 2	Y 3	T 4	D 5	L 13	L 15	E 17	G 18	S 19	N 20	V 21	K 24	P 46	K 47	E 49	S 50	H 51	D 55	F 56	E 57	E 58	I 59	P 60	Y 63	Q 65	Sum
F 34																				**3**							3
K 36																									3	(3)	6
E 39																		(6)									6
H 57	**7**																										7
Y 60A	**4**																										4
P 60C						**1**							3	**1**													5
W 60D	**2**														2	19											23
K 60F																(4)–1											4–1
L 65																							**1**		4		5
R 67																							3				3
R 73																			(7)–2	2							9–2
T 74																			**8**	**7**	4–1						19–1
R 75																					**10**						10
Y 76																					5	5–1		5			15–1
R 77A																						(8)					8
I 82																									**2**		2
W 96													3														3
L 99	**2**																										2
E 146								1	1																		2
K 149E																			(1)								1
R 173										1	**14**																15
I 174												**2**															2
C 191		2																									2
E 192		2		1													3										6
S 195	1–1																										1–1
S 214	1–1																										1–1
W 215	**3**																										3
G 216	2–1	1	5–1							1																	9–2
E 217		2	2							5	2	6–1															17–1
G 219			2	1						2–1																	5–1
C 220		**3**																									3
R 221A					(4)		**8**			1																	13
K 224										3–1																	3–1
Sum	22	10	9	2	4	1	8	1	1	12	16	8	6	1	2	23	3	6	16	12	19	13	4	5	9	3	216
H-bonds	3															1			2		1	1					12
Totals	1 → 41 → 3			4 → 6 → 5		13 → 53 → 24								46 → 35 → 51					55 → 60 → 58				59 → 21 → 65				216–12

[a] Numbers are intermolecular contacts < 4.0 Å; parentheses are ion pairs; dash followed by number is hydrogen bond; hydrophobic or neutral contacts in bold.

residues are involved in polar ion pair or hydrogen bond interactions: Glu-49' forms an ion pair with Lys-60F, and Glu-49' OE1 interacts with Arg-35 NH1 via a water molecule (Fig. 17). Moreover, OG of Ser-50'appears to interact in a bifurcated dipolar manner with Glu-192, and the imidazolium ion of His-51' forms an ion pair with Glu-39 (Fig. 17). The last three residues of the segment occupy a region of poorly defined electron density.

Although polar interactions between the second segment of the carboxy-terminal tail and thrombin persist, hydrophobic interactions are also important (Fig. 20 and Table IV). A robust total of 9 of the last 11 carboxy-terminal tail residues are involved in exosite interactions. The carboxylate of Asp-55' makes ion pairs with Arg-73 and Lys-149E; the position of the latter is also the γ-cleavage site of thrombin and is consistent with a successful lysine cross-linking experiment carried out using a synthetic dinitrofluorobenzyl-Gly-54'–Leu-64' peptide (Bourdon *et al.*, 1990). The amide nitrogen of Glu-57' forms a hydrogen bond with Thr-74 O, OE2 of Glu-57' forms another hydrogen bond with Tyr-76 N, and lastly, the carbonyl oxygen of Glu-57' makes a hydrogen bond with a water molecule which in turn is involved in a bifurcated hydrogen bond with the guanidinium group of Arg-67 (Fig. 20). Although Glu-57' also forms a double hydrogen-bonded ion pair with Arg-75 of *a twofold symmetry-related molecule* in the crystal, it is very possible that this electrostatic interaction can also occur intramolecularly in solution. The Glu-57'–Glu-58' residues are involved in an electrostatic interaction with Arg-77A: Glu-58' forms an ion pair with Arg-77A (at the β-cleavage site) directly while Glu-57' interacts with Arg-77A via a mediating water molecule (3.7 Å). Lastly, the electrostatic interactions conclude with the C-terminal carboxylate forming an ion pair with Lys-36 (Fig. 20) while both Glu-61' and Glu-62' are directed toward the solvent and are surprisingly not involved in direct salt bridge exosite interactions with the thrombin molecule.

An unexpected number of hydrophobic interactions also occur between the latter half of the extended tail and the fibrinogen-binding exosite (Table IV). For instance, Phe-56' penetrates into a depression on the thrombin surface and makes 11 close contacts with Phe-34, Leu-40, and Thr-74 collectively. In addition, the planes of Phe-56' and Phe-34 are nearly perpendicular to one another to give edge-on aromatic stacking (Fig. 20). Moreover, Ile-59' is close to Leu-65 and Ile-82, while Pro-60'abuts with Tyr-76 and Ile-82. The all-important Tyr-63', which is sulfated in native hirudin, is also involved in hydrophobic contacts with Leu-65, Ile-82, and Pro-60'. Although Tyr-63' is unsulfated in rHV2-K47, Lys-81, Lys-109, and Lys-110 are in the vicinity; so it is conceivable that they could form ion pairs with a sulfated residue via free bond rotations.

We have since solved the structure of the hirugen–thrombin complex at 2.2 Å resolution (Skrzypczak-Jankun *et al.*, 1991), hirugen being the Asn-53'–Leu-64' C-terminal of hirudin but with a sulfated Tyr-63'. The structure reveals that the oxygens of the sulfate of Tyr-63' make hydrogen bonds to Tyr-76 OH, Ile-82 N, and a water molecule; there are no ion pair interactions (Fig. 21). In addition, as a result of the 3_{10} helix beginning at Glu-61', Leu-64' makes close contacts with Ile-59', with main chain atoms of Glu-61', and with the side chain of Leu-65.

Thus, the C-terminal of hirudin is firmly anchored to thrombin at the end of the fibrinogen-binding exosite through hydrophobic interactions of the helical turn (Fig. 20 and Table IV) and the salt bridge formed by the terminal carboxylate of Gln-65' and Lys-36. Removal of Gln-65' has little effect on K_i (increases × 1.3) but deletion of Tyr-63'–Leu-64' raises K_i by a factor of 40 (Degryse, personal communication). The bulky 3_{10} helix might be important for recognition.

An important aspect of the hirudin carboxy-terminal and fibrinogen-binding exosite emerging from the structure of the complex is the extent of the nonpolar contribution. In the last half of the carboxy-terminal tail, nearly half of the residues (5 of 11) are hydrophobic or aromatic and *all 5 participate in nonpolar hirudin–thrombin interactions* (Table IV). The importance of hydrophobic interactions in the anticoagulant activity of the tail has already been suggested through the binding of synthetic polypeptides (Krstenansky and Mao, 1987) which indicate that the minimal peptide necessary for the detection of anticoagulant activity is Phe-56'–Gln-65',

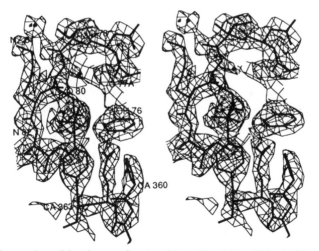

Figure 21. Stereoview of the electron density of the sulfated Tyr-63' in the hirugen–thrombin complex. Contours are at 2 σ(ρ).

and that the activity is sensitive to modification of residues Phe-56', Ile-59', Pro-60', Leu-64', as well as Glu-57'. From the structure of the complex, it is clear that the hydrophobic contribution in the last half of the tail is as important as the polar interactions. In all, 26 of the 65 residues of hirudin and 33 of the 259 B-chain residues of thrombin are involved in 216 contacts of < 4.0 Å of which 9 are ion pairs and 12 are hydrogen bonds; residues 1'–3', 13'–24', 46'–51', and 55'–58' make 41, 53, 35, and 60 of the contacts, respectively. A detailed summary of these interactions is presented in Table IV, from which it can be seen that hydrophobic contacts dominate the amino- and carboxy-terminus interactions.

Comparison of the amino-terminal hirudin active site interaction with that of PPACK (Figs. 11 and 18) and other serine proteinase–protein inhibitor complexes (Fig. 22) reveals clearly that hirudin is unique among serine proteinase inhibitors. Although Ile-1' and Tyr-3' occupy approximately the same positions as Pro-2I and D-Phe-3I of PPACK (Fig. 11), the chain direction of PPACK is *antiparallel* to Ser-214–Glu-217, while Ile-1'–Tyr-3' runs *parallel* to it. Moreover, the arginyl of PPACK occupies the specificity pocket fully in a substratelike manner (Fig. 12) but Thr-2' of hirudin does not (Fig. 18). All serine proteinase–inhibitor complexes show antiparallel alignment of a rigid loop segment in substratelike conformation and contain a residue corresponding to the specificity of the enzyme at the P_1 site (Fig. 22) (Huber and Bode, 1978; Laskowski and Kato, 1980; Read and James, 1986). These generalities coupled with the fact that hirudin displays no close sequence or topological homology to the existing

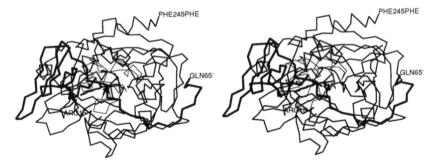

Figure 22. Stereoview of the CA structure of hirudin (bold) and thrombin (mediate connections), superimposed with BPTI (thin connections) as it would have to bind to thrombin according to the trypsin-BPTI complex (Huber et al., 1974). The carboxy-terminals of thrombin and hirudin are designated, the P_1-residue of BPTI, Arg15, is shown with dashed connections.

classes of canonical binding serine proteinase inhibitors, suggest that hirudin is a representative of a heretofore unknown family of inhibitors (Dodt *et al.*, 1985).

3.1.3. Comparison of Hirudin–Thrombin and PPACK Thrombin Structures

The structures of thrombin in the rHV2-K47–α-thrombin complex and in PPACK thrombin are very similar even to the extent of the disorder of the carbohydrate attached to Asn-60G. An optimal superposition of 818 CACN atoms between the two gives an r.m.s. deviation of 0.45 Å, which is about within the expected error, and 1.4 Å between side chains. By far the largest difference in the main-chain folding of the structures occurs in the decapeptide insertion loop between Glu-146 and Gly-150 (Fig. 23) which has an r.m.s. deviation of 5.7 Å for CA atoms. In PPACK-thrombin, Trp-148 is positioned close to the active site. A conformational change occurs in this insertion loop of the complex about the hinge points Glu-146 and Gly-150 which pivots Trp-148 into a totally different position: the loop now encircles the indole of this residue (Fig. 23).

The flexibility of this autolysis loop has already been alluded to (see above); the hirugen–thrombin and the hirulog–thrombin structures fur-

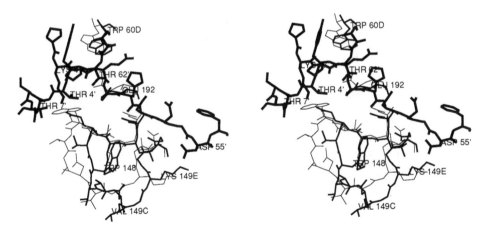

Figure 23. Stereoview of the middle part of the active-site cleft of the hirudin–thrombin complex: Hirudin segments Ile-1′–Thr-7′ and Pro-46′–Asp-55′ (bold connections) and Trp-60D, Glu-192, and the 148-insertion loop of the thrombin component (medium connections), superimposed with the same thrombin residues as observed in PPACK-thrombin (thin connections). Thrombin residue Trp-148 is very differently located in the thrombin complexes, and the whole insertion loop between Glu-146 and Gly-150 is differently arranged. Lys-47′ is shown with the hydrogen bond partners of its side chain NZ.

ther support this notion since about half of the loop is disordered in the structures of these complexes (Skrzypczak-Jankun *et al.*, 1991). The fact that the loop is ordered in the hirudin–thrombin complex might be the result of an extensive intermolecular interaction (eight hydrogen bonds) between the loop and the amino- and carboxy-terminals of a neighboring thrombin molecule.

Basically a two-pronged inhibitor, hirudin has two domains which could trigger the conformational change. From Fig. 23 it can be seen that the side chain of Trp-148 of PPACK thrombin occupies about the same space as the main chain of Thr-4′–Asp-5′ of hirudin suggesting that the binding of the amino-terminal three residues in the active-site cleft might be the source of the conformational change. The binding of the hirugen Gly-54′–Leu-65′ carboxy-terminal peptide of hirudin to thrombin also produces a change in CD (Mao *et al.*, 1988) but PPACK inhibition of thrombin does not (Konno *et al.*, 1988). These CD results thus suggest that the carboxy-terminal tail could be responsible for the conformational change. However, the ion pair interaction between Asp-55′ and Lys-149E (see before) which appears to be a result of the conformational change also appears to be too weak to help stabilize the position of the shifted loop, suggesting the hirudin and C-terminal peptide CD changes might correspond to somewhat different transitions.

Acknowledgments

This work was supported by the Sonderforschungsbereich 207 of the University of Munich (H-1 and H-2), the Fonds der Chemischen Industrie, by NIH grant HL43229, Transgene S.A., and Sanofi Recherche. We thank Dr. M. Stubbs for carefully reviewing the manuscript, Dr. K. G. Ravichandran for his contributions to the solution of the structure of the complex, Drs. Stuart R. Stone, Jan Hofsteenge, Carolyn Roitsch, and John W. Fenton II for supplying the proteins and for their interest and helpful discussions, Dr. G. Marius Clore for providing us with the NMR coordinates of the average solution structure of hirudin prior to their distribution by The Protein Data Bank, and Dr. John Maraganore of Biogen for providing us with a sample of Tyr-63′ sulfated hirugen.

4. REFERENCES

Apella, E., Weber, I. T., and Blasi, F., 1988, Structure and function of epidermal growth factor-like regions in proteins, *FEBS Lett.* **231**:1–4.

Ascenzi, P., Coletta, M., Amiconi, G., de Cristofaro, R., Bolognesi, M., Guarneri, M., and

Menegatti, E., 1988, Binding of the bovine basic pancreatic trypsin inhibitor (Kunitz) to human α- β- and γ-thrombin; a kinetic and thermodynamic study, *Biochim. Biophys. Acta* **956:**156–161.

Bagdy, D., Barabas, E., Graf, L., Peterson, T. E., and Magnusson, S., 1976, Hirudin, *Methods Enzymol.* **45:**669–678.

Bajusz, S., Barabas, E., Tolnay, P., Szell, E., and Bagdy, D., 1978, Inhibition of thrombin and trypsin by tripeptide aldehydes, *Int. J. Pept. Protein Res.* **123:**217–221.

Bajusz, S., Szell, E., Bagdy, D., Barabas, E., Horvath, G., Dioszegi, M., Fittler, Z., Szabo, G., Juhasz, A., Tomori, E., and Szilagy, E., 1990, Highly active and selective anticoagulants: D-Phe-Pro-Arg-H, a free tripeptide aldehyde prone to spontaneous inactivation, and its stable N-methyl derivative, D-Me-Phe-Pro-Arg-H, *J. Mol. Chem.* **33:**1729–1735.

Bar-Shavit, R., Kahn, A., Wilner, G. D., and Fenton, J. W., II, 1983, Monocyte chemotaxis: Stimulation by a specific exosite region in thrombin, *Science* **220:**728–731.

Bar-Shavit, R., Kahn, A., Mudd, M. S., Wilner, G. D., Mann, K. G., and Fenton, J. W., II, 1984, Localization of a chemotactic domain in human thrombin, *Biochemistry* **23:**397–400.

Bar-Shavit, R., Kahn, A., Mann, K. G., and Wilner, G. D., 1986, Growth-promoting effects of esterolytically inactive thrombin on macrophages, *Cell. Biochem.* **32:**261–272.

Bar-Shavit, R., Eldor, A., and Vlodavsky, I., 1989, Binding of thrombin to subendothelial extracellular matrix: Protection and expression of functional properties, *J. Clin. Invest.* **84:**1096–1104.

Berg, W., Hillvärn, B., Arwin, H., Stenberg, M., and Lundström, I., 1979, The isoelectric point of thrombin and its behaviour compared to prothrombin at some solid surfaces, *Thromb. Hemostasis* **42:**972–982.

Berliner, L. J., 1984, Structure–function relationship in human alpha- and gamma-thrombin, *Mol. Cell. Biochem.* **61:**159–172.

Berliner, L. J., and Shen, Y. Y. L., 1977, Physical evidence for an apolar binding site near the catalytic center of human α-thrombin, *Biochemistry* **16:**4622–4626.

Berliner, L. J., Sugawara, Y., and Fenton, J. W., II, 1985, Human α-thrombin binding to nonpolymerized fibrin–Sepharose: Evidence for an anionic binding region, *Biochemistry* **24:**7005–7009.

Berliner, L. J., Birktoft, J. J., Miller, T. L., Musci, G., Scheffler, J. W., Shen, Y. Y., and Sugawara, Y., 1986, Thrombin: Active-site topography, *Ann. N.Y. Acad. Sci.* **485:**80–95.

Bezeaud, A., and Guillin, M. C., 1988, Enzymic and non-enzymic properties of human β-thrombin, *J. Biol. Chem.* **263:**3576–3581.

Bing, D. H., Cory, M., and Fenton, J. W., II, 1977, Exosite affinity labeling of human thrombins: Similar labeling on the A chain and B chain (fragments of clotting α- and non-clotting γ/β thrombins), *J. Biol. Chem.* **252:**8027–8034.

Bing, D. H., Laura, L., Robinson, D. J., Furie, B. C., and Feldman, R. J., 1981, A computer-generated three dimensional model of the B chain of bovine α-thrombin, *Ann. N.Y. Acad. Sci.* **370:**496–510.

Bing, D. H., Feldmann, R. J., and Fenton, J. W., II, 1986, Structure and function relationships of thrombin based on the computer generated three-dimensional model of the B chain of bovine thrombin, *Ann. N.Y. Acad. Sci.* **485:**104–119.

Bizios, R., Lai, L., Fenton, J. W., II, and Malik, A. B., 1986, Thrombin-induced chemotaxis and aggregation of neutrophils, *J. Cell. Physiol.* **128:**485–490.

Björk, I., and Lindahl, V., 1982, Mechanism of the anticoagulant action of heparin, *Mol. Cell. Biochem.* **48:**161–182.

Blevins, R. A., and Tulinsky, A., 1985, The refinement and the structure of the dimer of α-chymotrypsin at 1.67 Å resolution, *J. Biol. Chem.* **260:**4264–4275.

Blombäck, B., Blombäck, M., Hessel, B., and Iwanaga, S., 1967, Structure of N-terminal fragments of fibrinogen and specificity of thrombin, *Nature* **215:**1445–1448.

Blombäck, B., Hessel, B., Hogg, D., and Claesson, G., 1977, Substrate specificity of thrombin on proteins and synthetic substrates, in: *Chemistry and Biology of Thrombin* (R. L. Lundblad, J. W. Fenton, II, and K. G. Mann, eds.), Ann Arbor Science, Ann Arbor, pp. 275–290.

Bode, W., 1979, The transition of bovine trypsinogen to a trypsin-like state upon strong ligand binding. II. The binding of the pancreatic trypsin inhibitor and of isoleucine-valine and of sequentially related peptides to trypsinogen and to p-guaninidobenzoate-trypsinogen, *J. Mol. Biol.* **127:**357–374.

Bode, W., and Huber, R., 1986, Crystal structures of pancreatic serine endopeptidases, in: *Molecular and Cellular Basis of Digestion* (P. Desnuelle, ed.), Elsevier, Amsterdam, pp. 213–234.

Bode, W., and Huber, R., 1991, Ligand binding: Proteinase–protein inhibitor interactions, *Curr. Op. Struct. Biol.* **1:**45–52.

Bode, W., and Schwager, P., 1975a, The refined crystal structure of bovine β-trypsin at 1.8 Å resolution. II. Crystallographic refinement, calcium binding site, benzamidine binding site and active site at pH 7, *J. Mol. Biol.* **98:**693–717.

Bode, W., and Schwager, P., 1975b, The single calcium binding site of crystalline bovine β-trypsin, *FEBS Lett.* **56:**139–143.

Bode, W., Schwager, P., and Huber, R., 1978, The transition of bovine trypsinogen to a trypsin-like state upon strong ligand binding. The refined crystal structures of the bovine trypsinogen–pancreatic trypsin inhibitor complex and of its ternary complex with Ile-Val at 1.9 Å resolution, *J. Mol. Biol.* **118:**99–112.

Bode, W., Chen, Z., Bartels, K., Kutzbach, C., Schmidt-Kastner, G., and Bartunik, H., 1983, Refined 2Å X-ray crystal structure of porcine pancreatic kallikrein A, a specific trypsin-like serine proteinase. Crystallization, structure determination, crystallographic refinement, structure and its comparison with bovine trypsin, *J. Mol. Biol.* **164:**237–282.

Bode, W., Walter, J., Huber, R., Wenzel, H. R., and Tschesche, H., 1984, The refined 2.2Å (0.22mm) X-ray crystal structure of the ternary complex formed by bovine trypsinogen, valine–valine, and the Arg15 analogue of bovine pancreatic trypsin inhibitor, *Eur. J. Biochem.* **144:**185–190.

Bode, W., Wei, A. Z., Huber, R., Meyer, E., Travis, J., and Neumann, S., 1986a, X-ray crystal structure of the complex of human leukocyte elastase (PMN elastase) and the third domain of the turkey ovomucoid inhibitor, *EMBO J.* **5:**2453–2458.

Bode, W., Papamokos, E., Musil, D., Seemüller, U., and Fritz, H., 1986b, Refined 1.2Å crystal structure of the complex formed between subtilisin Carlsberg and eglin C. Molecular structure of eglin and its detailed interaction with subtilisin, *EMBO J.* **5:**813–818.

Bode, W., Papamokos, E., and Musil, D., 1987, The high resolution X-ray crystal structure of the complex formed between subtilisin Carlsberg and the inhibitor eglin C, an elastase inhibitor from the leech Hirudo medicinalis. Structural analysis, subtilisin structure and interface geometry, *Eur. J. Biochem.* **166:**673–692.

Bode, W., Mayr, I., Baumann, U., Huber, R., Stone, S. R., and Hofsteenge, J., 1989a, The refined 1.9 Å crystal structure of human α-thrombin: Interaction with D-Phe-Pro-Arg chloromethylketone and significance of the Tyr-Pro-Pro-Trp insertion segment, *EMBO J.* **8:**3467–3475.

Bode, W., Greyling, H. J., Huber, R., Otlewski, J., and Wilusz, T., 1989b, The refined 2.0Å X-ray crystal structure of the complex formed between bovine β-trypsin and CMTI-I, a trypsin inhibitor from squash seeds *(Cucurbita maxima).* Topological similarity of the squash inhibitors with the carboxypeptidase A inhibitor from potatoes, *FEBS Lett.* **242**:285–292.

Bode, W., Turk, D., and Stürzebecher, J., 1990, Geometry of binding of the benzamidine- and arginine-based inhibitors NAPAP and MQPA to human α-thrombin. X-ray crystallographic determination of the NAPAP–trypsin complex and modeling of NAPAP–thrombin and MQPA-thrombin, *Eur. J. Biochem.* **193**:175–182.

Bode, W., Turk, D., and Karshikoff, A., 1992, The refined 1.9 Å X-ray crystal structure of D-Phe-Pro-Arg chloromethylketone-inhibited human α-thrombin. Structure analysis, overall structure, electrostatic properties, detailed active-site geometry, structure-function relationships, in press.

Boissel, J. P., LeBonniec, B., Rabiet, M. J., Labie, D., and Elion, J., 1984, Covalent structures of β and γ autolytic derivatives of human α-thrombin, *J. Biol. Chem.* **259**:5691–5697.

Bourdon, P., Fenton, J. W., II, and Maraganore, J. M., 1990, Affinity labeling of Lys-149 in the anion-binding exosite of human α-thrombin with an N^α (dinitrofluorobenzyl) hirudin C-terminal peptide, *Biochemistry* **29**:6379–6384.

Braun, P. J., Dennis, S., Hofsteenge, J., and Stone, S. R., 1988a, Use of site-directed mutagenesis to investigate the basis for the specificity of hirudin, *Biochemistry* **27**:6517–6522.

Braun, P. J., Hofsteenge, J., Chang, J. Y., and Stone, S. R., 1988b, Preparation and characterization of proteolyzed forms of human α-thrombin, *Thromb. Res.* **50**:273–283.

Brezniak, D. V., Brower, M. S., Witting, J. I., Walz, D. A., and Fenton, J. W., II, 1990, Human α- to ζ-thrombin cleavage occurs with neutrophil cathepsin G or chymotrypsin while fibrinogen clotting activity is retained, *Biochemistry* **29**:3526–3542.

Brower, M. S., Walz, D. A., Garry, E. E., and Fenton, J. W., II, 1987, Thrombin-induced elastase alters human α-thrombin function: Limited proteolysis near the γ-cleavage site results in decreased fibrinogen clotting and platelet stimulatory activity, *Blood* **69**:813–819.

Burley, S. K., and Petsko, G. A., 1985, Aromatic-aromatic interaction: A mechanism of protein structure stabilization, *Science* **229**:23–28.

Butkowski, R. J., Elion, J., Downing, M. R., and Mann, K. G., 1977, Primary structure of human prethrombin 2 and α-thrombin, *J. Biol. Chem.* **252**:4942–4957.

Cardin, A. D., and Weintraub, H. J. R., 1989, Molecular modeling of protein–glucosaminoglycan interactions, *Arteriosclerosis* **9**:21–32.

Carney, D. E., Herbosa, G. J., Stiernberg, J., Bergmann, J. S., Gordon, E. A., Scott, D. L., and Fenton, J. W., II, 1986, Double signal hypothesis for thrombin initiation of cell proliferation, *Semin. Thromb. Hemostas.* **12**:231–240.

Chang, J.-Y., 1983, The functional domain of hirudin, a thrombin-specific inhibitor, *FEBS Lett.* **164**:307–313.

Chang, J.-Y., 1985, Thrombin specificity. Requirement for apolar amino acid adjacent to the thrombin cleavage site of polypeptide substrates, *Eur. J. Biochem.* **151**:217–224.

Chang, J.-Y., 1986, The structures and proteolytic specificities of autolysed human thrombin, *Biochem. J.* **240**:797–802.

Chang, J.-Y., 1989, The hirudin-binding site of human α-thrombin: Identification of lysyl residues which participate in the combining site of hirudin–thrombin complex, *J. Biol. Chem.* **264**:7141–7146.

Chang, T. L., Feinman, R. D., Laudis, B. H., and Fenton, W., II, 1979, Antithrombin reactions with α- and γ-thrombins, *Biochemistry* **18**:113–119.

Charo, I. F., Bekeart, L.S., and Phillips, D. R., 1987, Platelet glycoprotein IIb-IIIa-like proteins mediate endothelial cell attachment to adhesive proteins and the extracellular matrix, *J. Biol. Chem.* **262**:9935–9938.

Chen, L. B., and Buchanan, J. M., 1975, Mitogenic activity of blood components. I. Thrombin and prothrombin, *Proc. Natl. Acad. Sci. USA* **72**:131–135.

Chow, M. M., Meyer, E. F., Jr., Bode, W., Kam, C.-M., Radhakrishnan, R., Vijayalakshmi, J., and Powers, J. C., 1990, The 2.2Å resolution X-ray crystal structure of the complex of trypsin inhibited by 4-chloro-3-ethoxy-7-guanidinoisocoumarin: A proposed model of the thrombin–inhibitor complex, *J. Am. Chem. Soc.* **112**:7783–7789.

Church, F. C., Pratt, C. W., Noyes, C. M., Kalayanamit, T., Sherril, G. B., Tobin, R. B., and Meade, B., 1989, Structural and functional properties of human α-thrombin, phosphopyridoxylated α-thrombin and γ_1-thrombin. Identification of lysyl residues in α-thrombin that are critical for heparin and fibrin(ogen) interactions, *J. Biol. Chem.* **264**:18419–18425.

Cohen, H. G., Silverton, E. W., and Davis, D., 1981, Refined crystal structure of γ-chymotrypsin at 1.9Å resolution. Comparison with other pancreatic serine proteases, *J. Mol. Biol.* **148**:449–479.

Cunningham, D. D., and Farrell, D. H., 1986, Thrombin interactions with cultured fibroblasts: Relationship to mitogenic stimulation, *Ann. N.Y. Acad. Sci.* **485**:240–248.

Degen, S. J. F., Gillivray, R. T. A., and Davie, E. W., 1983, Characterization of the complementary deoxyribonucleic acid and gene coding for human prothrombin. *Biochemistry* **22**:2087–2097.

Degryse, E., Acker, M., Defreyn, G., Bernat, A., Maffrand, J. P., Roitsch, C., and Courtney, M., 1989, Point mutations modifying the thrombin inhibition kinetics and antithrombin activity in vivo of recombinant hirudin, *Protein Engin.* **2**:459–465.

Dodt, J., Müller, H., Seemüller, U., and Chang, J.-Y., 1984, The complete amino acid sequence of hirudin, a thrombin specific inhibitor. *FEBS Lett.* **165**:180–183.

Dodt, J., Seemüller, U., Maschler, R., and Fritz, H., 1985, The complete covalent structure of hirudin. Localisation of the disulfide bonds, *Biol. Chem. Hoppe-Seyler* **366**:379–385

Dodt, J., Köhler, S., and Baici, A., 1988, Interaction of site specific hirudin variants with α-thrombin, *FEBS Lett.* **229**:87–90.

Doyle, M. F., and Mann, K. G., 1990, Multiple active forms of thrombin IV. Relative activities of meizothrombins, *J. Biol. Chem.* **265**:10693–10710.

Elion, J., Boissel, J. P., LeBonniec, B., Bezeaud, A., Jandrot-Perrus, M., Rabiet, M. J., and Guillin, M. C., 1986, Proteolytic derivatives of thrombin, *Ann. N.Y. Acad. Sci.* **485**:16–26.

Esmon, N. L., Owen, W. G., and Esmon, C. T., 1982, Isolation of membrane-bound cofactor for thrombin-catalyzed activation of protein C, *J. Biol. Chem.* **257**:859–864.

Esmon, N. L., Carrol, R. C., and Esmon, C. T., 1983, Thrombomodulin blocks the ability of thrombin to activate platelets, *J. Biol. Chem.* **258**:12238–12242.

Fenton, J. W., II, 1981, Thrombin specificity, *Ann. N.Y. Acad. Sci.* **370**:468–495.

Fenton, J. W., II, 1986, Thrombin, *Ann. N.Y. Acad. Sci.* **485**:5–15.

Fenton, J. W., II, 1988a, Regulation of thrombin generation and functions, *Semin. Thromb. Hemostas.* **14**:234–240.

Fenton, J. W., II, 1988b, Thrombin bioregulatory functions, *Adv. Clin. Enzymol.* **6**:186–193.

Fenton, J. W., II, 1989, Thrombin interactions with hirudin, *Semin. Thromb. Hemostas.* **15**:265–268.

Fenton, J. W., II, and Bing, D. H., 1986, Thrombin active-site regions, *Semin. Thromb. Hemostas.* **12**:200–208.

Fenton, J. W., II, Fasco, M. J., Stackrow, A. B., Arouson, D. L., Young, A. M., and Finlayson, J. S., 1977a, Human thrombins: Production, evaluation and properties of α-thrombin, *J. Biol. Chem.* **252**:3587–3598.

Fenton, J. W., II, Landis, B. H., Walz, D. H., and Finlayson, J. S., 1977b, Human thrombins, in: *Chemistry and Biology of Thrombin* (R. L. Lundblad, J. W. Fenton, II, and K. G. Mann, eds.), Ann Arbor Science, Ann Arbor, pp. 43–70.

Fenton, J. W., II, Zabinski, M. P., Hsieh, K., and Wilner, G. D., 1981, Thrombin non-covalent protein binding and fibrin(ogen) recognition, *Thromb. Haemostas.* **46:**177.

Fenton, J. W., II, Olson, T. A., Zabinski, M. P., and Wilner, G. D., 1988, Anion binding exosite of human α-thrombin and fibrin(ogen) recognition, *Biochemistry* **27:**7106–7112.

Fenton, J. W., II, Witting, J. I., Pouliott, C., and Fareed, J., 1989, Thrombin anion-binding exosite interactions with heparin and various polyanions. *Ann. N.Y. Acad. Sci.* **556:**158–165.

Finzel, B. C., 1987, Incorporation of fast Fourier transforms to speed restrained least-squares refinement of protein structures, *J. Appl. Crystallogr.* **20:**53–55.

Folkers, P. J. M., Clore, G. M., Driscoll, P. C., Dodt, J., Köhler, S., and Gronenborn, A. M., 1989, Solution structure of recombinant hirudin and the Lys-47–Glu mutant: A nuclear magnetic resonance and hybrid distance geometry–dynamical simulated annealing study, *Biochemistry* **28:**2601–2617.

Furie, B., Bing, D. H., Feldmann, R. J., Robinson, D. J., Burnier, J. P., and Furie, B. C., 1982, Computer-generated models of blood coagulation factor Xa, factor IXa and thrombin based upon structural homology with other serine proteases, *J. Biol. Chem.* **257:**3875–3882.

Glenn, K. C., Frost, G. H., Bergmann, J. S., and Carney, D. H., 1988, Synthetic peptides bind to high-affinity thrombin receptors and modulate thrombin mitogenesis, *Pept. Res.* **1:**65–73.

Greer, J., 1981a, Comparative model-building of the mammalian serine proteinases, *J. Mol. Biol.* **153:**1027–1042.

Greer, J., 1981b, Model of a specific interaction. Salt-bridges form between prothrombin and its activating enzyme blood clotting factor X_a, *J. Mol. Biol.* **153:**1043–1053.

Griffith, M. J., 1979, Covalent modification of human α-thrombin with pyridoxal 5′-phosphate, *J. Biol. Chem.* **254:**3401–3406.

Griffith, M. J., 1982, Kinetics of the heparin-enhanced antithrombin III/thrombin reaction. Evidence for a template model for the mechanism of action of heparin, *J. Biol. Chem.* **257:**7360–7365.

Griffith, M. J., Kingdon, H. S., and Lundblad, R. L., 1979, The interaction of heparin with human α-thrombin: Effect on the hydrolysis of anilide tripeptide substrates, *Arch. Biochem. Biophys.* **195:**378–384.

Gronke, R. S., Bergman, B. L., and Baker, J. B., 1987, Thrombin interaction with platelets. Influence of a platelet protease nexin, *J. Biol. Chem.* **262:**3030–3036.

Grütter, M. G., Priestle, J. P., Rahuel, J., Grossenbacher, H., Bode, W., Hofsteenge, J., and Stone, S. R., 1990, Crystal structure of the thrombin–hirudin complex: A novel mode of serine protease inhibitor, *EMBO J.* **9:**2361–2365.

Hageman, T. C., and Scheraga, H. A., 1974, Mechanism of action of thrombin on fibrinogen: Reaction of the N-terminal CNBr fragment from the Aα chain of human fibrinogen with bovine thrombin, *Arch. Biochem. Biophys.* **164:**707–715.

Hageman, T. C., Endres, G. F., and Scheraga, H. A., 1975, Mechanism of action of thrombin on fibrinogen; on the role of the A chain of bovine thrombin in specificity and differentiating between thrombin and trypsin, *Arch. Biochem. Biophys.* **171:**327–336.

Hartley, B. S., and Kauffman, D., 1966, Corrections to the amino acid sequence of bovine chymotrypsinogen A, *Biochem. J.* **101:**229.

Hartley, B. S., and Shotton, D. M., 1971, Pancreatic elastase, in: *The Enzymes* (P. D. Boyer, ed.), Academic Press, New York, Vol. 3, pp. 323–373.

Haruyama, H., and Wüthrich, K., 1989, The conformation of recombinant desulfatohirudin in aqueous solution determined by nuclear magnetic resonance, *Biochemistry* **28:**4301–4312.

Harvey, R. P., Degryse, E., Stefani, L., Schamber, F., Cazenave, J.-P., Courtney, M., Tolstoshev, P., and Lecocq, J.-P., 1986, Cloning and expression of a cDNA coding for the anticoagulant hirudin from the bloodsucking leech, *Hirudo medicinalis, Proc. Natl. Acad. Sci, USA* **83:**1084–1088.

Henriksen, R. A., and Mann, K. G., 1988, Identification of the primary structural defect in the dysthrombin thrombin Quick I: Substitution of cysteine for arginine-382, *Biochemistry* **27:**9160–9165.

Henriksen, R. A., and Mann, K. G., 1989, Substitution of valine for glycine-558 in the congenital dysthrombin thrombin Quick II alters primary substrate specificity, *Biochemistry* **28:**2078–2082.

Heuck, C. C., Schiele, V., Horn, D., Fronda, D., and Ritz, E., 1985, The role of surface charge on the accelerating action of heparin on the antithrombin III-inhibited activity of α-thrombin, *J. Biol. Chem.* **260:**4598–4603.

Hofsteenge, J., and Stone, S. R., 1987, The effect of thrombomodulin on the cleavage of fibrinogen fragments by thrombin, *Eur. J. Biochem.* **168:**49–56.

Hofsteenge, J., Taguchi, H., and Stone, S. R., 1986, Effect of thrombomodulin on the kinetics of the interaction of thrombin with substrates and inhibitors, *Biochem. J.* **237:**243–251.

Hofsteenge, J., Braun, P. J., and Stone, S. R., 1988, Enzymatic properties of proteolytic derivatives of human α-thrombin, *Biochemistry* **27:**2144–2151.

Hogg, D. H., and Blombäck, B., 1978, The mechanism of the fibrinogen–thrombin reaction, *Thromb. Res.* **12:**953–964.

Hogg, P. J., and Jackson, C. M., 1989, Fibrin monomer protects thrombin from inactivation by heparin–antithrombin III: Implications for heparin efficacy, *Proc. Natl. Acad. Sci. USA* **86:**3619–3623.

Horne, M. K., and Gralnick, H. R., 1983, The oligosaccharide of human thrombin: Investigations of functional significance, *Blood* **63:**188–194.

Huber, R., and Bode, W., 1978, Structural basis of the activation and action of trypsin, *Acc. Chem. Res.* **11:**114–122.

Huber, R., and Carrell, R. W., 1989, Implications of the three-dimensional structure of α_1-antitrypsin for structure and function of serpins, *Biochemistry* **28:**8951–8966.

Huber, R., Kukla, D., Rühlmann, A., Epp, O., and Formanek, H., 1970, The basic trypsin inhibitor of bovine pancreas. I. Structure analysis and conformation of the polypeptide chain, *Naturwissenschaften* **57:**389.

Huber, R., Kukla, D., Bode, W., Schwager, P., Bartels, K., Deisenhofer, J., and Steigemann, W., 1974, Structure of the complex formed by bovine trypsin and bovine pancreatic trypsin inhibitor. II. Crystallographic refinement at 1.9Å resolution, *J. Mol. Biol.* **89:**73–101.

Jack, A., and Levitt, M., 1978, Refinement of large structures by simultaneous minimization of energy and R- factor, *Acta Crystallogr.* **A34:**931–935.

Jackson, C. M., and Nemerson, Y., 1980, Blood coagulation, *Annu. Rev. Biochem.* **49:**765–811.

Johnson, P. J., Sze, P., Winant, R., Payne, P. W., and Lazar, J. B., 1989, Biochemistry and genetic engineering of hirudin, *Semin. Thromb. Hemostas.* **15:**302–315.

Kabsch, W., and Sander, C., 1983, Dictionary of protein secondary structure: Pattern, recognition of hydrogen-bonded and geometrical features, *Biopolymers* **22:**2577–2637.

Kaczmarek, E., and McDonagh, J., 1988, Thrombin binding to the Aα- Bβ- and γ-chains of fibrinogen and to their remnants contained in fragment E, *J. Biol. Chem.* **263:**13896–13900.

Kaminski, M., and McDonagh, J., 1987, Inhibited thrombins: Interactions with fibrinogen and fibrin, *Biochem. J.* **242:**881–887.

Karpatkin, S., and Karpatkin, M., 1974, Inhibition of the enzymatic activity of thrombin by concanavalin A, *Biochem. Biophys. Res. Commun.* **57:**1111–1118.

Karshikoff, A. D., Engh, R., Bode, W., and Atanasov, B. P., 1989, Electrostatic interactions in proteins: Calculations of the electrostatic term of free energy and the electrostatic potential field, *Eur. Biophys. J.* **17:**287–297.

Kawabata, S., Morita, T., Iwanaga, S., and Igarashi, H., 1985, Staphylocoagulase-binding region in human prothrombin, *J. Biochem.* **97:**325–331.

Kettner, C., and Shaw, E., 1979, D-Phe-Pro-Arg CH_2Cl—a selective affinity label for thrombin, *Thromb. Res.* **14:**969–973.

Kettner, C., and Shaw, E., 1981, Inactivation of trypsin-like enzymes with peptides of arginine chloromethyl ketone, *Methods Enzymol.* **80:**826–842.

Konno, S., Fenton, J. W., II, and Villanueva, G. B., 1988, Analysis of the secondary structure of hirudin and the mechanism of its interaction with thrombin, *Arch. Biochem. Biophys.* **267:**158–166.

Krstenansky, J. L., and Mao, S. J. T., 1987, Antithrombin properties of C-terminus of hirudin using synthetic unsulfated N-α-acetyl-hirudin 45–65, *FEBS Lett.* **211:**10–16.

Laskowski, M., and Kato, I., 1980, Protein inhibitors of proteinases, *Annu. Rev. Biochem.* **49:**593–626.

Laskowski, M., Kato, F., Ardelt, W., Cook, J., Denton, A., Empie, M. W., Kohr, W. J., Park, S. J., Parks, K., Schatzley, B. L., Schoenberger, O. L., Tashiro, M., Vichot, G., Whatley, H. E., Wieczorek, A., and Wieczorek, M., 1987, Protein third domains from 100 avian species: Isolation, sequences and hypervariability of enzyme–inhibitor contact residues, *Biochemistry* **26:**202–221.

Lewis, S. D., Lorand, L., Fenton, J. W., II, and Shafer, J. A., 1987, Catalytic competence of human α- and γ-thrombin in the activation of fibrinogen and factor XIII, *Biochemistry* **26:**7597–7603.

Li, E. H. H., Fenton, J. W., II, and Feinman, R. D., 1976, The role of heparin in the thrombin–antithrombin III reaction, *Arch. Biochem. Biophys.* **175:**153–159.

Liem, R. K. H., and Scheraga, H. A., 1974, Mechanism of action of thrombin on fibrinogen IV. Further mapping of the active sites of thrombin and trypsin. The binding of thrombin and fibrin, *Arch. Biochem. Biophys.* **160:**333–339.

Liu, C. Y., Nossel, H. L., and Kaplan, K. L., 1979, The binding of thrombin by fibrin, *J. Biol. Chem.* **254:**10421–10425.

Löbermann, H., Tokuoka, R., Deisenhofer, J., and Huber, R., 1984, Human α₁-proteinase inhibitor. Crystal structure analysis of two crystal modifications, molecular model and preliminary analysis of the implications for function, *J. Mol. Biol.* **177:**531–556.

Lottenberg, R., Hall, J. A., Blinder, M., Binder, E. P., and Jackson, C. M., 1983, The action of thrombin on peptide p-nitroanilide substrates. Substrate selectivity and examination of hydrolysis under different reaction conditions, *Biochim. Biophys. Acta* **742:**539–557.

Lundblad, R. L., Noyes, C. M., Mann, K. G., and Kingdon, H. S., 1979, The covalent differences between bovine α- and β-thrombin: A structural explanation for the changes in catalytic activity, *J. Biol. Chem.* **254:**8524–8528.

Lundblad, R. L., Noyes, C. M., Featherstone, G. L., Harrison, J. H., and Jenzano, J. W., 1988, The reaction of bovine α-thrombin with tetranitromethane. Characterization of the modified protein, *J. Biol. Chem.* **263:**3729–3734.

McGillivray, R. T. A., and Davie, E. W., 1984, Characterization of bovine prothrombin mRNA and its translation product, *Biochemistry* **23**:1626–1634.

McGowan, E. B., and Detwiler, T. C., 1986, Modified platelet responses to thrombin. Evidences for two types of receptors or coupling mechanisms, *J. Biol. Chem.* **261**:739–746.

McKay, D. B., Kay, L. M., and Stroud, R. M., 1977, Preliminary crystallization and X-ray diffraction studies of human thrombin, in: *Chemistry and Biology of Thrombin* (R. L. Lundblad, J. W. Fenton, II, and K. G. Mann, eds.), Ann Arbor Science, Ann Arbor, pp. 113–121.

Magnusson, S., 1971, Thrombin and prothrombin, *Enzymes* (3rd ed.), **3**:277–321.

Magnusson, S., Peterson, T. E., Sottrup-Jensen, L., and Claeys, H., 1975, Complete primary structure of prothrombin: Isolation, structure and reactivity of ten carboxylated glutamic acid residues and regulation of prothrombin activation by thrombin, in: *Proteases and Biological Control* (E. Reich, D. B. Rifkin, and E. Shaw, eds.), Cold Spring Harbor Laboratory Press, Cold Spring Harbor, N.Y., pp. 123–149.

Mao, S. J. T., Yates, M. T., Owen, T. J., and Krstenansky, J. L., 1988, Interaction of hirudin with thrombin: Identification of a minimal binding domain of hirudin that inhibits clotting activity, *Biochemistry* **27**:8170–8173.

Markwardt, F., 1970, Hirudin as an inhibitor of thrombin, *Methods Enzymol.* **19**:924–932.

Marsh, H. C., Meinwald, Y. C., Lee, S., and Scheraga, H. A., 1983, Mechanism of action of thrombin on fibrinogen: NMR evidence for a β-bend at or near fibrinogen Aα Gly (P_5)–Gly (P_4), *Biochemistry* **21**:6167–6171.

Martin, P. D., Robertson, W., Turk, D., Huber, R., Bode, W., and Edwards, B. F. P., 1992, The structure of residues 7–16 of the Aα chain of human fibrinogen bound to bovine thrombin at 2.3Å resolution, *J. Biol. Chem.*, Submitted.

Matsuzaki, T., Sasaki, C., Okumura, C., and Umeyama, H., 1989, X-ray analysis of a thrombin inhibitor–trypsin complex, *J. Biochem.* **105**:949–952.

Meinwald, Y. C., Martinelli, R. A., van Nispen, J. W., and Scheraga, H. A., 1980, Mechanism of action of thrombin on fibrinogen. Size of the Aα fibrinogen-like peptide that contacts the active site of thrombin, *Biochemistry* **19**:3820–3825.

Meloun, B., Kluh, I., Kostka, V., Moravek, L., Prusik, Z., Vanecek, J., Keil, B., and Sorm, F., 1966, Covalent structure of bovine chymotrypsin A, *Biochim. Biophys. Acta* **130**:543.

Meyer, E., Cole, G., Radhakrishnan, R., and Epp, O., 1988, Structure of native porcine pancreatic elastase at 1.65 Å resolution, *Acta Crystallogr.* **B44**:26–38.

Mikes, O., Holeysoksky, V., Tomasek, V., and Sorm, F., 1966, Covalent structure of bovine trypsinogen. The position of the remaining amides, *Biochem. Biophys. Res. Commun.* **24**:346–352.

Miyata, T., Morita, T., Inomoto, T., Kawauchi, S., Shirakami, A., and Iwanaga, S., 1987, Prothrombin Tokushima, a replacement of arginine-418 by tryptophan that impairs the fibrinogen clotting activity of derived thrombin Tokushima, *Biochemistry* **26**:1117–1122.

Nagy, J. A., Meinwald, Y. C., and Scheraga, H. A., 1982, Immunochemical determination of conformational equilibria for fragments of the Aα chain of fibrinogen, *Biochemistry* **21**:1794–1806.

Naski, M. C., Fenton, J. W., II, Maraganore, J. M., Olson, S. T., and Shafer, J. A., 1990, The COOH-terminal domain of hirudin. An exosite-directed competitive inhibitor of the action of α-thrombin on fibrinogen, *J. Biol. Chem.* **265**:13484–13489.

Nesheim, M. E., 1983, A simple rate law that describes the kinetics of the heparin-catalyzed reaction between antithrombin III and thrombin, *J. Biol. Chem.* **258**:14708–14717.

Ni, F., Konishi, Y., Frazier, R. B., and Scheraga, H. A., 1989a, High-resolution NMR studies of fibrinogen-like peptides in solution: Interaction of thrombin with residues 1–23 of the Aα chain of human fibrinogen, *Biochemistry* **28:**3082–3094.

Ni, F., Konishi, Y., Bullock, L. D., Rivetna, M. N., and Scheraga, H. A., 1989b, High resolution NMR studies of fibrinogen-like peptides in solution: Structural basis for the bleeding disorder caused by the single mutation of Gly (12) to Val (12) in the Aα chain of human fibrinogen Rouen, *Biochemistry* **28:**3106–3119.

Ni, F., Konishi, Y., and Scheraga, H. A., 1990, Thrombin-bound conformation of the C-terminal fragments of hirudin determined by transferred nuclear Overhauser effects, *Biochemistry* **29:**4479–4489.

Nilsson, B., Horne, M. K., and Gralnick, H. R., 1983, The carbohydrate of human thrombin: Structural analysis of glycoprotein oligosaccharides by mass spectrometry, *Arch. Biochem. Biophys.* **224:**127–133.

Noé, G., Hofsteenge, J., Rovelli, G., and Stone, S. R., 1988, The use of sequence specific antibodies to identify a secondary binding site in thrombin, *J. Biol. Chem.* **263:**11729–11735.

Okamoto, S., Hijikata, A., Kikumoto, R., Tonomura, S., Hara, N., Ninomiya, K., Maruyama, A., Sugano, M., and Tamao, Y., 1981, Potent inhibition of thrombin by the newly synthesized arginine derivative No. 805. The importance of stereostructure of its hydrophobic carboxamide portion, *Biochem. Biophys. Res. Commun.* **101:**440–446.

Olson, S. T., and Shore, J. D. 1982, Demonstration of a two-step reaction mechanism for inhibition of α-thrombin by antithrombin III and identification of the step affected by heparin, *J. Biol. Chem.* **257:**14891–14895.

Olson, T. A., Sonder, S. A., Wilner, G. D., and Fenton, J. W., II, 1986, Heparin binding in proximity to the catalytic site of human α-thrombin, *Ann. N.Y. Acad. Sci.* **485:**96–103.

Pomerantz, M. W., and Owen, W. G., 1978, A catalytic role for heparin: Evidence for a ternary complex of heparin cofactor thrombin and heparin, *Biochim. Biophys. Acta* **535:**66–77.

Preissner, K. T., DelVos, V., and Müller-Berghaus, G., 1987, Binding of thrombin to thrombomodulin accelerates inhibition of the enzyme by antithrombin III. Evidence for a heparin-independent mechanism, *Biochemistry* **26:**2521–2528.

Prescott, S. M., Seeger, A. R., Zimmerman, G. A., McIntyre, T. M., and Maraganore, J. M., 1990, Hirudin-based peptides block the inflammatory effects of thrombin on endothelial cells, *J. Biol. Chem.* **265:**9614–9616.

Priestle, J. P., 1988, Ribbon: A stereo cartoon drawing program for proteins, *J. Appl. Crystallogr.* **21:**572–576.

Read, R. J., and James, M. N. G., 1986, Introduction to the proteinase inhibitors: X-ray crystallography, in: *Proteinase Inhibitors* (A. J. Barret and G. Salvesen, eds.), Elsevier, Amsterdam, pp. 301–336.

Richardson, J. S., and Richardson, D. C., 1990, Principles and patterns of protein conformation, in: *Prediction of Protein Structure and the Principles of Protein Conformation* (G. Fasman, ed.), Plenum Press, New York, pp. 1–98.

Rosenberg, R. D., 1977, Chemistry of the hemostatic mechanism and its relationship to the action of heparin, *Fed. Proc. Fed. Am. Soc. Exp. Biol.* **36:**10–18.

Rosenberg, R. D., and Damus, P. S., 1973, The purification and mechanism of action of human antithrombin–heparin cofactor, *J. Biol. Chem.* **248:**6490–6505.

Rossmann, M. G., and Argos, P., 1975, A comparison of the heme binding pocket in globins and cytochrome b$_5$, *J. Biol. Chem.* **250:**7525–7532.

Rydel, T. J., Ravichandran, K. G., Tulinsky, A., Bode, W., Huber, R., Roitsch, C., and Fenton, J. W., II, 1990, The structure of a complex of recombinant hirudin and human α-thrombin, *Science* **249**:277–280.

Rydel, T. J., Tulinsky, A., Bode, W., and Huber, R., 1991, Refined structure of the hirudin–thrombin complex, *J. Mol. Biol.* **221**:583–601.

Schechter, I., and Berger, A., 1967, On the size of the active site in proteases. I. Papain, *Biochem. Biophys. Res. Commun.* **27**:157.

Scheraga, H. A., 1977, Active site mapping of thrombin, in: *Chemistry and Biology of Thrombin* (R. L. Lundblad, J. W. Fenton, II, and K. G. Mann, eds.), Ann Arbor Science, Ann Arbor, pp. 145–158.

Shuman, M. A., 1986, Thrombin–cellular interactions, *Ann. N.Y. Acad. Sci.* **485**:228–239.

Skaug, K., and Christenson, T. B., 1971, The significance of the carbohydrate constituents of bovine thrombin for the clotting activity, *Biochim. Biophys. Acta* **230**:627.

Skrzypczak-Jankun, E., Rydel, T. J., Tulinsky, A., Fenton, J. W., II, and Mann, K. G., 1989, Human D-Phe-Pro-Arg-CH_2-α-thrombin: Crystallization and diffraction data, *J. Mol. Biol.* **206**:755–757.

Skrzypczak-Jankun, E., Carperos, V. E., Ravichandran, K. G., Tulinsky, A., Westbrook, M., and Maraganore, J. M., 1991, Structure of the hirugen and hirulog1 complexes of α-thrombin, *J. Mol. Biol.* **221**:1379–1393.

Sonder, S. A., and Fenton, J. W., II, 1984, Proflavin binding within the fibrinopeptide groove adjacent to the catalytic site of human α-thrombin, *Biochemistry* **23**:1818–1823.

Stein, P. E., Leslie, A. G. W., Finch, J. T., Turnell, W. G., McLaughlin, P. J., and Carrell, R., 1990. Crystal structure of ovalbumin as a model for the reactive center of serpins, *Nature* **347**:99–102.

Stone, S. R., and Hofsteenge, J., 1986, Kinetics of the inhibition of thrombin by hirudin, *Biochemistry* **25**:4622–4628.

Stone, S. R., Braun, P. J., and Hofsteenge, J., 1987, Identification of regions of α-thrombin involved in its interaction with hirudin, *Biochemistry* **26**:4617–4624.

Stubbs, M., Oschkinat, H., Mayr, I., Huber, R., Angliker, H., Stone, S. R., and Bode, W., 1992, The interaction of thrombin with fibrogen—a structural basis for its specificity, *Eur. J. Biochem., in press.*

Stürzebecher, J., Markwardt, F. Voigt, B., Wagner, G., and Walsmann, P., 1983, Cyclic amides of $N^α$-arylsulfonylaminoacylated 4-amidinophenylalanine—Tight binding inhibitors of thrombin, *Thromb. Res.* **29**:635–642.

Sugawara, Y., Birktoft, J. J., and Berliner, L. J., 1986, Human α- and γ-thrombin inhibition by trypsin inhibitors supports predictions from molecular graphics experiments, *Semin. Thromb. Hemostas.* **12**:209–210.

Suzuki, K., Nishioka, J., and Hayashi, T., 1990, Localization of thrombomodulin-binding site within human thrombin, *J. Biol. Chem.* **265**:13263–13267.

Svendsen, L., Blombäck, B., Blombäck, M., and Olsson, P. I., 1972, Synthetic chromogenic substrates for determination of trypsin, thrombin and thrombin-like enzymes, *Thromb. Res.* **1**:267–278.

Thomas, K. A., Smith, G. M., Thomas, T. B., and Feldmann, R. J., 1982, Electronic distributions within protein phenylalanine aromatic rings are reflected by the three-dimensional oxygen atom environments, *Proc. Natl. Acad. Sci. USA* **79**:4843–4847.

Thompson, A. R., 1976, High affinity binding of human and bovine thrombin to p-chlorobenzylamido-ε-aminocaproyl-agarose, *Biochim. Biophys. Acta* **422**:200–209.

Tollefsen, D. M., Feagler, J. R., and Majerus, P. W., 1974, The binding of thrombin to the surface of human platelets, *J. Biol. Chem.* **249**:2646–2651.

Tollefsen, D. M., Majerus, P. W., and Blank, M. K., 1982, Heparin cofactor II. Purification and properties of a heparin-dependent inhibitor of thrombin in human plasma, *J. Biol.*

Chem. **257:**2162–2169.

Toma, K., and Suzuki, K., 1989, Mapping active sites of blood coagulation serine proteinases—activated protein C and thrombin—on simple graphics models, *J. Mol. Graphics* **7:**146–149.

Tsernoglou, D., Walz, D. A., McCoy, L. E., and Seegers, W. H., 1974, Crystallization and X-ray diffraction studies of bovine thrombin, *J. Biol. Chem.* **249:**999.

Tsukuda, H., and Blow, D., 1985, Structure of α-chymotrypsin refined at 1.68Å resolution, *J. Mol. Biol.* **184:**703–711.

Turk, D., Sturzebecher, J., and Bode, D., 1991, Geometry of binding of the Nα-tosylated piperidides of m-amidino, p-amidino and p-guanidino phenylalanine to thrombin and trypsin, *FEBS Lett.* **1:**133–138.

Vali, Z., and Scheraga, H. A., 1988, Localization of the binding site on fibrin for the secondary binding site of thrombin, *Biochemistry* **27:**1956–1963.

van Nispen, J. W., Hageman, T. C., and Scheraga, H. A., 1977, Mechanism of action of thrombin on fibrinogen: The reaction of thrombin with fibrinogen-like peptides containing 11, 14, and 16 residues, *Arch. Biochem. Biophys.* **182:**227–243.

Wallace, A., Rovelli, G., Hofsteenge, J., and Stone, S. R., 1989, Effect of heparin on the glia-derived nexin–thrombin interaction, *Biochem. J.* **257:**191–196.

Walsh, K. A., and Neurath, H., 1964, Trypsinogen and chymotrypsinogen as homologous proteins, *Proc. Natl. Acad. Sci. USA* **52:**884.

Walsmann, P., and Markwardt, F., 1981, Biochemische und pharmakologische Aspekte des Thrombininhibitors Hirudin, *Pharmazie* **36:**653–660.

Walz, D. A., Hewett-Emmett, D., and Seegers, W. H., 1977, Amino acid sequence of human prothrombin fragments 1 and 2, *Proc. Natl. Acad. Sci. USA* **74:**1962–1972.

Wang, D., Bode, W., and Huber, R., 1985, Bovine chymotrypsinogen A. X-ray crystal structure analysis and refinement of a new crystal form at 1.8 Å resolution, *J. Mol. Biol.* **185:**595–624.

White, G. C., Lundblad, R. L., and Griffith, M. J., 1981, Structure–function relations in platelet–thrombin reactions, *J. Biol. Chem.* **256:**1763–1766.

Wright, H. T., Qian, H. X., and Huber, R., 1990, Crystal structure of plakalbumin, a proteolytically nicked form of ovalbumin. Its relationship to the structure of cleaved α$_1$-proteinase inhibitor, *J. Mol. Biol.* **213:**513–528.

Chapter 2

NUCLEAR MAGNETIC RESONANCE STUDIES OF THROMBIN–FIBRINOPEPTIDE AND THROMBIN–HIRUDIN COMPLEXES

Feng Ni, Kenneth D. Gibson, and Harold A. Scheraga

1. INTRODUCTION

The thrombin–fibrinogen interaction to form the fibrin clot proceeds in three reversible steps (Scheraga and Laskowski, 1957), with thrombin being involved in only the first one:

Step 1: Proteolysis $\quad$ F $\overset{T}{\rightleftharpoons}$ f + P

Step 2: Polymerization $\quad nf \rightleftharpoons f_n$

Step 3: Clotting $\quad mf_n \rightleftharpoons$ fibrin

Feng Ni • Biotechnology Research Institute, National Research Council of Canada, Montréal, Québec, Canada H4P 2R2. **Kenneth D. Gibson and Harold A. Scheraga** • Baker Laboratory of Chemistry, Cornell University, Ithaca, New York 14853-1301.

Thrombin: Structure and Function, edited by Lawrence J. Berliner. Plenum Press, New York, 1992.

where F is fibrinogen, T is thrombin, f is the fibrin monomer, P represents the fibrinopeptides A and B (two each) released by proteolytic cleavage of the Arg–Gly bonds in the Aα and Bβ chains, respectively, of fibrinogen, f_n is a series of intermediate, rodlike staggered overlapped polymers of variable length (i.e., varying n), and fibrin is the clotted product formed by the incorporation of m rodlike polymers in an organized networklike structure.

Considerable chemical and physicochemical evidence has been accumulated about the equilibrium of Step 1 (Laskowski *et al.*, 1960) and about the energetics of the polymerization of fibrin monomer in Step 2 (Sturtevant *et al.*, 1955; Endres *et al.*, 1966). It appears that one of the polymerization sites is near the C-terminus of the γ chains of fibrinogen (Varadi and Scheraga, 1986) but that the other one, near the N-termini of the α and β chains, becomes available only after the proteolytic release of fibrinopeptides A and B by thrombin in Step 1 (Laudano and Doolittle, 1980). Polymerization of the fibrin monomer by means of hydrogen-bond formation between tyrosyl donors and histidyl acceptors (possibly in cooperation with hydrophobic interactions) accounts for the pH range of clotting, the pH dependence of the release and uptake of protons during polymerization, and the pH dependence of the enthalpy of polymerization (Sturtevant *et al.*, 1955; Endres *et al.*, 1966; Scheraga, 1983).

Our more recent work on the thrombin–fibrinogen interaction has focused on Step 1. For this purpose, we have carried out active-site mapping studies by examining the kinetics of hydrolysis of the Arg–Gly bonds of fibrinogen and of fibrinogenlike peptides, with sequences from the N-termini of the Aα and Bβ chains (Scheraga, 1986, and references therein), and have determined the immunochemical properties of these oligopeptides (Nagy *et al.*, 1982, 1985). It appears that the Arg–Gly bonds of *both* the Aα and Bβ chains of fibrinogen are accessible to thrombin and to antibodies, and that hydrolysis of both chains proceeds simultaneously and competitively, with no lag period, when thrombin is added to fibrinogen, but that the rate of hydrolysis of the Arg–Gly bond of the Aα chain is much greater than that of the Bβ chain (Martinelli and Scheraga, 1980; Hanna *et al.*, 1984).

Synthetic fibrinogenlike peptides containing the Arg 16–Gly 17 peptide bond are also good substrates for thrombin, but only if the peptide is long enough to include Phe 8 (the numbering scheme being that for the Aα chain of human fibrinogen) (van Nispen *et al.*, 1977). The remoteness of Phe 8 from the Arg 16–Gly 17 bond led to the hypothesis that this portion of the Aα chain may be folded in such a way that the active site of thrombin could accommodate Phe 8 as well as Arg 16 and Gly 17 (Blombäck *et al.*, 1969; Marsh *et al.*, 1982). This hypothesis was examined by carrying out transferred NOE measurements on complexes of thrombin with fibrinogenlike peptides in order to determine the structure of such peptides when bound to thrombin (Ni *et al.*, 1989b,c). Presumably, such structure (aside from that of Arg 16, for which no NOEs were observed) is the same as that of the corresponding portion of the whole fibrinogen molecule since the peptides are also good substrates for thrombin.

As a related problem, there are numerous functionally deficient fibrinogens that arise because of single-site mutations in the fibrinopeptide A portion of fibrinogen; hydrolysis of the corresponding Arg–Gly bond proceeds extremely slowly, if at all (Southan, 1988). Consequently, similar transferred NOE measurements on complexes of thrombin with such mutated variants of fibrinopeptide A can provide information as to how an altered conformation of fibrinopeptide A prevents optimal orientation in the active site of thrombin, thereby rendering hydrolysis very inefficient (Ni *et al.*, 1989a).

Another interaction of thrombin that can be examined by transferred NOE measurements concerns the complexes of thrombin with peptides having the amino acid sequence of the C-terminal portion of hirudin. Hirudin is a very potent anticoagulant found in the salivary glands of medicinal leeches (Markwardt, 1970). It is a small protein of 65 residues and forms a tight complex with α-thrombin ($K_d \sim 10^{-13}$ M), thereby preventing the thrombin-catalyzed conversion of fibrinogen to the fibrin clot (Bagdy *et al.*, 1976; Dodt *et al.*, 1984; Stone and Hofsteenge, 1986; Stone and Maragonore, this volume). In solution, the hirudin molecule has a disulfide-linked core structure (residues 4–29 and 38–47) and loose strands in the chain segments of residues 1–3, 30–37, and 48–65 (Folkers *et al.*, 1989; Haruyama and Wüthrich, 1989). These flexible strands, particularly residues 1–3 and 48–65, may be directly involved in the interaction with thrombin since modifications of residues in both the N- and the C-terminal regions of hirudin resulted in drastically reduced affinity for thrombin (Chang, 1983; Dodt *et al.*, 1987; Braun *et al.*, 1988; Wallace *et al.*, 1989; Stone and Maraganore, this volume). Furthermore, the C-terminal residues 52–65 of hirudin may represent an independent functional domain since peptides derived from this segment inhibit the thrombin-induced polymerization of fibrinogen, but they do not bind to the (primary) active site of thrombin (Mao *et al.*, 1988; Maraganore *et al.*, 1989; Stone and Maraganore, this volume). Transferred NOE measurements have helped define the nature of the interaction of thrombin with these hirudin peptides (Ni *et al.*, 1990).

The discovery of the anticoagulant activities of synthetic fibrinogenlike peptides has led to the development of potent active-site-directed inhibitors of thrombin (Blombäck *et al.*, 1969; Kettner and Shaw, 1979;

Stürzebecher *et al.*, 1983; Kikumoto *et al.*, 1984). Similar approaches have also led to antithrombotic peptides of 10–11 residues in length based on the C-terminal region of hirudin (Maraganore *et al.*, 1989; Krstenansky *et al.*, 1990; Stone and Maraganore, this volume). Knowledge of the thrombin-bound conformation of the fibrinogen and hirudin peptides will be useful for the design of novel molecules of lower molecular weight that are specific inhibitors of the thrombin–fibrinogen interaction. This chapter is concerned with the transferred NOE measurements made on complexes of bovine α-thrombin with wild-type fibrinogenlike peptides, genetically altered fibrinogenlike peptides, and peptides derived from the C-terminal region of hirudin, respectively.

2. NMR STUDIES

2.1. NMR Introduction

Linear peptides generally do not adopt a single native conformation in solution as they presumably do when bound to thrombin. Most of the peptides related to fibrinopeptide A bind to thrombin rather weakly. In other words, the dissociation of a peptide from the complex is rather fast, as illustrated in Fig. 1. This made it rather difficult to prepare crystals of the thrombin–peptide complexes for x-ray structure determination, but success in this regard has recently been achieved (Martin *et al.*, 1992). Prior to this crystallographic breakthrough, we initiated an NMR investigation of these complexes involving the measurement of transferred NOEs of the bound peptides (Fig. 1).

In the transferred NOE experiment, we concentrate on the peptides in solution. Because of their small sizes (up to 20 amino acid residues), it is relatively easy to assign all the proton resonances of the peptides by using two-dimensional NMR spectroscopy (Ni *et al.*, 1988). Moreover, these small peptides tumble relatively fast in aqueous solution. This fast tumbling results in negligible NOEs for the peptides themselves if the NOE experiment is carried out by using the normal NOESY technique (Bothner-By *et al.*, 1984; Brown and Farmer, 1989). The presence of a much less than stoichiometric amount of thrombin briefly shifts the bulk peptide away from the NOE null toward the negative NOE regime when the peptides associate to and dissociate from thrombin rapidly. Therefore, information about the spatial proximity of peptide protons in the bound state is transferred to the free peptides whose proton resonance signals are monitored and studied.

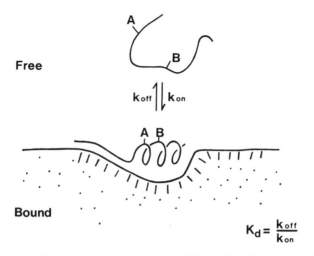

Figure 1. A schematic illustration of the interaction of thrombin with peptides derived from proteins such as fibinogen and hirudin. The peptides in solution can assume an ensemble of conformations. Protons A and B are, on average, far away in the free peptide but, in the bound state, they may be nearby in space. The proximity between protons A and B in the bound state can be established by transferred NOE experiments.

2.2.   Thrombin–Fibrinopeptide Interaction

Fibrinopeptide A (FpA) is found to bind to thrombin after the cleavage of the Arg 16–Gly 17 peptide bond since thrombin induces line broadening of the proton resonances of specific residues such as Gly 6, Glu 11, and Gly 12 (Fig. 2) among others of the chain segment from Asp 7 to Arg 16. The resonances of the entire region from Ala 1 to Glu 5 in FpA, on the other hand, are not much affected by thrombin binding. This result suggests that residues Asp 7 to Arg 16 constitute an essential structural element in the interaction of thrombin with fibrinogen, a finding that is in agreement with results from active-site mapping studies (van Nispen *et al.*, 1977; Meinwald *et al.*, 1980; Marsh *et al.*, 1983).

Transferred NOEs were observed for the complex of thrombin with FpA (Fig. 2). In particular, the ring protons of Phe 8 are in close proximity (< 5 Å) with the side-chain protons of Leu 9 and Val 15 (Fig. 3A). Furthermore, the same ring protons of Phe 8 are close to the $C^\alpha H$ protons of Gly 13 and/or Gly 14. If all these NOEs arise from the same complex of FpA with thrombin, there must be a hydrophobic cluster formed by the side chains of Phe 8, Leu 9, and Val 15 in that complex. Such a hydrophobic

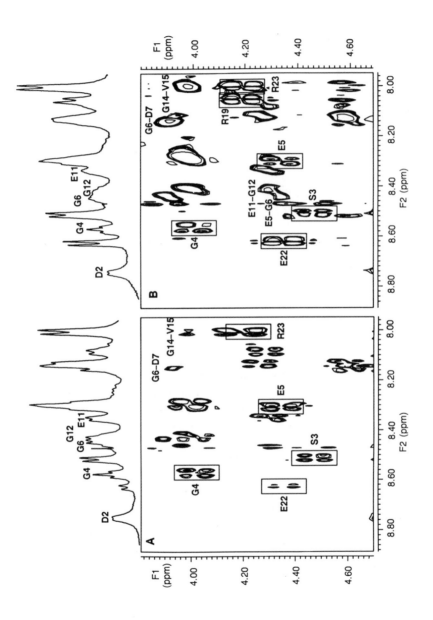

cluster could be in direct contact with thrombin since the side-chain proton resonances of Phe 8, Leu 9, and Val 15 are all selectively broadened upon thrombin binding (Ni *et al.*, 1989b,c).

Other studies have demonstrated the importance of a hydrophobic pocket in the active site of thrombin (Sonder and Fenton, 1984; Berliner, this volume). A hydrophobic pocket is occupied by the side chains of the D-Phe and Pro residues in the active-site-directed inhibitor, D-Phe-Pro-Arg-CH$_2$Cl (PPACK) as seen in the recently determined crystal structure of PPACK–thrombin (Bode *et al.*, this volume); it is not yet certain that this is exactly the same hydrophobic pocket. PPACK prevents the binding of FpA to thrombin since no transferred NOEs can be observed when thrombin is blocked by this inhibitor (Fig. 3B). Furthermore, all the thrombin-induced line broadening of FpA disappeared when the experiments were repeated with PPACK–thrombin (F. Ni and H. A. Scheraga, unpublished results). These observations suggest either (1) that FpA binds to the same site of thrombin as PPACK or (2) that the conformation around the active site is significantly altered by PPACK so that FpA can no longer bind.

To determine the thrombin-bound structure of FpA, shorter peptides (residues Asp 7 to Arg 16 and Phe 8 to Arg 16 of FpA) were used to identify transferred NOEs obscured because of peak overlaps with resonances from residues Ala 1 to Gly 6 in FpA (Ni *et al.*, 1989c). Apart from the transferred NOEs found with FpA, other NOEs found include that between the NH proton of Asp 7 and the C$^\beta$H protons of Ala 10 and that between the C$^\alpha$H proton of Glu 11 and the NH proton of Gly 13 (Fig. 4). These NOEs, combined with those involving the backbone protons of neighboring residues, indicate that there is a turn of helical structure for residues Asp 7 to Glu 11 followed by a chain reversal at residues Glu 11 and Gly 12, which helps bring residues Phe 8 and Leu 9 close to Val 15, and presumably next to the active-site residue Arg 16 (Ni *et al.*, 1989c). The proximity between the side chains of Phe and Val was also seen from ^{1}H

← ──

Figure 2. NMR spectra of fibrinopeptide A and Gly–Pro–Arg–Val–Val–Glu–Arg (the latter resulting from the thrombin cleavage of the Arg 16–Gly 17 bond of a peptide corresponding to fibrinogen Aα 1–23). (A) Thrombin concentration, 0.14 mM; mole ratio, FpA:thrombin ~ 40:1. (B) Thrombin concentration, 0.6 mM; mole ratio, FpA:thrombin 10:1. It is seen that the NH resonances of Gly 6, Glu 11, and Gly 12 are progressively broadened as the concentration of thrombin increases (top spectra). The two-dimensional spectra are the NOE and zero-quantum correlations between the NH and the C$^\alpha$H proton resonances of the peptides. The pattern of the boxed cross peaks remains the same in both spectra, suggesting that these residues do not bind to thrombin. NOE peaks, such as those between the C$^\alpha$H and NH protons of Gly 6 and Asp 7 and between Gly 14 and Val 15, become stronger at higher concentration of thrombin, indicating that the observed NOEs are primarily transferred NOEs from the thrombin–peptide complex (from Ni *et al.*, 1989b).

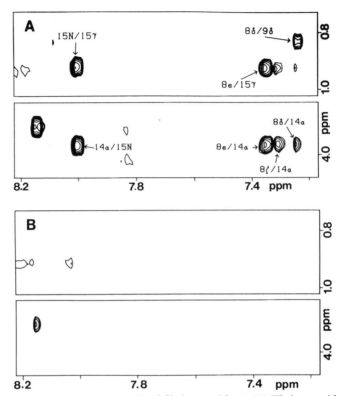

Figure 3. Long-range transferred NOEs of fibrinopeptide A. (A) Fibrinopeptide A in the presence of thrombin; (B) fibrinopeptide A in the presence of PPACK–thrombin.

and ^{13}C spectra of D-Phe-Val-Arg-pNA (pNA, p-nitroaniline) which is a good substrate for thrombin (Rae and Scheraga, 1979); in the L-Phe analog, which is not a good substrate, the side chain of Phe is not near that of Val.

A tentative model for the structure of the peptide was provided by distance geometry calculations incorporating the interproton distances determined from transferred NOEs (Ni *et al.*, 1989c). Sets of conformations for the decapeptide (N-acetyl-Asp-Phe-Leu-Ala-Glu-Gly-Gly-Gly-Val-Arg-NHMe) were generated that were free from steric overlaps and that were consistent with the NMR information. A prominent feature in these structures is the hydrophobic cluster formed by the side chains of residues Phe 8, Leu 9, and Val 15. The hydrophobic interactions between Phe 8 and Leu 9 could account for the requirements for Phe–Leu in maintaining the catalytic efficiency of the thrombin cleavage of the Arg–Gly peptide bond in synthetic fibrinogenlike peptides (van Nispen *et al.*, 1977). It also may explain why the placement of a Phe residue at position P3 (in place of Gly 14) enhances the inhibitory effects of arginine methyl esters on thrombin action (Blombäck *et al.*, 1969). The relative orientation of the arginine side chain could not be defined in this study because of the absence of NOE constraints.

The NOEs that are observed in the complex with thrombin can also be seen in solutions of the free peptide at low temperature (Ni *et al.*, 1989b), where the tumbling of the peptide is slower, and at ambient temperatures using the rotating frame NOE technique (E. T. Fossel and J. McDonagh, personal communication; F. Ni and H. A. Scheraga, unpublished results) that is less sensitive to the tumbling rate (Bothner-By *et al.*, 1984; Brown and Farmer, 1989). This suggests that the NOEs may arise from conformations that are intrinsically stable, but may be further stabilized by contact with thrombin. Accordingly, energy-minimized conformations of the decapeptide have been generated by a combination of the buildup and Monte Carlo-minimization procedures developed in this laboratory (Pincus *et al.*, 1982; Gibson and Scheraga, 1987; Li and Scheraga, 1987) based on the ECEPP/2 (Empirical Conformational Energy Pro-

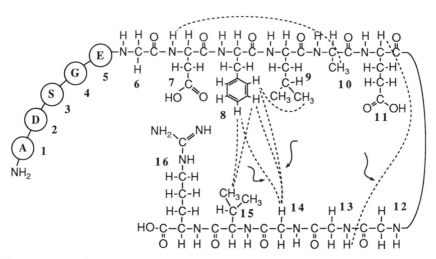

Figure 4. Proximity relations of protons of fibrinopeptide A in the thrombin-bound state. The circled residues are those whose proton resonances were not affected by thrombin binding. The arrows indicate the NOEs that are absent when Gly 12 is replaced by Val 12 in fibrinopeptide A. The NOEs between the ring H atoms of Phe 8 and the H^α atoms of Gly 14 may also be attributed to interactions with the H^α atoms of Gly 13 because of the spectral overlap between the H^α resonances of Gly 13 and Gly 14 (Ni *et al.*, 1989c).

gram for Peptides) algorithm (Momany *et al.*, 1975; Némethy *et al.*, 1983; Sippl *et al.*, 1984), together with numerical tests and visual examination using graphics to ensure that the NOE constraints were never grossly violated. Because no sequential NOEs were observed between Gly 13 and Gly 14 as a result of resonance overlap, the conformation of this part of the chain could not be predicted; therefore, it was necessary to generate and examine very many conformations of the decapeptide. Out of a total of 9705 conformations that were generated, 111 were found to satisfy the NOE constraints of Ni *et al.* (1989c) with an RMS deviation of less than 0.2 Å. In many of the conformations that are consistent with the NOE constraints, the backbone dihedral angles at position 12 are such (viz., $\phi > 0°$) that only a glycine residue could be placed there; however, in other conformations, these dihedral angles (viz., $\phi < 0°$) adopt values compatible with *any* L-amino acid. Thus, although the almost universal preference for a Gly residue at position 12 in FpA probably reflects the conformational preference of this part of the chain, it may also be dictated by some as yet unknown packing requirements in the complex with thrombin.

The lowest-energy conformation of the decapeptide that is consistent with the NOE constraints is depicted in Fig. 5A. As can be seen, its overall appearance is much like the conformations in Fig. 7 of Ni *et al.* (1989c) (whose ECEPP/2 energy, after minimization, was more than 10 kcal/mol greater). Certain extra requirements are satisfied by the conformation in Fig. 5A; for example, the H^β of Val 15 is more than 5 Å from the ring protons of Phe 8, in keeping with the fact that no NOEs are observed between these protons, the H^γ atoms of Val 15 are at least 4.3 Å from the H^δ atoms of Phe 8, in agreement with the data in Fig. 3A which show at most a very weak NOE for this interaction, and the side chain of Arg 16 is extended, as would be necessary if this residue is to fit into the specificity pocket of thrombin.

With the recent availability of an x-ray structure of a complex of bovine thrombin with the peptide consisting of residues 7–16 of FpA (Martin *et al.*, 1992), it is of interest to compare the structure in Fig. 5A, and the previously published structure of Ni *et al.* (1989c), with the x-ray structure. The C^α atoms of the first five residues (Asp–Phe–Leu–Ala–Glu) from the structure in Fig. 5A could be superimposed on the crystal structure of the bound peptide with an RMS deviation of 0.26 Å, and the same C^α atoms from the structure of Ni *et al.* (1989c) could be similarly superimposed with an RMS deviation of 0.36 Å. Thus, both of these computed conformations predicted the helical portion of the bound decapeptide satisfactorily. These two computed conformations also predicted that the dihedral angle ϕ for Gly 12 would be positive, again in agreement with the x-ray observation (although, as pointed out above, other, higher energy,

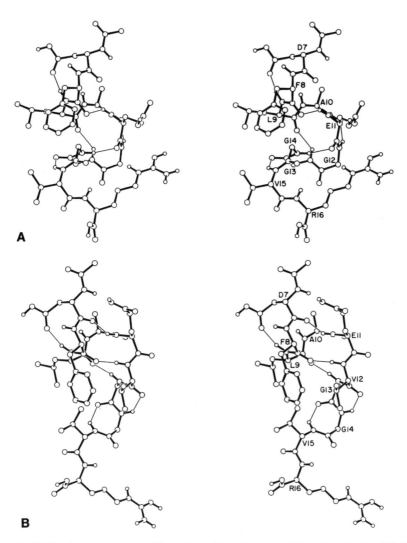

Figure 5. (A) The lowest-energy conformation of the decapeptide (residues 7–16) of fibrin-opeptide A that is consistent with the transferred NOE constraints. (B) Conformation of fibrinopeptide A (residues 7–16) when Gly 12 is replaced by Val 12 as found in the abnormal fibrinogen Rouen. The coordinates of the model structures displayed here along with those of the hirudin peptide in Fig. 8 have been deposited in the Brookhaven Protein Data Bank.

conformations of the peptide in which $\phi < 0°$ for Gly 12 also fit the observed NOEs reasonably well). Further, all the computed conformations that satisfied the NOEs contained a hydrophobic cluster, consisting of the side chains of Phe 8, Leu 9, and Val 15 in approximately the same orientation as is observed in the crystal. However, none of the computed conformations were able to reproduce the elongated structure of the three contiguous glycine residues that is seen in the crystal structure. A possible explanation for this disagreement is that the strong NOEs that were attributed entirely to interactions between the H^α atoms of Gly 14 and the ring H atoms of Phe 8 (Ni *et al.*, 1989c) actually contained a contribution from the H^α atoms of Gly 13 as well; this possibility cannot be ruled out because the resonances of the H^α atoms of Gly 13 and Gly 14 overlap and no NOE experiments were carried out for a peptide in which the H^α atoms of Gly 14 (rather than Gly 13) were replaced with deuterium. If the strong NOEs between the glycine H^α atoms and the Phe 8 side chain arose from transfer of magnetization from Gly 13 as well as Gly 14, the conformations deduced from the NMR data should certainly have been more extended in this region. In support of this possibility, the conformation of the peptide in the crystal structure has been generated by first measuring the dihedral angles (using the program FRODO) and then generating the structure using ECEPP/2 geometry. In this structure, both Gly 13 and Gly 14 had at least one H^α atom whose distance from at least one of the ring H atoms of Phe 8 was < 3 Å.

Molecular defects in genetically abnormal human fibrinogens cause delayed formation of fibrin clots. Alterations in the amino acid sequence of some of the abnormal fibrinogens have been identified and, in most of the fibrinogens analyzed so far, mutations of single amino acids have been located in the regions from residue Asp 7 to Arg 19 in the Aα chain. In particular, a bleeding disorder is caused by a single mutation of Gly 12 to Val 12 in FpA (Southan, 1988). This mutation reduces the rate of cleavage of the Arg–Gly bond in synthetic peptides derived from residues Ala 1 to Val 20 of the Aα chain of the abnormal fibrinogen (Ni *et al.*, 1989a).

The mutated FpA still binds to thrombin, on the basis of the observation of line broadening of the proton resonances of Val 12–FpA in the presence of thrombin (Ni *et al.*, 1989a). The hydrophobic cluster postulated for normal FpA still exists in the complex of thrombin with the Val 12-substituted FpA, but with an increase in size since Val 12 is now also part of the cluster (Ni *et al.*, 1989a). The Val 12 substitution seems to affect the conformation of only residues at positions 12, 13, and 14 compared to normal FpA. In particular, transferred NOEs between Phe 8 and Gly 13/14 and between Glu 11 and Gly 13 observed with FpA are absent in the mutated complex (Fig. 4). Since sequential NOEs are observed for every pair of residues of the mutant peptide (Ni *et al.*, 1989a), it was possible to build an energy-minimized conformation of this decapeptide in a straightforward manner by sequential addition of amino acids starting from the C-terminus, using computer graphics to generate starting conformations for energy minimization. In this way, a nearly unique conformation, consistent with all the observed NOEs, was generated. The main uncertainty was in the conformation adopted by the side chain of Arg 16, for which values of χ^1 corresponding to *trans, gauche*$^+$, or *gauche*$^-$ were equally consistent with the observed constraints. One of these conformations is shown in Fig. 5B. As pointed out above, the nonpolar side chains of Phe 8, Leu 9, Val 12, and Val 15 form a cluster, which presents a hydrophobic face that should bind tightly to any exposed hydrophobic sites on the surface of thrombin. The marked difference between the conformations in Fig. 5A and B could easily account for the large difference in the rates of hydrolysis of the normal and mutant fibrinogens by thrombin, since it is unlikely that two such dissimilar molecules would bind to any site on thrombin with comparable affinities.

As in the case of the native decapeptide, it is instructive to compare the computed conformation of the Val-12 decapeptide with the crystal structure of the native decapeptide complexed with thrombin (Martin *et al.*, 1992). In this case, the C^α atoms of the first six residues (Asp–Phe–Leu–Ala–Glu–Val) of the computed conformation can be superimposed on the C^α atoms of the first six residues of the native peptide in the crystal with an RMS deviation of 0.24 Å. Thus, the backbone of the mutant peptide follows the backbone of the native peptide over this region very closely. However, thereafter, the backbones of the two peptides diverge strongly, and the positions of the Val-15 C^α atoms in the two conformations are well separated, even though the side chain of this residue is close to the side chain of Phe 8 in both cases. The divergence of the backbones in these two structures has significant consequences for the relative placement of the Arg-16 side chain.

2.3. Thrombin–Hirudin Interaction

The interaction of thrombin with hirudin was investigated by using synthetic peptides having the sequence of residues 52 to 65 of the hirudin C-terminus. Thrombin induces significant line broadening in the peptides, particularly on the NH proton resonances of residues Phe 56 to Ile 59 and residues Glu 62 to Gln 65 (Ni *et al.*, 1990). The side-chain proton resonances of Phe 56, Ile 59, Tyr 63, and Leu 64 are broadened by thrombin binding, indicating that they all participate in the interaction with thrombin. Figure 6A shows the aliphatic region of a two-dimensional spectrum

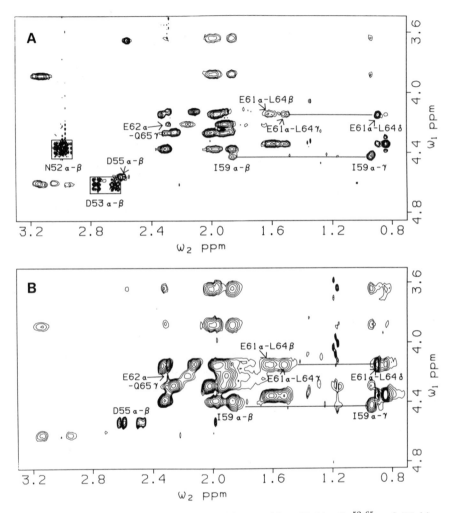

Figure 6. Regions of the NOESY spectra of the peptides, (A) hirudin[52–65] and (B) hirudin[55–65], in the presence of thrombin. The boxed cross peaks are from the interactions (zero-quantum coherence or NOE) between the C$^\alpha$H and C$^\beta$H protons of residues Asn 52 and Asp 53. Also indicated are the medium-range NOEs between the C$^\alpha$H proton of Glu 61 and the side-chain protons of Gln 65 (from Ni *et al.*, 1990).

(J-correlation + transferred NOE) of the hirudin peptide (52–65) inter-acting with thrombin. Because the side-chain ($C^\beta H$) protons of Asn 52 and Asp 53 are not much affected by thrombin binding, the correlations be-tween the $C^\alpha H$ and the $C^\beta H$ protons of these two residues are dominated by the so-called zero-quantum coherence that shows up as fine structures within one peak compared with other residues such as those for Asp 55 and Ile 59. These observations suggest that residues Asn 52 and Asp 53 (in-cluding Gly 54, see Ni *et al.*, 1990) are free to rotate in the thrombin–peptide complex and can be dispensed with. Indeed, the shorter peptide, Asp 55–Gln 65, has similar transferred NOEs as the longer peptide (Fig. 6).

The NMR experiments were also carried out with thrombin inacti-vated with PPACK. Thrombin-induced line broadening still persists in these spectra in contrast to what was found with FpA in the presence of PPACK–thrombin. Furthermore, all of the transferred NOEs observed for the hirudin peptide interacting with PPACK–thrombin were identical with those identified with unmodified thrombin. These results show that the C-terminal peptides of hirudin do not bind to the (primary) catalytic site of thrombin in agreement with the finding that these peptides do not inhibit the thrombin-catalyzed proteolysis of tripeptide chromogenic sub-strates (Krstenansky and Mao, 1987).

The thrombin-bound conformation of the hirudin peptides is char-acterized by the following transferred NOEs involving protons of residues remote in the sequence. First, thrombin induces NOEs between the C^α-proton of Glu 61 and the side-chain protons of Leu 64 and between the C^α-proton of Glu 62 and the C^γ-protons of Gln 65 (Fig. 6). These NOEs indicate that residues Glu 61 to Gln 65 of the hirudin peptides form a short stretch of helix in the complex. This conclusion is supported by the ob-servation of other NOEs involving residues from Ile 59 to Gln 65 (Fig. 7). There is an NOE between the C^α-proton of Glu 61 and the NH proton of Leu 64. And there are also NOEs between the side-chain protons of Ile 59 and Tyr 63, between those of residues Pro 60 and Tyr 63, and between residues Tyr 63 and Leu 64. An NOE between the side-chain protons of Phe 56 and Ile 59 combined with the NOEs mentioned above shows that there is also a hydrophobic cluster in the complex of thrombin with the hirudin peptides involving residues Phe 56, Ile 59, Pro 60, Tyr 63, and Leu 64.

The structural features can be seen more clearly in the computed model (Fig. 8) that satisfies all the interproton distances deduced from the transferred NOEs (Ni *et al.*, 1990). Residues 56–65 of the hirudin peptides form an amphiphilic structure where one side of the molecule is polar and charged and the other side is hydrophobic. The phenyl ring of Phe 56 is

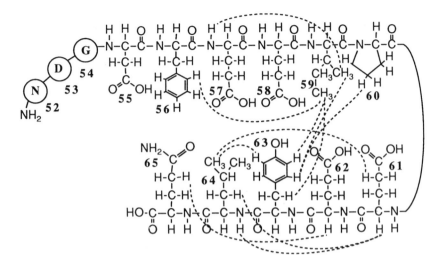

Figure 7. Proximity relations of protons of residues 52–65 of the hirudin C-terminus in the thrombin-bound state. The circled residues are those whose proton resonances were not affected by thrombin binding (Fig. 6 and Ni *et al.*, 1990).

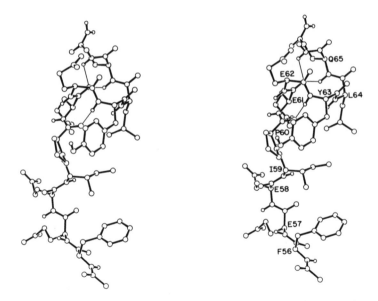

Figure 8. A model structure of residues Phe 56 to Gln 65 of the hirudin C-terminus on the basis of NOE-constrained molecular modeling (Ni *et al.*, 1990). The position of Phe 56 is not unique and deviates from that observed in the x-ray structure of the thrombin–hirudin complex (Table I).

brought back by a chain reversal at residues Glu 57 and Glu 58 to form a hydrophobic cluster with the side chains of residues Ile 59, Pro 60, Tyr 63, and Leu 64. Residues Ile 59 to Gln 65 form a very compact structure with Ile 59 and Pro 60 in extended, and Glu 61 to Gln 65 in distorted, helical conformations.

It is of interest to compare the NMR results on the interaction of the hirudin C-terminal peptides with *bovine* thrombin *in solution* with x-ray results obtained for the crystal of the complex of *human* thrombin with the entire hirudin molecule (see Chapter 1). While a detailed comparison must wait for the refinement of both the NMR and x-ray structures, a preliminary comparison revealed that the NMR and x-ray results are consistent with each other. Table I shows the deviations of the C^α atoms between the NMR and the x-ray structures of residues 56–65 of the hirudin C-terminus. The fit is excellent for residues Ile 59 to Gln 65 that are well characterized by NOE constraints (Fig. 7 and Ni *et al.*, 1990). Furthermore, the electron density of residues Asn 52 and Asp 53 of hirudin was found to be poorly defined in the crystal of the complex, in agreement with our conclusion that they are not involved in binding to thrombin since no thrombin-induced line broadening was observed for the side-chain protons of these residues (Figs. 6 and 7).

2.4. Preliminary Docking Experiments

The active site of thrombin contains two surface loops; one of these contains 12 residues and runs from residues Leu 60 to Asn 62, and the other contains 10 residues and runs from Glu 146 to Gly 150, in the

Table I. Differences[a] in the Positions of the C^α Atoms of Hirudin[56-65] between the NMR Structures and Those Determined by the X-ray Analysis[b] of the Thrombin–Hirudin Complex

Hirudin[56-65]	56	57	58	59	60	61	62	63	64	65
	F	E	E	I	P	E	E	Y	L	Q
(A)	5.44	0.93	0.64	0.35	0.13	0.20	0.27	0.22	0.41	0.53
	−9.86	−3.92	−2.78	−1.31	−1.16	0.76	−0.82	−0.89	−1.44	−1.45
(B)	5.75	0.93	0.91	0.58	0.28	0.40	0.28	0.38	1.08	1.45

[a] All of the NMR structures were superimposed on the x-ray structure by a fit based on the C^α atoms of residues Ile 59 to Gln 65. Residue Phe 56 deviates significantly from the x-ray structure since very few NOEs pertaining to this residue were observed (Fig. 7). Values listed in row (A) represent the ranges of the deviation (in Å) between the x-ray structure and all of the structures generated by distance geometry calculations (Ni *et al.*, 1990). Numbers in row (B) are the deviations of the x-ray structure from the NMR structure displayed in Fig. 8.
[b] Chapter 1.

terminology of Bode *et al.* (this volume) (this nomenclature was used because insertions were labeled with numbers and letters, e.g., 149A to 149E; hence, the length of the loop may not bear a direct relationship to the numbering of the first and last residues in the loop). These loops lie on either side of the specificity pocket, and contain several sites at which hydrophobic contacts can be made with fibrinogen. The second loop, which consists of ten residues, takes very different conformations in PPACK–thrombin and in the hirudin–thrombin complex. This loop appears to be quite flexible, and its conformation in complexes with fibrinogen or fibrinopeptides may be quite different from either of the conformations observed in the crystal structures of PPACK–thrombin or the hirudin–thrombin complex (Chapter 1).

On the other hand, in these two crystal structures, the conformation of the first loop is essentially identical; thus, this loop appears to be stable, and it probably retains the same conformation when bound to fibrinogen. The same is true of the specificity pocket into which the side chain of Arg 16 of the Aα chain of fibrinogen fits. It is reasonable to demand, in any attempt to pack FpA into thrombin, that the conformations of the last two residues, Val 15 and Arg 16, should resemble those of the Pro and Arg residues in PPACK, and that the Leu 60-to-Asn 62 loop (which contains ten residues) should not be greatly displaced by the peptide.

Eleven heavy atoms from the Val 15 and Arg 16 residues of the computed FpA decapeptide conformations (see Section 2.2) were superimposed on the corresponding heavy atoms of PPACK. Out of 9705 conformations, 799 could be superimposed in this way with an RMS deviation of less than 0.4 Å. However, none of the 111 conformations that came near to satisfying all the NOE constraints of Ni *et al.* (1989c) was included among this number. Furthermore, preliminary attempts to pack some of the conformations that satisfy the constraints into the crystal structure of PPACK–thrombin, with no atomic overlap involving the Leu 60-to-Asn 62 loop, have been unsuccessful because of the rather large volume occupied by the cluster of side chains from Phe 8, Leu 9, and Val 15 of FpA. The situation is even worse in the case of the Val 12 mutant, in which the hydrophobic cluster occupies a larger volume. If the conformations of Gly 13 and Gly 14 are allowed to change somewhat, it is possible to accommodate Val 15 and Arg 16 in the space occupied by the Pro and Arg residues of PPACK in the crystal; however, this causes the side chain of Val 15 to move some distance from the phenyl ring of Phe 8, in seeming contradiction to the observed transferred NOEs between the side-chain protons of these two residues.

These results, and other computations involving the native and mutant decapeptides, raise some questions concerning the experimental results, which must be answered before the binding of fibrinogen to thrombin is fully understood. (1) Is the conformation of the Glu 146-to-Gly 150 loop in PPACK–thrombin specific for that complex, or does it bear some relation to the conformation adopted by that loop when fibrinogen binds to thrombin? (2) Are the relative spatial positions of the Phe 8 and the Arg 16 side chains in the fibrinogen Aα chain bound to thrombin significantly different from the relative positions of the side chains of D-Phe and Arg in PPACK? (3) Do fibrinopeptides such as FpA and the Asp 7–Arg 16 decapeptide bind to thrombin in more than one conformation, only one of which represents the productive complex? The questions posed here may be answered by inspection of the crystal structure of the complex of the fibrinogenlike peptide (residues 7–16 of the Aα chain) with thrombin (Martin *et al.*, 1992). Further NMR studies are also in progress to attempt to answer these questions.

3. CONCLUSIONS AND FUTURE WORK

The NMR work discussed here, and the x-ray work discussed in Chapter 1, provide considerable information about the thrombin–fibrinogen and thrombin–hirudin complexes. Further important details of the interactions in these binary complexes are being obtained by mutating the residues of the peptide substrates (by direct chemical synthesis) and of thrombin (by recombinant DNA methods). The effect of such mutations on the kinetics of hydrolysis of the Arg–Gly bonds and on the structure obtained by transferred NOE measurements is being examined. Thus, the role of specific residues in both the enzyme and the substrate in the formation of the complex is being assessed.

Since thrombin is a two-chain disulfide-linked molecule, the recombinant DNA methodology is being carried out on prothrombin and on prethrombin-2, rather than on thrombin. Prethrombin-2 is the last single-chain precursor of thrombin in the cascade from prothrombin to thrombin. Prethrombin-2 can be folded (L. S. Hanna and H. A. Scheraga, unpublished results), and converted to thrombin by one proteolytic cleavage by Factor Xa with no loss of any amino acids. Thus, specific mutations can be made in the genes for prothrombin and/or prethrombin-2, and the effect of such mutations on the kinetics of hydrolysis of the Arg–Gly bond of the fibrinogenlike peptide by these altered thrombins can be assessed.

The following are some examples of thrombin mutants that are under consideration for these studies. Replacement of Ser 195 (in the terminology of Bode *et al.*, this volume) by Ala in recombinant thrombin, or by dehydroalanine in chemically modified thrombin, will prevent hydrolysis of the Arg 16–Gly 17 peptide bond in substrates; thus, transferred NOE experiments can lead to structural information about the Arg–Gly–Pro–Arg–Val portion of the fibrinogen Aα chain. Replacement of Cys 1 and Cys 122 by Ala will eliminate the disulfide bond connecting the A and B chains of thrombin, thereby enabling us to determine whether the folded B chain, lacking the A chain, is as effective an enzyme as intact thrombin. Finally, incorporation of ^{15}N- and ^{13}C-labeled amino acids in recombinant thrombin will lead to information about thrombin–peptide interactions in NMR spectroscopic studies.

The coordinates of the computed structures in Figs. 5A, 5B, and 8 have been deposited with the Brookhaven Protein Data Bank, with entry names assigned as follows: 1FPA, fibrinopeptide A fragment (Fig. 5A); 2FPA, fibrinopeptide A fragment mutant (Fig. 5B); 1HRG, hirudin fragment (Fig. 8).

Acknowledgments

We thank Professor R. Huber for providing the coordinates of the human PPACK–thrombin, Professor A. Tulinsky for the C^α coordinates of the human thrombin–hirudin complex, and Dr. P. Martin and Professor B. Edwards for the coordinates of the bovine thrombin–fibrinopeptide complex. This work was supported at Cornell University by grant HL-30616 from the National Heart, Lung and Blood Institute, National Institutes of Health. F. N. acknowledges support from the National Research Council of Canada.

4. REFERENCES

Bagdy, D., Barabas, E., Graf, L., Petersen, T. E., and Magnusson, S., 1976, Hirudin, *Methods Enzymol.* **45:**669–678.

Blombäck, B., Blombäck, M., Olsson, P., Svendsen, L., and Åberg, G., 1969, Synthetic peptides with anticoagulant and vasodilating activity, *Scand. J. Clin. Lab. Invest. Suppl.* **24**(107):59–64.

Bothner-By, A. A., Stephens, R. L., Lee, J.-M., Warren, C. D., and Jeanloz, R. W., 1984, Structure determination of a tetrasaccharide: Transient nuclear Overhauser effects in the rotating-frame, *J. Am. Chem. Soc.* **106:**811–813.

Braun, P. J., Dennis, S., Hofsteenge, J., and Stone, S. R., 1988, Use of site-directed mutagenesis to investigate the basis for the specificity of hirudin, *Biochemistry* **27:**6517–6521.

Brown, L. R., and Farmer, B. T., II, 1989, Rotating-frame nuclear Overhauser effect, *Methods Enzymol.* **176:**199–217.

Chang, J.-Y., 1983, The functional domain of hirudin: A thrombin-specific inhibitor, *FEBS Lett.* **164:**307–313.

Dodt, J., Müller, H.-P., Seemüller, U., and Chang, J.-Y., 1984, The complete amino acid sequence of hirudin, a thrombin specific inhibitor: Application of colour carboxymethylation, *FEBS Lett.* **165**:180–183.

Dodt, J., Seemüller, U., and Fritz, H., 1987, Influence of chain shortening on the inhibitory properties of hirudin and eglin c, *Biol. Chem. Hoppe-Seyler* **368**:1447–1453.

Endres, G. F., Ehrenpreis, S., and Scheraga, H. A., 1966, Equilibria in the fibrinogen–fibrin conversion. VI. Ionization changes in the reversible polymerization of fibrin monomer, *Biochemistry* **5**:1561–1567.

Folkers, P. J. M., Clore, G. M., Driscoll, P. C., Dodt, J., Köhler, S., and Gronenborn, A. M., 1989, Solution structure of recombinant hirudin and the Lys-47 to Glu mutant: A nuclear magnetic resonance and hybrid distance geometry–dynamical simulated annealing study, *Biochemistry* **28**:2601–2617.

Gibson, K. D., and Scheraga, H. A., 1987, Revised algorithms for the build-up procedure for predicting protein conformations by energy minimization, *J. Comput. Chem.* **8**:826–834.

Hanna, L. S., Scheraga, H. A., Francis, C. W., and Marder, V. J., 1984, Comparison of structures of various human fibrinogens and a derivative thereof by a study of the kinetics of release of fibrinopeptides, *Biochemistry* **23**:4681–4687.

Haruyama, H., and Wüthrich, K., 1989, Conformation of recombinant desulfatohirudin in aqueous solution determined by nuclear magnetic resonance, *Biochemistry* **28**:4301–4311.

Kettner, C., and Shaw, E., 1979, D-Phe-Pro-ArgCH$_2$Cl—a selective affinity label for thrombin, *Thromb. Res.* **14**:969–973.

Kikumoto, R., Tamao, Y., Tezuka, T., Tonomura, S., Hara, H., Ninomiya, K., Hijikata, A., and Okamoto, S., 1984, Selective inhibition of thrombin by (2R,4R)-4-methyl-1-[N^2-[(3-methyl-1,2,3,4-tetrahydro-8-quinolinyl)sulfonyl]-L-arginyl]-2-piperidinecarboxylic acid, *Biochemistry* **23**:85–90.

Krstenansky, J. L., and Mao, S. J. T., 1987, Antithrombin properties of C-terminus of hirudin using synthetic unsulfated N$^\alpha$-acetyl-hirudin$_{45-65}$, *FEBS Lett.* **211**:10–16.

Krstenansky, J. L., Broersma, R. J., Owen, T. J., Payne, M. H., Yates, M. T., and Mao, S. J. T., 1990, Development of MDL 28,050, a small stable antithrombin agent based on a functional domain of the leech protein, hirudin, *Thromb. Haemostasis* **63**:208–214.

Laskowski, M., Jr., Ehrenpreis, S., Donnelly, T. H., and Scheraga, H. A., 1960, Equilibria in the fibrinogen–fibrin conversion. V. Reversibility and thermodynamics of the proteolytic action of thrombin on fibrinogen, *J. Am. Chem. Soc.* **82**:1340–1348.

Laudano, A. P., and Doolittle, R. F., 1980, Studies on synthetic peptides that bind to fibrinogen and prevent fibrin polymerization. Structural requirements, number of binding sites, and species differences, *Biochemistry* **19**:1013–1019.

Li, Z., and Scheraga, H. A., 1987, Monte Carlo-minimization approach to the multiple-minima problem in protein folding, *Proc. Natl. Acad. Sci. USA* **84**:6611–6615.

Mao, S. J. T., Yates, M. T., Owen, T. J., and Krstenansky, J. L., 1988, Interaction of hirudin with thrombin: Identification of a minimal binding domain of hirudin that inhibits clotting activity, *Biochemistry* **27**:8170–8173.

Maraganore, J. M., Chao, B., Joseph, M. L., Jablonski, J., and Ramachandran, K. L., 1989, Anticoagulant activity of synthetic hirudin peptides, *J. Biol. Chem.* **264**:8692–8698.

Markwardt, F., 1970, Hirudin as an inhibitor of thrombin, *Methods Enzymol.* **19**:924–932.

Marsh, H. C., Jr., Meinwald, Y. C., Lee, S., and Scheraga, H. A., 1982, Mechanism of action of thrombin on fibrinogen. Direct evidence for the involvement of phenylalanine at position P$_9$, *Biochemistry* **21**:6167–6171.

Marsh, H. C., Jr., Meinwald, Y. C., Thannhauser, T. W., and Scheraga, H. A., 1983, Mechanism of action of thrombin on fibrinogen. Direct evidence for the involvement of aspartic acid at position P_{10}, *Biochemistry* **22**:4170–4174.

Martin, P. D., Robertson, W., Turk, D., Huber, R., Bode, W., and Edwards, B. F. P., 1992, The structure of residues 7–16 of the Aα chain of human fibrinogen bound to bovine thrombin at 2.3 Å resolution, *J. Biol. Chem.*, in press.

Martinelli, R. A., and Scheraga, H. A., 1980, Steady-state kinetics study of the bovine thrombin-fibrinogen interaction, *Biochemistry* **19**:2343–2350.

Meinwald, Y. C., Martinelli, R. A., van Nispen, J. W., and Scheraga, H. A., 1980, Mechanism of action of thrombin on fibrinogen. Size of the Aα fibrinogen-like peptide that contacts the active site of thrombin, *Biochemistry* **19**:3820–3825.

Momany, F. A., McGuire, R. F., Burgess, A. W., and Scheraga, H. A., 1975, Energy parameters in polypeptides. VII. Geometric parameters, partial atomic charges, nonbonded interactions, hydrogen bond interactions, and intrinsic torsional potentials for the naturally occurring amino acids, *J. Phys. Chem.* **79**:2361–2381.

Nagy, J. A., Meinwald, Y. C., and Scheraga, H. A., 1982, Immunochemical determination of conformational equilibria for fragments of the Aα chain of fibrinogen, *Biochemistry* **21**:1794–1806.

Nagy, J. A., Meinwald, Y. C., and Scheraga, H. A., 1985, Immunochemical determination of conformational equilibria for fragments of the Bβ chain of fibrinogen, *Biochemistry* **24**:882–887.

Némethy, G., Pottle, M. S., and Scheraga, H. A., 1983, Energy parameters in polypeptides. 9. Updating of geometric parameters, nonbonded interactions, and hydrogen bond interactions for the naturally occurring amino acids, *J. Phys. Chem.* **87**:1883–1887.

Ni, F., Scheraga, H. A., and Lord, S. T., 1988, High-resolution NMR studies of fibrinogen-like peptides in solution: Resonance assignments and conformational analysis of residues 1–23 of the Aα chain of human fibrinogen, *Biochemistry* **27**:4481–4491.

Ni, F., Konishi, Y., Bullock, L. D., Rivetna, M. N., and Scheraga, H. A., 1989a, High-resolution NMR studies of fibrinogen-like peptides in solution: Structural basis for the bleeding disorder caused by a single mutation of Gly(12) to Val(12) in the Aα chain of human fibrinogen Rouen, *Biochemistry* **28**:3106–3119.

Ni, F., Konishi, Y., Frazier, R. B., Scheraga, H. A., and Lord, S. T., 1989b, High-resolution NMR studies of fibrinogen-like peptides in solution: Interaction of thrombin with residues 1–23 of the Aα chain of human fibrinogen, *Biochemistry* **28**:3082–3094.

Ni, F., Meinwald, Y. C., Vásquez, M., and Scheraga, H. A., 1989c, High-resolution NMR studies of fibrinogen-like peptides in solution: Structure of a thrombin-bound peptide corresponding to residues 7–16 of the Aα chain of human fibrinogen, *Biochemistry* **28**:3094–3105.

Ni, F., Konishi, Y., and Scheraga, H. A., 1990, Thrombin-bound conformation of the C-terminal fragments of hirudin determined by transferred nuclear Overhauser effects, *Biochemistry* **29**:4479–4489.

Pincus, M. R., Klausner, R. D., and Scheraga, H. A., 1982, Calculation of the three-dimensional structure of the membrane-bound portion of melittin from its amino acid sequence, *Proc. Natl. Acad. Sci. USA* **79**:5107–5110.

Rae, I. D., and Scheraga, H. A., 1979, ^{1}H and ^{13}C nuclear magnetic resonance spectra of some peptides with fibrinogen-like activity, *Int. J. Pept. Protein Res.* **13**:304–314.

Scheraga, H. A., 1983, Interaction of thrombin and fibrinogen and the polymerization of fibrin monomer, *Ann. N.Y. Acad. Sci.* **408**:330–343.

Scheraga, H. A., 1986, Chemical basis of thrombin interactions with fibrinogen, *Ann. N.Y. Acad. Sci.* **485**:124–133.

Scheraga, H. A., and Laskowski, M., Jr., 1957, The fibrinogen–fibrin conversion, *Adv. Protein Chem.* **12**:1–131.

Sippl, M. J., Némethy, G., and Scheraga, H. A., 1984, Intermolecular potentials from crystal data. 6. Determination of empirical potentials for O—H · · · O=C hydrogen bonds from packing configurations, *J. Phys. Chem.* **88**:6231–6233.

Sonder, S. A., and Fenton, J. W., II, 1984, Proflavin binding within the fibrinopeptide groove adjacent to the catalytic site of human α-thrombin, *Biochemistry* **23**:1818–1823.

Southan, C., 1988, Molecular and genetic abnormalities of fibrinogen, in: *Fibrinogen, Fibrin Stabilization, and Fibrinolysis: Clinical, Biochemical and Laboratory Aspects* (J. L. Francis, ed.), Ellis Horwood VCH, England, pp. 65–99.

Stone, S. R., and Hofsteenge, J., 1986, Kinetics of the inhibition of thrombin by hirudin, *Biochemistry* **25**:4622–4628.

Sturtevant, J. M., Laskowski, M., Jr., Donnelly, T. H., and Scheraga, H. A., 1955, Equilibria in the fibrinogen–fibrin conversion. III. Heats of polymerization and clotting of fibrin monomer, *J. Am. Chem. Soc.* **77**:6168–6172.

Stürzebecher, J., Markwardt, F., Voigt, B., Wagner, G., and Walsmann, P., 1983, Cyclic amides of N^{α}-arylsulfonyl-aminoacylated 4-amidinophenylalanine—tight binding inhibitors of thrombin, *Thromb. Res.* **29**:635–642.

van Nispen, J. W., Hageman, T. C., and Scheraga, H. A., 1977, Mechanism of action of thrombin on fibrinogen: The reaction of thrombin with fibrinogen-like peptides containing 11, 14, 16 residues, *Arch. Biochem. Biophys.* **182**:227–243.

Varadi, A., and Scheraga, H. A., 1986, Localization of segments essential for polymerization and for calcium binding in the γ-chain of human fibrinogen, *Biochemistry* **25**:519–528.

Wallace, A., Dennis, S., Hofsteenge, J., and Stone, S. R., 1989, Contribution of the N-terminal region of hirudin to its interaction with thrombin, *Biochemistry* **28**:10079–10084.

Chapter 3

ESR AND FLUORESCENCE STUDIES OF THROMBIN ACTIVE SITE CONFORMATION

Lawrence J. Berliner

1. INTRODUCTION

The chapters in this book have introduced thrombin from a detailed primary structure viewpoint, from the three-dimensional x-ray structure, and from NMR studies of ligands, such as hirudin and fibrinopeptide analogs, which complex with this enzyme. It should be noted that the highly impressive accomplishments of the three-dimensional structure determination of PPACK-human α-thrombin and its complexes with hirudin and hirudin analogs are quite recent reports, within one year of the publication of this tome. Prior to that time we had available several spectroscopic techniques allowing us to probe aspects of protein structure and conformation *in solution*. While most of the spectroscopic techniques cannot distinguish individual amino acid residues and their precise conformation (except high-resolution proton NMR), they are, nonetheless, sensitive

Lawrence J. Berliner • Department of Chemistry, The Ohio State University, Columbus, Ohio 43210-1173.

Thrombin: Structure and Function, edited by Lawrence J. Berliner. Plenum Press, New York, 1992.

methods for examining aspects of conformation and dynamics. That is, the physical bases, which contribute to the spectral output from techniques such as electron spin resonance (ESR), fluorescence, circular dichroism (CD), and other optical spectroscopies, are highly sensitive to small changes or movements within a protein structure. This chapter addresses two biophysical techniques, ESR (spin-labeling) and fluorescence spectroscopy, with examples demonstrating their power and simplicity in unraveling conformational states and binding modes in various thrombin species.

2. BIOPHYSICAL APPROACHES—ESR SPIN LABELING AND FLUORESCENCE

2.1. The Spin Label Technique

The application of ESR or electron paramagnetic resonance (EPR) to biology and biochemistry grew from the need for complementary techniques to fluorescence, NMR, and other optical spectroscopies which could monitor and report dynamic information, such as macromolecular tumbling rates or local segmental flexibility, about a system. The technique utilizes a stable organic free radical, which can be synthetically manipulated to mimic substrates, inhibitors, cofactors, and effectors for a broad variety of biological systems. The requirement is beautifully met by the nitroxide moiety, which is most suitably found as a five- and six-membered heterocyclic ring structure, such as the pyrrolinyl, pyrrolidinyl, piperidinyl, and oxazolidinyl structures shown in Fig. 1.

The information derived from the ESR experiment is dynamic, i.e., a direct monitor of the rotational freedom of the nitroxide moiety about the single bonds, which connect it (covalently or noncovalently) to the protein backbone. Thus, the restricted mobility of a nitroxide moiety reflects steric hindrance at the binding locus on the protein surface or, more likely, strong hydrophobic interactions between the predominantly hydrocarbon structure of the nitroxide and a complementary apolar region on the protein. These interactions are potentially modulated by small changes in protein structure, which the (ESR) spin label technique can observe. While these changes must occur locally, they can, nonetheless, be mediated over much longer distances within the protein. Such changes are qualitative, but may be correlated quantitatively with binding equilibria or similar changes measured by other techniques. One of the quantitative aspects of the spin label method relates to the actual tumbling rate of the nitroxide

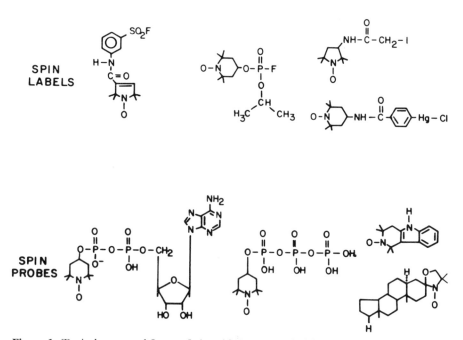

Figure 1. Typical structural forms of nitroxide groups. Spin labels can covalently modify a protein, while spin probes bind noncovalently. From Berliner (1980) with permission.

and the direction of changes in motion accompanying selective perturbations of the protein. The actual numbers are useful when the nitroxide label is rigid with respect to the macromolecule or when unique segmental motion occurs at the nitroxide site. In most other cases, however, the qualitative changes in motion (i.e., faster or slower, more or less mobility) are of more interest than the actual rotational correlation time (Berliner, 1980).

Instrumentally, the technique requires an X-band EPR spectrometer with a rectangular resonant cavity and a quartz flat sample cell. Due to the inherent problems of microwave absorption by aqueous samples, the particular geometry of the cavity and flat cell described above is critical. Sensitivity requirements are dictated primarily by the line shape and linewidth of the resultant ESR spectrum. That is, the broader the lines, the smaller the signal height, the lower the signal-to-noise. In practice this translates to labeled protein concentrations of 1 mM (preferred) down to 10 to 20 μM with time averaging of multiple acquisitions. Very recent technology, which utilizes a microwave device called the loop gap resonator, promises to reduce sample size to 1 μl (from 50 to 200 μl) and concentrations toward submicromolar.

All of the ESR studies reported to date with thrombin have utilized a series of sulfonylphenyl nitroxide analogs such as those depicted in Fig. 2. They all contain the unique Ser 195 specific fluorosulfonyl moiety, which guarantees a common "labeling point" between the enzyme and inhibitor (Berliner and Wong, 1974). They differ topographically in the position and structure of the nitroxide moiety on the aromatic ring. The most success has been with *meta-* and *para-*substituted sulfonyl fluoride spin labels. The *ortho-*substituted derivatives were usually unreactive due to steric hindrance from the adjacent nitroxide ring.

2.2. Fluorescence Spectroscopy

This highly sensitive, yet complex, technique has been a tremendous boon to the biochemical community by virtue of the low concentrations of fluorophore needed and the many changes in spectral characteristics, which result from small perturbations of the environment of the fluorescent moiety. Basically, in some molecules (including Trp and Tyr in proteins), the absorption of a photon is followed by light emission of lower energy (longer wavelength) than the exciting light. As is found with conventional absorption spectroscopy, there are many environmental factors that affect a fluorescence spectrum, including the efficiency of the fluorescence. With proteins and macromolecules, the scope of information covers conformation, binding sites, solvent interactions, degree of flexibility, intramolecular distances, and overall rotational diffusion coefficients. Unfortunately, the theory is not yet adequate to permit a positive correlation between a fluorescence spectrum and the properties of the immediate environment of the fluorophore; however, many researchers have attempted to draw conclusions from studies with model compounds in varying solvent environments. This latter approach with proteins is rather naive, and sometimes dangerous (as we learn more from fluorescence studies on high-resolution, well-defined x-ray structures). Thus, the principal benefits—and indeed the great power of the method—come from the use of fluorescence as a spectroscopic "monitor" of structure/conformation/environment changes, even when the "direction" of the effect cannot be correlated specifically to polarity, solvent, electronic, motional, or other physical changes.

The two most common uses of fluorescence in enzyme systems are intrinsic fluorescence (Trp or Tyr residues in the protein itself) or extrinsic fluorophores (covalent labels or noncovalent probes), which are added to

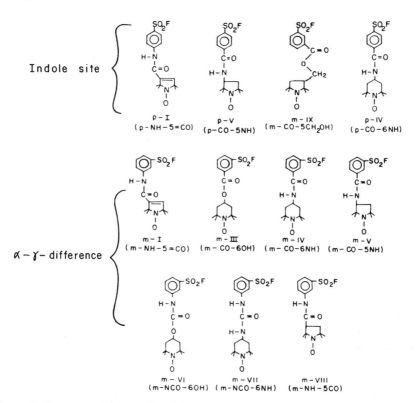

Figure 2. Structures of fluorosulfonylphenyl nitroxide spin labels. Compounds which are isomers of the same structure are designated by the same roman numeral. The two classes of labels, indole site and α–γ difference, respectively, are related to the tumbling domains in space of each nitroxide group relative to the sulfonyl substituent on the ring.

The parenthetical designation completely describes the structure: o, m, or p specifies the position of the SO_2F group on the ring. The second term specifies the linking functional group to the nitroxide: -NH, amido; -CO, acyl; -NCO, carbamoyl; or -SO_2, sulfonyl. The third term specifies the nitroxide ring: 5, the five-membered (saturated) pyrrolidinyl ring; 5=, the five-membered (unsaturated) pyrrolinyl ring; 6, the six-membered piperidinyl ring. The last term describes the functional group derived from the nitroxide moiety in the covalent linkage: CO, acyl (from the caroxylic acid); OH, an ester (from the alcohol); NH, amido (from the amine); and CH_2OH, ester (from the primary alcohol). For example, label m-I (m-NH-5=CO) is the m-fluorosulfonylanilide of the unsaturated pyrrolinecarboxylic acid nitroxide. Adapted from Berliner *et al.* (1981) with permission.

the system. We will discuss results with a variety of active-site-directed fluorescent probes (Ser and His directed inhibitor analogs), noncovalent inhibitor analogs, and less specific hydrophobic probes.

3. COMPARING THE ACTIVE SITES OF HUMAN AND BOVINE THROMBINS

If one examines the amino acid sequences (i.e., the primary structures of bovine and human thrombins in Chapter 1), the differences are few and, with contemporary prediction methods, virtually impossible to discriminate structurally. However, we know that bovine and human thrombin differ in some kinetic properties, especially in their proteolyzed derivatives (i.e., α, β, γ, ϵ, ζ human thrombin; α, β bovine thrombin). For example, β bovine thrombin retains about 10% fibrinogen clotting activity with virtually no change in K_m while human β- and γ-thrombin are essentially inactive toward fibrinogen (Lundblad *et al.*, 1984). The latter observations will become clear throughout this chapter, also with reflections on the crystal structure reported in Chapter 1. Second, in complexes with trypsin inhibitors such as BPTI (bovine pancreatic trypsin inhibitor), it has been shown that bovine thrombin is inhibited with a K_I of about 1 μM, while human α-thrombin–BPTI has a K_I in the millimolar range (Sugawara *et al.*, 1986). Other studies, such as the kinetics of inhibition by exosite affinity labels, have also distinguished bovine from human thrombin (Sonder and Fenton, 1983, 1986).

3.1. Human Thrombins

Looking solely at one species, i.e., human thrombin, we note that the various proteolyzed thrombin forms—β, γ, ϵ, ζ—differ in the position of a single proteolytic cleavage (save γ-thrombin, which has two cleavages), yet all of the same "primary structure" as that of human α-thrombin. The substantial decrease in fibrinogen clotting activity noted for β- and γ-thrombin undoubtedly contains secrets as to the mechanism and function by which thrombin effects fibrinogen clotting. There are also differences reported by Hofsteenge *et al.* (1986, 1988) as to the interactions with thrombomodulin. Thus, a thorough comparison of the extended active site structures of the various proteolyzed species of human thrombin offers much insight into the differences in structure and conformation in solution, which may be compared with functional differences. This comparison may also be extended to the bovine species, which has minor amino acid substitutions and insertions/deletions in sequence, versus the human form.

Such a study with spectroscopic methods such as spin labeling is quite informative, since any doubts as to the spectral contributions posed by the label itself are canceled out since the same label is placed on the same serine in all of the analogous derivatives. The approach using spin labels for both the human and bovine thrombin forms was a thorough comparison of the ESR spectra of most of the entire series of fluorosulfonylphenyl nitroxides shown in Fig. 2. An example of the type of ESR spectral results that one obtains is shown in Fig. 3, where spectrum 3a is that of the spin label m-V on the active serine 195 of human α-thrombin. This is to be compared with the same label on human γ-thrombin in spectrum 3b. The arrows on the left side of the spectrum denote one of the outer hyperfine extrema, which is a direct monitor of the rotational tumbling motion of the nitroxide moiety.

The larger splitting for m-V-labeled γ-thrombin indicates that the nitroxide moiety was more immobilized in the γ- versus α-thrombin derivative. On the other hand, the *para*-substituted label, p-I, gave *identical* spectra for the α- and γ-thrombin labeled derivative, respectively, as shown in Fig. 4a and b. Note that upon exposing p-I-labeled α- or γ-thrombin to the apolar ligand indole, a marked shift in the nitroxide mobility occurred (Fig. 4d), reflecting a conformational change at the active site upon complexing this apolar ligand. On the other hand, the m-V-labeled thrombins displayed *no* spectral shifts upon addition of indole. In fact, the results shown for m-V were consistent for *all* seven labels denoted "*α-γ-difference*" (Fig. 2) and the results shown for p-I were consistent for *all* four labels denoted "*indole site.*"

The results obtained comparing α- versus γ-thrombin were remarkable. That is, one class of nitroxide labels, designated indole site labels, produced identically superimposable spectra when any nitroxide was on the serine 195 of α- and γ-thrombin, respectively. This identity in tumbling motion for three different labels strongly suggests an identity in active site structure, conformation, and interactions in the region of the nitroxide moiety. On the other hand, a series of mostly *meta*-substituted phenylsulfonyl nitroxides clearly distinguished the α- from γ-thrombin active site conformation. Since we know the differences between α and γ thrombin are profound, functionally, as to fibrinogen hydrolytic activity and interactions with other macromolecular clotting factors (e.g., thrombomodulin), yet kinetically identical with tripeptide substrates, we have undoubtedly detected one of the structural alterations, which occurs upon conversion of α- to γ- human thrombin. In a more quantitative sense, we note that the mobilities of all of the γ-thrombin derivatives were *less* than those with corresponding α-thrombin analogs. Such immobilization is potentially caused by two factors. One is steric hindrance, the other increased

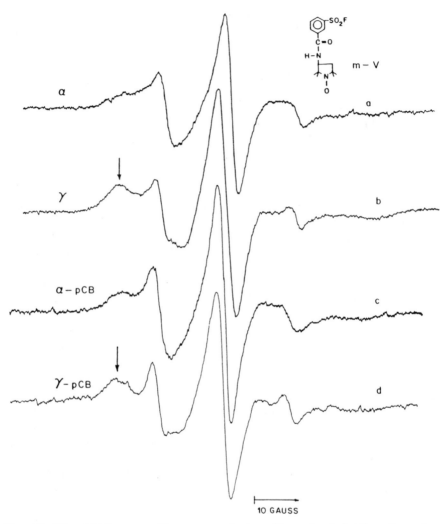

Figure 3. X-band ESR spectra of human α- and γ-thrombin conjugated at the active serine with *m*-V (m-CO-5NH): (a) spin-labeled α-thrombin; (b) spin-labeled γ-thrombin; (c) spectrum a in the presence of 50 mM *p*-chlorobenzylamine; (d) spectrum b in the presence of 50 mM *p*-chlorobenzylamine. All spectra were measured at pH 6.5, 0.05 M sodium phosphate, and 0.75 M NaCl, 26 ± 2°C. Protein concentration was typically 70–80 μM. From Berliner *et al.* (1981) with permission.

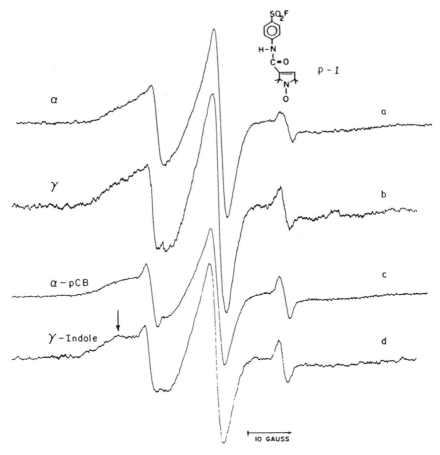

Figure 4. X-band ESR spectra of human α- and γ-thrombin spin-labeled at the active serine with p-I (p-NH-5=CO). All conditions were identical to those in Fig. 3. (a) p-I-α-thrombin; (b) p-I-γ-thrombin; (c) p-I-α plus 50 mM p-chlorobenzylamine (p-I-γ under the same conditions gave an identical effect); (d) p-I-γ in saturated (~20 mM) indole. From Berliner *et al.* (1981) with permission.

apolar interactions. Perhaps both contributed in the case of γ-thrombin, since we are aware that the fibrinogen K_m is substantially increased in γ-thrombin, that the two insertion loops of thrombin (versus the corresponding regions in the chymotrypsin or trypsin structures) must be important in interactions with fibrinogen. We have undoubtedly observed some "collapse" of these important loop structures upon proteolytic conversion. Figure 5 shows CPK (Corey–Pauling–Koltun) space-filling mod-

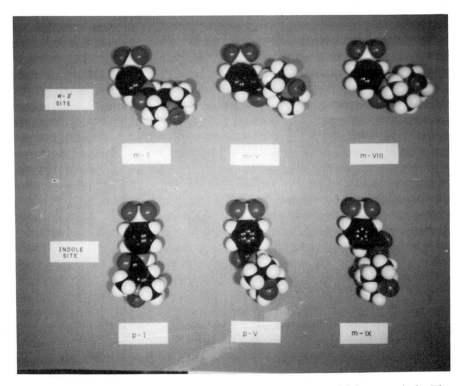

Figure 5. CPK models of α–γ *difference site* and *indole site sensitive labels,* respectively. The labels shown are (top row) *m*-I (*m*-NH-5=CO), *m*-V (*m*-CO-5NH), and *m*-VIII (*m*-NH-5CO); and (bottom row) *p*-I (*p*-NH-5=CO), *p*-V (*p*-CO-5NH), and *m*-IX (*m*-CO-5CH₂OH). The sulfonylphenyl moiety is fixed in the same orientation in all cases while the nitroxide moieties are placed in a common orientation for each label class. Where amide bonds existed between moieties, a planar, *trans* conformation was chosen. From Berliner *et al.* (1981) with permission.

els of these two groups of labels. Note how the *indole site* labels, which could not distinguish α- from γ-thrombin, are all linear, while the second group, designated α-γ, were all bent. Since these, of course, can undergo tumbling motion in the thrombin active site, they were allowed to rotate and the generated volumes of rotation are shown in Fig. 6. The model shown provides two oblate ellipsoids of revolution of the dimensions shown. The fixed reference points in this map are the sulfonyl group, covalently attached to Ser 195, and a *fixed* orientation of the benzene ring. This latter assumption is based on our knowledge of the previous crystal structures of tosyl-chymotrypsin and elastase, respectively, where somewhat unique orientations were observed, particularly involving the oxygen atoms of the sulfonyl group embedding themselves in the "oxyanion hole" where the carbonyl oxygen of a typical substrate resides. The validity of this assumption is confirmed by the self-consistency of our results; that is, we would not have found that all of the bent, *meta*-substituted labels are sensitive to, and rotating in, one unique region of space while the linear indole-sensitive labels resided in another region of space. Other aspects of this model, particularly those that implicate ligand binding, will be discussed in a later section.

While these spectra may have influences from anisotropic tumbling motion (i.e., rotation of the nitroxide about one axis preferentially over the

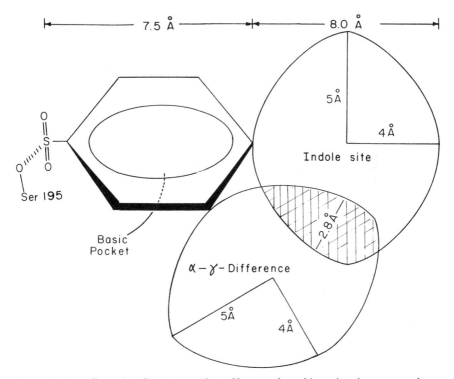

Figure 6. Two-dimensional representation of human thrombin active-site topography as a "map" constructed from the accessible volume of rotation of each nitroxide moiety with respect to the phenylsulfonyl group according to the constraints detailed in the text and Fig. 5 legend. These distances, which include van der Vaals radii, were derived from CPK models of the spin labels in Fig. 2. The distance marker at the phenyl ring bisects the covalent bond between the ring and the nitroxide moiety. The two oblate ellipsoids of revolution approximate those for the smallest spin label which detects each structural feature. Their volume of intersection thus *excludes* either the indole or α-γ difference sites. From Berliner *et al.* (1981) with permission.

other), these are usually minor contributions that do not affect the analyses, especially where multiple spin labels are examined. In this direct comparison of "dynamic active site conformations" of serine proteinases we analyze the spectra carefully by both overlapping them on a light box to see if they superimpose, as well as measuring these hyperfine extrema, called $2T_{||}$. There are occasionally small contributions due to free undialyzed label, or labeled thrombin which was subsequently proteolyzed to a structurally damaged form (which is impossible to dialyze in the presence of intact thrombin). These species would give the narrow three line components which are centered about the middle line in the spectrum with a splitting of ca. 15–16 G.

3.2. Bovine versus Human Thrombin

The A chain of bovine α-thrombin contains an additional 13 amino acids versus the human species (Lundblad *et al.*, 1979). However, the differences in their B-chain structure, which contains the catalytic site and most (or all) of the binding interactions with macromolecular substrates, are at first glance difficult to identify from their sequences (shown in Figure 2 of Chapter 1). Furthermore, as of this writing, we do not have the benefit of a published, completed bovine α-thrombin structure, except for information from the solution NMR work of Scheraga, Gibson, and Ni in Chapter 2.

An attempt at describing these differences structurally allows one to compare the efficacy of the spin label method versus crystallographic data. The differences shown between the bovine α- and human α-thrombin structures from both ESR spin labeling and fluorescence are subtle yet real. Specifically, Nienaber and Berliner (1991) found that the nitroxide moiety was consistently more immobilized in the bovine α-thrombin active site versus the human α-thrombin derivative. Furthermore, two active serine- directed fluorophores were both blue shifted and of decreased quantum yield with bovine versus human α-thrombin, respectively.

If one examines the amino acid sequences in more detail around the Trp 60 and Leu 144–Gly 150 insertion loops, the "apparent" differences which one might utilize in secondary or tertiary structure prediction are few. Focusing on the 144–150 loop, we note one stretch of residues from 149A to 150 which differ significantly between the bovine and human structures. The crystal work of Bode *et al.* (1989) and Rydel *et al.* (1990) reveals some flexibility in the 144–150 loop between the structures of PPACK–human α-thrombin and the hirudin complex of native thrombin. (The details are described in Chapter 1). We note that the bovine species contains slightly bulkier sidechains in this region (149A–150) with a marked substitution of Gly 149E versus Lys 149E in the human structure. Since we know that the spin labels are sensitive to the γ cleavage in this region, it is quite likely that the nitroxide interactions are sensitive to the sequence substitutions noted above. While the question arises from crystal structures as to whether the observed conformation of a movable loop is the most stable conformation vis-à-vis the structure or simply reflects crystal packing phenomena, we must take the data at their face value. Of course, the crystal structure of the spin-labeled thrombin complexes would add much to our interpretation of these results. On the other hand, since the crystal structure gives a static picture, even of the spin label derivative, we may or may not see the "dynamic" conformational differences found from the ESR studies.

3.3. Fluorescent Probes of the Thrombin Active Site

Figures 7A and B show the structures of several fluorescent probes suitable for active site (and exosite) studies of thrombin. Two of these labels, dansylfluoride and *p*-nitrophenylanthranilate, are substrate-inhibitor analogs, which form stable sulfonylated or acylated intermediates, respectively. In previous work, Berliner and Shen (1977a) examined the fluorescence emission properties of dansyl and anthraniloyl α- and γ-human thrombins, respectively. Remarkably, the fluorescence parameters for corresponding α- and γ-thrombin derivatives were identical! For example, the emission spectrum of dansyl thrombin has a λ_{max} of 510 nm and an enhancement factor of 60 (versus free label in water). Anthraniloyl thrombin has a λ_{max} (emission) of 415 nm with an enhancement factor of 300. Both labeled species were coupled to one (or more) intrinsic Trp residues as evidenced by a distinct excitation energy transfer band (at 286 nm in all cases). It was also observed that these labels resided over the basic substrate sidechain binding pocket since addition of the reversible inhibitor, benzamidine, resulted in an apparent quenching of the dansyl or anthraniloyl emission, which was consistent with a *shift* of the fluorophore to a more solventlike environment. These probes revealed *identical* environments close to the immediate region of Ser 195 and the basic sidechain binding, where we know that chromogenic tripeptide substrates exhibit essentially identical kinetic behavior (Sonder and Fenton, 1986; Lottenberg *et al.*, 1982). Thus, the structural *differences* between α- and γ-human thrombin are manifested somewhat *away* from the catalytic center where the dansyl or anthraniloyl moieties reside.

The ε- and ζ- forms of human thrombin have also been examined as the dansyl and anthraniloyl derivatives. In the case of ζ-thrombin a single cleavage has occurred between Trp 148 and Thr 149, resulting in a de-

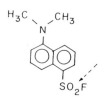

dansyl fluoride

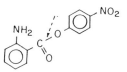

p-nitrophenylanthranilate

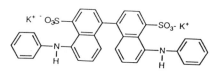

bisANS

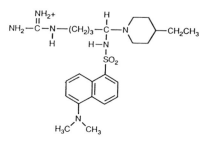

DAPA

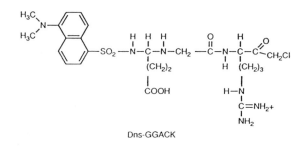

A   Dns-GGACK

Figure 7. Structures of fluorescent probes for thrombin studies. (A) The two active site covalent labels are shown with an arrow denoting where Ser 195 attacks the reagent. The remaining probes, with the exception of the chloromethyl ketone analogs, bind noncovalently. (B) A peptide chloromethyl ketone is synthesized containing a thioester at the aminoterminus and subsequently reacted at the catalytic His of a serine protease (step 1). After gentle removal of the acetyl moiety (step 2), the exposed thiol group may be reacted with a thiol-reactive fluorescent probe such as iodoacetamido-fluorescein. The chloromethyl ketone analogs reported by Bock (1988) incorporated D-Phe-Pro-Arg as the peptide segment $(NH-CHR-CO)_n$. From Bock (1988) with permission.

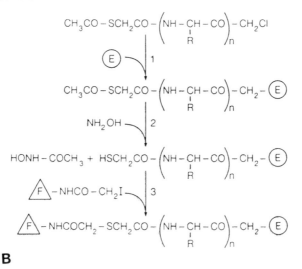

B

Figure 7. *continued*

rivative with almost complete retention of fibrinogen clotting activity (Brezniak *et al.*, 1989). On the other hand, ε-thrombin results from a cleavage at Ala 149A–Asn 149B, with a 30–40% loss of clotting activity (Kawabata *et al.*, 1985; Brower *et al.*, 1987). In both cases, ε- and ζ-dansyl and anthraniloyl thrombins showed five- to sixfold enhanced emission quantum yields versus the corresponding α-thrombin derivatives (J. K. Rowand and L. J. Berliner, submitted for publication). Thus, the structural changes resulting from the ε and ζ cleavages are significant enough to alter the fluorescence of these two active site probes, in contrast to the doubly cleaved γ- form, which was essentially identical in fluorescence properties (Berliner and Shen, 1977a).

A hydrophobic probe which binds noncovalently to nonpolar "patches" in proteins is the dimeric fluorophore, bis-ANS, shown in Fig. 7A. The presence of two sulfonate groups affords this probe a reasonable degree of water solubility; however, it is an avid, sensitive, hydrophobic ligand. In fact, it is critical that the binding stoichiometry of bis-ANS be determined, as with any new protein; it has been reported recently to bind at multiple sites on various proteins (J. E. Scheffler, personal communication; V. L. Nienaber and L. J. Berliner, unpublished results). In the case of α- and γ-thrombins, bis-ANS binds in a 1:1 complex with K_d values of 14.8 ± 2.2 and 5.8 ± 1.0 mM, respectively, at pH 7.0, 25 mM Tris, 0.15 M NaCl (Musci *et al.*, 1985). The fluorescence emission maximum is significantly blue shifted, from 515 to 485 nm, when complexed with either thrombin form. However, it was apparent that γ-thrombin was somewhat more hydrophobic than α-thrombin as evidenced both by its stronger dissociation constant and fluorescence emission enhancement factor (220 versus 70 for γ- versus α-thrombin, respectively). While the precise *location* of the bis-ANS binding site awaits future crystallographic studies, it is likely that the binding site is homologous in all of the thrombin species, especially since the λ_{max}^{em} was identical for both α- and γ-thrombin. The β, ϵ, and ζ forms appear to have significantly reduced binding affinity for bis-ANS versus the α and γ forms; in fact, the K_d values must be greater than 65 μM (J. K. Rowand and L. J. Berliner, unpublished results).

The two fluorescent, active-site-directed substrate-inhibitor analogs discussed above fell in the class of "poor substrates," i.e., they formed intermediates that were analogs of acyl enzyme intermediates found in the typical serine protease mechanism with ester substrates. This gives us a picture of a "transition state intermediate" conformer of the enzyme. Other active-site-directed fluorophores have been designed which target either the active site His 57 or are strong noncovalent, reversible inhibitors which strongly fluoresce when complexed with thrombin versus little fluorescence intensity in water.

Bock (1988) has developed a promising series of chloromethyl ketone analogs containing a free thiol group that can be derivatized *in situ* on a His-blocked thrombin derivative. The general structures of these labels and their chemistry are shown in Fig. 7b. Bock (1988) labeled human α-thrombin stoichiometrically with N^α-[(acetylthio)acetyl]-D-Phe-Pro-Arg-CH_2Cl followed by deacetylation (NH_2OH) and alkylation of the thiol group with 5-(iodoacetamido)fluorescein to yield the fluorescent thrombin product. Another (commercially available) analog is DEGRK or DNS-GGACK (dansyl-glutamyl-glycyl-arginine chloromethyl ketone), which has been employed with a variety of basic proteases including thrombin, and tissue plasminogen activator (Kettner and Shaw, 1981; Higgins and Lamb, 1986). A consideration with these labels is that the fluorophore moiety may reside *away* from the peptide binding region, rotating freely in the solution environment. In these cases a negligible to small change in fluorescence parameters may occur upon complexing a labeled thrombin sample with ligands which modulate its conformation implying if the probe lies far away from the effector or from the region where the conformational changes are propagated.

A novel series of strong *reversible* inhibitors, several of which are also fluorescent, were developed by Okamoto and co-workers (1979) and subsequently exploited in fluorescence studies by Nesheim *et al.* (1979). In particular, the inhibitor DAPA (dansyl-arginine-N-[3-ethyl-1,5-pentanediyl] amide) was highly fluorescent in the thrombin binary complex and almost nonfluorescent in water. Its strong affinity for thrombin ($K_d = 10^{-7}$ μM) suggested that this probe could be used as a highly sensitive monitor of thrombin conversion in the prothrombinase reaction (Krishnaswamy et al., 1986, 1987). There were further advantages in that the active thrombin(s) produced were immediately blocked (inhibited) from participating in further autoproteolytic steps.

Lastly, a small, basic thrombin inhibitor, para-aminobenzamidine, was also found to show enhanced fluorescence in the thrombin binary complex (Evans et al., 1982). This ligand binds at the basic sidechain binding pocket, exhibiting an emission blue shift from 376 to 368 nm and a 230-fold fluorescence enhancement versus the label in neutral aqueous solution. This probe was used to monitor the kinetics of thrombin:antithrombin III (AT-III) reactions, since p-aminobenzamidine is displaced upon thrombin:AT-III complexation, which is accompanied by a decrease in fluorescence emission (Evans et al., 1982). See also the chapter by Olson and Bjork which describes thrombin:AT-III interactions using this probe.

4. THROMBIN COMPLEXES WITH REGULATORY METABOLITES

The interactions of thrombin with its various effectors, clotting factors, and other regulatory factors in solution are complex as they all lend to the highly regulated process of blood coagulation and fibrinolysis to which thrombin plays a central role. We already know that much more than the catalytic residues of the thrombin structure play important parts in its overall function in thrombosis and hemostasis. The molecular picture of a thrombin–ligand complex is difficult for macromolecular complexes. To date we have only a relatively small complex, i.e., hirudin–thrombin, as discussed in earlier chapters. How then does one approach larger complexes, where crystal structures are difficult to attain, even at low resolution? Furthermore, how does one unravel and understand the regulatory sites on the thrombin surface? An approach utilized with the spin label and fluorescence methods was to examine small ligands which contain some physical chemical property that was complementary to a particular thrombin regulatory site.

4.1. Apolar Ligands

The first report by Berliner and Shen (1977b) confirmed the existence of at least one apolar binding site on the thrombin surface, to which indole was an effective ligand, which resulted in conformational change mediated in the catalytic region. It is important to point out that the structural change, i.e., the change in environment of the nitroxide moiety of the active serine spin label, can be mediated either by an indirect conformational change or by an overlapping binding interaction between the nitroxide moiety and the indole ligand at the same (i.e., active site) region. Refining this model further requires additional experiments such as dye binding (UV-visible), fluorescence, and activity measurements. In fact, all of the evidence to date for apolar indole-like ligand binding points to a site somewhat distant from the active center. This is demonstrably shown by the unusual ability of apolar ligands, particularly tryptamines, to activate thrombin macromolecular catalytic activity such as fibrinogen clotting. A marked activation of fibrinogen clotting rate was noted by Berliner *et al.* (1986) using 5-fluorotryptamine. Even more striking was the ability of indole and tryptamine analogs to activate thrombin activation of protein PC (Musci and Berliner,1987), which will be expanded in a later section on thrombin–thrombomodulin interactions.

The indole–apolar ligand binding interactions are to be contrasted with the binding of small ligands which are designed to be reversible competitive inhibitors for thrombin substrates, such as the basic ligands benzamidine and *para*-chlorobenzylamine; here it is well known that both basic ligands bind specifically to the basic sidechain binding pocket (normally for an Arg residue), thus blocking all thrombin catalytic functions. For example, the (active site directed) dye, proflavin, is completely displaced upon titrating the thrombin:proflavin complex with benzamidine (Koehler and Magnusson, 1974). Direct physical evidence of overlap at this region is found from the entire series of fluorosulfonylphenyl nitroxide spin labels discussed above. In *every* case of every spin label with *all* of the thrombin derivatives in species, an environment shift (i.e., a shift in mobility) was noted upon complexing the spin-labeled thrombin derivative with either benzamidine or *para*-chlorobenzylamine under saturating conditions. In most of the cases the label became more immobilized, although the specific explanation for this reduction in mobility is not understood. Needless to say, the interpretation of these results is the binding of the aromatic basic inhibitor in the substrate binding pocket, which protrudes far enough into the catalytic triad region so as to obligate the phenylsulfonyl group to shift to an environment that will tolerate both spin label and basic ligand binding simultaneously. Such a "conformational shift" was observed earlier for α-chymotrypsin with a similar series of fluorosulfonyl spin labels (Berliner and Wong, 1974). In the case of serine proteases, which have a similar specificity pocket, we note from the crystal structures that the basic binding pocket in trypsin is somewhat deeper than that in thrombin. This was manifested in spin label experiments by *no* conformational shift in nitroxide spin labels upon binding benzamidine to fluorosulfonylphenyl nitroxide spin-labeled trypsins.

4.2. Nucleotide and Pyrophosphate Analogs

An unusual series of ligands were found to bind to thrombin which, at first glance, were completely different. Since it was known that substantial amounts of ATP are released as the "secretion reaction" in thrombin-induced platelet aggregation, and also since nucleotide and inorganic phosphate levels are relatively high in blood, there was an interest in whether this modulated thrombin structure and function. Indeed, Conery and Berliner (1983) found that ATP and other pyrophosphate-containing nucleotides were effective inhibitors of fibrinogen clotting. The most specific was ATP, although other nucleotides such as UDP were effective, presumably by virtue of the pyrophosphate moiety. The adenine nucleotide binding was allosteric, with at least two types of sites. The binding was

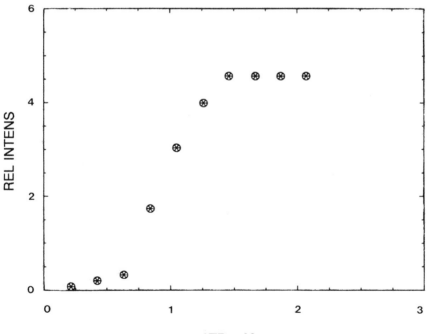

Figure 8. ATP titration of α-thrombin–bis-ANS complex at pH 7.0. α-Thrombin concentration was 0.54 μM; bis-ANS, 1.83 μM. Experimental conditions: λ_{ex} = 375 nm, λ_{em} = 485 nm, and 25°C. From Musci *et al.* (1985) with permission.

most exemplary in studies of the thrombin:bis-ANS complex by fluorescence spectroscopy. Figure 8 shows the emission intensity for a 0.54 μM human α-thrombin:1.83 μM bis-ANS complex, pH 7.0, with increasing ATP. Note the strong sigmoidal behavior in the low millimolar concentration region. What was even more striking were the unusual effects of adenine nucleotides on thrombin catalytic behavior with chromogenic tripeptide anilide substrates. That is, binding of ATP at low submillimolar concentrations induced an *activation* of amidase or esterase activity followed by a linked, second inhibition phase at higher concentrations. The kinetic phenomena were first reported by Berliner *et al.* (1986) and then further refined in more detail by DeCristofaro *et al.* (1990). More recently, we have shown similar behavior with esterase activity with Tos-Arg-OMe or Bz-Arg-OMe (L. J. Berliner, A. Suehiro, V. L. Nienaber, and J. S. Yu, unpublished results). Remarkably, in the presence of adenosine, no kinetic effects are observed at all. Since the physical binding of adenosine and ATP was observed from studies of fluorescent labeled or fluorescent complexed thrombin, we are aware, however, of some type of, perhaps apolar, binding site for the adenosyl moiety.

This nucleotide binding phenomenon and the associated nucleotide binding region(s) may overlap partially with the anionic binding exosite found for hirudin C-terminal fragments.

5. PROBING CONFORMATIONAL CHANGES IN THROMBIN–HIRUDIN COMPLEXES

The natural anticoagulant polypeptide hirudin has been described in detail in this book, from the crystal structure work by Bode, Huber, Rydel, and Tulinsky (Chapter 1) to the chapter on biochemical aspects by Stone and Maraganore (Chapter 6). An important consequence of the incisive kinetic studies of Stone and Hofsteenge (1986) was the notion that the binding interactions of hirudin with thrombin involved in a two-step process. One model suggested that part of the hirudin molecule binds at some exosite followed by a second binding step over the catalytic center. Furthermore, the kinetic studies fit a two-step model, but this does not rule out other, more complex mechanisms. Unfortunately, such a dynamic binding process may not be observable from the x-ray crystallographic studies. The goal of biophysical studies was to see if stepwise binding interactions might be detected by complexing thrombin with hirudin and hirudin C-terminal analogs (Krstenansky *et al.*, 1987). Both the sulfonylphenyl nitroxide labels, as well as the fluorescent anthraniloyl and dansyl probes were sensitive to the binding of both C-terminal hirudin fragments as well as recombinant hirudins (Berliner and Rowand, 1992; Rowand and Berliner, 1992). What was noteworthy in these studies was the distinct active site conformational changes observed upon binding a dodecapeptide C-terminal hirudin fragment (Mao et al., 1988) which should bind well away from any active site label. That is, these active site probes may be detecting one of the two-step binding interactions of hirudin, this step involving a conformational change propagated to the catalytic site.

6. PROBING CONFORMATIONAL CHANGES IN THROMBIN–THROMBOMODULIN COMPLEXES

When thrombin interacts with endothelium it can interact with thrombomodulin (TM) to activate protein C, which ultimately serves to inhibit clot formation (Esmon and Owen, 1981). When thrombin interacts with TM, the rate of protein C activation may be increased up to ca. 20,000-fold (Esmon and Owen, 1981). This rate enhancement is manifested by a decreased K_m and increased k_{cat} for protein C, yet little change is observed in the hydrolysis of synthetic substrates (Esmon et al., 1983a,b). Although activation requires Ca(II), complex formation is essentially independent of free Ca(II). Several experiments suggested that a conformational change occurs in thrombin upon complex formation with TM; in particular, dansyl-thrombin undergoes a 47% decrease in fluorescence emission intensity in the TM complex (Johnson et al., 1983). Musci et al. (1988) provided further insight as to whether these conformational changes occur near the active site using nitroxide spin labels. These studies examined both intact TM and a proteolyzed 50,000-dalton derivative, ε-TM. Figure 9A and B show ESR spectra of p-I human α-thrombin in the absence and presence of ε-TM, respectively. When comparing these ESR spectra, i.e., of labeled α-thrombin alone versus the ε-TM:thrombin complex, one must account for the change in macromolecular rotation since the overall molecular mass shifts from 36,600 to 86,600 daltons, respectively. Therefore, Musci et al. (1988) measured ESR spectra for labeled, uncomplexed thrombin alone under viscosity conditions where the overall rotational tumbling rate would be equivalent to that for an 86,600-dalton molecule (i.e., isokylindric). This was accomplished in sucrose-containing buffers of 2.4-poise viscosity assuming that the molecular mass of a spherical globular protein is directly related to the radius cubed, since the partial specific volumes of most proteins are quite similar. Spectrum 9C depicts the isokylindric control of ε-TM:p-I-α-thrombin in 24% (w/v) sucrose; however, note the *decreased* immobilization of the nitroxide in ε-TM:p-I-α-thrombin (Fig. 9B), as reflected by the position of the low-field

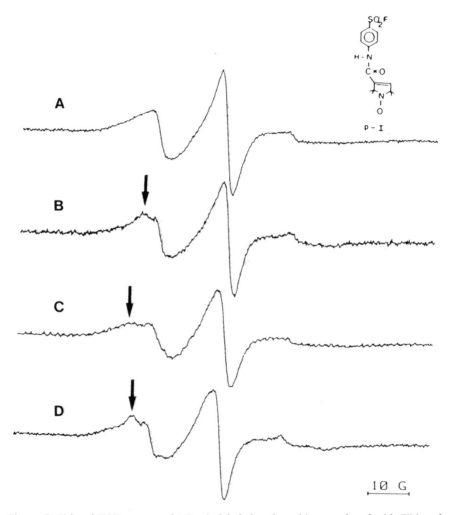

Figure 9. X-band ESR spectra of *p*-I spin-labeled α-thrombin complexed with TM and ε-TM. (A) p-I-labeled thrombin in buffer. (B) p-I spin-labeled α-thrombin complexed with ε-TM. (C) an isokylindric control of thrombin spin-labeled with *p*-I in 24% (w/v) sucrose. (D) *p*-I labeled thrombin complexed with intact TM. Experimental conditions: 98 μM labeled thrombin (A); 32 μM labeled thrombin and 35 μM ε-TM (B); 150 μM labeled thrombin (C); 60 μM labeled thrombin and 98 μM TM (D); 20 mM Tris–HCl buffer, 0.16 M NaCl, pH 7.5. Scan range 100 G, field set 3395 G, modulation amplitude 2 G, microwave power 20 mW, scan time 2 min. (four scans were averaged for each spectrum). From Musci *et al.* (1988) with permission.

shoulder (arrows). Furthermore, the larger intact TM:p-I-α-thrombin complex also showed a *decreased* immobilization (Fig. 9D), which is clearly evidence for a TM-induced conformational change since decreased macro-molecular tumbling would cause precisely the reverse spectral changes.

TM (or ε-TM) interaction with thrombin clearly results in several alterations of the thrombin active site region, which may contribute not only to the increased rate of protein C activation, but also to the decreased rates of procoagulant reactions such as fibrinogen clotting (Esmon *et al.*, 1982; Hofsteenge *et al.*, 1986). It has been shown that the binding of thrombin to TM is competitively inhibited by fibrinogen and is consistent with TM sterically hindering the thrombin binding interaction with fibrinogen (Hofsteenge *et al.*, 1986). Upon titrating Ca(II) into the samples in Fig. 9B and D, no change in the nitroxide mobility was detected. When Gd(III) was substituted for Ca(II), no paramagnetic broadening of the nitroxide was observed indicating a distance exceeding ca. 12 Å for the separation between the cation and the free electron on the N–O moiety. These metal ion binding experiments, while yielding negative results, were nonetheless consistent with the fact that the calcium binding site simply functions as a structural site in stabilizing the TM conformation. Certainly, this site is not manifested in the TM-induced conformational change of α-thrombin at the active site. While 12 Å is small relative to the molecular sizes of α-thrombin, the contact region was sufficiently distant from the thrombin active site.

7. CONCLUSIONS FROM SOLUTION VERSUS CRYSTAL STRUCTURES

7.1. Efficacy of Labeling Techniques

The x-ray crystal structure, of course, provides one of the most precise maps of the three-dimensional structure of the enzyme in the static crys-talline state. But the high-resolution x-ray crystal structure is not the final word, serving as a valuable starting point for further detailed struc-ture/function studies. In fact, several findings from spectroscopic ap-proaches have contributed to a reasonably detailed map of the active site structure of thrombin. More precise identification requires crystal struc-tures, putative structures, or a large number of solution distance measure-ments. A good example of where an x-ray structure served as a starting point was the thrombin–BPTI studies (Sugawara *et al.*, 1986) which were suggested by the computer graphics model analysis, while previous literature dictated otherwise. Strong arguments for continuing biophysical solution distance measurements are seen from the following:

1. Some thrombin binding loci found from solution studies may be inaccessible to a ligand analog in the crystal due to crystal packing restrictions or insufficient ligand diffusion into the crystal. Unfortunately, thrombin has been difficult to crystallize unless blocked with a strong irreversible inhibitor.

2. Ligand binding phenomena which induce protein conformational changes (e.g., indole or ATP) may be difficult or impossible to resolve e.g., in the crystal. Subtle changes, such as those manifested in catalytic rate activation or fluorophore emission changes, may not be detectable as defined residue movements, even within the resolution (ca. 2–2.5%) of an x-ray study.

3. While the accuracy of absolute distances from most spectroscopic techniques is not as precise (ca. 10–20%) as x-ray derived distances, distance *differences* are more accurate at times than those obtainable from crystal studies.

It is obvious that the results obtained from labeling techniques are reported by the *label* and, hence, may affect the label as well as the protein structure. However, when comparing two forms of the same protein (i.e., γ- and ε-thrombin), the structural differences are real for each label. Furthermore, if several other physicochemical measurements indicate that the derivative is (essentially) identical to the native form, the contribution from the label itself is at most very local. Lastly, labels and probes are designed to mimic substrates, inhibitors and a substrate/inhibitor/effector interacts with the protein structure. The overall value of labeling methods is measured by what they add to our knowledge about a protein or enzyme.

7.2. Precision of Distance Measurements and the Significance of Mapping Binding Regions by Solution Techniques

All of the spectroscopic distance measurement techniques outlined in this proposal are based on dipolar interactions, which are sensitive to the inverse sixth power of the separation between the two groups. Furthermore, while it is, of course, desirable to have the precise position, dimensions, and depth of a binding site, it is first important to identify and localize new binding regions. A complete description of the geometry of a site itself requires x-ray study with a series of ligands of varying structure. An important point to reiterate about the value of the spectroscopic methods *per se* is that many binding regions were discovered directly from these uniquely sensitive spectroscopic experiments. Summarizing, the accuracy of spectroscopic experiments is quite good in localizing a binding region, although less precise than x-ray crystallography in identifying the specific amino acid residues which define the site.

8. REFERENCES

Berliner, L. J., 1980, Using the spin label method in enzymology, in: *Spectroscopy in Biochemistry* (J. E. Bell, ed.), Vol. 2, CRC Press, Boca Raton, Fla., pp. 1–56.

Berliner, L. J., and Shen, Y. Y. L., 1977a, Active site fluorescent labeled dansyl and anthraniloyl human thrombins, *Thromb. Res.* **12:**15–25.

Berliner, L. J., and Shen, Y. Y. L., 1977b, Physical evidence for an apolar binding site near the catalytic center of human α-thrombin, *Biochemistry* **16:**4622–4626.

Berliner, L. J., and Wong, S. S., 1974, Spin-labeled sulfonyl fluorides as active site probes of protease structure, *J. Biol. Chem.* **249:**1668–1677.

Berliner, L. J., Bauer, R. S., Chang, T.-L., Fenton, J. W., II, and Shen, Y. Y. L., 1981, Active site topography of human coagulant (α) and noncoagulant (γ) thrombins, *Biochemistry* **20:**1831–1837.

Berliner, L. J., and Rowand, J. K., 1992, Biophysical studies of interactions of hirudin analogs with bovine and human thrombin by ESR and fluorescence labeling studies, in: *Design and Synthesis of Thrombin Inhibitors* (M. Scully, G. Claeson, and V. Kakkar, eds.), Plenum Press, New York (submitted).

Berliner, L. J., Birktoft, J. J., Miller, J. L., Musci, G., Scheffler, J. E., Shen, Y. Y., and Sugawara, Y., 1986, Thrombin: Active site topography, *Ann. N.Y. Acad. Sci.* **488:**80–95.

Bock, P. E., 1988, Active site selective labeling of serine proteases with spectroscopic probes using thioester peptide chloromethyl ketones, *Biochemistry* **27:**6633–6639.

Bode, W., Mayr, I., Baumann, U., Huber, R., Stone, S. R., and Hofsteenge, J., 1989, The refined 1.9 Å crystal structure of human α-thrombin: Interaction with D-Phe-Pro-Arg chloromethylketone and significance of the Tyr-Pro-Pro-Trp insertion segment, *EMBO J.* **8:**3467–3475.

Brezniak, D. V., Brower, M. J., Witting, J. I., Walz, D. A., and Fenton, J. W., II, 1989, Human α- to ζ-thrombin cleavage occurs with neutrophil cathepsin G or chymotrypsin while retaining fibrinogen clotting activity, *Biochemistry* **28:**3536–3542.

Brower, M. S., Walz, D. A., Garry, K. E., and Fenton, J. W., II, 1987, Human neutrophil elastase alters human α-thrombin function: Limited proteolysis near the γ-cleavage site results in decreased fibrinogen clotting and platelet-stimulatory activity, *Blood* **69:**813–373.

DeCristofaro, R., Landolfi, R., and Di Cera, E., 1990, The linkage between adenosine nucleotide binding and amidase activity in human α-thrombin, *Biophys. Chem.* **36:**77–84.

Esmon, C. T., and Owen, W. G., 1981, Identification of an endothelial cell factor for thrombin-catalyzed activation of protein C, *Proc. Natl. Acad. Sci. USA* **78:**2249–2252.

Esmon, C. T., Esmon, N. L., and Harris, K. W., 1982, Complex formation between thrombin and thrombomodulin inhibits both thrombin-catalyzed fibrin formation and factor V activation, *J. Biol. Chem.* **257:**7944–7947.

Esmon, N. L., Carroll, R. C., and Esmon, C. T., 1983a, Thrombomodulin blocks the ability of thrombin to activate platelets, *J. Biol. Chem.* **258:**12238–12242.

Esmon, N. L., DeBault, L. E., and Esmon, C. T., 1983b, Proteolytic formation and properties of γ-carboxyglutamic acid-domainless protein C, *J. Biol. Chem.* **258**:5548–5553.

Evans, S. A., Olson, S. T., and Shore, J. D., 1982, p-Aminobenzamidine as a fluorescent probe for the active site of serine proteases, *J. Biol. Chem.* **257**:3014–3017.

Higgins, D. L., and Lamb, M. C., 1986, The incorporation of a fluorescent probe into the active sites of one- and two-chain tissue-type plasminogen activator, *Arch. Biochem. Biophys.* **249**:418–426.

Hofsteenge, J., Taguchi, H., and Stone, S. R., 1986, Effect of thrombomodulin on the kinetics of the interaction of thrombin with substrates and inhibitors, *Biochem. J.* **237**:243–251.

Hofsteenge, J., Braun, P. J., and Stone, S. R., 1988, Enzymatic properties of proteolytic derivatives of human α-thrombin, *Biochemistry* **27**:2144–2151.

Johnson, A. E., Esmon, N. L., Laue, T. M., and Esmon, C. T., 1983, Structural changes required for activation of protein C are induced by Ca (II) binding to a high affinity site that does not contain γ-carboxyglutamic acid, *J. Biol. Chem.* **258**:5554–5560.

Kawabata, S., Morita, T., Iwanaga, S., and Igarashi, H., 1985, Staphylocoagulase-binding region in human prothrombin, *J. Biochem.* **97**:325–331.

Kettner, C., and Shaw, E., 1981, Inactivation of trypsin-like enzymes with peptides of arginine chloromethylketone, *Methods Enzymol.* **80**:826–842.

Koehler, K. A., and Magnusson, S., 1974, The binding of proflavin to thrombin, *Arch. Biochem. Biophys.* **160**:175–184.

Krishnaswamy, S., Mann, K. G., and Nesheim, M. E., 1986, The prothrombinase-catalyzed activation of prothrombin proceeds through the intermediate meizothrombin in an ordered, sequential reaction, *J. Biol. Chem.* **261**:8977–8984.

Krishnaswamy, S., Church, W. R., Nesheim, M. E., and Mann, K. G., 1987, Activation of human prothrombin by human prothrombinase. Influence of factor Va on the reaction mechanism, *J. Biol. Chem.* **262**:3291–3299.

Krstenansky, J. L., Owen, T. J., Yates, M. T., and Mao, S. J. T., 1987, Anticoagulant peptides: Nature of the interaction of the C-terminal region of hirudin with a noncatalytic binding site on thrombin, *J. Med. Chem.* **30**:1688–1691.

Lottenberg, R., Hall, J. A., Fenton, J. W., and Jackson, C. M., 1982, The action of thrombin on peptide p-nitroanilide substrates: Hydrolysis of tos-gly-pro-arg-pNA and D-phe-pip-arg-pNA by human α- and γ- and bovine α- and β-thrombins, *Thromb. Res.* **28**:313–332.

Lundblad, R. L., Noyes, C. M., Mann, K. G., and Kingdon, H. S., 1979, The covalent differences between bovine α- and β-thrombin. A structural explanation for the changes in catalytic activity, *J. Biol. Chem.* **254**:8524–8528.

Lundblad, R. L., Nesheim, M. E., Straight, D. L., Sailor, S., Bowie, J., Jenzano, J. W., Robert, J. D., and Mann, K. G., 1984, Bovine α- and β-thrombin is not a consequence of reduced affinity for fibrinogen, *J. Biol. Chem.* **259**:6991–6995.

Mao, S. J. T., Yates, M. T., Owen, T. J., and Krstenansky, J. L., 1988, Interaction of hirudin with thrombin: Identification of a minimal binding domain of hirudin that inhibits clotting activity, *Biochemistry* **27**:8170–8173.

Musci, G., and Berliner, L. J., 1987, Ligands which effect human protein C activation by thrombin, *J. Biol. Chem.* **262**:13889–13891.

Musci, G., Metz, G. D., Tsunematsu, H., and Berliner, L. J., 1985, Bis-ANS binding to human thrombins. A sensitive exosite fluorescent affinity probe, *Biochemistry* **24**:2034–2039.

Musci, G., Berliner, L. J., and Esmon, C. T., 1988, Evidence for multiple conformational changes in the active center of thrombin induced by complex formation thrombomodulin: An analysis employing nitroxide spin labels, *Biochemistry* **27**:769–773.

Nesheim, M. E., Prendergast, F. G., and Mann, K. G., 1979, Interactions of a fluorescent active-site-directed inhibitor of thrombin: Dansylarginine N-(3-ethyl-1,5-pentane-diyl)amide, *Biochemistry* **18**:996–1003.

Nienaber, V. L., and Berliner, L. J., 1991, Conformational differences between human and bovine thrombins as detected by electron spin resonance and fluorescence spectroscopy. *Thromb. Haemostasis* **65**:40–45.

Okamoto, S., Hijikata, A., Kikumoto, R., and Tamao, Y., 1979, A synthetic thrombin inhibitor taking extremely active stereo-structure, *Thromb. Haemostas.* **42**:205.

Rowand, J. K., and Berliner, L. J., 1992, Active site labeled thrombin forms can distinguish between hirudin isoinhibitors, *J. Prot. Chem.* **11** (submitted).

Rydel, T. J., Ravichandran, K. G., Tulinsky, A., Bode, W., Huber, R., Roitsch, C., and Fenton, J. W., II, 1990, The structure of a complex of recombinant hirudin and human α-thrombin, *Science* **249**:277–280.

Sonder, S. A., and Fenton, J. W., II, 1983, Differential inactivation of human and bovine α-thrombins by exosite affinity-labeling reagents, *Thromb. Res.* **32**:623–629.

Sonder, S. A., and Fenton, J. W., II, 1986, Thrombin specificity with tripeptide chromogenic substrates: Comparison of human and bovine thrombins with and without clotting activities, *Clin. Chem.* **32**:934–937.

Stone, S. R., and Hofsteenge, J., 1986, Kinetics of the inhibition of thrombin by hirudin, *Biochemistry* **25**:4622–4628.

Sugawara, Y., Birktoft, J. J., and Berliner, L. J., 1986, Human α- and γ-thrombin inhibition by trypsin inhibitors supports predictions from molecular graphics experiments, *Semin. Thromb. Hemostas.* **12**:209–212.

Part 2

BIOCHEMISTRY

Chapter 4

SYNTHETIC SUBSTRATES AND INHIBITORS OF THROMBIN

James C. Powers and Chih-Min Kam

1. INTRODUCTION

Thrombin has a central regulatory role in hemostasis and is formed by both the intrinsic and extrinsic pathways of blood coagulation (Fenton, 1986, 1981). The major function of this serine protease is the cleavage of fibrinogen to form fibrin clots, but thrombin also activates factors V, VIII, XIII and protein C which are important in the control of hemostasis and thrombosis. In addition, thrombin can stimulate platelet secretion and aggregation in blood, and mediate other nonhemostatic cellular events. Our understanding of the various biological roles of thrombin has been facilitated by the development of convenient chromogenic and fluorogenic assays for the measurement of thrombin activity; and synthetic, low-molecular-weight inhibitors which are important probes for the study of thrombin's mechanism, specificity, and function in the coagulation system.

Thrombin is a member of the trypsin family of serine proteases, a family which encompasses many other important plasma enzymes including coagulation enzymes (factors Xa, VIIa, IXa, XIa, XIIa, protein C),

James C. Powers and Chih-Min Kam • School of Chemistry and Biochemistry, Georgia Institute of Technology, Atlanta, Georgia 30332.

Thrombin: Structure and Function, edited by Lawrence J. Berliner. Plenum Press, New York, 1992.

117

fibrinolytic enzymes (plasmin, plasminogen activators), and complement enzymes (C1s, C1r, protein B, protein C2, protein D, factor I). In addition to the catalytic triad (Asp 102, His 57, and Ser 195), a feature common to the active site of all serine proteases, Asp 189 in the primary substrate binding site (S_1)* of the trypsin family plays an important role in the recognition and binding of substrates and inhibitors.

Thrombin is a well-recognized target for the design of new antithrombotic agents since this serine protease is such a powerful trigger in thrombus formation in the blood. At present, one of the most widely used anticoagulant drugs is heparin, a large sulfonate polysaccharide which inhibits thrombin by increasing the activity of the natural plasma thrombin inhibitor antithrombin III. Considerable effort is being devoted to the development of new antithrombotic agents since the currently available drugs such as heparin often cause bleeding when used therapeutically and are unable to prevent the occlusive complications in atherosclerotic vascular disease or reocclusion following successful thrombolysis. Although much of this activity has been focused on high-molecular-weight thrombin inhibitors such as hirudin or low-molecular-weight platelet aggregation inhibitors, increasing research is being devoted to the development of suitable small synthetic thrombin inhibitors. In the future, it is likely that synthetic thrombin inhibitors will supplement or even replace currently used antithrombotic agents in a variety of therapeutic situations. In this review, we will limit our attention to small synthetic inhibitors and substrates of thrombin.

2. SUBSTRATES

Synthetic peptide substrates for the coagulation enzymes have existed for more than 30 years and are widely used for both research and clinical diagnosis. Most assays use specific peptide sequences for thrombin to which are attached a chromogenic or fluorogenic leaving group. The most widely used substrates are chromogenic p-nitroanilide substrates (pNA or NA) and fluorogenic peptide derivatives of 7-amino-4-methylcoumarin (AMC), β-naphthylamide (β-NA), and aminoisophthalic acid dimethyl ester (AIE).[†] Chromogenic peptide thioesters have been studied with many

---

*The nomenclature used for the individual amino acid residues (P_1, P_2, etc.) of a substrate and the subsites (S_1, S_2, etc.) of the enzyme is that of Schechter and Berger (1967).

[†]Standard abbreviations are used for amino acid residues and blocking groups in peptide substrates and inhibitors. Other abbreviations used include: ACITIC, 7-amino-4-chloro-3-(3-isothiureidopropoxy) isocoumarin; AMC, 7-amino-4-methylcoumarin; APPA, 4-ami-

coagulation enzymes including thrombin and are extremely reactive toward this enzyme. Thioester substrates are usually more reactive than corresponding nitroanilides and aminomethylcoumarins toward various serine proteases, but they are less specific than the amide substrates. Thus, each group of substrates has its own advantages and disadvantages. Several peptide nitroanilides, thiobenzyl esters, and fluorogenic substrates for thrombin are commercially available. Several reviews of synthetic substrates for thrombin have been published (Izquierdo and Burguillo, 1989; Lottenberg et al., 1981).

2.1. Peptide 4-Nitroanilides

Specific peptide nitroanilides were first developed as substrates of thrombin, plasmin, and trypsin by Svendsen et al. (1972). The substrate Bz-Phe-Val-Arg-pNA was based on the known cleavage site of the α chain of fibrinopeptide A by thrombin. Since the initial peptide substrate was prepared in 1972, many new and improved chromogenic and fluorogenic substrates have been developed with specificity for thrombin and many other coagulation enzymes. Substrates such as Bz-Phe-Val-Arg-pNA·HCl (S-2160), H-D-Phe-Pip-Arg-pNA·2HCl (S-2238), Z-Gly-Pro-Arg-pNA·HCl (Cbz-Chromozym TH) are specific for thrombin (Bang and Mattler, 1977). Tos-Gly-Pro-Arg-pNA·HCl (Tos-Chromozym TH) has a greater sensitivity for thrombin but a lower specificity than its relative Cbz-Chromozym TH. Many of these substrates are widely used since they are commercially available.

The kinetic constants for the hydrolysis of Tos-Gly-Pro-Arg-pNA·HCl and H-D-Phe-Pip-Arg-pNA by human thrombins (α and γ forms) and bovine thrombins (α and β forms) have been studied and all forms of thrombin hydrolyze these substrates very efficiently with k_{cat}/K_M values of 10^6–10^7 M^{-1} s^{-1} (Lottenberg et al., 1982). The small synthetic substrates are thus less able to distinguish the various forms of thrombin than clotting assays which use fibrinogen as the substrate.

The various peptide substrates have been invaluable for the subsite mapping of the active site of thrombin. Human thrombin has been studied with many blocked and unblocked tripeptidyl pNA substrates (Pozsgay et

---

dinophenylpyruvic acid; FOY, ε-guanidinocaproic acid p-carboxyethylphenyl ester (Gabexate Mesilate); FUT-175, 6'-amidino-2-naphthyl-4-guanidinobenzoate (Nafamostat mesylate); MQPA, 4-methyl-1-[N^2-[(3-methyl-1,2,3,4-tetrahydro-8-quinolinyl)sulfonyl]-L-arginyl]- 2-piperidinecarboxylic acid (Argatroban); pNA, p-nitroanilide; NAPAP, N^α-(β-naphthylsulfonyl-glycyl)-DL-p-amidinophenylalanyl-piperidine; ONO-3307, 4-sulfamoyl-phenyl 4-guanidinobenzoate; Z, benzyloxycarbonyl.

al., 1981). H-D-Phe-Pro-Arg-pNA was found to be the most sensitive substrate in this series with a k_{cat}/K_M value of $9 \times 10^6 \, M^{-1} \, s^{-1}$. Bovine α-thrombin was tested with 24 commercially available peptide p-nitroanilides and the corresponding kinetic parameters were measured (Lottenberg *et al.*, 1983). The kinetic constant k_{cat}/K_M ranges from 33 to $1.1 \times 10^8 \, M^{-1} \, s^{-1}$ for the poorest and the best substrates, respectively. The best substrates were those with Arg in the P_1 position, Pro or a Pro homolog in the P_2 position, and an apolar amino acid residue in the P_3 position (kinetic constants are shown in Table I). Thrombin will also hydrolyze substrates with Lys in the P_1 position, but the k_{cat}/K_M values are significantly lower than those with Arg in this position. The substitution of Lys for Arg in Tos-Gly-Pro-Arg-pNA, D-Val-Leu-Arg-pNA, and D-Val-Phe-Arg-pNA drops the k_{cat}/K_M value by 25- to 95-fold. A series of 14 tripeptide pNA substrates of Z-AA-Gly-Arg-pNA and Z-AA-Phe-Arg-pNA with various amino acids in the P_3 position were used to map the S_3 subsite of several coagulation enzymes including thrombin (Cho *et al.*, 1984). The best substrate in this series is Z-Lys-Gly-Arg-pNA with a k_{cat}/K_M value of $10^4 \, M^{-1} \, s^{-1}$, which is ca. 10^3- to 10^4-fold lower than those p-nitroanilides with the optimum peptide sequence for thrombin.

Recently, the new peptide pNA substrate Bz-Phe-Pro-GPA-pNA, which contains the p-guanidino-L-phenylalanine (GPA) residue in the P_1 site, was synthesized and found to have a smaller specificity constant (k_{cat}/K_M) for hydrolysis by thrombin than the corresponding arginine derivative Bz-Phe-Pro-Arg-pNA (Tsunematsu *et al.*, 1986). Although this substrate binds better to thrombin (two-fold lower K_M value) than Bz-Phe-Pro-Arg-pNA, its k_{cat} value was remarkably low.

Table I. Kinetic Constants for the Hydrolysis of Peptide Nitroanilides by Bovine Thrombin[a]

Substrate[b]	k_{cat} (s^{-1})	K_M (μM)	k_{cat}/K_M ($M^{-1} \, s^{-1}$)
H-D-Phe-Aze-Arg-pNA	48	0.49	1.1×10^8
H-D-Phe-Pip-Arg-pNA	98	1.5	6.6×10^7
H-D-Ile-Pro-Arg-pNA	74	1.2	6.2×10^7
H-D-Val-Pro-Arg-pNA	89	2.0	4.5×10^7
Tos-Gly-Pro-Arg-pNA	100	3.6	2.8×10^7
<Glu-Pro-Aze-Arg-pNA	150	38	4.1×10^6
Z-Gly-Pro-Arg-pNA	83	25	3.3×10^6
Tos-Gly-Pro-Lys-pNA	23	21	1.1×10^6

[a]Data from Lottenberg *et al.* (1983).
[b]Aze, 2-azetidinecarboxylic acid; <Glu, pyroglutamic acid; Pip, pipecolic acid.

Peptide derivatives of 5-amino-2-nitrobenzoic acid (ANBA) have been developed as alternatives of the pNA leaving group (Kolde et al., 1986). The chromophoric group (ANBA) had spectroscopic properties similar to p-nitroaniline and the carboxyl group on the phenyl ring can easily be modified by introducing various substituents to interact with thrombin's S'subsites. For example, the new substrate H-D-Phe-Pro-Arg-ANBA-iso-propylamide had greater specificity toward thrombin than H-D-Phe-Pro-Arg-pNA; however, the kinetic constant k_{cat}/K_M for enzymatic hydrolysis was 2.4-fold lower than the latter.

2.2. Amino Acid and Peptide Thioesters

Amino acid and peptide thioesters are very sensitive substrates for serine proteases since they have high k_{cat}/K_M values and the thiol leaving group can be detected at fairly low concentrations. Cleavage of the thioester bond can be monitored continuously by reaction with a thiol reagent such as 4,4'-dithiodipyridine or 5,5'-dithiobis(2-nitrobenzoic acid) present in the assay mixture to produce a chromogenic compound (Grassetti and Murray, 1967; Farmer and Hageman, 1975). Z-Lys-SBzl was the first thioester substrate synthesized for trypsinlike enzymes and has a k_{cat}/K_M value of $8.8 \times 10^5 \, M^{-1} \, s^{-1}$ for thrombin (Green and Shaw, 1979). Several Arg-containing amino acid and peptide thioesters have been used to map the active site of blood coagulation enzymes including thrombin and their kinetic constants are shown in Table II (McRae et al., 1981; Cook et al., 1984). The best substrate for thrombin is H-D-Phe-Pro-Arg-SBzl with a k_{cat}/K_M value of $9.9 \times 10^6 \, M^{-1} \, s^{-1}$. Z-Gly-Arg-SBzl is also a reactive substrate, but with a slightly smaller k_{cat}/K_M value. Z-Lys-SBzl and a few Arg thioester substrates are commercially available.

Table II. Kinetic Constants for the Hydrolysis of Amino Acid and Peptide Thioesters by Bovine Thrombin

Substrate	k_{cat} (s^{-1})	K_M (μM)	k_{cat}/K_M $(M^{-1} s^{-1})$
Boc-Arg-SBzl[a]	8.3	5.2	1.6×10^6
Z-Arg-SBzl[b]	7.0	1.9	3.7×10^6
Z-Gly-Arg-SBzl[a]	27	6.6	4.1×10^6
Boc-Phe-Phe-Arg-SBzl[a]	13	11	1.2×10^6
D-Phe-Pro-Arg-SBzl[b]	7.1	0.72	9.9×10^6

[a]Data from McRae et al. (1981).
[b]Data from Cook et al. (1984).

2.3. Fluorogenic Substrates

The most widely studied group of fluorogenic substrates for thrombin are peptide derivatives of 7-amino-4-methylcoumarin (AMC). These substrates have been used both for subsite mapping and in routine assays. Twenty Arg-containing peptide-AMC derivatives were first tested with thrombin and other trypsinlike enzymes by Morita *et al.* (1977). Boc-Val-Pro-Arg-AMC was found to be the most specific for thrombin among this group of peptides. This substrate was not hydrolyzed by kallikrein, urokinase, and only slowly hydrolyzed by factor Xa and plasmin. A series of Boc-AA-AA-Arg-AMC (where AA = amino acid residue) were synthesized and tested to find specific substrates for blood coagulation enzymes and trypsin (Kawabata *et al.*, 1988; kinetic constants are shown in Table III). The most reactive substrate for human α-thrombin among these compounds was Boc-Asp(OBzl)-Pro-Arg-AMC with an extremely high k_{cat}/K_M value of 1.5×10^7 M^{-1} s^{-1}. The k_{cat}/K_M value of this substrate is 2.7-fold higher than the commercially available substrate Boc-Val-Pro-AMC. Thrombin effectively hydrolyzes substrates containing an apolar residue (e.g., Pro) or nonaromatic residue (e.g., Ala) in the P_2 site. The importance of the apolar residue at P_2 has also been demonstrated with peptide p-nitroanilides (Lottenberg *et al.*, 1983; Cho *et al.*, 1984; Pozsgay *et al.*, 1981) and peptide thioesters (Cho *et al.*, 1984; McRae *et al.*, 1981). The AMC peptide with a Phe residue at P_2 was less reactive, although it was still a good substrate. Recently, a new water-soluble fluorogenic substrate, Z-Gly-Gly-Arg-7-aminocoumarin-4-methanesulfonic Acid (ACMS), has

Table III. Kinetic Constants for the Hydrolysis of Peptide Aminomethylcoumarins (AMC) by Human Thrombin[a]

Substrate	k_{cat} (s^{-1})	K_M (μM)	k_{cat}/K_M (M^{-1} s^{-1})
Boc-Asp(OBzl)-Pro-Arg-AMC	160	11	1.5×10^7
Boc-Ala-Pro-Arg-AMC	130	13	1.0×10^7
Boc-Val-Pro-Arg-AMC	120	22	5.5×10^6
Boc-Asp(OBzl)-Ala-Arg-AMC	52	44	1.2×10^6
Boc-Gly-Ala-Arg-AMC	20	48	4.2×10^5
Boc-Gly-Gly-Arg-AMC	2.5	56	4.5×10^4
Boc-Gly-Asn-Arg-AMC	1.5	110	1.4×10^4
Boc-Phe-Phe-Arg-AMC	2.7	43	6.3×10^4
Boc-Gly-Val-Arg-AMC	5.6	120	4.7×10^4
Boc-Gly-Phe-Arg-AMC	0.23	39	5.9×10^3

[a]Data from Kawabata *et al.* (1988).

been reported (Sato *et al.*, 1988). However, this substrate has a lower k_{cat}/K_M value (1.9×10^2 M^{-1} s^{-1}) when compared to Z-Gly-Gly-Arg-AMC.

Other fluorogenic substrates such as H-D-Phe-Pro-Arg-AIE and Bz-Leu-Ala-Arg-α-naphthylester (NE) are also very sensitive and quite specific substrates for thrombin (Mitchell *et al.*, 1978a,b; Hitomi *et al.*, 1981). A new class of fluorogenic substrates for trypsinlike enzymes, (Z-Arg-NH)$_2$-rhodamine, and (Z-AA-Arg-NH)$_2$-rhodamine, employ rhodamine as the fluorophoric leaving group (Leytus *et al.*, 1983a,b). Cleavage of one of the amide bonds by the enzyme converts the nonfluorescent bisamide substrate into a highly fluorescent monoamide product. Some kinetic constants for the hydrolysis of amino acid and dipeptide substrates by trypsin, plasmin, and thrombin are shown in Table IV. The dipeptide derivative (Z-Pro-Arg-NH)$_2$-rhodamine exhibits high selectivity for thrombin with a k_{cat}/K_M value of 3.7×10^5 M^{-1} s^{-1}. Another derivative, (Z-Phe-Arg-NH)$_2$-rhodamine, was not hydrolyzed by human thrombin, although it is the best substrate for plasmin among this series. Comparison of the kinetic constants for hydrolysis of dipeptide substrates, (Z-AA-Arg-NH)$_2$-rhodamine, with the single amino acid derivative, (Z-Arg-NH)$_2$-rhodamine, indicates that selection of the proper amino acid residue in the P$_2$ site can affect the substrate specificity.

2.4. Other Types of Substrates

Besides the three common types of substrates (peptide *p*-nitroanilides, thioesters, and fluorogenic substrates), Arg or Lys esters such as tosylar-

Table IV. Kinetic Constants for the Hydrolysis of Amino Acid and Dipeptide Rhodamine Derivatives by Bovine Trypsin, Human Plasmin, and Human Thrombin[a]

Substrate	k_{cat}/K_M (M^{-1} s^{-1})		
	Bovine trypsin	Human plasmin	Human thrombin
(Arg-NH)$_2$-rhodamine	450	No hydrolysis	No hydrolysis
(Z-Arg-NH)$_2$-rhodamine	170,000	1,200	4,400
(Z-Ala-Arg-NH)$_2$-rhodamine	1,600,000	8,100	15,000
(Z-Gly-Arg-NH)$_2$-rhodamine	1,700,000	6,500	7,000
(Z-Phe-Arg-NH)$_2$-rhodamine	760,000	30,000	No hydrolysis
(Z-Pro-Arg-NH)$_2$-rhodamine	1,000,000	8,800	370,000
(Z-Trp-Arg-NH)$_2$-rhodamine	250,000	27,000	No hydrolysis
(Z-Val-Arg-NH)$_2$-rhodamine	690,000	15,000	23,000

[a]Data from Leytus *et al.* (1983b).

ginine methyl ester (TAME, Tos-Arg-OMe), and benzoylarginine ethyl ester (BAEE, Bz-Arg-OEt) have been used as substrates for thrombin and other trypsinlike enzymes (Sherry and Troll, 1954). The enzymatic hydrolysis rates of these substrates can be measured by a spectrometric method (Hummel, 1959), a potentiometric method, or by release of radioactively labeled alcohol (Roffman *et al.*, 1970). A more sensitive chromogenic ester substrate *p*-nitrobenzyl-*p*-toluene sulfonyl-L-arginine (NBTA) has also been developed (Aogichi and Plaut, 1977; Plaut, 1978). However, all of the ester substrates generally had low selectivity toward thrombin, and the analysis for the hydrolysis products is not as convenient compared to other chromogenic or fluorogenic substrates.

Another type of thrombin substrate is H-D-Phe-Pip-Arg-*p*-aminodiphenylamide·2HCl (S-2497, Pip = pipecolic acid), which can be detected electrochemically (Daraio de Peuriot *et al.*, 1980). Hydrolysis of the substrate is measured using the oxidation current of the leaving group (*p*-aminodiphenylamide) determined by its potentiodynamic signal. Thrombin concentration can be evaluated up to 0.01 NIH u/ml by this method, which is more sensitive than the corresponding spectrometric method. In addition, the enzymatic activity can be analyzed in heterogeneous solution although the assay is not as convenient as the spectrometric method.

3. INHIBITORS

Synthetic inhibitors for serine proteases are generally divided into five categories (Powers and Harper, 1986): simple substrate analogs (e.g., benzamidine and arginine derivatives), transition state analogs (e.g., peptide aldehydes, peptide trifluoromethyl ketones, peptide α-ketoesters, and boronic acids), alkylating agents which react with the active site histidine (e.g., peptide chloromethyl ketones), acylating agents which react with the active site serine forming stable acyl enzymes (e.g., ester derivatives of *p*-guanidinobenzoic acid, sulfonyl fluorides, phosphoryl fluorides), and mechanism-based (suicide) inhibitors. The first two classes of compounds are reversible inhibitors, and the last three are generally irreversible inhibitors.

Reversible inhibitors usually contain substratelike features and their potency depends on binding interactions with the enzyme. Since the enzyme–inhibitor complex formed with reversible inhibitors can dissociate, restoration of enzymatic activity may occur as the inhibitor is destroyed or cleared from the system. The strength of the binding is reflected in K_I, the dissociation constant for the enzyme–inhibitor complex E·I. Medicinal chemists often report inhibitor potency with IC_{50} (I_{50}) values, which are easier to determine than K_I values. However, IC_{50} values are not directly related to K_I values and can depend on both the mechanism of inhibition and the concentrations of substrate (Cheng and Prusoff, 1973). Comparison of IC_{50} values obtained in different laboratories is only valid if the assay conditions are the same and the mechanism of inhibition is identical. Irreversible inhibitors usually inactivate serine proteases by first forming a reversible E·I complex followed by covalent bond formation. The potency of the inhibitor depends on the strength of reversible binding to the enzyme (K_I), the rate of the inactivation step (k_2), or both (k_2/K_I). The most suitable parameter for comparing irreversible inhibitors is k_2/K_I, which is identical with $k_{obsd}/[I]$ in most cases (Knight, 1986; Powers and Harper, 1986).

Thrombin in common with the other plasma trypsinlike enzymes cleaves peptide bonds following Arg residues and thus all potent synthetic inhibitors incorporate a guanidino or amidino group into their structures. The acidic side chain of Asp 189 in the S_1 pocket of thrombin hydrogen bonds with basic functional groups including guanidine and amidine moieties. With the exception of Asp 189, the remainder of the S_1 pocket of thrombin is quite hydrophobic and inhibitors containing aromatic amidino or guanidino functional groups are often more potent than those which more closely resemble arginine. Since the same features are found in other plasma trypsinlike enzymes, most specific thrombin inhibitors incorporate structural features which interact with subsites beyond the primary specificity site. Thrombin in particular contains a unique hydrophobic insertion loop near the active site which is an attractive target in the design of new specific inhibitors (Bode et al., 1989). However, construction of small synthetic inhibitors which are specific for thrombin still remains a formidable challenge for the medicinal chemist.

Other reviews of thrombin inhibitors include Stürzebecher (1984).

3.1. Reversible Inhibitors

3.1.1. Arginine Derivatives

N^α-substituted L-arginine ester and amide derivatives are inhibitors of thrombin with IC_{50} values in the range of 5.0×10^{-3} to 7.5×10^{-8} M (Okamoto et al., 1980; Kikumoto et al., 1980a,b, 1984). The inhibitory potency depends on the nature of the N^α substituent, and the ester functional group or the amide derivative. Ester derivatives of arginine are readily hydrolyzed by thrombin and other trypsinlike enzymes, and therefore are not suitable for use as antithrombotic agents in vivo (Okamoto et al., 1980). However, amide derivatives of N^α-substituted L-arginine are

more stable and primary amides of N^α-dansyl-L-arginine are good thrombin inhibitors with the n-butyl derivative being the most potent ($IC_{50} = 2.1$
μM in a clotting time assay). Cyclic amides of N^α-dansyl-L-arginine are
better inhibitors and the piperidine amide has an IC_{50} value of 1.0 μM.
Inhibition of substituents onto the piperidine ring further increases
thrombin inhibitory potency and the 4-methylpiperidine and 4-ethylpi-
peridine derivatives are the most potent inhibitors in the series with
IC_{50} values of 0.3 and 0.1 μM, respectively. Changing the dansyl substituent into a naphthalene-1-sulfonyl group, exchanging it for other heterocyclic structures, or introduction of substituents on the naphthalene
ring results in small improvements in the inhibitory potency (Kikumoto et
al., 1980a).

The potent reversible thrombin inhibitor MQPA (4-methyl-1-
[N^2-[(3-[methyl-1,2,3,4 - tetrahydro-8-quinolinyl)-sulfonyl]-L-arginyl]-
2-piper-idinecarboxylic acid) was an outgrowth of the extensive structure–
activity studies with piperidine amide derivatives of N^α-dansyl-L-arginine.
The parent compounds were too toxic for use in man, but introduction of
a carboxyl group on the piperidine amide reduced the toxicity while
having little effect on the inhibitory potency. The inhibitory potency of
MQPA depends on the stereochemistry of the 2-piperidinecarboxylic acid
moiety, and the K_I values obtained with the four stereoisomers of MQPA
are shown in Table V (Kikumoto et al., 1984). (2R,4R)-MQPA (MD-805,
MCI-9038, Fig. 1) was the most potent inhibitor for bovine α-thrombin
with a K_I value of 0.019 μM while the other stereoisomers were 10- to
10^5-fold less potent. This isomer also inhibited trypsin with a K_I value of
5.0 μM, but was much less effective with other trypsinlike enzymes. MQPA
is currently being developed for therapeutic use by Genentech under the
name of Argatroban.

**Table V. Inhibition Constants of Four Stereoisomers
of MQPA for Trypsinlike Proteases[a]**

Stereoisomers of MQPA	K_I (μM)				
	Bovine thrombin	Trypsin	Factor Xa	Plasmin	Bovine plasma kallikrein
(2R,4R)-MQPA	0.019	5.0	210	800	1500
(2R,4S)-MQPA	0.24	30	>500	>500	>500
(2S,4R)-MQPA	1.9	630	>1500	>1500	>1500
(2S,4S)-MQPA	280	>500	>500	>500	>500

[a]Data from Kikumoto et al. (1984).

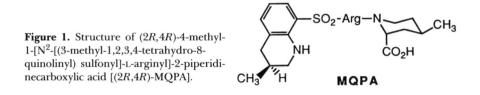

Figure 1. Structure of (2*R*,4*R*)-4-methyl-1-[N²-[(3-methyl-1,2,3,4-tetrahydro-8-quinolinyl) sulfonyl]-L-arginyl]-2-piperidinecarboxylic acid [(2*R*,4*R*)-MQPA].

The x-ray structure of (2*R*,4*R*)-MQPA bound to trypsin has been determined and used to explain the high potency of (2*R*,4*R*)-MQPA toward thrombin (Matsuzaki *et al.*, 1989). The arginine side chain of the inhibitor extends into the primary binding pocket S_1, the quinoline ring lies on Trp 215, while the piperidine ring is close to His 57 and Ser 195. There are two hydrogen bonding sites between the inhibitor and trypsin: one site involves the guanidino group of the inhibitor and the carboxyl group of the Asp 189 in the primary binding pocket S_1, the second one is formed between the NH of the arginine residue of MQPA and the peptide backbone carbonyl oxygen of Gly 216 in trypsin. The latter hydrogen bond is observed in almost all complexes of peptides with serine proteases. Molecular modeling with thrombin suggests that the binding mode should be quite similar except that the quinoline ring is in a more hydrophobic environment in thrombin (Bode *et al.*, 1990).

The potent thrombin inhibitor 1-(N^α-dansylarginyl)-4-ethylpiperidine [DAPA, frequently called *N*-dansylarginine *N*-(3-ethyl-1,5-pentanediyl)amide] had been used as a fluorescent probe. Although this compound is toxic and cannot be used for *in vivo* studies, it is a potent competitive thrombin inhibitor ($K_I = 10^{-7}$ M) and exhibits a change in the fluorescence properties of the dansyl moiety when it is bound to thrombin. Thus, DAPA has proved to be a useful tool for study of thrombin and its interactions with cofactors (Nesheim *et al.*, 1979; Hibbard *et al.*, 1982), and the mechanism of activation of prothrombin by prothrombinase (Krishnaswamy *et al.*, 1986, 1987).

Peptide derivatives of arginine such as H-D-Phe-Pro-Arg-isopropylester and H-D-Phe-Pro-Arg-4-methyl-piperidine amide are also moderate inhibitors of thrombin with K_I values of 6 and 500 μM respectively (Mattson *et al.*, 1982). The isopropylester inhibited plasmin, factor Xa, and kallikrein 20- to 40-fold less potently than thrombin.

3.1.2. Aromatic Amidines and Guanidines

Aromatic compounds containing amidino or guanidino groups are poor to good inhibitors for thrombin. Simple aryl guanidines such as phenylguanidine, 4-guanidinobenzoic acid, and 4-guanidinophenylpy-ruvic acid are poor thrombin inhibitors with K_I values of 9.0, 4.0, and 1.5 mM, respectively (Markwardt and Walsmann, 1968; Markwardt et al., 1974). However, aromatic amidines, particularly derivatives of naphtha-lene and indole, are much better inhibitors of thrombin and have greater inhibitory potency than arylalkyl amines or aryl guanidines (Table VI). Benzamidine derivatives with aliphatic or arylaliphatic moieties at the 3- or 4-position of the benzamidine ring have slightly increased inhibitory activ-ities toward thrombin (e.g., 4-AmPh-$(CH_2)_4CH_3$, K_I = 19 μM; 4-AmPh-$O(CH_2)_{10}CH_3$, K_I = 10 μM; 4-AmPh = 4-amidinophenyl; Walsmann et al., 1974, 1975). Derivatives with 3- and 4-substituents differ only slightly in their inhibitory activity (4-AmPh-CH_2O-β-naphthyl, K_I = 6.6 μM; 3-AmPh-$(CH_2)_2CO_2CH_2$Ph, K_I = 18 μM).

The transition state inhibitor 4-amidinophenylpyruvic acid (APPA, Fig. 2) is a much more potent thrombin inhibitor than simple amidines. APPA inhibits thrombin and trypsin with K_I values of 6.5 and 1.6 μM, respectively. The interactions observed in the x-ray structure of the APPA–trypsin complex at 1.4-Å resolution are shown in Fig. 2 (Walter and Bode, 1983). In addition to the interaction of the amidinophenyl group of the inhibitor with Asp 189 of trypsin, a tetrahedral intermediate is formed by the addition of Ser 195 Oγ to the carbonyl carbon of the pyruvate. The carboxylate oxygens of the inhibitor form hydrogen bonds with Nε2 of His 57. The structure with thrombin is likely to be very similar to that observed with trypsin.

Table VI. Inhibition of Thrombin by Aromatic Compounds Substituted with Basic Functional Groups

Inhibitor	K_I (μM)
Benzylamine	12,000[a]
Phenylguanidine	9,000[b]
Benzamidine	220[a]
4-Amidinopyridine	550[c]
β-Naphthylamidine	85[d]
6-Amidinoquinoline	130[c]
5-Amidinoindole	8[c]

[a]Data from Markwardt et al. (1968a).
[b]Data from Markwardt et al. (1968b).
[c]Data from Gertz et al. (1979).
[d]Data from Markwardt et al. (1969).

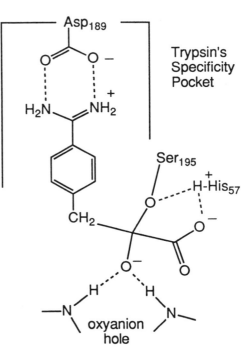

Figure 2. Structure of the complex between 4-amidinophenyl pyruvic acid (APPA) and the active site of trypsin.

A large number of derivatives of N^α-arylsulfonyl-4-amidinophenyla-lanine have been studied as inhibitors of thrombin. Simple amides of tosyl 4-amidinophenylalanine such as the n-butyl derivative have K_I values of 23 μM, while cyclic amides are much more potent with the piperidine, 4-methylpiperidine, and morpholine derivatives having K_I-values of 2.3, 4.1, and 5.4 μM, respectively (Markwardt *et al.*, 1980). Replacement of the tosyl group with a β-naphthylsulfonyl group increases the inhibitory activity slightly. Varying the position of the amidino group (*para* or *meta*), changing the separation between the amidinophenyl group and the α carbon, and varying the N^α substituent led to several potent thrombin inhibitors including Tos-3-AmPhGly-piperidine (K_I = 1.7 μM, 3-AmPhGly = 3-amidinophenylglycine), Tos-NH[4-AmPh(CH$_2$)$_4$]CH-CO-piperidine (0.36 μM), and Tos-NH[3-AmPh(CH$_2$)$_3$]CH-CO-piper-idine (0.06 μM; Stürzebecher *et al.*, 1983).

Replacing the N^α-sulfonyl group with an N^α-sulfonylglycyl residue resulted in a substantial improvement in inhibitor potency and produced one of the most potent reversible thrombin inhibitors N^α-(β-naphthylsulfo-nyl-glycyl)-DL-p-amidinophenylalanyl-piperidine (NAPAP) which has a K_I-

value of 0.006 μM with thrombin (Table VII; Stürzebecher *et al.*, 1983; Voigt *et al.*, 1988). Elongation of NAPAP by adding an additional glycine residue or replacing the piperidine moiety with pyrrolidine, proline, or morpholine resulted in 2- to 2200-fold less potent thrombin inhibitors.

The x-ray crystal structure of the trypsin–NAPAP complex has recently been determined at 1.8-Å resolution (Bode *et al.*, 1990). The D form of NAPAP binds compactly to the active site of trypsin with the 4-amidino-phenylalanine moiety in the S_1 pocket, the glycyl group of the inhibitor hydrogen bonding with Gly 216, the naphthyl group standing perpendicular to the indole ring of Trp 215, and the piperidine enclosed by the naphthyl moiety, Leu 99, and His 57. Molecular modeling has been used to transform the conformation and position of NAPAP in the trypsin–inhibitor complex to the thrombin active site. NAPAP fits quite well into the more restricted thrombin active site. The naphthyl binding site in thrombin is more hydrophobic than in trypsin because Ile 174 is closer. The piperidine moiety is also tightly packed into the S_2 subsite of thrombin which leaves little space for substitution due to the hydrophobic side chains of Tyr 60A and Trp 60D in thrombin's insertion loop. The restricted nature of the piperidine binding site in thrombin is indicated by the observation that the binding affinity of the L-proline analog of NAPAP which has a carboxylate in the equatorial 2-position is significantly reduced when compared to NAPAP, and the D-proline derivative binds to thrombin rather weakly with a K_I value of 13 μM (Table VII).

Table VII. Inhibition Constants of Thrombin by Derivatives of N^α-Substituted 4-Amidinophenylalanine Amides (R₁-AmPhe-R₂)[a]

R₁	R₂	K_I (μM)
Tosyl	Piperidine	2.3
α-Naphthylsulfonyl	Piperidine	2.8
β-Naphthylsulfonyl	Piperidine	0.42
Tosyl-glycyl	Piperidine	0.048
α-Naphthylsulfonyl-glycyl	Piperidine	0.014
β-Naphthylsulfonyl-glycyl	Piperidine	0.006
β-Naphthylsulfonyl-glycyl-glycyl	Piperidine	3.8
β-Naphthylsulfonyl-glycyl	Pyrrolidine	0.013
β-Naphthylsulfonyl-glycyl	L-Proline	0.51
β-Naphthylsulfonyl-glycyl	D-Proline	13
β-Naphthylsulfonyl-glycyl	n-Butylamine	2.3
β-Naphthylsulfonyl-glycyl	Morpholine	0.23

[a]Data from Stürzebecher *et al.* (1983) and Voigt *et al.* (1988). AmPhe = 4-amidinophenylalanine.

Bis-arylamidine derivatives are also potent inhibitors of thrombin (Geratz *et al.*, 1973; Geratz and Tidwell, 1977; Tidwell *et al.*, 1978, 1980). Two effective inhibitors are shown in Fig. 3. The most potent thrombin inhibitor was α,α′-bis(4-amidino-2-iodophenoxy)-*p*-xylene) which had K_I values of 0.38 and 30 μM with thrombin and factor Xa, respectively. The heterocyclic bisamidine 1,2-bis(5-amidino-2-benzofuranyl)ethane was a less potent thrombin inhibitor (K_I = 15 μM), but was a better factor Xa inhibitor (0.57 μM). The inhibitory potency of bisamidine increases with the chain length of the alkyl or arylalkyl group separating the two benzamidines and with halogen substitution. The enhanced inhibitory activity of bis(amidinoaryl) compounds, compared to simple benzamidines, suggests that multiple binding interactions with the enzyme are involved. Tris-benzamidine derivatives are also potent thrombin inhibitors and have K_I values of 0.65–1.4 μM (Tidwell *et al.*, 1976, 1978).

3.1.3. Peptide Fluoroalkyl Ketones

Peptide fluoroalkyl ketones (monofluoromethyl, difluoromethyl, and trifluoromethyl) have been widely used as transition-state and irreversible inhibitors for serine proteases. Several lysine and arginine monofluoroketones including Lys-Ala-Lys-CH₂F (McMurray and Dyckes, 1987), Bz-Phe-Lys-CH₂F and Ala-Phe-Lys-CH₂F (Angliker *et al.*, 1987), Bz-Phe-Arg-CH₂F and H-D-Phe-Pro-Arg-CH₂F (Angliker *et al.*, 1988) have been synthesized as inhibitors for trypsin, thrombin, and other trypsinlike en-

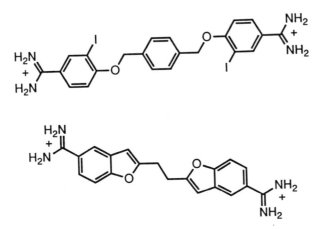

Figure 3. Structures of the bisamidine inhibitors, α,α′-bis(4-amidino-2-iodophenoxy)-*p*-xylene (top) and 1,2-bis(5-amidino-2-benzofuranyl)ethane (bottom).

zymes. These lysine monofluoromethyl ketones, like their corresponding chloromethyl ketones, were found to be active-site-directed irreversible inhibitors of trypsin. H-D-Phe-Pro-Arg-CH_2F inhibited thrombin with a k_2/K_1 of 6000 $M^{-1}s^{-1}$, which is 1800-fold less than H-D-Phe-Pro-Arg-CH_2Cl (Angliker et al., 1988).

More highly fluorinated fluoroketones (trifluoromethyl or difluoromethylene) are potent transition-state inhibitors for serine proteases and in most cases do not alkylate the enzyme. Upon binding to the serine protease, several fluoroketone inhibitors have been shown to react with the γ-OH of Ser195 to form a hemiketal structure resembling the transition state for peptide bond hydrolysis (Takahashi et al., 1988). These inhibitors include Ac-Leu-ambo-Phe-CF_3 for chymotrypsin, Ac-Ala-Ala-Pro-ambo-Ala-CF_3 for porcine pancreatic elastase (Imperiali and Abeles, 1986; ambo, racemic amino acid residue), and Z-Lys(Z)-Val-Pro-Val-CF_3 for human leukocyte elastase (Stein et al., 1987; Dunlap et al., 1987). The peptide lysine trifluoromethyl ketone H-D-Phe-Pro-Lys-CF_3 is a potent inhibitor of thrombin with a K_1 of 1 nM and is an orally active anticoagulant (R. Abeles, 1988 June ACS meeting). Recently, several arginine fluoroalkyl ketones have been synthesized as inhibitors of trypsin and blood coagulation enzymes (Ueda et al., 1990). These compounds were found to be slow-binding inhibitors for bovine trypsin with K_1 values of 0.2–56 μM. They also inhibited thrombin, kallikrein, factors Xa, XIa, and XIIa less potently. Benzoyl-Arg-CF_2CF_3 was the best inhibitor in the series for bovine thrombin and inhibited thrombin competitively with a K_1 of 13 μM. Benzoyl-Arg-CF_2CF_3 was shown to exist primarily as a hydrate or cyclic carbinolamine in solution. Although the inhibitors contain only a single amino acid residue (arginine), these fluoroalkyl ketones showed some selectivity toward the various coagulation enzymes. The potency and selectivity of the fluoroalkyl ketones would be expected to be increased by introducing systematic changes in the structure and by the addition of longer peptide sequences.

3.1.4. Peptide Aldehydes and Peptide α-Ketoamides

Peptide aldehydes are potent transition-state inhibitors for serine and cysteine proteases. The mechanism of inhibition involves attack of the active site serine on the aldehyde carbonyl group to form a tetrahedral hemiacetal adduct, which can be stabilized by subsite interactions between the enzyme and the peptide chain of the inhibitor (Thompson, 1973; Kuramochi et al., 1979). The tripeptide aldehydes, Boc-D-Phe-Pro-Arg-H, and H-D-Phe-Pro-Arg-H possess the highest antithrombin activity in a series of compounds studied by Bajusz et al. (1978). The K_1 values differed by four orders of magnitude and the inhibition mechanism could be either competitive or noncompetitive depending on the substrate used. For example, H-D-Phe-Pro-Arg-H was a noncompetitive inhibitor when fibrinogen (K_1 = 0.075 µM) or Z-D-Phe-Pro-Arg-pNA (K_1 = 0.3 mM) was used as the substrate, while competitive inhibition and K_1 values of 0.45 µM and 0.7 mM were obtained respectively with Z-Phe-Val-Arg-pNA and H-D-Phe-Pro-Arg-pNA. The natural arginine peptide aldehydes antipain and leupeptin are also inhibitors of thrombin (K_1 values of 8.4 and 45 µM, respectively), but are less potent than H-D-Phe-Pro-Arg-H (0.0072 µM) and its Boc derivative (0.0065 µM) using chromogenic substrates or clotting assays (Witting et al., 1988). Leupeptin also prolonged the thrombin time and the activated partial thromboplastin time (APTT) of normal human citrated plasma (Fareed et al., 1981). Both H-D-Phe-Pro-Arg-H and its Boc derivative also inhibit trypsin, plasmin, and other blood coagulation enzymes, with H-D-Phe-Pro-Arg-H being the more selective toward thrombin. A newer argininal analog, D-MePhe-Pro-Arg-H, which contains an N-methyl-D-Phe residue, is as potent and selective toward thrombin as its parent H-D-Phe-Pro-Arg-H, but is more stable than H-D-Phe-Pro-Arg-H (Bajusz et al., 1990).

Recently, two new potent thrombin inhibitors, cyclotheonamide A and B which contain an arginine α-ketoamide functional group (Fig. 4), have been isolated from a marine sponge *Theonella sp.* (Fusetani et al., 1990). Cyclotheonamide A was found to inhibit thrombin, trypsin, and plasmin

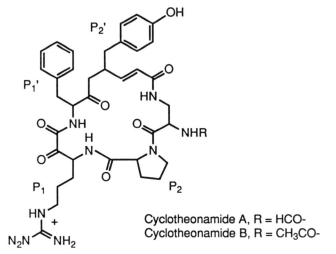

Cyclotheonamide A, R = HCO-
Cyclotheonamide B, R = CH₃CO-

Figure 4. Structures of cyclotheonamide A and B showing the proposed interaction with the subsites of thrombin.

quite potently with IC_{50} values of 0.1, 0.27, and 0.41 µM, respectively. It is likely that the arginine α-ketoamide functional group of cyclotheonamide is forming a tetrahedral adduct similar to that observed with APPA (Fig. 2). Peptide α-ketoesters have previously been reported to be transition-state inhibitors for other serine proteases (Hori *et al.*, 1985; Peet *et al.*, 1990). If this is the case with cyclotheonamide, then the proline residue would fit in the S_2 subsite, and the Phe and Tyr-like residues could fit respectively in the S_1' and S_2' subsites.

3.1.5. Peptide Boronic Acids

Peptide boronic acids containing a C-terminal α-aminoalkyl boronic acid residue [-NH-CH(R)-B(OR)$_2$] are potent inhibitors of serine proteases including chymotrypsin, cathepsin G, porcine pancreatic elastase (PPE), and human leukocyte elastase (HLE) (Kettner and Shenvi, 1984). The trigonal boron can react with either the nucleophilic hydroxyl group of Ser 195 or the imidazole of His 57 in the active site to give a tetrahedral boron adduct. Crystallographic studies have indicated that both types of adducts are formed with different boronic acids. In the case of the subtilisin BPN′complex with benzeneboronic acid, the hydroxyl of Ser 221 is covalently bound to the inhibitor boron atom to form a tetrahedral structure (Matthews *et al.*, 1975). Similar structures have been observed in complexes of the α-lytic protease and five peptide boronic acid inhibitors, MeO-Suc-Ala-Ala-Pro-boroAA, where AA is the α-aminoboronic acid analog of Ala, Val, Ile, Nle, or Phe (Bone *et al.*, 1989). In the case of the peptidyl-boroPhe derivatives, a trigonal adduct is formed between the active site serine and the boronic acid moiety in which the catalytic histidine occupies a position axial to the plane of the trigonal adduct, with the distance between Nε2 of the histidine and boron suggesting a covalent bond. Similarly, both the histidine and the serine form covalent bonds with the boron in the crystal structure of PPE complexed with Z-Ala-boroIle-OH (Takahashi *et al.*, 1989).

Peptide boroarginine derivatives have recently been synthesized and found to be potent inhibitors for trypsinlike enzymes (Kettner *et al.*, 1988, 1990). Ac-D-Phe-Pro-boroArg-OH, Boc-D-Phe-Pro-boroArg-OH, and H-D-Phe-Pro-boroArg-OH are highly effective slow tight-binding inhibitors of thrombin. The final inhibition constants and association rate constants for these compounds are 41 pM, $5.5 \times 10^6 \, M^{-1} \, s^{-1}$; 3.6 pM, $9.3 \times 10^6 \, M^{-1} \, s^{-1}$; < 1 pM, $8.0 \times 10^6 \, M^{-1} \, s^{-1}$, respectively. These inhibitors have been studied with other enzymes such as plasma kallikrein, factor Xa, plasmin, tissue plasminogen activator, and were found to be quite selective for thrombin.

3.2. Irreversible Inhibitors

3.2.1. Organophosphorus Inhibitors

Diisopropylphosphofluoridate (Dip-F) is one of the most extensively used serine protease inhibitors and can inactivate thrombin (Gladner and Laki, 1958) by reaction with the active site Ser 195 (Laki *et al.*, 1958). Other simple organophosphorus compounds also inhibit thrombin, factors VIIa, IXa, and Xa, but they are less potent than Dip-F. Organophosphorus inhibitors are extremely toxic and most are therefore not very useful for *in vivo* testing.

3.2.2. Sulfonyl Fluorides and Nitrophenylsulfonates

Phenylmethanesulfonyl fluoride (PMSF), one of first inhibitors shown to inactivate trypsin and chymotrypsin (Fahrney and Gold, 1963; Gold and Fahrney, 1964), also inhibits thrombin (Lundblad, 1971). However, the inactivation is slow and a high concentration of inhibitor is required. PMSF and other similar sulfonylating reagents are general serine protease inhibitors since they can react with a wide variety of enzymes including chymases, tryptases, elastases, and blood coagulation proteases. Reaction occurs with the active site serine to give a sulfonyl enzyme derivative which is usually stable under neutral conditions (Gold, 1965; James, 1978). Desulfonylation or elimination can occur at higher temperatures or under nonneutral conditions.

Introduction of an amidino group onto the aromatic nucleus of PMSF enhances the inhibitory activity for trypsinlike enzymes (Walsmann *et al.*, 1972; Markwardt *et al.*, 1973; Laura *et al.*, 1980; Tanaka *et al.*, 1983). One of the most potent inhibitors is 4-amidinophenylmethanesulfonyl fluoride (APMSF, Fig. 5), which can totally inactivate thrombin, trypsin, plasmin, factor Xa, complement enzymes C1r and C1s at micromolar concentration in a few minutes, with binding constants in the range of 1 to 2.4 μM (Laura *et al.*, 1980). APMSF stoichiometrically inactivates trypsin and thrombin by reaction with the active site serine. 4-Amidinobenzenesulfonyl fluoride (Fig. 5), which lacks the methylene group of APMSF, is only a moderate inhibitor of thrombin and trypsin with $k_{obsd}/[I]$ values of 12 and 40 M^{-1} s^{-1}, respectively (Walsmann *et al.*, 1972; Markwardt *et al.*, 1973). 4-(2-Amino-ethyl)benzenesulfonyl fluoride is about 10-fold less reactive toward both enzymes than APMSF. The distance between the cationic group and the sulfonyl fluoride group is crucial for inhibitory potency with trypsin and thrombin, and appears to be optimal with APMSF (Laura *et al.*, 1980) since 4-amidinobenzensulfonyl fluoride is 15-fold less reactive toward trypsin than APMSF.

Exosite inhibitors containing sulfonyl fluoride groups have been reported for thrombin, trypsin, plasmin, and kallikrein (Geratz, 1972). The inactivation constants (k_2/K_I) of these enzymes by 4-[3-(3-fluorosulfonylphenylureido)phenoxyethoxy]benzamidine (Fig. 5) are respectively 0.3, 6900, 0.9, and 180 $M^{-1} s^{-1}$. Thrombin is inactivated 22,000-fold less effectively than trypsin. Human thrombin is also inactivated by 3-[2-(2-chloro-5-fluorosulfonylphenylureido)phenoxybutoxy]benzamidine with K_I and k_2 values of 73 μM and 0.003 s^{-1} (Bing et al., 1977). Labeling experiments indicate that the inhibitor reacts with both the B chain, which contains the active site serine (80%) and the A chain.

Nitrophenyl esters of benzenesulfonic and phenylmethane sulfonic acid substituted with a cationic group can sulfonylate the active site serine of the thrombin, although their inhibitory activity is low (Wong and Shaw, 1974; Wong et al., 1978). 4-Nitrophenyl 4-amidinophenylmethane sulfonate inactivates thrombin specifically and stoichiometrically with k_2 of 0.019 s^{-1} and K_I of 0.12 mM (Wong and Shaw, 1974). The inactive sulfonyl thrombin does not have any coagulant activity. This inhibitor does not sulfonylate trypsin, chymotrypsin, plasmin, and kallikrein, but forms reversible complexes with K_I values in the range of 0.036–500 mM. The 3-amidino compound does not inactivate any of the enzymes tested. The position of the nitro group on the phenol leaving plays an important role in the specificity of sulfonylation (Wong and Shaw, 1976). Thus, the 4-nitrophenyl ester of 4-amidinophenylmethanesulfonate sulfonylates only thrombin, while the 3- and 2-nitrophenyl esters will also inactivate trypsin.

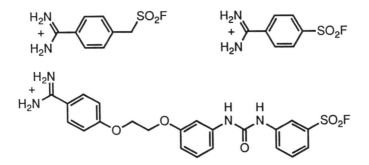

Figure 5. Structures of 4-amidinophenylmethanesulfonyl flouride (APMSF, top left), 4-amidinobenzenesulfonyl flouride (top right), and 4-[3-(3-fluorosulfonylphenylureido)phen-oxyethoxy]benzamidine.

3.2.3. Peptide Chloromethyl Ketones

Peptide chloromethyl ketones are a well-understood class of substrate-derived irreversible inhibitors for serine proteases which alkylate the active site His 57 and form a tetrahedral adduct with Ser195. Tos-Lys-CH$_2$Cl (TLCK), which was first developed as a specific inactivator of trypsin (Shaw et al., 1965), inhibits thrombin by alkylation of the active site histidine (Glover and Shaw, 1971). Tripeptide Lys and Arg chloromethyl ketones are more potent thrombin inhibitors with the arginine derivatives being more effective than the lysine derivatives (Kettner and Shaw, 1977, 1979; Kettner et al., 1978). Chloromethyl ketones with a P$_2$ Pro have a greater affinity for thrombin since the Pro-Arg sequence corresponds to the cleavage sites of prothrombin and factor XIII by thrombin and is also found in many synthetic peptide substrates. H-D-Phe-Pro-Arg-CH$_2$Cl is one of the most potent and selective thrombin inhibitors (Kettner and Shaw, 1979, 1981). The inactivation constants for thrombin, plasmin, factor Xa, and kallikrein are shown in Table VIII. H-D-Phe-Pro-Arg-CH$_2$Cl is 3 to 4 orders of magnitude more reactive toward thrombin than the other trypsinlike enzymes.

A crystal structure of human α-thrombin complexed with H-D-Phe-Pro-Arg-CH$_2$Cl has been determined at 1.9-Å resolution (Bode et al., 1989). The D-Phe-Pro-Arg backbone of the inhibitor forms an antiparallel β-sheet structure with Ser 214 to Gly 216 backbone residues of the enzyme while the Arg side chain of the inhibitor interacts with the S$_1$ subsite of thrombin via a salt link with Asp 189. The S$_1$ pocket of thrombin appears more hydrophobic than trypsin. The proline of the P$_2$ residue of the inhibitor is encapsulated in a very hydrophobic cage formed by the side chains of Trp215, Leu99, His57, Tyr60A, and Trp60D. This cage is closed by the benzyl group of the P$_3$ D-Phe residue and is a major factor for the

Table VIII. Inhibition Constants of Plasma Proteases by Peptide Chloromethyl Ketones[a]

	$k_{obsd}/[I]$ (M^{-1} s^{-1})			
Inhibitor	Human plasma kallikrein	Bovine factor Xa	Bovine thrombin	Human plasmin
H-D-Phe-Phe-Arg-CH$_2$Cl	390,000	3,200	7,500	6,100
DNS-Ala-Phe-Arg-CH$_2$Cl	94,000	8,200	—	16,000
DNS-Glu-Gly-Arg-CH$_2$Cl	24,000	370,000	4,300	4,800
H-D-Phe-Pro-Arg-CH$_2$Cl	8,000	4,500	12,000,000	700

[a]Data from Kettner and Shaw (1981).

exceptional specificity of this compound toward thrombin. The catalytic Ser and His residues of thrombin have reacted with chloromethyl ketone functional group and formed two covalent bonds with the inhibitor.

3.2.4. Acylating Agents

A variety of simple acyl esters, amides, isocyanates, azapeptides, and heterocyclic compounds will acylate the active site serine of serine proteases to form acyl enzymes. Some acyl enzymes are very unstable and deacylate rapidly, whereas others are quite stable due to various geometric and electronic features contained in the acyl enzyme structure.

Thrombin is acylated by various esters of guanidino and amidinobenzoic acids. 4-Nitrophenyl 4'-guanidinobenzoate (NPGB) acylates thrombin rapidly to release stoichiometric amounts of nitrophenol and can be used as an active-site titrant (Chase and Shaw, 1969). The 3-guanidinobenzoate derivative has a lower affinity for thrombin than NPGB. Thioester and fluorescein derivatives of 4-guanidinobenzoate have been studied with thrombin (Cook and Powers, 1983; Melhado et al., 1982) and the kinetic constants are shown in Table IX. All four compounds acylate thrombin rapidly (high k_2 values) but deacylate more slowly (low k_3 values) since the 4-guanidinobenzoyl derivative of thrombin is quite stable. In each case the leaving group can be detected spectrophotometrically or fluorometrically, and the reagents can be used as active-site titrants.

Other aromatic amidine and guanidines which inhibit thrombin effectively are benzyl 4-guanidinobenzoate (Markwardt et al., 1970), aromatic esters of ε-guanidinocaproic acid (Muramatsu and Fujii, 1972; Ohno et al., 1980), substituted phenyl 4-guanidinobenzoates (Tamura et al., 1977), amidinoaryl 4-guanidinobenzoates (Fujii and Hitomi, 1981), amino and guanidino substituted naphthoates and tetrahydronaphthoates (Nakayama et al., 1984; Oda et al., 1990). Several inhibitors including ε-guani-

Table IX. Kinetic Constants for Reaction of Bovine Thrombin with Esters of 4-Guanidinobenzoate[a]

Compound	k_2 (s^{-1})	K_s (μM)	k_3 (s^{-1})	K_m (μM)
4-Nitrophenyl 4'-guanidinobenzoate	0.13	3.9	0.00098	0.03
4-Nitrophenyl 3'-guanidinobenzoate	0.12	16	0.0055	17
Benzyl 4'-guanidinothiobenzoate	0.13	66	0.00098	0.48
Fluorescein mono-4-guanidinobenzoate	0.06	22	0.0018	0.64

[a]Data from Chase and Shaw (1969), Cook and Powers (1983), and Melhado et al. (1982).

dinocaproic acid p-carboxyethylphenyl ester (FOY, Gabexate Mesilate, Ohno *et al.*, 1980), 6'-amidino-2-naphthyl-4-guanidinobenzoate (FUT-175, Nafamostat mesylate; Fujii and Hitomi, 1981), and 4-sulfamoylphenyl 4-guanidinobenzoate (ONO-3307, Matsuoka *et al.* 1989) have been developed as anticoagulants (Fig. 6). FOY inhibits thrombin with a K_I value of 0.97 µM (Ohno *et al.*, 1980), and also inhibits trypsin, plasmin, plasma kallikrein equally well with K_I values of 0.07–3.6 µM (Matsuoka *et al.*, 1989). FUT-175 inhibits thrombin (K_I 0.8 µM) and other trypsinlike enzymes including trypsin, plasmin, plasma kallikrein, C1r, and C1s with IC_{50} values of 0.02–0.4 µM. ONO-3307 is also a good thrombin inhibitor (K_I 0.18 µM), and inhibits trypsin ca. 4-fold better, but is 2- to 20-fold less potent with other trypsinlike enzymes.

Inverse esters containing an amidinoaryl group as the leaving group are potent irreversible inhibitors of thrombin, but are not very specific and also inhibit other trypsinlike enzymes (Yagashi *et al.*, 1984; Fujii *et al.*, 1986; Turner *et al.*, 1986). The active site serine of thrombin was also acylated by N-acylimidazoles (Lundblad, 1975), 5-acyloxyoxazoles (Valenty *et al.*, 1979), substituted benzoxazinones (Spencer *et al.*, 1986), and substituted isatoic anhydrides (Gelb and Abeles, 1986). 2-Phenyl-4-(3-nitroguanidino-propyl)-5-pivaloyloxyoxazole inactivates thrombin as effectively (k_2/K_I = 4300 $M^{-1} s^{-1}$) as 4'-nitrophenyl 4-guanidinobenzoate, while 7-(amino-methyl)-1-benzylisatoic anhydride inhibits thrombin more slowly ($k_{obsd}/[I]$ = 600 $M^{-1} s^{-1}$). The isatoic anhydride derivative reacts preferentially with thrombin compared to trypsin and plasmin, but is rapidly degraded in serum, has a half-life of less than 1 min, and thus is not suitable for use as an anticoagulant.

3.3. Mechanism-Based Inhibitors

Mechanism-based inhibitors, suicide inhibitors, k_{cat} inhibitors, or enzyme-activated inhibitors often contain a masked functional group which is unmasked upon reaction with the active site of a serine protease. Usually the initial step in the reaction involves formation of an acyl enzyme with Ser 195 with simultaneous unmasking of the reactive group which can then react with another active site residue to form an irreversibly inactivated enzyme.

Halomethylcoumarins rapidly inactivate chymotrypsin since the 4-hydroxybenzyl halides, proposed to be intermediates formed upon acylation of the enzyme, decompose rapidly at neutral pH to give 4-quinone methides which are powerful alkylating agents (Bechet *et al.*, 1977). 3,4-Dihydro-3-benzyl-6-chloromethylcoumarin also irreversibly inactivated thrombin, plasmin, and tissue plasminogen activator (one- and two-chain

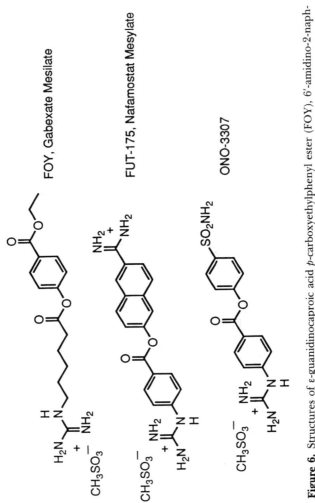

FOY, Gabexate Mesilate

FUT-175, Nafamostat Mesilate

ONO-3307

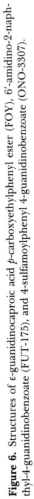

Figure 6. Structures of ε-guanidinocaproic acid *p*-carboxyethylphenyl ester (FOY), 6′-amidino-2-naphthyl-4-guanidinobenzoate (FUT-175), and 4-sulfamoylphenyl 4-guanidinobenzoate (ONO-3307).

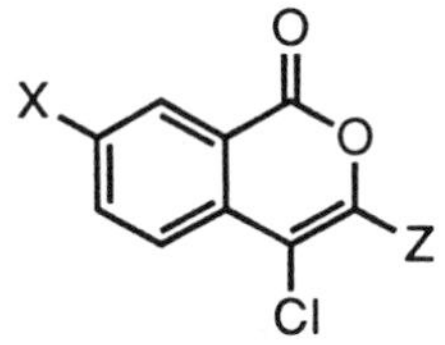

Figure 7. Structures of substituted isocoumarins: DCI, 3,4-dichloroisocoumarin, X = H, Z = Cl; 3-alkoxy-4-chloro-7-guanidinoisocoumarins, X = NH–CH($=$NH$_2^+$)NH$_2$, Z = OR; 7-amino-4-chloro-3-(3-isothiureidopropoxy)isocoumarin (ACITIC), X = NH$_2$, Z = OCH$_2$CH$_2$CH$_2$SC($=$NH$_2^+$)NH$_2$.

forms) with apparent second-order rate constants of 31,000, 316, 187, and 250 $M^{-1} s^{-1}$, respectively (Mor *et al.*, 1990). This compound was tested as anticoagulant in human plasma and was effective at prolonging the prothrombin time.

3,4-Dichloroisocoumarin (DCI, Fig. 7) is a general serine protease inhibitor (Harper *et al.*, 1985) which contains a masked acid chloride group. It will inactivate thrombin, but the rate is slow compared to the inactivation rate of elastases or chymotrypsinlike serine proteases.

Substituted isocoumarins containing basic functional groups (aminoalkoxy, guanidino, and isothiureidoalkoxy) effectively inactivate thrombin and other trypsinlike enzymes (Kam *et al.*, 1988). 3-Alkoxy-4-chloro-7-guanidinoisocoumarins (Fig. 7) are very potent inhibitors of thrombin with k_{obsd}/[I] values in the range of $10^4–10^5 M^{-1} s^{-1}$. 7-Amino-4-chloro-3-(3-isothiureidopropoxy)isocoumarin (ACITIC, Fig. 7) also inhibited thrombin, but only moderately with a k_{obsd}/[I] value of 630 $M^{-1} s^{-1}$. However, ACITIC is more stable than the 7-guanidinoisocoumarin derivatives in buffer and in plasma. Thrombin inactivated by these isocoumarins was very stable and regained less than 20% activity after long standing in the buffer or in the presence of hydroxylamine.

The mechanism of inhibition of trypsinlike enzymes by 3-alkoxy-4-chloro-7-guanidinoisocoumarins and ACITIC involves initial formation of an acyl enzyme intermediate with the unmasking of the 4-aminobenzyl chloride functional group in the case of ACITIC (not shown) or a 4-guanidinobenzyl chloride functional group in the case of 3-alkoxy-4-chloro-7-guanidinoisocoumarins (Fig. 8; Kam *et al.*, 1988; Harper and Powers, 1985). These structures can eliminate chloride to give a quinone imine methide intermediate, which can react further either with an enzyme nucleophile such as His 57 to give an alkylated enzyme or with a solvent molecule to give a new acyl enzyme that can regenerate active enzyme upon deacylation.

The x-ray structure of trypsin complexed with 4-chloro-3-ethoxy-7-guanidinoisocoumarin has provided structural evidence for the inhibition mechanism, since both the covalent adduct with His57 and the chloroacyl enzyme intermediate at Ser 195 are present in the crystal (Chow *et al.*,

1990). The guanidinium group of the inhibitor extends into the S_1 pocket and forms a hydrogen bond with a water molecule which also serves as a H-bond donor/acceptor to Asp189. Molecular modeling with human thrombin indicates that the guanidinium group of the inhibitor should also interact with Asp 189 of thrombin via a water molecule and the 3-alkoxy group should be directed toward thrombin's insertion loop 60A–60H.

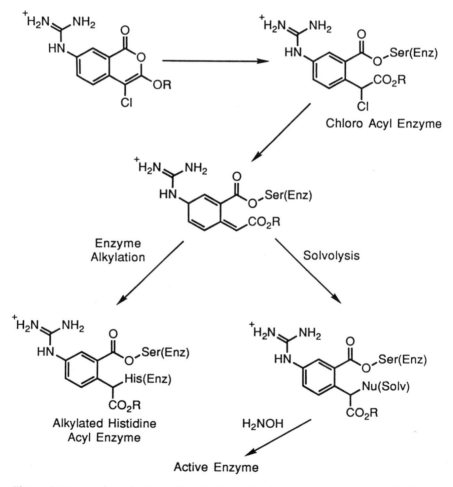

Figure 8. Proposed mechanism of inactivation of serine proteases by 3-alkoxy-4-chloro-7-guanidinoisocoumarins (Kam *et al.*, 1988).

4. SPECIFICITY OF THROMBIN INHIBITORS

A major problem in the design of low-molecular-weight inhibitors for thrombin is specificity. A large number of trypsinlike enzymes are present in mammalian systems and include pancreatic trypsin, the blood coagulation enzymes, fibrinolytic enzymes, complement enzymes, acrosin and other enzymes involved in fertilization, and mast cell and lymphocyte tryptases. For some uses, a highly specific thrombin inhibitor may be necessary, while for others, a less specific inhibitor may be acceptable. The design of specificity into any inhibitor presents a considerable challenge to the medicinal chemist, since the various trypsinlike enzymes share many active site structural features.

The specificity of some of the more potent reversible and irreversible thrombin inhibitors are summarized respectively in Tables X and XI. Among these inhibitors, NAPAP, (2R,4R)-MQPA, and H-D-Phe-Pro-boro-Arg-OH are the most specific reversible inhibitors for thrombin and bind more tightly to thrombin than any of other enzymes studied by at least 115-, 263-, and 150-fold, respectively. The irreversible inhibitor H-D-Phe-Pro-Arg-CH$_2$Cl reacted with thrombin more rapidly than any of the other enzymes studied by at least 1450-fold. It should be noted that many new compounds are claimed to be specific thrombin inhibitors without widespread testing of their activity toward other trypsinlike enzyme.

5. THERAPEUTIC USE OF THROMBIN INHIBITORS

Thrombin is involved in many thrombotic disorders and it is often desirable to control thrombin activity *in vivo*. The major anticoagulant drug in use today is heparin, which promotes the binding of the serpin anti-thrombin III to thrombin. Heparin's drawbacks include the lack of oral activity and the frequent occurrence of hemorrhage following its use. A number of other protein inhibitors of thrombin are being developed including hirudin, a naturally occurring thrombin inhibitor isolated from medicinal leeches (Markwardt and Walsmann, 1958; Markwardt, 1970). Protein inhibitors are unlikely to be orally active and therefore, low-molecular-weight thrombin inhibitors have been widely tested *in vivo* since they are potentially orally active and directly acting anticoagulants.

One of the most widely tested thrombin inhibitors is the irreversible peptide chloromethyl ketone H-D-Phe-Pro-Arg-CH$_2$Cl. Although it was found to be eliminated rapidly in rabbits with a half-life of 2.9 min, this inhibitor is able to prevent intravascular coagulation caused by thrombin or tissue thromboplastin (Collen *et al.*, 1982). It also appears to prevent

Table X. Inhibition Constants of Various Reversible Inhibitors toward Trypsinlike Enzymes

Compound	K_I (μM)					References
	Bovine thrombin	Bovine factor Xa	Human plasmin	Porcine plasma kallikrein	Bovine trypsin	
4-Amidinophenylpyruvic acid (APPA)	6.5	9.4	60	3.0	1.6	Markwardt et al. (1974)
α,α'-Bis(4-amidino-2-iodophenoxy)-p-xylene	0.11	30			2.3	Geratz et al. (1976)
NAPAP	0.0060	7.9	30		0.69	Stürzebecher et al. (1983)
FOY	1.4	2.1	1.6	0.2	2.6	Ohno et al. (1980)
(2R,4R)-MQPA	0.019	210	800	1500[a]	5.0	Kikumoto et al. (1984)
FUT-175	1.9[b]		0.31	0.042[c]	0.084	Matsuoka et al. (1989)
ONO-3307	0.18[b]		0.31	0.29[c]	0.048	Matsuoka et al. (1989)
H-D-Phe-Pro-Arg-H	0.075					Bajusz et al. (1978)
	0.1[d]	4[e]		3[f]	2	Mattson et al. (1982)
H-D-Phe-Pro-boroArg- OH	>0.000004[b]	0.0082[e]	0.0023	0.0006[c]		Kettner et al. (1990)

[a]Bovine plasma kallikrein was used.
[b]Human thrombin was used.
[c]Human plasma kallikrein was used.
[d]Enzyme and inhibitor incubated for 30 sc.
[e]Human factor Xa was used.
[f]Bovine glandular kallikrein was used.

Table XI. Inactivation Constants of Various Irreversible Inhibitors toward Trypsinlike Enzymes

Compound	$k_{obsd}/[I]$ (M^{-1} s^{-1})					References
	Bovine thrombin	Bovine factor Xa	Human plasmin	Porcine plasma kallikrein	Bovine trypsin	
H-D-Phe-Pro-Arg-CH_2Cl	12,000,000	4,500	720	7,900		Kettner and Shaw (1981)
p-Amidinophenyl benzoate	24,000	3,700			800	Turner et al. (1986)
4-Chloro-3-ethoxy-7-guanidinoisocoumarin	55,000	27,000		<500,000		Kam et al. (1988)

thrombin-mediated disseminated intravascular coagulation (DIC) in dogs (Schaffer *et al.*, 1986). The continuous infusion of H-D-Phe-Pro-Arg-CH$_2$Cl (100 nmol/kg per min) abolishes platelet accumulation and the occlusion of thrombogenic segments in baboon models of thrombosis (Hanson and Harker, 1988). Heparin and aspirin are ineffective in this model of high-shear, platelet-dependent thrombosis. The results with H-D-Phe-Pro-Arg-CH$_2$Cl suggest that low-molecular-weight thrombin inhibitors would be useful in the prevention of reocclusion in patients with acute myocardial infarction who have undergone successful reperfusion of occluded coronary arteries by thrombolytic therapy, or for the prevention of occlusion in patients undergoing angioplasty or endarterectomy (Hanson and Harker, 1988; Krupski *et al.*, 1990). Furthermore, H-D-Phe-Pro-Arg-CH$_2$Cl can prevent heparin-resistant thrombotic occlusion in hollow-fiber hemodialyzers (Kelly *et al.*, 1989).

The potent reversible thrombin inhibitor (2R,4R)-MQPA (MD-805, MCI-9038) has low toxicity, good anticoagulant effect *in vitro* (Green *et al.*, 1985), and prevents thrombin formation in experimental anti-thrombin III-deficient animals, whereas heparin is ineffective (Kumada and Abiko, 1981). MQPA is effective in preventing DIC and platelet-rich arterial thrombosis in experimental models (Hara *et al.*, 1987; Jang *et al.*, 1990), and has also been tested in humans for treatment of progressing cerebral thrombosis in Japan (Kobayashi *et al.*, 1989).

The benzamidine derivative NAPAP is one of the most potent reversible inhibitor of thrombin, and has excellent anticoagulant activity *in vitro* and *in vivo* (Hauptmann *et al.*, 1985). NAPAP has low toxicity in mice and has an excellent antithrombotic effect in various experimental animal thrombosis models such as stasis-induced venous thrombosis, electrically induced arterial thrombosis, and extracorporeal arteriovenous shunt thrombosis, but it has a short half-life (9 min) in rabbit plasma (Kaiser *et al.*, 1985; Kaiser and Markwardt, 1986). Although the related N^α-(2-naphthylsulfonylglycyl)-4-amidinophenylalanyl-proline is a less potent thrombin inhibitor than NAPAP (Table VII), this compound has a significantly longer plasma half-life in rabbits ($t_{1/2}$ = 25 min) and is better tolerated in mice and rats than NAPAP (Hauptmann *et al.*, 1989). Lower toxicity and better pharmacokinetics may compensate for the loss of thrombin inhibitory potency of this inhibitor. The NAPAP analog DL-1-[N^α-methyl-N^α-(N'-2-naphthylsulfonylglycyl)-4-amidinophenylalanyl]-piperidine (TA-PAP) have been compared with other benzamidine derivatives which are better factor Xa inhibitors (Stürzebecher and Stürzebecher, 1988; Hauptmann *et al.*, 1990). *In vitro*, the factor Xa inhibitors N^α-tosylglycyl-3-amidinophenylalanine methylester (TAPAM) and 2,7-bis-(4-amidinobenzylidene)-cycloheptanone (BABCH) were more effective in prolonging the prothrombin time (PT) compared to the APTT, while the opposite is true with the thrombin inhibitors. *In vivo*, the thrombin inhibitors NAPAP and TAPAP were effective as antithrombotic agents in a venous stasis thrombosis model and thromboplastin-induced microthombosis model, while factor Xa inhibitors TAPAM (K_I of 0.84 µM) and BABCH (K_I of 0.022 µM) were not effective at equimolar doses. However, this may not be a fair test of the potency of factor Xa inhibitors since both models were initiated with thrombin or thromboplastin. The recently reported tick anticoagulant peptide (TAP), a potent factor Xa inhibitor (K_I = 0.59 nM), appears to have promising activity *in vivo* (Waxman *et al.*, 1990).

The peptide aldehyde H-D-Phe-Pro-Arg-H, a potent transition-state inhibitor, is orally active and can significantly reduce the thrombus weight in an experimental thrombosis model in rabbits (Bagdy *et al.*, 1988). Both H-D-Phe-Pro-Arg-H and D-MePhe-Pro-Arg-H also reduce thrombus formation in an arteriovenous shunt model in rabbits when administered by i.v. infusion, s.c. injection, or oral application (Bajusz *et al.*, 1990). H-D-Phe-Pro-Arg-OiPr had less anticoagulant effect *in vitro* and *in vivo* when compared to H-D-Phe-Pro-Arg-H (Mattson *et al.*, 1982). The related transition-state fluoroketone inhibitors benzoyl-Arg-CF_2CF_3 and 1-naphthoyl-Arg-CF_3 have moderate anticoagulant activity *in vitro* (Ueda *et al.*, 1990). Bz-Arg-CF_2CF_3 prolongs both the PT and APTT at 90–120 µM, while 1-naphthoyl-Arg-CF_3 only prolongs the APTT at 100 µM.

Peptide boronic acid derivatives of arginine including Ac-D-Phe-Pro-boroArg-OH, Boc-D-Phe-Pro-boroArg-$C_{10}H_{16}$, H-D-Phe-Pro-boroArg-OH, and H-D-Phe-Pro-boroArg-$C_{10}H_{16}$ are effective thrombin inhibitors in rabbit plasma against physiological substrates and also prolong APTT *in vitro* at concentrations of 50–200 nM (Kettner *et al.*, 1990). In rabbits, Ac-D-Phe-Pro-boroArg-OH prolongs APTT after i.v. or s.c. injections which suggests that peptide boronic acids have good potential for development as antithrombotic agents.

The guanidinobenzoate derivative FOY (Fig. 6) is very unstable in blood with half-lives of 7–8 and 70–80 s in rat and human plasma, respectively. However, it does prevent clot formation in thrombin-induced DIC in rats when administered by continuous infusion (Ohno *et al.*, 1981). FOY also has some anticoagulant activity in extracorporeal blood circuits where the state of anticoagulation is maintained by continuous infusion of this compound (Mottaghy *et al.*, 1985). However, this anticoagulant effect is not shown outside of the circuit due to the short half-life of FOY. The related guanidinobenzoate FUT-175 (Fig. 6, Nafamostat mesylate) has also been used as an anticoagulant for hemodialysis in Japan even though it has a short half-life (Oka *et al.*, 1986). ONO-3307 (Fig. 6) completely inhibits the deposition of radioactive fibrin in rat kidney and lung (continuous infusion of 10 mg/kg per h) in an experimental thrombosis model for DIC (Matsuoka *et al.*, 1989). FOY was also effective in this model at a higher dose of 50 mg/kg per h, while nafamostat mesylate did not show a significant effect at a dose of 10 mg/kg per h. *p*-Amidinophenyl esters have also shown moderate *in vitro* and *in vivo* anticoagulant activity (Pizzo *et al.*, 1986).

The isocoumarins ACITIC and 4-chloro-3-ethoxy-7-guanidino-isocoumarin (Fig. 7) have potent *in vitro* anticoagulant activity in both human and rabbit plasma (Kam *et al.*, 1988, 1990). With human plasma, these two compounds prolonged the PT ca. 2-fold and prolonged the APTT more than 4.5-fold at 20–30 μM. Both compounds have smaller anticoagulant effects in rabbit plasma. ACITIC has been shown to be an active anticoagulant in rabbits and can inhibit the formation of venous thrombi (Oweida *et al.*, 1990). A short-acting, controllable anticoagulant could have considerable utility in surgery where it is desirable to turn off coagulation for only a defined period of time.

6. SUMMARY

A number of potent reversible and irreversible low-molecular-weight thrombin inhibitors are now available for use in the study of the physiological roles of thrombin. Considerable progress has also been made in designing specificity into the inhibitor structures, although this still remains a formidable challenge for the medicinal chemist. Several thrombin inhibitors appear to be therapeutically useful or have potential application in thrombotic disorders. Although many of the compounds have good potency, low toxicity, and good selectivity, and are anticoagulants *in vivo*, all of the structures appear to have short half-lives in animal models. Nevertheless, it is clear that potent, orally active, low-molecular-weight inhibitors can be developed for a wide variety of uses including hemodialysis and heart–lung bypass. In the future, it is likely that these compounds will replace or supplement heparin and coumarin anticoagulants in many clinical situations.

NOTE ADDED IN PROOF

Crystal structures of the complexes of human thrombin with four inhibitors [benzamidine, H-D-Phe-Pro-Arg-CH$_2$Cl, NAPAP, and (2R,4R)-MQPA] were determined at 3.0 Å resolution (Banner and Hadváry, 1991).

NAPAP and (2R,4R)-MQPA have similar binding modes to thrombin even though they require different stereochemistries for potent inhibition. Both inhibitors bind to thrombin similar to their trypsin binding mode, but the extra Tyr-Pro-Pro-Trp loop in thrombin provides additional binding interactions with inhibitors.

Several new inhibitors, such as phosphonopeptides (Cheng *et al.*, 1991), fluoromethyleneketone retroamides (Altenburger and Schirlin, 1991), and the natural product Nazumamide A (Fusetani *et al.*, 1991) have been reported recently.

Acknowledgment

We gratefully acknowledge financial support from the NIH (grants HL39035 and GM42212).

7. REFERENCES

Altenburger, J. M. and Schirlin, D., 1991, General synthesis of polyfunctionalized fluoromethylene retroamides as pootential inhibitors of thrombin, *Tetrahedron Lett.* **32**: 7255–7258.

Angliker, H., Wikström, P., Rauber, P., and Shaw, E., 1987, The synthesis of lysylfluoromethane and their properties as inhibitors of trypsin, plasmin, and cathepsin B, *Biochem. J.* **241**:871–875.

Angliker, H., Wikström, P., Rauber, P., Stone, S., and Shaw, E., 1988, Synthesis and properties of peptidyl derivatives of arginylfluoromethane, *Biochem. J.* **256**:481–486.

Aogichi, T., and Plaut, G. W. E., 1977, Assay of the esterase activity of thrombin, plasmin, and trypsin with chromogenic substrate *p*-nitrobenzyl *p*-toluenesulfonyl-L-arginine, *Thromb. Haemostas.* **37**:253–261.

Bagdy, D., Barabas, E., Fittler, Z., Orban, E., Rabloczky, G., Bajusz, S., and Szell, E., 1988, Experimental oral anticoagulation by a direct acting thrombin inhibitor (RGH-2958), *Folia Haematol. (Leipzig)* **115**:136–140.

Bajusz, S., Barabas, E., Tolnay, P., Szell, E., and Bagdy, D., 1978, Inhibition of thrombin and trypsin by tripeptide aldehydes, *Int. J. Pept. Protein Res.* **12**:217–221.

Bajusz, S., Szell, E., Bagdy, D., Barabas, E., Horvath, G., Dioszegi, M., Fittler, Z., Szabo, G., Juhasz, A., Tomori, E., and Szilagyi, G., 1990, Highly active and selective anticoagulants: D-Phe-Pro-Arg-H, a free tripeptide aldehyde prone to spontaneous inactivation, and its stable N-methyl derivative, D-MePhe-Pro-Arg-H, *J. Med. Chem.* **33**:1729–1735.

Bang, N. U., and Mattler, L. E., 1977, Thrombin sensitivity and specificity of three chromogenic peptide substrates, in: *Chemistry and Biology of Thrombin* (R. L. Lundblad, J. W. Fenton, II, and K. G. Mann, eds.), Ann Arbor Science, Ann Arbor, pp. 305–311.

Banner, D. W., and Hadvary, P., 1991, Crystallographic analysis at 3.0-Å resolution of the binding to human thrombin of four active site-directed inhibitors, *J. Biol. Chem.* **266**: 20085–20093.

Bechet, J. J., Dupaix, A., and Blagoeva, I., 1977, Inactivation of α-chymotrypsin by new bifunctional reagents: Halomethylated derivatives of dihydrocoumarins, *Biochimie* **59**:231–239.

Bing, D. H., Cory, M., and Fenton, J. W., II, 1977, Exo-site affinity labeling of human thrombins, *J. Biol. Chem.* **252**:8027–8034.

Bode, W., Mayr, I., Baumann, U., Huber, R., Stone, S. R., and Hofsteenge, J., 1989, The refined 1.9 Å crystal structure of human α-thrombin: Interaction with D-Phe-Pro-Arg-chloromethylketone and significance of Tyr-Pro-Pro-Trp insertion segment, *EMBO J.* **8**:3467–3475.

Bode, W., Turk, D., and Stürzebecher, J., 1990, Geometry of binding of the benzamidine- and arginine-based inhibitors N^{α}-(2-naphthyl-sulphonyl-glycyl)-DL-*p*-amidinophenyla-lanyl-piperadine (NAPAP) and (2R,4R)-4-methyl-1-[N^{α}-(3-methyl-1,2,3,4-tetrahydro-8-quinolinesulphonyl)-L-arginyl]-2-piperidine carboxylic acid (MQPA) to human α-thrombin. X-ray crystallographic determination of the NAPAP–trypsin complex and modeling of NAPAP–thrombin and MQPA–thrombin, *Eur. J. Biochem.* **193**:175–182.

Bone, R., Frank, D., Kettner, C. A., and Agard, D. A., 1989, Structural analysis of specificity: α-lytic protease complexes with analogues of reaction intermediates, *Biochemistry* **28**:7600–7609.

Chase, T., Jr., and Shaw, E., 1969, Comparison of esterase activities of trypsin, plasmin and thrombin on guanidinobenzoate esters. Titration of the enzymes, *Biochemistry* **8**:2212–2224.

Cheng, L., Goodwin, C. A., Scully, M. F., Kakkar, V. V., and Claeson, G., 1991, Substrate-related phosphonopeptides, a new class of thrombin inhibitors, *Tetrahedron Lett.* **32**:7333–7336.

Cheng, Y.-C., and Prusoff, W. H., 1973, Relationship between the inhibition constant (K_I) and the concentration of inhibitor which causes 50 per cent inhibition (I_{50}) of an enzymatic reaction, *Biochem. Pharmacol.* **22**:3099–3108.

Cho, K., Tanaka, T., Cook, R. R., Kisiel, W., Fujikawa, K., Karachi, K., and Powers, J. C., 1984, Active site mapping of bovine and human blood coagulation serine proteases using synthetic tripeptide 4-nitroanilide and thioester substrates, of the relative re-activities of human and bovine factor XIIa, *Biochemistry* **23**:644–650.

Chow, M. M., Meyer, E. F., Bode, W., Kam, C.-M., Radhakrishnan, R., Vijayalakshmi, J., and Powers, J. C., 1990, The 2.2 Å resolution x-ray crystal structure formed by trypsin with 4-chloro-3-ethoxy-7-guanidinoisocoumarin, a thrombin inhibitor, *J. Am. Chem. Soc.* **112**:7783–7789.

Collen, D., Matsuo, O., Stassen, J. M., Kettner, C., and Shaw, E., 1982, *In vivo* studies of a synthetic inhibitor of thrombin, *J. Lab. Clin. Med.* **99**:76–83.

Cook, R. R., and Powers, J. C., 1983, Benzyl *p*-guanidinothiobenzoate hydrochloride, a new active-site titrant for trypsin and trypsin-like enzymes, *Biochem. J.* **215**:287–294.

Cook, R. R., McRae, B. J., and Powers, J. C., 1984, Kinetics of hydrolysis of peptide thioester derivatives of arginine by human and bovine thrombin, *Arch. Biochem. Biophys.* **234**:82–88.

Daraio de Peuriot, M., Nigretto, J.-M. and Jozefowicz, M., 1980, Electrochemical activity determination of trypsin-like enzymes. II. Thrombin, *Thromb. Res.* **19**:647–654.

Dunlap, R. P., Stone, P. J., and Abeles, R. H., 1987, Reversible, slow, tight-binding inhibition of human leukocyte elastase, *Biochem. Biophys. Res. Commun.* **145**:509–515.

Fahrney, D. E., and Gold, A., 1963, Sulfonyl fluorides as inhibitors of esterases. 1. Rates of reaction with acetylcholinesterase, α-chymotrypsin and trypsin, *J. Am. Chem. Soc.* **85**:997–1000.

Fareed, J., Messmore, H. L., Kindle, G., and Balis, J. U., 1981, Inhibition of serine proteases by low molecular weight peptides and their derivatives, *Ann. N.Y. Acad. Sci.* **370:**765–784.

Farmer, D. A., and Hageman, J. H., 1975, Use of N-benzoyl-L-tyrosine thiobenzyl ester as a protease substrate. Hydrolysis by α-chymotrypsin and subtilisin BPN', *J. Biol. Chem.* **250:**7366–7371.

Fenton, J. W., II, 1981, Thrombin specificity, *Ann. N.Y. Acad. Sci.*, **370:**468–495.

Fenton, J. W., II, 1986, Thrombin, *Ann. N.Y. Acad. Sci.* **485:**5–15.

Fujii, S., and Hitomi, Y., 1981, New synthetic inhibitors of C1r, C1s esterase, thrombin, plasmin, kallikrein and trypsin, *Biochim. Biophys. Acta* **661:**342–345.

Fujii, S., Nakayama, T., Nunomura, S., Sudo, K., Watanabe, S.-I., Okutome, T., Sakurai, Y., Kurami, M., and Aoyama, T., 1986, Amidine compounds, U.S. Patent 4,563,527.

Fusetani, N., Matsunaga, S., Matsumoto, H., and Takebayashi, Y., 1990, Cyclotheonamides, potent thrombin inhibitors, from a marine sponge *Theonella sp.*, *J. Am. Chem. Soc.* **112:**7053–7054.

Fusetani, N., Nakao, Y., and Matsunaga, S., 1991, Nazumamide A, a thrombin-inhibitory tetrapeptide from a marine sponge, *Theonella* sp., *Tetrahedron Lett.* **32:** 7073–7074.

Gelb, M. H., and Abeles, R. H., 1986, Substituted isatoic anhydride: Selective inactivators of trypsin-like serine proteases, *J. Med. Chem.* **29:**585–589.

Geratz, J. D., 1972, Kinetic aspects of the irreversible inhibition of trypsin and related enzymes by *p-[m-(m*-fluorosulfonylphenylureido)phenoxyethoxy]benzamidine, *FEBS Lett.* **20:**294–296.

Geratz, J. D., and Tidwell, R. R., 1977, The development of competitive reversible thrombin inhibitors, in: *Chemistry and Biology of Thrombin* (R. L. Lundblad, J. W. Fenton, II, and K. G. Mann, eds.), Ann Arbor Science, Ann Arbor, pp. 179–197.

Geratz, J. D., Whitmore, A. C., Cheng, M. C.-F., and Piantadosi, C., 1973, Diamino-α-ω-diphenoxy-alkanes. Structure–activity relationships for the inhibition of thrombin, pancreatic kallikrein, and trypsin, *J. Med. Chem.* **16:**970–975.

Geratz, J. D., Cheng, M. C.-F., and Tidwell, R. R., 1976, Novel bis(benzamidino) compounds with aromatic central link. Inhibitors of thrombin, pancreatic kallikrein, trypsin, plasmin, and complement, *J. Med. Chem.* **19:**634–639.

Geratz, J. D., Stevens, F. M., Polakoski, R. L., Parrish, R. F., and Tidwell, R. R., 1979, Amidino-substituted aromatic heterocycles as probes of the specificity pocket of trypsin-like proteases, *Arch. Biochem. Biophys.* **197:**551–559.

Gladner, J. A., and Laki, K., 1958, Active site of thrombin, *J. Am. Chem. Soc.* **80:**1263–1264.

Glover, G., and Shaw, E., 1971, Purification of thrombin and isolation of a peptide containing the active center histidine, *J. Biol. Chem.* **246:**4594–4601.

Gold, A. M., 1965, Sulfonyl fluorides as inhibitors of esterases. III. Identification of serine as the site of sulfonylation in phenylmethanesulfonyl fluoride, *Biochemistry* **4:**897–902.

Gold, A. M., and Fahrney, D. E., 1964, Sulfonyl fluorides as inhibitors of esterases. II. Formation and reactions of phenylmethanesulfonyl-α-chymotrypsin, *Biochemistry* **3:**783–791.

Grassetti, D. R., and Murray, J. F., Jr., 1967, Determination of sulfhydryl groups with 2,2'- or 4,4'-dithiodipyridine, *Arch. Biochem. Biophys.* **119:**41–49.

Green, G. D. J., and Shaw, E., 1979, Thiobenzyl benzyloxycarbonyl-L-lysinate, substrate for a sensitive colorimetric assay for trypsin-like enzymes, *Anal. Biochem.* **93:**223–226.

Green, D., Ts'ao, C., Reynolds, N., Kahn, D., Kohl, H., and Cohen, I., 1985, *In vitro* studies of a new synthetic thrombin inhibitor, *Thromb. Res.* **37:**145–153.

Hanson, S. R., and Harker, L. A., 1988, Interruption of acute platelet-dependent thrombosis by the synthetic antithrombin D-phenylalanyl-L-prolyl-L-arginyl chloromethyl ketone, *Proc. Natl. Acad. Sci. USA* **85:**3184–3188.

Hara, H., Tamao, Y., Kikumoto, R., and Okamoto, S., 1987, Effect of a synthetic thrombin inhibitor MCI-9038 on experimental models of disseminated intravascular coagulation in rabbits, *Thromb. Haemostas.* **57:**165–170.

Harper, J. W., and Powers, J. C., 1985, Reaction of serine proteases with substituted 3-alkoxy-4-chloroisocoumarins and 3-alkoxy-7-amino-4-chloroisocoumarins. New reactive mechanism based inhibitors, *Biochemistry* **24:**7200–7213.

Harper, J. W., Hemmi, K., and Powers, J. C., 1985, Reaction of serine proteases with substituted isocoumarins: Discovery of 3,4-dichloroisocoumarin, a new general mechanism based serine protease inhibitor, *Biochemistry* **24:**1831–1841.

Hauptmann, J., Kaiser, B., and Markwardt, F., 1985, Anticoagulant action of synthetic tight binding inhibitors of thrombin *in vitro* and *in vivo*, *Thromb. Res.* **39:**771–775.

Hauptmann, J., Kaiser, B., Paintz, M., and Markwardt, F., 1989, Pharmacological characterization of a new structural variant of 4-amidinophenylalanine amide-type synthetic thrombin inhibitor, *Pharmazie* **44:**282–284.

Hauptmann, J., Kaiser, B., Nowak, G., Stürzebecher, J., and Markwardt, F., 1990, Comparison of the anticoagulant and antithrombotic effects of synthetic thrombin and factor Xa inhibitors, *Thromb. Haemostas.* **63:**220–223.

Hibbard, L. S., Nesheim, M. E., and Mann, K. G., 1982, Progressive development of a thrombin inhibitor binding site, *Biochemistry* **21:**2285–2292.

Hitomi, Y., Kanda, T., Niinobe, M., and Fujii, S., 1981, A sensitive colorimetric assay for thrombin, prothrombin and antithrombin III in human plasma using a new synthetic substrate, *Clin. Chem. Acta* **119:**157–164.

Hori, H., Yasutake, A., Minematsu, Y., and Powers, J. C., 1985, Inhibition of human leukocyte elastase, porcine pancreatic elastase and cathepsin G by peptide ketones, in: *Peptides: Synthesis–Structure–Function. Proceeding of the Ninth American Peptide Symposium* (C. M. Deber, V. J. Hruby, and K. D. Kopple, eds.), Pierce Chemical Co., Rockford, Ill., pp. 819–822.

Hummel, B. C. W., 1959, Modified spectrophotometric determination of chymotrypsin, trypsin, and thrombin, *Can. J. Biochem. Physiol.* **37:**1393–1399.

Imperiali, B., and Abeles, R. H., 1986, Inhibition of serine proteases by peptidyl fluoromethyl ketones, *Biochemistry* **25:**3760–3767.

Izquierdo, C., and Burguillo, F. J., 1989, Synthetic substrates for thrombin, *Int. J. Biochem.* **21:**579–592.

James, G. T., 1978, Inactivation of the proteinase inhibitor phenylmethanesulfonyl fluoride in buffers, *Anal. Biochem.* **86:**574–579.

Jang, I.-K., Gold, H. K., Ziskind, A. A., Leinbach, R. C., Fallon, J. T., and Collen, D., 1990, Prevention of platelet-rich arterial thrombosis by selective thrombin inhibition, *Circulation* **81;**219–225.

Kaiser, B., and Markwardt, F., 1986, Experimental studies on the antithrombotic action of a highly effective synthetic thrombin inhibitor, *Thromb. Haemostas.* **55:**194–196.

Kaiser, B., Hauptmann, J., Weiss, A., and Markwardt, F., 1985, Pharmacological characterization of a new highly effective synthetic thrombin inhibitor, *Biomed. Biochim. Acta* **44:**1201–1210.

Kam, C.-M., Fujikawa, K., and Powers, J. C., 1988, Mechanism based isocoumarin inhibitors for trypsin and blood coagulation serine proteases: New anticoagulants, *Biochemistry* **27:**2547–2557.

Kam, C.-M., Vlasuk, G. P., Smith, D. E., Arcuri, K. E., and Powers, J. C., 1990, Thioester chromogenic substrates for human factor VIIa. Substituted isocoumarins are inhibitors of factor VIIa and *in vitro* anticoagulants, *Thromb. Haemostas.* **64:**133–137.

Kawabata, S.-I., Miura, T., Morita, T., Kato, H., Fujikawa, K., Iwanaga, S., Takada, K.,

Kimura, T., and Sakakibara, S., 1988, Highly sensitive peptide-4-methylcoumaryl-7-amide substrates for blood-clotting proteases and trypsin, *Eur. J. Biochem.* **172**:17–25.

Kelly, A. B., Hanson, S. R., Henderson, L. W., and Harker, L. A., 1989, Prevention of heparin-resistant thrombotic occlusion of hollow-fiber hemodialyzers by synthetic antithrombin, *J. Lab. Clin. Med.* **114**:411–418.

Kettner, C., and Shaw, E., 1978, Synthesis of peptides of Arginine chloromethyl ketones: Selective inactivation of human plasma kallikrein, *Biochemistry* **17**:4778–4784.

Kettner, C., and Shaw, E., 1977, The selective inactivation of thrombin by peptides of arginine chloromethyl ketone, in: *Chemistry and Biology of Thrombin* (R. L. Lundblad, J. W. Fenton, II, and K. G. Mann, eds.), Ann Arbor Science, Ann Arbor, pp. 129–143.

Kettner, C., and Shaw, E., 1979, D-Phe-Pro-Arg-CH$_2$Cl—A selective label for thrombin, *Thromb. Res.* **14**:969–973.

Kettner, C., and Shaw, E., 1981, Inactivation of trypsin-like enzymes with peptides of arginine chloromethylketones, *Methods Enzymol.* **80**:826–842.

Kettner, C. A., and Shenvi, A. B., 1984, Inhibition of the serine proteases leukocyte elastase, pancreatic elastase, cathepsin G, and chymotrypsin by peptide boronic acids, *J. Biol. Chem.* **259**:15106–15114.

Kettner, C., Mersinger, L., and Siefring, G., 1988, Peptides of boroarginine as inhibitors of trypsin-like enzymes, *J. Cell. Biochem.* Suppl. **12B**:292.

Kettner, C., Mersinger, L., and Knabb, R., 1990, The selective inhibition of thrombin by peptides of boroarginine, *J. Biol. Chem.* **265**:18289–18297.

Kikumoto, R., Tamao, Y., Ohkubo, K., Tezuka, T., Tonomura, S., Okamoto, S., Funahara, Y., and Hijikata, A., 1980a, Thrombin inhibitors. 2. Amide derivatives of N$^\alpha$-substituted L-arginine, *J. Med. Chem.* **23**:830–836.

Kikumoto, R., Tamao, Y., Ohkubo, K., Tezuka, T., Tonomura, S., Okamoto, S., and Hijikata, A., 1980b, Thrombin inhibitors. 3. Carboxyl-containing amide derivatives of N$^\alpha$-substituted L-arginine, *J. Med. Chem.* **23**:1293–1299.

Kikumoto, R., Tamao, Y., Tezuka, T., Tonomura, S., Hara, H., Ninomiya, K., Hijikata, A., and Okamoto, S., 1984, Selective inhibition of thrombin by (2R,4R)-4-methyl-1-[N$^\alpha$-[(3-methyl-1,2,3,4-tetrahydro-8-quinolinyl)-sulfonyl]-L-arginyl]-2-piperidinecarboxylic acid, *Biochemistry* **23**:85–90.

Knight, C. G., 1986, The characterization of enzyme inhibition, in: *Proteinase Inhibitors* (A. J. Barrett, and G. Salvensen, eds.), Elsevier, Amsterdam, pp. 23–51.

Kobayashi, S., Kitani, M., Yamaguchi, T., Suzuki, T., Okada, K., and Tsunematsu, T., 1989, Effects of an antithrombotic agent (MD-805) on progressing cerebral thrombosis, *Thromb. Res.* **53**:305–317.

Kolde, H.-J., Eberle, R., Heber, H., and Heimburger, N., 1986, New chromogenic substrate for thrombin with increased specificity, *Thromb. Haemostas.* **56**:155–159.

Krishnaswamy, S., Mann, K. G., and Nesheim, M. E., 1986, The prothrombinase-catalyzed activation of prothrombin proceeds through the intermediate meizothrombin in an ordered, sequential reaction, *J. Biol. Chem.* **261**:8977–8984.

Krishnaswamy, S., Church, W. R., Nesheim, M. E., and Mann, K. G., 1987, Activation of human prothrombin by human prothrombinase. Influence of factor Va on the reaction mechanism, *J. Biol. Chem.* **262**:3291–3299.

Krupski, W. C., Bass, A., Kelly, A. B., Marzed, U. M., Hanson, S. R., and Harker, L. A., 1990, Heparin-resistant thrombus formation by endovascular stents in baboons. Interruption by a synthetic antithrombin, *Circulation* **82**:570–577.

Kumada, T., and Abiko, Y., 1981, Comparative study on heparin and a synthetic thrombin inhibitor No. 805 (MD-805) in experimental antithrombin III-deficient animals, *Thromb. Res.* **24**:285–298.

Kuramochi, H., Nakada, H., and Ishii, S.-I., 1979, Mechanism of association of a specific aldehyde inhibitor, leupeptin, with bovine trypsin, *J. Biochem.* **80:**1403–1410.

Laki, A., Gladner, J. A., Folk, J. E., and Kominz, D. R., 1958, The mode of action of thrombin, *Thromb. Diath. Haemorrh.* **2:**205.

Laura, R., Robinson, D. J., and Bing, D. H., 1980, (*p*-Amidinophenyl)methanesulfonyl fluoride, an irreversible inhibitor of serine proteases, *Biochemistry* **19:**4859–4864.

Leytus, S. P., Melhado, L. L., and Mangel, W. F., 1983a, Rhodamine-based compounds as fluorogenic substrates for serine proteinases, *Biochem. J.* **209:**299–307.

Leytus, S. P., Patterson, W. L., and Mangel, W. F., 1983b, New class of sensitive and selective fluorogenic substrates for serine proteinases, *Biochem. J.* **215:**253–260.

Lottenberg, R., Christensen, U., Jackson, C. M., and Coleman, P. L., 1981, Assay of coagulation proteases using peptide chromogenic and fluorogenic substrates, *Methods Enzymol.* **80:**341–361.

Lottenberg, R., Hall, J. A., Fenton, J. W., II, and Jackson, C. M., 1982, The action of thrombin on peptide *p*-nitroanilide substrates: Hydrolysis of Tos-Gly-Pro-Arg-*p*NA and D-Phe-Pip-Arg-*p*NA by human α and γ and bovine α and β-thrombins, *Thromb. Res.* **28:**313–332.

Lottenberg, R., Hall, J. A., Blinder, M., Binder, E. P., and Jackson, C. M., 1983, The action of thrombin on peptide *p*-nitroanilide substrates, substrate selectivity and examination of hydrolysis under different reaction conditions, *Biochim. Biophys. Acta* **742:**539–557.

Lundblad, R. L., 1971, A rapid method for the purification of bovine thrombin and the inhibition of the purified enzyme by phenylmethanesulfonyl fluoride, *Biochemistry* **10:**2501–2506.

Lundblad, R. L., 1975, The reaction of bovine thrombin with N-butylimidazole. Two different reactions resulting in the inhibition of catalytic activity, *Biochemistry* **14:**1033–1037.

McMurray, J. S., and Dyckes, D. F., 1986, Evidence of hemiketals as intermediates in the activation of serine proteinases with halomethyl ketones, *Biochemistry* **25:**2298–2301.

McRae, B. J., Kurachi, K., Heimark, R. L., Fujikawa, K., Davie, E. W., and Powers, J. C., 1981, Mapping the active sites of bovine thrombin, factor IXa, factor Xa, factor XIa, factor XIIa, plasma kallikrein, and trypsin with amino acid and peptide thioesters: Development of new sensitive substrates, *Biochemistry* **20:**7196–7206.

Markwardt, F., and Walsmann, P., 1968a, Inhibition of the clotting enzyme, thrombin, by benzamidine, *Experientia* **24:**25–26.

Markwardt, F., Landmann, H., and Walsmann, P., 1968b, Inhibition of trypsin, plasmin, and thrombin by derivatives of benzylamine and benzamidine, *Eur. J. Biochem.* **6:**502–506.

Markwardt, F., Walsmann, P., and Kazmirowski, H. G., 1969, Effects of ring substitution on the thrombin-inhibiting effect of benzylamine and benzamidine derivatives, *Pharmazie* **24:**400–402.

Markwardt, F., 1970, Hirudin as an inhibitor of thrombin. *Methods Enzymol.* **19:**924–932.

Markwardt, F., and Walsmann, P., 1958, The reaction of hirudin and thrombin, *Hoppe-Seylers Z. Physiol. Chem.* **312:**85–98.

Markwardt, F., Richter, P., Walsmann, P., and Landmann, H., 1970, The inhibition of trypsin, plasmin and thrombin by benzyl 4-guanidinobenzoate and 4'-nitrobenzyl 4-guanidinobenzoate, *FEBS Lett.* **8:**170–172.

Markwardt, F., Hoffman, J., and Korbs, E., 1973, The influence of the synthetic thrombin inhibitors on thrombin–antithrombin reaction, *Thromb. Res.* **2:**343–348.

Markwardt, F., Richter, P., Stürzebecher, J., Wagner, G., and Walsmann, P., 1974, Synthetic inhibitors of serine proteinases. 6. Inhibition of trypsin, plasmin and thrombin by phenylpyruvic acids with several basic substituents, *Acta Biol. Med. Ger.* **33**:K1–K7.

Markwardt, F., Wagner, G., Stürzebecher, J., and Walsmann, P., 1980, N^{α}-Arylsulfonyl-ω-(4-amidino-phenyl)-α-aminoalkylcarboxylic acid amides–Novel selective inhibitors of thrombin, *Thromb. Res.* **17**:425–431.

Matsuoka, S., Futagami, M., Ohno, H., Imaki, K., Okegawa, T., and Kawasaki, A., 1989, Inhibitory effects of ONO-3307 on various proteases and tissue thromboplastin *in vitro* and experimental thrombosis *in vivo*, *Jpn. J. Pharmacol.* **51**:455–463.

Matsuzaki, T., Sasaki, C., Okumura, C., and Umeyama, H., 1989, X-ray analysis of a thrombin inhibitor–trypsin complex, *J. Biochem.* **105**:949–952.

Matthews, D. A., Adler, R. A., Birktoft, J. J., Freer, S. T., and Kraut, J., 1975, X-ray crystallographic study of boronic acids adducts with subtilisin BPN' (Novo), *J. Biol. Chem.* **250**:7120–7126.

Mattson, C., Eriksson, E., and Nilsson, S., 1982, Anti-coagulant and anti-thrombotic effects of some protease inhibitors, *Folia Haematol. (Leipzig)* **1**(Suppl):43–51.

Melhado, L. L., Peltz, S. W., Leytus, S. P., and Mangel, W. F., 1982, *p*-Guanidinobenzoic acid esters of fluorescein as active site titrants of serine proteases, *J. Am. Chem. Soc.* **104**:7299–7306.

Mitchell, G. A., Gargiulo, R. J., Huseby, R. M., Lawson, D. E., Porchron, S. P., and Sehuanes, J. A., 1978a, Assay for plasma heparin using a synthetic peptide substrate for thrombin: Introduction of the fluorophore aminoisophthalic acid dimethylester, *Thromb. Res.* **13**:47–52.

Mitchell, G. A., Hudson, P. M., Huseby, R. M., Porchron, S. P., and Gargiulo, R. J., 1978b, Fluorescent substrate assay for antithrombin III, *Thromb. Res.* **12**:219–225.

Mor, A., Mailard, J., Favreau, C., and Reboud-Ravaux, M., 1990, Reaction of thrombin and proteinases of the fibrinolytic system with a mechanism-based inhibitor, 3,4-dihydro-3-benzyl-6-chloromethylcoumarin, *Biochim. Biophys. Acta* **1038**:119–124.

Morita, T., Kato, H., Iwanaga, S., Takada, K., Kimura, T., and Sakakibara, S., 1977, New fluorogenic substrates for α-thrombin, factor Xa, kallikrein and urokinase, *J. Biochem.* **82**:1495–1498.

Mottaghy, K., Oedekoven, B., Bey, R., and Schmid-Schonbein, H., 1985, Extracorporeal anticoagulation using a serine protease inhibitor in hemodialysis in sheep, *Trans. Am. Soc. Artif. Intern. Organs* **31**:534–536.

Muramatsu, M., and Fujii, S., 1972, Inhibitory effects of ω-guanidino acid esters on trypsin, plasmin, plasma kallikrein and thrombin, *Biochim. Biophys. Acta* **268**:221–224.

Nakayama, T., Okutome, T., Matsui, R., Kurumi, M., Sakurai, Y., Aoyama, T., and Fujii, S., 1984, Synthesis and structure–activity study of protease inhibitor. II. Amino- and guanidino-substituted naphthoates and tetrahydronaphthoates, *Chem. Pharm. Bull.* **32**:3968–3980.

Nesheim, M. E., Prendergast, F. G., and Mann, K. G., 1979, Interactions of a fluorescent active-site-directed inhibitor of thrombin: Dansylarginine N-(3-ethyl-1,5-pentanediyl)amide, *Biochemistry* **18**:996–1003.

Oda, M., Ino, Y., Nakamura, K., Kuramoto, S., Shimamura, K., Iwaki, M., and Fujii, S., 1990, Pharmacological studies on 6-amidino-2-naphthyl[4-(4,5-dihydro-1H-imidazole-2-yl)amino]benzoate dimethane sulfonate (FUT-187). I. Inhibitory activities on various kinds of enzymes *in vitro* and anticomplement activity *in vivo*, *Jpn. J. Pharmacol.* **52**:23–34.

Ohno, H., Kosaki, G., Kambayashi, J., Imaoka, S., and Hirata, F., 1980, FOY:(ethyl *p*-(6-guanidinohexanoyloxy)benzoate)methanesulfonate as a proteinase inhibitor. I. Inhibition of thrombin and factor Xa *in vitro*, *Thromb. Res.* **19**:579–588.

Ohno, H., Kambayashi, J., Chang, S. W., and Kosaki, G., 1981, FOY:(ethyl *p*-(guanidino-hexanoyloxy)benzoate) methanesulfonate as a proteinase inhibitor. II. *In vivo* effect on coagulofibrinolytic system in comparison with heparin and aprotinin, *Thromb. Res.* **24**:445–452.

Oka, T., Hanasawa, K., Yoshika, T., Endo, Y., Matsuda, K., Tani, T., and Kodama, M., 1986, An inhibitor of the coagulation, complement, and contact system as an anticoagulant, *Life Support Syst.* **4**(Suppl. 2):195–197.

Okamoto, S., Kinjo, K., and Hijikata, A., 1980, Thrombin inhibitors, 1. Ester derivatives of N^{α}-(arylsulfonyl)-L-arginine, *J. Med. Chem.* **23**:827–830.

Oweida, S. W., Ku, D. N., Lumsden, A. B., Kam, C.-M., and Powers, J. C., 1990, *In vivo* determination of the anticoagulant effect of a substituted isocoumarin (ACITIC), *Thromb. Res.* **58**:191–197.

Peet, N. P., Burkhart, J. P., Angelastro, M. R., Giroux, E. L., Mehdi, S., Bey, P., Kolb, M., Neises, B., and Schirlin, D., 1990, Synthesis of peptidyl fluoromethyl ketones and petidyl α-keto esters as inhibitors of porcine pancreatic elastase, human neutrophil elastase, and rat and human neutrophil cathepsin G, *J. Med. Chem.* **33**:394–407.

Pizzo, S. V., Turner, A. D., Porter, N., and Gonias, S., 1986, Evaluation of *p*-amidinophenyl esters as potential antithrombotic agents, *Thromb. Haemostas.* **56**:387–390.

Plaut, G. W. E., 1978, *p*-Nitrobenzyl *p*-toluenesulfonyl-L-arginine a chromogenic substrate for thrombin, plasmin, and trypsin, *Haemostasis* **7**:105–108.

Powers, J. C., and Harper, J. W., 1986, Inhibitors of serine proteinases, in: *Proteinase Inhibitors* (A. J. Barrett, and G. Salvensen, eds.), Elsevier, Amsterdam, pp. 55–152.

Pozsgay, M., Szabo, G., Bajusz, S., Simonsson, R., Gaspar, R., and Elodi, P., 1981, Study of the specificity of thrombin with tripeptidyl-*p*-nitroanilide substrates, *Eur. J. Biochem.* **115**:491–495.

Roffman, S., Sanocka, U., and Troll, W., 1970, Sensitive proteolytic enzyme assay using differential solubilities of radioactive substrates and products in biphasic systems, *Anal. Biochem.* **36**:11–17.

Sato, E., Matsuhisa, A., Sakashita, M., and Kanaoka, Y., 1988, New water-soluble fluorogenic amine. 7-Aminocoumarin-4-methanesulfonic acid (ACMS) and related substrates for proteinases, *Chem. Pharm. Bull.* **36**:3496–3502.

Schaffer, R. C., Briston, C., Chilton, S.-M., and Carlson, R. W., 1986, Disseminated intravascular coagulation following Echiscarinatus venom in dogs: Effects of a synthetic thrombin inhibitor, *J. Lab. Clin. Med.* **107**:488–497.

Schechter, I., and Berger, A., 1967, On the size of the active site in protease. 1. Papain, *Biochem. Biophys. Res. Commun.* **27**:157–162.

Shaw, E., Mares-Guia, M., and Cohen, W., 1965, Evidence for an active site histidine in trypsin through use of a specific reagent, 1-chloro-3-tosylamido-7-amino-2-heptanone, the chloromethyl ketone derived from N^{α}-tosyl-L-lysine, *Biochemistry* **4**:2219–2224.

Sherry, S., and Troll, W., 1954, The action of thrombin on synthetic substrates, *J. Biol Chem.* **208**:95–105.

Spencer, R. W., Copp, L. J., Bonaventura, B., Tam, T. F., Liak, T. J., Billedeau, R. J., and Krantz, A., 1986, Inhibition of serine proteases by benzoxazinones: Effects of electron withdrawal and 5-substitution, *Biochem. Biophys. Res. Commun.* **140**:928–933.

Stein, R. L., Strimpler, A. M., Edwards, P. D., Lewis, J. J., Mauger, R. C., Schwartz, J. A., Stein, M. M., Trainer, A., Wildonger, R. A., and Zottola, M. A., 1987, Mechanism of slow-binding inhibition of human leukocyte elastase by trifluoromethyl ketones, *Biochemistry* **26**:2682–2689.

Stürzebecher, J., 1984, Inhibitors of thrombin, in: *The Thrombin* (R. Machovich, ed.), CRC Press, Boca Raton, Vol. 1, pp. 131–160.

Stürzebecher, J., and Stürzebecher, U., 1988, Are factor Xa inhibitors superior to thrombin inhibitors in anticoagulation? *Folia Haematol. (Leipzig)* 115:152–156.

Stürzebecher, J., Markwardt, F., Voigt, B., Wagner, G., and Walsmann, P., 1983, Cyclic amides of N^α-arylsulfonylaminoacylated 4-amidinophenylalanine—Tight binding inhibitors of thrombin, *Thromb. Res.* 29:635–642.

Svendsen, L., Blombäck, B., Blombäck, M., and Olsson, P., 1972, Synthetic chromogenic substrates for the determination of trypsin, thrombin, and thrombin-like enzymes, *Thromb. Res.* 1:267–278.

Takahashi, L. H., Radhakrishnan, R., Rosenfield, R. E., Meyer, E. F., Trainor, D. A., and Stein, M., 1988, X-ray diffraction analysis of the inhibition of porcine pancreatic elastase by a peptidyl trifluoromethylketone, *J. Mol. Biol.* 201:423–428.

Takahashi, L. H., Radhakrishnan, R., Rosenfield, R. E., and Meyer, E. F., 1989, Crystallographic analysis of the inhibition of porcine pancreatic elastase by a peptidyl boronic acid: Structure of a reaction intermediate, *Biochemistry* 28:7610–7617.

Tamura, Y., Hirado, M., Okamura, K., Yoshihiro, M., and Fujii, S., 1977, Synthetic inhibitors of trypsin, plasmin, kallikrein, thrombin, C1r, and C1s esterase, *Biochim. Biophys. Acta* 484:417–422.

Tanaka, T., McRae, B. J., Cho, K., Cook, R., Fraki, J. E., Johnson, J. E., and Powers, J. C., 1983, Mammalian tissue trypsin-like enzymes: Comparative reactivities of human skin tryptase, human lung tryptase, and bovine trypsin with peptide 4-nitroanilide and their thioester substrates, *J. Biol. Chem.* 258:13552–13557.

Thompson, R. C., 1973, Use of peptide aldehydes to generate transition-state analogs of elastase, *Biochemistry* 12:47–51.

Tidwell, R. R., Fox, L. L., and Geratz, J. D., 1976, Aromatic tris-amidines. A new class of highly active inhibitors of trypsin-like proteases, *Biochim. Biophys. Acta* 445:729–738.

Tidwell, R. R., Geratz, J. D., Dann, O., Volz, G., Zeh, D., and Loewe, H., 1978, Diarylamidine derivatives with one or both of the aryl moieties consisting of an indole or indole-like ring. Inhibitors of arginine-specific esteroproteases, *J. Med. Chem.* 21:613–623.

Tidwell, R. R., Webster, W. P., Shaver, S. R., and Geratz, J. D., 1980, Strategies for anticoagulation with synthetic protease inhibitors: Xa inhibitors versus thrombin inhibitors, *Thromb. Res.* 19:339–349.

Tsunematsu, H., Mizusaki, K., Hatanaka, Y., Kamahori, M., and Makisumi, S., 1986, Kinetics of hydrolysis of a new peptide substrate containing p-guanidino-L-phenylalanine by trypsin and thrombin, *Chem. Pharm. Bull.* 34:1351–1354.

Turner, A. D., Monroe, D. M., Roberts, H. R., Porter, N. A., and Pizzo, S. V., 1986, p-Amidino esters are irreversible inhibitors of factor IXa and Xa and thrombin, *Biochemistry* 25:4929–4935.

Ueda, T., Kam, C.-M., and Powers, J. C., 1990, The synthesis of arginylfluoroalkanes, their inhibition of trypsin and blood-coagulation serine proteinases and their anticoagulant activity, *Biochem. J.* 265:539–545.

Valenty, V. B., Wos, J. D., Lobo, A. P., and Lawson, W. B., 1979, 5-Acyloxyoxazoles as serine protease inhibitors. Rapid inactivation of thrombin in contrast to plasmin, *Biochem. Biophys. Res. Commun.* 88:1375–1381.

Voigt, B., Stürzebecher, J., Wagner, G., and Markwardt, F., 1988, Syntheses of N^α-(arylsulfonyl)-4-amidino-phenylalanine-proline and of N^α-(arylsulfonylglycyl)-4-amidino-phenylalanine-proline and investigation on inhibition of serine proteinases 33. Synthetic inhibitors of serine *Pharmazie* 43:412–414.

Walsmann, P., Markwardt, F., and Landmann, H., 1970, Inhibition of thrombin, plasmin, and trypsin by alkyl- and alkoxybenzamidine, *Pharmazie* **25**:551–554.

Walsmann, P., Richter, P., and Markwardt, F., 1972, Inactivation of trypsin and thrombin by 4-amidinobenzenesulfonyl fluoride and 4-(2-aminoethyl)-benzenesulfonyl fluoride, *Acta Biol. Med. Ger.* **28**:577–585.

Walsmann, P., Markwardt, F., Richter, P., Stürzebecher, J., Wagner, G., and Landmann, H., 1974, Synthetic inhibitors of serine proteinases. 2. Inhibitory effect of homologs of amidinobenzoic acids and their esters against trypsin, plasmin and thrombin, *Pharmazie* **29**:333.

Walsmann, P., Horn, H., Landmann, H., Markwardt, F., Stürzebecher, J., and Wagner, G., 1975, Synthetic inhibitors of serine proteinases. 5. Inhibition of trypsin, plasmin and thrombin by arylaliphatic amidino compounds with ether structure, as well as esters of 3- and 4-amidinophenoxyacetic acid, *Pharmazie* **30**:386–389.

Walter, J., and Bode, W., 1983, The x-ray crystal structure analysis of the refined complex formed by bovine trypsin and *p*-amidinophenyl pyruvate at 1.4 Å resolution, *Hoppe-Seyler's Z. Physiol. Chem.* **364**:949–959.

Waxman, L., Smith, D. E., Arcuri, K. E., and Vlasuk, G. P., 1990, Tick anticoagulant peptide (TAP) is a novel inhibitor of blood coagulation factor Xa, *Science* **248**:593–595.

Witting, J. I., Pouliott, C., Catalfamo, J. A., Fareed, J., and Fenton, J. W., II, 1988, Thrombin inhibition with dipeptidyl argininals, *Thromb. Res.* **50**:461–468.

Wong, S.-C., and Shaw, E., 1974, Differences in active center reactivity of trypsin homologs. Specific inactivation of thrombin by nitrophenyl *p*-amidinophenylmethane sulfonate, *Arch. Biochem. Biophys.* **161**:536–543.

Wong, S.-C., and Shaw, E., 1976, Inactivation of trypsin-like proteinases by active-site-directed sulfonation. Ability of departing group to confer selectivity, *Arch. Biochem. Biophys.* **176**:113–118.

Wong, S.-C., Green, G. D. J., and Shaw, E., 1978, Inactivation of trypsin-like proteases by sulfonation. Variation of positively charged group and inhibitor length, *J. Med. Chem.* **21**:456–459.

Yagashi, T., Nunomura, S., Okutome, T., Nakayama, T., Kurumi, M., Sakurai, Y., Aoyama, T., and Fujii, S., 1984, Synthesis and structure–activity study of protease inhibitors. III. Amidinophenols and their benzoyl esters, *Chem. Pharm. Bull.* **32**:4466–4477.

Chapter 5

REGULATION OF THROMBIN BY ANTITHROMBIN AND HEPARIN COFACTOR II

Steven T. Olson and Ingemar Björk

1. INTRODUCTION

Thrombin plays a pivotal role in blood coagulation. It cleaves fibrinogen, a reaction that initiates the formation of the fibrin gel, which constitutes the framework of the blood clot. It also activates the cofactors factor V and factor VIII of the clotting system, thereby greatly accelerating the coagulation process. When bound to thrombomodulin on the endothelial cell surface, thrombin also activates protein C, which then inactivates the two cofactors and impedes blood clotting. In addition, thrombin activates factor XIII, leading to cross-linking of the fibrin gel. It is thus apparent that accurate regulation of thrombin activity is important in maintaining normal hemostasis.

The activity of thrombin generated during blood clotting is primarily

Steven T. Olson • Division of Biochemical Research, Henry Ford Hospital, Detroit, Michigan 48202.   **Ingemar Björk** • Department of Veterinary Medical Chemistry, Swedish University of Agricultural Sciences, Uppsala Biomedical Center, S-751 23 Uppsala, Sweden.

Thrombin: Structure and Function, edited by Lawrence J. Berliner. Plenum Press, New York, 1992.

regulated by inactivation of the enzyme by plasma proteinase inhibitors. Four such inhibitors, α_1-proteinase inhibitor, α_2-macroglobulin, antithrombin, and heparin cofactor II, have been shown to be able to inhibit thrombin. The first two of these are general proteinase inhibitors that can inactivate several different enzymes, while antithrombin and heparin cofactor II are more specific for thrombin and, in the case of antithrombin, also for other proteinases of the blood clotting system. The latter two inhibitors also share the property that they bind heparin and certain similar glycosaminoglycans and that the rate of their inactivation of thrombin is greatly accelerated by these interactions.

In this chapter we will review current knowledge of the structure and function of the two specific thrombin inhibitors, antithrombin and heparin cofactor II. All evidence indicates that antithrombin is the physiologically most important of the two, and particular emphasis will therefore be given to this inhibitor. We will first describe the properties of antithrombin and discuss the mechanism of its inactivation of thrombin and other clotting proteinases. The binding of heparin to antithrombin and thrombin or other proteinases, as well as the role of these interactions in the mechanisms by which heparin accelerates the antithrombin–proteinase reaction will then be considered. We will also examine current evidence regarding the physiological role of antithrombin and of heparinlike polysaccharides and discuss how other proteins in plasma may modulate the heparin-accelerated antithrombin–proteinase reaction. Finally, we will review the properties of heparin cofactor II, the mechanism of its inhibition of thrombin, the effect of glycosaminoglycans on this reaction, and the possible physiological role of the inhibitor.

2. ANTITHROMBIN

2.1. Structure and Function

2.1.1. Structure

Antithrombin is a plasma glycoprotein that has been isolated from a large number of species and presumably occurs in all vertebrates (Jordan, 1983). Its concentration in human plasma is ~0.125 mg/ml, i.e., ~2.3 µM (Conard et al., 1983). It is synthesized in the liver and circulates in plasma with a half-life of ~3 days (Collen et al., 1977; Léon et al., 1983).

Antithrombin can be isolated from plasma by affinity chromatography on matrix-linked heparin (Miller-Andersson et al., 1974). Recombinant human antithrombins with properties similar or identical to those of the plasma protein, except for differences in glycosylation, have been expressed in several mammalian cell lines (Stephens *et al.*, 1987b; Wasley *et al.*, 1987; Zettlmeissl *et al.*, 1988, 1989).

Determinations of protein and cDNA sequences have shown that human antithrombin has 432 amino acid residues and three disulfide bridges (Petersen *et al.*, 1979; Bock *et al.*, 1982; Prochownik *et al.*, 1983; Chandra *et al.*, 1983). The antithrombin sequence is homologous to that of several other plasma proteinase inhibitors with related function, such as α_1-proteinase inhibitor, α_2-antiplasmin, α_1-antichymotrypsin, and heparin cofactor II, and also to many proteins with no known proteinase inhibitory activity, such as ovalbumin, angiotensinogen, and certain carriers of lipophilic molecules (Hunt and Dayhoff, 1980; Huber and Carrell, 1989). The proteins of this superfamily have been called serpins, an acronym for serine proteinase inhibitors (Carrell and Boswell, 1986). All serpins have a common central region, comprising about 350 amino acids with 25–50% sequence homology, to which amino-terminal or carboxy-terminal extensions specific for each individual member may be attached.

All four potential N-glycosylation sites with the structure Asn-X-Ser/Thr in the antithrombin sequence carry oligosaccharide side chains in the predominant form of the protein (Petersen *et al.*, 1979; Bock *et al.*, 1982). However, about 10% of the inhibitor in plasma lack the carbohydrate side chain on Asn-135, leading to a slightly higher heparin affinity (Carlson and Atencio, 1982; Peterson and Blackburn, 1985; Brennan *et al.*, 1987). The four oligosaccharide side chains of the predominant form have a largely identical, biantennary structure but differ in their terminal sialic acids (Franzén *et al.*, 1980; Mizuochi *et al.*, 1980), resulting in a charge heterogeneity of the protein (Nordenman *et al.*, 1977; Borsodi and Bradshaw, 1977).

The molecular weight of human antithrombin, calculated from the amino acid sequence and carbohydrate composition, is 58,200, in agreement with sedimentation equilibrium determinations (Nordenman *et al.*, 1977). Hydrodynamic and low-angle x-ray diffraction measurements have shown that the shape of antithrombin is slightly more elongated than that of a typical globular protein (Nordenman *et al.*, 1977; Furugren *et al.*, 1977). The existence of two structural domains, differing in stability, in antithrombin has been inferred from the denaturation behavior of the protein in guanidinium hydrochloride (Villanueva and Allen, 1983a; Fish *et al.*, 1985; Gettins and Wooten, 1987).

The three-dimensional structure of bovine antithrombin, although in a modified form cleaved one residue away from the reactive bond (see Section 2.1.2), has been described by Mourey *et al.*, (1990). The structure is highly similar to that of an analogous modified form of human α_1-

proteinase inhibitor that is cleaved at the reactive bond (Loebermann *et al.*, 1984). This latter structure has been reported in greater detail. The structures of both modified inhibitors are well ordered. In α_1-proteinase inhibitor, 80% of the amino acids are located in eight α-helices, denoted A to H, and the large β-sheets, denoted A to C (Loebermann *et al.*, 1984; Fig. 1). Analogous structural elements are present in antithrombin (Mourey *et al.*, 1990). The major differences between the two structures generally occur in connecting loops between secondary structural elements. An additional difference is the presence of a 44-residue amino-terminal extension in antithrombin that is not found in α_1-proteinase inhibitor and which contains helices and loops (Mourey *et al.*, 1990). The two disulfide bonds (linking Cys-8 with Cys-128 and Cys-21 with Cys-95) that join this segment to the rest of the molecule result in its being localized close to helix D

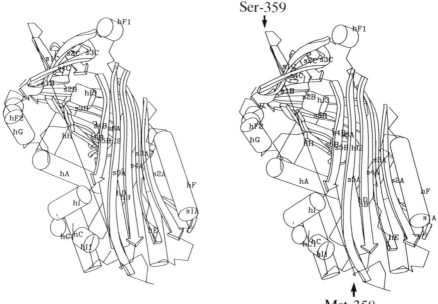

Figure 1. Polypeptide chain folding of reactive-bond cleaved α_1-antitrypsin with the secondary structural elements represented by arrows (β-sheet strands) and cylinders (α-helixes). β-Sheets (s) are denoted A–C, with numbers distinguishing individual strands, and α-helixes (h) are denoted A–H. The A sheet is the major sheet on the "front" side of the molecule. The oblique straight line joins residues Met-358 and Ser-359, which form the reactive bond of the intact inhibitor. These residues correspond to Arg-393 and Ser-394, respectively, in antithrombin. (Reprinted, with permission, from Huber and Carrell, 1989. Copyright 1989 by the American Chemical Society. Figure courtesy of Dr. Robert Huber.)

(Mourey *et al.*, 1990). The structure inferred for intact antithrombin from the known structure of the reactive-site cleaved forms of antithrombin and α_1-proteinase inhibitor will be discussed in conjunction with an account of the reactive site of antithrombin and the nature of the interaction of the inhibitor with proteinases in Section 2.1.2.

2.1.2. Reaction with Thrombin and Other Clotting Proteinases

Antithrombin can inhibit all enzymes of the intrinsic coagulation pathway and also some noncoagulation serine proteinases, such as plasmin, trypsin, and the complement proteinase, C1s̄ (Ogston *et al.*, 1976; Rosenberg, 1977). The reaction with the proteinase leads to the formation of an equimolar, tight complex, in which the active site of the enzyme is blocked (Abildgaard, 1969; Rosenberg and Damus, 1973). Antithrombin does not bind proteinases in which the reactive serine residue has been covalently modified with inactivating reagents (Rosenberg and Damus, 1973), indicating that an intact active site of the enzyme is required for formation of a tight complex.

Antithrombin inactivates its target enzymes rather slowly in the absence of heparin, although the rate increases dramatically in the presence of the polysaccharide (see Section 2.2). The bimolecular rate constant for the antithrombin–thrombin reaction at near-neutral pH in the absence of heparin is 7×10^3–$1.1 \times 10^4 \, M^{-1} \, s^{-1}$ at 25°C (Olson and Shore, 1982; Danielsson and Björk, 1982; Wong *et al.*, 1983; Latallo and Jackson, 1986) and $\sim 1.4 \times 10^4 \, M^{-1} \, s^{-1}$ at 37°C (Jesty, 1979b), corresponding to a half-life of thrombin of ~ 20 s at 37°C under pseudo-first-order conditions at plasma concentrations of antithrombin. The actual half-life of exogenous thrombin in plasma is somewhat longer, ~ 40 s, primarily due to the inhibitory effect of fibrinogen on the antithrombin–thrombin reaction (Jesty, 1986). The half-life of thrombin formed during prothrombinase-catalyzed prothrombin activation in plasma is even longer, due to the effect of the activation fragments produced concurrently (see Section 2.4.2). The proteolytically cleaved derivative of human thrombin, γ-thrombin, is inactivated about four times slower than the intact proteinase (Latallo and Jackson, 1986). All other coagulation proteinases are inactivated slower than thrombin. The bimolecular rate constant for the reaction with factor Xa is thus 3–$4 \times 10^3 \, M^{-1} \, s^{-1}$ at 37°C (Jesty, 1978; Jordan *et al.*, 1980b), and the rate constants for inhibition of other coagulation proteinases, and also of kallikrein and plasmin, appear to be at least about tenfold lower than this value (Jordan *et al.*, 1980b; Schapira *et al.*, 1982; Scott *et al.*, 1982; Pixley *et al.*, 1985; Beeler *et al.*, 1986; Scott and Colman, 1989). In contrast, trypsin is inactivated appreciably faster than thrombin, with a rate constant of 1.5–4.5 × 10^5 M^{-1} s^{-1} at 25°C (Wong *et al.*, 1982; Danielsson and Björk, 1982). Inhibition rates somewhat faster than those cited above have been reported for several enzymes but may have been caused by a minute contamination of heparin in the antithrombin preparations.

Rapid-kinetics studies have shown that the antithrombin–thrombin reaction proceeds by a two-step mechanism, in analogy with other proteinase–proteinase inhibitor reactions (Laskowski and Kato, 1980). A weak, Michaelis–Menten-type complex with a dissociation constant of ~1.4 × 10^{-3} M is formed initially, and this complex is subsequently converted to a stable complex with a rate constant of ~10 s^{-1} (Olson and Shore, 1982). The relatively low bimolecular inhibition rate constant of the antithrombin–thrombin reaction is primarily due to the high dissociation constant of the initial complex.

All available evidence suggests that the formation of the antithrombin–proteinase complex involves interaction between the enzyme and a reactive bond of antithrombin, the Arg-393–Ser-394 bond in the carboxy-terminal region of the inhibitor (Fig. 2). The localization of this bond was originally proposed from identification of the single cleavage site in the modified form of antithrombin that can be dissociated by hydroxylamine from complexes with several proteinases (Jörnvall *et al.*, 1979; Björk *et al.*, 1981, 1982; Danielsson and Björk, 1982) and has subsequently been amply verified. Analogous cleavage sites at homologous positions have thus been identified in several serpins (Wiman and Collen, 1979; Carrell *et al.*, 1980; Morii and Travis, 1983; Griffith *et al.*, 1985a,b). Moreover, overwhelming support has come from studies of a number of congenital antithrombin variants isolated from patients with thrombotic disorders. Mutation of the P_1 (Schechter and Berger, 1967) residue, Arg-393, to His (Erdjument *et al.*, 1988a, 1989; Owen *et al.*, 1988; Lane *et al.*, 1989a; Erdjument *et al.*, 1989), Cys (Erdjument *et al.*, 1988a,b), or Pro (Lane *et al.*, 1989b) thus results in an inactive antithrombin that is unable to form a complex with clotting proteinases, demonstrating that this residue is essential for the interaction. In mutants with Cys in this position, this residue can form a disulfide bond with the thiol group of albumin, thereby also restricting the access to the reactive bond (Erdjument *et al.*, 1987). The requirement for an arginine at the P_1 position of the reactive bond is consistent with the known substrate specificity of the enzymes that are inhibited by antithrombin. The essential role of the P_1 residue of the inhibitor is further illustrated by a mutation of this residue in α_1-proteinase inhibitor from Met to Arg, which converted the protein from an inhibitor of elastase to a thrombin inhibitor (Owen *et al.*, 1983). The P_1' residue of antithrombin has also been shown to be important for the interaction with the target proteinase, although the requirement for Ser in this position is not absolute. A congenital anti-

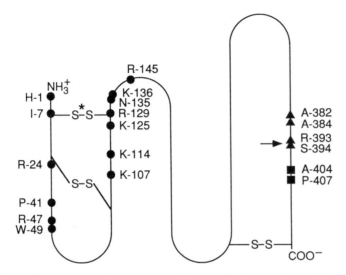

Figure 2. Residues suggested to be involved in or affecting antithrombin function. ▲, proteinase binding; ●, heparin binding; ■, maintenance of normal circulating levels of the inhibitor. The arrow marks the reactive bond and the asterisk the disulfide bond proposed to be of importance for the heparin-binding conformation.

thrombin variant with a mutation of this residue to Leu had a greatly decreased ability to inhibit thrombin (Stephens *et al.*, 1987a), as had mutant recombinant antithrombins with Pro, Met, Cys, Val, or Leu at this site (Stephens *et al.* 1988). However, recombinant antithrombins with Gly, Ala, or Thr in the P_1' position retained near-normal activity, indicating that there is a size optimum which is modulated by hydrophobic effects for the side chain in this position (Stephens *et al.*, 1988).

In the crystal structures of cleaved antithrombin and α_1-proteinase inhibitor, the two residues of the cleavage site, i.e., the reactive bond in α_1-proteinase inhibitor and the adjacent bond in antithrombin (corresponding to the Ser-394-Leu-395 bond in the human inhibitor), are at opposite ends of the molecule, separated by 60–70 Å (Loebermann *et al.*, 1984; Mourey *et al.*, 1990; Fig. 1). The cleavage thus must have been followed by a conformational change of the inhibitors. However, substantial evidence indicates that this change is moderate and that the intact molecules may be reconstructed from those of the cleaved forms, as originally proposed for α_1-proteinase inhibitor (Loebermann *et al.*, 1984; Huber and Carrell, 1989; Engh *et al.*, 1990). This reconstruction can be done by extraction of the central 4A strand, which has the amino-terminal of the cleaved bond as its carboxy-terminus, from the major β-sheet of the inhibitors, the six-stranded, antiparallel A sheet (Fig. 1), and joining it with the residue carboxy-terminal of the cleaved bond, located at the other end of this sheet. This removal can be done by suitable rotations of the backbone at a "hinge" residue, corresponding to Glu-378 in antithrombin (Engh *et al.*, 1990). In the case of α_1-proteinase inhibitor, molecular dynamics simulations have shown that the gap after the removed strand can be closed with only limited overall structural perturbations, resulting in formation of a five-stranded A sheet with the two middle strands in a parallel arrangement (Engh *et al.*, 1990). In this way, the reactive bond will be located in an exposed region of the molecule and will be easily accessible to the target proteinases of the inhibitor. Such an exposed localization of the reactive-bond segment of antithrombin has been suggested to be required for this segment to fit into the narrow active-site "canyon" in thrombin (Bode *et al.* 1989; see Chapter 1). Moreover, an exposure of the extracted 4A strand is consistent with observations that certain enzymes which are not inhibited cleave, and thereby inactivate, antithrombin in a region immediately amino-terminal to the reactive bond, i.e., at residues located in this strand (Kress and Catanese, 1981; Carrell and Owen, 1985). The proposed reconstruction of the active conformations of α_1-proteinase inhibitor and antithrombin is strongly supported by the recently determined three-dimensional structure of ovalbumin, a related serpin with no known inhibitory activity, crystallized in its intact form (Stein *et al.*, 1990). In this structure, the region corresponding to the reactive-bond segments in α_1-proteinase inhibitor and antithrombin forms an exposed, somewhat mobile, three-turn α-helix, supported on two short peptide "stalks," and separated from the protein core by a solvent channel. Moreover, the structure of the remainder of the molecule, including that of the five-stranded A sheet, is highly similar to the reconstructed structures of native α_1-proteinase inhibitor and antithrombin.

The data thus strongly suggest that inactivation of a proteinase by antithrombin is initiated by the enzyme attacking the reactive bond of the inhibitor as in a regular substrate. However, it is apparent that subsequent cleavage does not proceed normally, since a stable complex is formed. Instead, antithrombin presumably is activated to trap the enzyme and arrest cleavage at some intermediate stage of the normal proteolytic reaction. The trapping most likely involves a conformational change of the inhibitor, as indicated by the changes in antigenic, heparin-binding, and spectroscopic properties of antithrombin that accompany complex formation (Carlström *et al.*, 1977; Collen and DeCock, 1978; Villanueva and Danishefsky, 1979; Wallgren *et al.*, 1981; Wong *et al.*, 1983; Peterson and Blackburn, 1987a). A conformational change is also consistent with the two-step mechanism, involving a stabilization of an initial weak antithrombin–thrombin complex, that has been demonstrated by rapid-kinetics studies (Olson and Shore, 1982). The conformational change may involve insertion of at least part of the exposed reactive-bond loop into the A sheet to a position related to that in the cleaved form of the inhibitor (Engh *et al.*, 1990). This reaction has been suggested to be initiated by residues just carboxy-terminal to the hinge residue filling a hole present between the two central, parallel strands of the A sheet in the reconstructed structure of α_1-proteinase inhibitor. This proposal is supported by the fact that serpins which are not proteinase inhibitors, such as ovalbumin and angiotensinogen, have amino acids with bulky, charged side chains in this position of the reactive loop, which would prevent the incorporation of these residues into the A sheet (Wright *et al.*, 1990; Engh *et al.*, 1990).

Circumstantial evidence indicates that the trapping may be triggered at the acyl-intermediate stage of proteolysis. At this stage, the peptide fragment carboxy-terminal to the reactive bond of antithrombin would be released with the P_1' Ser residue as its amino-terminus, but an acyl bond would still link the active-site Ser of the proteinase with the P_1 Arg residue that now forms the carboxy-terminus of the major fragment of the inhibitor. Cessation of the proteolytic reaction at the acyl-intermediate stage is thus consistent with the demonstration of a new amino-terminal Ser in the antithrombin–thrombin complex (Ferguson and Finlay, 1983a) and also with the finding that this complex is stable under strong denaturing conditions but can be dissociated by hydroxylamine or high pH (Rosenberg and Damus, 1973; Owen, 1975; Jesty, 1979a; Fish and Björk, 1979), indicative of a covalent ester bond joining the two proteins in the complex. However, recent NMR studies of the stable complex formed between elastase and α_1-proteinase inhibitor labeled with ^{13}C at the P_1 Met residue have suggested that an arrest of the proteolytic reaction occurs at the tetrahedral intermediate stage (Matheson *et al.*, 1991. The antithrombin conformational change may trap the proteinase by drastically reducing the rate at which the acyl or tetrahedral intermediate can deacylate to regenerate free proteinase and cleaved inhibitor. This possibility is indicated by the demonstration that the antithrombin–thrombin complex is not thermodynamically, but only kinetically, stable and slowly (i.e., with a half-life of ~3 days) dissociates to intact enzyme and cleaved inhibitor (Danielsson and Björk, 1980, 1983). The slow deacylation rate could be due to a noncovalent interaction so tight that water is efficiently excluded. It has been suggested (Björk *et al.*, 1981) that such an interaction might involve residues in a hydrophobic hexapeptide, Ala-Val-Val-Ile-Ala-Gly, located immediately amino-terminal to the reactive bond of antithrombin, and a complementary hydrophobic site in thrombin (Bode *et al.*, 1989).

A peculiar feature of the reaction between antithrombin and thrombin or certain other proteinases is that a small amount of a noncomplexed, modified form of antithrombin, in which the reactive bond is cleaved, is formed concurrent with the stable antithrombin–proteinase complex (Fish et al., 1979; Jörnvall et al., 1979; Björk and Fish, 1982; Björk et al., 1982; Olson, 1985; Gettins and Harten, 1988). This modified antithrombin, which is analogous to the modified form of α_1-proteinase inhibitor for which the crystal structure is known (Loebermann et al., 1984), is inactive, has a greatly decreased heparin affinity (Björk and Fish, 1982), and may have a conformation similar to that of antithrombin in the stable complex, as indicated by the similar antigenic properties and heparin affinities of the two forms (Carlström et al., 1977; Jordan et al., 1979; Wallgren et al., 1981). It must be formed by a pathway which branches off the inhibition pathway at some intermediate stage and in which antithrombin does not trap the proteinase in a stable complex but simply acts as a substrate for the enzyme (Björk and Fish, 1982; Olson, 1985). The partitioning between the inhibition and substrate pathways is affected by heparin, which thereby greatly increases the amount of modified antithrombin formed (Björk and Fish, 1982; Olson, 1985; see Section 2.2.3d). A preferential production of the modified antithrombin by the substrate pathway, rather than formation of a complex, has been shown in reactions of thrombin with antithrombin to which is bound a monoclonal antibody directed against an epitope comprising residues 382 to 386, Ala-Ala-Ala-Ser-Thr, located ~10 residues amino-terminal to the reactive bond (Asakura et al., 1990). Similarly, congenital antithrombin variants in which Ala-384 (Fig. 2) is substituted by Pro have also been shown to be substrates rather than inhibitors of thrombin (Aiach et al., 1985; Molho-Sabatier et al., 1989; Caso et al., 1991), and substitution of Ala-382 by Thr possibly causes the same effect (Devraj-Kizuk et al., 1988). It thus appears that the conformation of the region of antithrombin ~10 residues amino-terminal to the reactive bond can determine whether antithrombin can trap its target enzymes in a stable complex or whether it only acts as a normal substrate of these enzymes. This region is at the amino-terminal end of the reactive-bond loop, near the "hinge" residue discussed earlier. A different three-dimensional structure of this region therefore could influence the insertion of the reactive-bond loop into the A sheet (Engh et al., 1990) in such a manner that the noncovalent interaction with the proteinase becomes much weaker, allowing access of water to the acyl or tetrahedral intermediate and resulting in an appreciable deacylation rate.

2.2. Effect of Heparin on the Reaction with Thrombin and Other Clotting Proteinases

Heparin is a sulfated glycosaminoglycan produced by a certain type of mast cells. Its long-known anticoagulant activity has led to its widespread clinical use for the prophylaxis and treatment of thrombosis. Commercial preparations used for this purpose are usually isolated from porcine intestinal mucosa or bovine lung. The polysaccharide consists of alternating hexuronic acid residues, which can be either D-glucuronic acid or L-iduronic acid, and D-glucosamine residues. The resulting polymer is heavily sulfated, carrying N-sulfate groups on the glucosamine units as well as O-sulfate groups in different positions on both the hexuronic and glucosamine units. The distribution of glucuronic acid and iduronic acid units as well as the sulfation pattern is variable, albeit far from random, resulting in a considerable structural heterogeneity of the polysaccharide (see further Björk and Lindahl, 1982; Lane and Lindahl, 1989). Commercial heparin preparations are polydisperse with molecular weights varying from ~5000 to ~30,000, i.e., with ~15 to 100 monosaccharide residues per chain.

Heparin exerts its anticoagulant activity by greatly accelerating the rate at which antithrombin inactivates its target proteinases. The apparent second-order rate constant for the inactivation of thrombin in the presence of optimal heparin concentrations is thus $1.5-4 \times 10^7 \, M^{-1} s^{-1}$ at 37°C (Jordan et al., 1979, 1980b; Griffith, 1982a,b; Olson and Björk, 1991), i.e., the reaction is accelerated ~2000-fold (see Section 2.1.2). This rate constant would correspond to a half-life for thrombin of ~10 ms under pseudo-first-order conditions at plasma concentrations of antithrombin. However, due to the two-step mechanism of the heparin-accelerated antithrombin–thrombin reaction, the pseudo-first-order rate constant approaches a limiting value of $~5 \, s^{-1}$ at these antithrombin concentrations (Olson and Shore, 1982; Olson and Björk, 1991), resulting in an actual thrombin half-life of ~140 ms. Extents of acceleration comparable to that seen with thrombin have been shown for the reactions of antithrombin with factors Xa and IXa, although the maximum second-order rate constants for these reactions in the presence of heparin are almost 10-fold lower than that of the reaction with thrombin (Jordan et al., 1980b). In contrast, the acceleration of the inactivation of enzymes participating in the contact activation phase of blood clotting appears to be much smaller (Beeler et al., 1986; Soons et al., 1987; Scott and Colman, 1989; Olson, 1989; Olson and Shore, 1989). Most evidence indicates that binding of both inhibitor and enzyme to heparin in a ternary complex is a prerequisite for the ability of the polysaccharide to accelerate the inactivation of thrombin and factor IXa by antithrombin, whereas acceleration of the inactivation of factor Xa only appears to require binding of antithrombin to heparin (see further Section 2.2.3).

2.2.1. Binding of Heparin to Antithrombin

Commercial heparin can be fractionated into two distinct fractions with different affinities for antithrombin by affinity chromatography on matrix-linked inhibitor (Lam *et al.*, 1976; Höök *et al.*, 1976; Andersson *et al.*, 1976). A high-affinity fraction which constitutes about one-third of the polysaccharide, has high anticoagulant activity and accounts for ~90% of the activity of the starting material, whereas the remaining low-affinity fraction has low activity. A unique pentasaccharide sequence (Fig. 3), not found in low-affinity heparin, has been shown to constitute the specific antithrombin binding region of high-affinity heparin (Lindahl *et al.*, 1980; Casu *et al.*, 1981; Thunberg *et al.*, 1982; Choay *et al.*, 1983; Atha *et al.* 1984, 1985). Most high-affinity chains contain one such sequence, although some longer chains may contain two (Danielsson and Björk, 1981; Jordan *et al.*, 1982; Nesheim *et al.*, 1986). This pentasaccharide has been synthesized chemically (Sinaÿ *et al.*, 1984).

High-affinity heparin binds to a single site on antithrombin with a dissociation constant of ~2 × 10^{-8} M at physiological pH and ionic strength (Nordenman *et al.*, 1978; Jordan *et al.*, 1979; Olson *et al.*, 1981). This affinity is essentially independent of heparin chain length for naturally occurring high-affinity heparin chains and for high-affinity fragments, produced by chemical degradation of longer chains, that contain a complete pentasaccharide sequence, i.e., from the size of an octasaccharide upwards (Danielsson and Björk, 1981; Lindahl *et al.*, 1984). The synthetic pentasaccharide also binds with virtually the same affinity, demonstrating

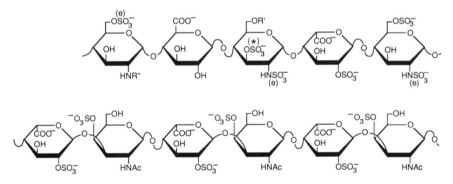

Figure 3. The binding regions for antithrombin in heparin (upper structure; Lindahl *et al.*, 1984) and for heparin cofactor in dermatan sulfate (lower structure; Maimone and Tollefsen, 1990). Structural variants in the antithrombin-binding region are indicated by R' (-H or -SO$_3^-$) and R'' (-SO$_3^-$ or -COCH$_3$). The 3-*O*-sulfate group marked by an asterisk is a distinguishing structural feature of the antithrombin-binding region of heparin and is required for high-affinity binding of the polysaccharide to antithrombin. The three sulfate groups marked (e) are also essential for this interaction. (Reproduced, with permission, from *The Annual Review of Biochemistry*, Volume 60, Copyright 1991 by Annual Reviews Inc. Figure courtesy of Dr. Ulf Lindahl.)

that this sequence in high-affinity heparin contributes all, or the major part, of the free energy of binding to antithrombin (Atha *et al.*, 1985; Olson *et al.*, 1992). Low-affinity heparin binds to the same site on antithrombin as high-affinity heparin but with ~1000-fold lower affinity (Nordenman and Björk, 1978b; Danielsson and Björk, 1978; Jordan *et al.*, 1979). The affinity of the binding of high-affinity heparin to antithrombin decreases markedly with increasing ionic strength, consistent with ionic interactions between the negatively charged polysaccharide and positive charges on the protein (Nordenman *et al.*, 1978; Jordan *et al.*, 1979; Olson *et al.*, 1981). A quantitative analysis of this effect has suggested that at most five to six charged groups on each molecule participate in the interaction (Nordenman and Björk, 1981; Olson and Björk, 1991), in good agreement with the number of charged groups in the pentasaccharide sequence shown to be essential for high affinity (Fig. 3). However, this analysis also suggests that a major fraction of the binding energy is further derived from nonionic interactions, consistent with the specificity of the binding.

The binding of high-affinity heparin to antithrombin induces changes in the ultraviolet absorption near ultraviolet circular dichroism, and tryptophan fluorescence of the inhibitor, suggesting perturbations of the environment of aromatic residues (Einarsson and Andersson, 1977; Villanueva and Danishefsky, 1977; Nordenman and Björk, 1978b; Nordenman *et al.*, 1978). Most of these changes are identical or similar for all high-affinity chains of the size of an octasaccharide or larger, indicating a similar mode of binding of these chains (Lindahl *et al.*, 1984). However, the small differences in the circular dichroism changes that can be observed between high-affinity chains containing 14 or fewer monosaccharide units and longer chains have been taken to reflect a secondary, weak interaction between antithrombin and a region of the longer chains outside the pentasaccharide sequence (Oosta *et al.*, 1981; Stone *et al.*, 1982). Binding of high-affinity heparin also induces perturbations of the proton NMR spectrum of antithrombin which indicate that His residues, other aromatic residues, and Lys and/or Arg residues are affected by the binding (Gettins, 1987). These changes are similar for high-affinity heparin chains containing at least eight monosaccharides but somewhat different for the synthetic pentasaccharide (Gettins and Choay, 1989), indicating that the environment of residues of antithrombin in the vicinity of the heparin binding site may be affected by the monosaccharide units immediately adjacent to the pentasaccharide region in the longer heparin chains.

The spectroscopic changes observed on binding of high-affinity heparin are consistent with, but not proof of, a conformational change in antithrombin being induced by the binding. More conclusive evidence for such a change has been provided by the observation that the fluorescence enhancement of antithrombin on interaction with heparin arises from buried Trp residues rather than merely from a perturbation of surface residues (Olson and Shore, 1981). The exposure of a new cleavage site for a snake-venom proteinase in antithrombin on heparin binding (Kress and Catanese, 1981) and the increased reactivity of Lys-236 to chemical modification in the presence of the polysaccharide (Chang, 1989) also support a conformational change. Furthermore, rapid-kinetics studies have shown that heparin binding is a two-step process, involving an initial binding in a rapid equilibrium with a dissociation constant of 3–4 × 10^{-5} M under physiological conditions, followed by a conformational change at a rate constant of ~500 s^{-1} (Olson *et al.*, 1981, 1992). The altered conformation induced by the latter step increases the affinity of heparin for antithrombin ≥ 300-fold and is responsible for the tight binding.

Although the heparin binding site of antithrombin has not been definitely located, considerable evidence indicates that two portions of the antithrombin sequence participate in the binding, namely the amino-terminal region and a region comprising residues ~100 to ~145. Evidence that the amino-terminal region (Fig. 2) is involved was first provided by studies showing that chemical modification of Trp-49 resulted in decreased heparin binding (Blackburn *et al.*, 1984). Further support has come from the characterization of congenital antithrombin variants with amino acid substitutions in this region. Mutations of Arg-47 to Cys (Koide *et al.*, 1984; Duchange *et al.*, 1987; Brunel *et al.* 1987; Owen *et al.*, 1989b), His (Owen *et al.*, 1987; Caso *et al.*, 1990), or Ser (Borg *et al.*, 1988) and of Arg-24 to Cys (Borg *et al.*, 1990) have been shown to abolish or greatly decrease the affinity of antithrombin for heparin. This effect presumably is due to the loss of the positive charge, although steric hindrance by disulfide-linked groups may also have contributed to the decreased affinity in the case of mutations to Cys. Substitution of Pro-41 by Leu similarly reduced the heparin affinity, probably as a result of an altered conformation of the amino-terminal region (Chang and Tran, 1986; Daly *et al.*, 1989). Mutation of Ile-7 to Asn also decreased heparin binding, presumably because of interference by the bulky oligosaccharide side chain bound to the new carbohydrate attachment site introduced by this substitution (Brennan *et al.*, 1988). Additional data corroborate that the amino-terminal region of antithrombin is involved in heparin binding. Proton NMR studies have shown that resonances from two histidines, suggested to be His-1 and, more tentatively, His-65, are perturbed by this binding (Gettins and Wooten, 1987). Moreover, a disulfide bond, first proposed to be the carboxy-terminal disulfide bond but later identified as that between Cys-8 and Cys-128, has been shown to be important for the integrity of the heparin binding site (Ferguson and Finlay, 1983b; Sun and Chang, 1989a). Also, the specific cleavage by V8 protease of three Glu-X bonds in the amino-terminal region of the inhibitor was inhibited by binding of heparin (Liu and Chang, 1987a). Finally, the heparin binding site of antithrombin has been suggested to be lost on unfolding of a separate domain of the protein, which may be the amino-terminal region (Villanueva and Allen, 1983a,b).

Evidence that the antithrombin segment from residues ~100 to ~145 (Fig. 2) is involved in binding of heparin has mainly come from chemical modification studies. An initial investigation showed that reaction of antithrombin with pyridoxal 5'-phosphate produced a non-heparin-binding protein in which Lys-125 was modified, indicating that the positive charge contributed by this residue is involved in the interaction (Peterson et al., 1987b). Work with other lysine-specific reagents showed that modification of Lys-107, Lys-114, Lys-125, and Lys-136 reduced the heparin-accelerated rate of inactivation of thrombin and that the extent of modification was greatly decreased in the presence of the polysaccharide (Liu and Chang, 1987b; Chang, 1989). Further studies with arginine-specific modification reagents similarly implicated Arg-129 and Arg-145 in the binding (Sun and Chang, 1990). However, all of these residues presumably do not participate directly in heparin binding; some may be located only in the vicinity of the binding site, their reactivity with the modifying reagents being decreased by the long flexible heparin chains used in these studies. The best evidence for a direct involvement in heparin binding has been provided for Lys-125. More definite proof that the antithrombin region from residues ~100 to ~145 participates in binding of heparin has recently emerged from characterization of a hereditary abnormal antithrombin with decreased heparin affinity, in which Arg-129, a residue also implicated by chemical modification, was replaced by Gln (Gandrille et al., 1990). Further evidence is provided by the finding that the oligosaccharide side chain on Asn-135 interferes with binding of the polysaccharide (Peterson and Blackburn, 1985; Brennan et al., 1987). The requirement for an intact Cys-8–Cys-128 disulfide bond for heparin binding discussed above may also reflect the importance of the region around the latter cysteine residue for the interaction (Sun and Chang, 1989a). The isolation of heparin-binding antithrombin fragments comprising residues 104–251 (Rosenfeld and Danishefsky, 1986) or 114–156 (Smith and Knauer, 1987) following degradation of the protein with cyanogen bromide or V8 protease, respectively, is also consistent with residues in the region of antithrombin extending from residues ~100 to ~145 contributing to the heparin binding site. Moreover, polyclonal antibodies directed against a synthetic peptide comprising residues 124–145 decrease heparin binding to antithrombin and may accelerate complex formation with thrombin (Smith et al., 1990).

The bulk of the data presented above, together with the identification of those positively charged residues that are conserved in heparin-binding proteinase inhibitors, has led to a proposal of where the heparin binding site may be located in a model of antithrombin based on the three-dimensional structure of the reactive-site-cleaved α_1-proteinase inhibitor (Borg *et al.*, 1988; Huber and Carrell, 1989). The binding site is suggested to comprise a band of positively charged residues stretching from the base of the A helix, where Arg-47 and Ile-7 presumably are located, along the underside of the D helix, with Lys-125 and Arg-129 as well as several of the conserved basic residues, and further toward the A sheet in a region containing Asn-135 and Lys-136 (Fig. 4). This binding site is of an appropriate size to allow binding of the heparin pentasaccharide region. Although attractive, this hypothesis has the limitation that it is modeled on the inferred structure of reactive-site-cleaved antithrombin, which binds heparin very weakly (Björk and Fish, 1982). It is possible, however, that the large decrease in heparin affinity is due to only limited conformational changes of the heparin binding region following reactive-bond cleavage of the inhibitor. This possibility is supported by the high degree of similarity between the three-dimensional structure of reactive-bond-cleaved α_1-proteinase inhibitor and that of ovalbumin, in which the region corresponding to the reactive-bond loop is intact (Stein *et al.*, 1990).

2.2.2. Binding of Heparin to Thrombin and Other Clotting Proteinases

An interaction between thrombin and heparin has been demonstrated qualitatively by affinity chromatography (Nordenman and Björk, 1978a), crossed immunoelectrophoresis (Holmer *et al.*, 1979), differential heat denaturation (Machovich and Aranyi, 1977), and aqueous two-phase partitioning (Petersen and Jørgensen, 1983). Thrombin elutes from an immobilized heparin affinity matrix at ~0.5 M NaCl (Nordenman and Björk, 1978a), consistent with the enzyme binding to heparin through ionic interactions and with a weaker affinity than that of antithrombin. A similar binding of heparin to an immobilized thrombin affinity column has also been reported (Nordenman and Björk, 1980). The polysaccharide is eluted from the immobilized proteinase by a salt gradient as a single, broad band rather than as discrete fractions differing in affinity, suggesting that thrombin does not bind to a specific saccharide sequence in heparin as does antithrombin.

Most quantitative studies of the interaction of thrombin with size-fractionated heparins have used changes in the polarization or the intensity of the fluorescence of extrinsic probes covalently bound to either the enzyme or the polysaccharide to measure binding (Jordan *et al.*, 1980a; Oshima *et al.*, 1986; Nesheim *et al.*, 1986; Evington *et al.*, 1986b). Such

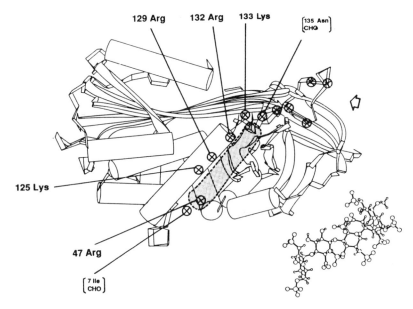

Figure 4. Proposed heparin binding site of antithrombin. The structure of antithrombin is modeled on the known structure of reactive-bond-cleaved α_1-antitrypsin (Loebermann *et al.*, 1984). The proposed primary binding site (shaded) is formed by Arg-47, Lys-125, Arg-129, Arg-132, and Lys-133. A secondary site may extend around the molecule toward the reactive bond along residues Lys-136, Lys-228, Arg-235, and Lys-236, which are shown but not labeled. The estimated position of the reactive bond is marked with an arrow. The primary binding site is flanked by residues 135 and 7, which carry oligosaccharide side chains in normal antithrombin and in a congenital variant of the protein with Ile-7 replaced by Asn, respectively. The size of the pentasaccharide is indicated in the lower right of the figure. (Reprinted, with permission, from Huber and Carrell, 1989. Copyright 1989 by the American Chemical Society. Figure courtesy of Dr. Robert Huber.)

studies have reported similar dissociation constants ranging from 0.86 to 2.4×10^{-6} M for the interaction at near physiological ionic strengths (I 0.15–0.22). However, the stoichiometry of binding measured in these studies has varied greatly. Thus, binding of one to two heparins per thrombin molecule was found in studies which used fluorescence polarization changes of labeled heparins of molecular weight 11,000 and 6500, respectively, to monitor binding (Jordan *et al.*, 1980a; Oshima *et al.*, 1986). In contrast, two studies in which fluorescence intensity or polarization changes of a dansyl-labeled thrombin were used to monitor binding demonstrated a progressive linear increase of from 2 to 7 thrombin molecules bound per heparin chain as the heparin molecular weight increased from ~6000 to 30,000 (Nesheim *et al.*, 1986; Evington *et al.*, 1986b). Such discrepancies in stoichiometry may reflect the multivalency of both heparin and thrombin suggested by later studies (Olson *et al.*, 1991), together with the different manner in which titrations were conducted in these studies, i.e., titration of thrombin into heparin or heparin into thrombin. Alternatively, the assumption of the former studies of a two-state interacting system in which a single polarization change characterizes all thrombin–heparin complexes (Jordan *et al.*, 1980a; Oshima *et al.*, 1986) may not be valid, since the molecular weight, and thereby the polarization change, of such complexes would be expected to be variable.

One problem inherent in the analysis of the thrombin–heparin interaction in all of these studies was the assumption that thrombin binds to discrete sites on the heparin chain with identical affinities. However, the failure to detect any specifically binding polysaccharide species by affinity chromatography of heparin on immobilized thrombin and the correlation of thrombin binding stoichiometry with heparin chain length have suggested that thrombin binds to heparin mainly through nonspecific electrostatic interactions. A consequence of nonspecific electrostatic binding of a protein to a linear polyelectrolyte chain is that protein binding sites will overlap, thereby resulting in an apparent negative cooperativity of binding, due to each protein binding event eliminating several binding sites for subsequent binding events (Kowalczykowski *et al.*, 1986). Analyses of the binding assuming a discrete site interaction would thus be expected to overestimate the binding affinity and underestimate the stoichiometry. Recognition of this problem led to efforts in one study to discriminate between specific and nonspecific binding models for the thrombin–heparin interaction (Olson *et al.*, 1991). Equilibrium binding studies were conducted over a wide range of thrombin and heparin concentrations using fluorescence intensity changes of the active-site-bound probe, *p*-aminobenzamidine, or quantitative affinity chromatography to monitor binding. Analysis of the binding by the nonspecific binding model developed for protein–polynucleotide interactions (McGhee and von Hippel, 1974; Epstein, 1978) or the discrete binding site model that had been previously used to analyze the interaction (Jordan *et al.*, 1980a; Nesheim *et al.*, 1986; Evington *et al.*, 1986b), revealed that the two models could not be distinguished over the range of thrombin binding densities accessible to measurement. This was due to thrombin–heparin complexes precipitating when thrombin binding densities exceeded 2 thrombin molecules bound per heparin chain, a phenomenon that also has been observed in previous studies (Nesheim *et al.*, 1986; Evington *et al.*, 1986b). However, analyses of the ionic strength dependence of the interaction using polyelectrolyte theory (Record *et al.*, 1976, 1978; Manning 1978) demonstrated that the binding energy could be almost completely accounted for by ionic interactions. This finding, together with the minimal dependence of the binding affinity on temperature (Record *et al.*, 1978), and the ability of the nonspecific, but not the specific, binding model to account for the apparent increase in binding affinity of thrombin for heparin chains of increasing size in terms of a single intrinsic dissociation constant, strongly argued in favor of the nonspecific binding model (Olson *et al.*, 1991). This model indicated that thrombin interacts with 5–6 ionic groups in a 3-disaccharide binding site of heparin with an intrinsic dissociation constant of 6–10 μM at physiological ionic strength and pH. Thrombin binding to heparin was thus suggested to be ~1000-fold weaker than antithrombin binding. However, this intrinsic binding affinity represents an upper limit due to the possible underestimation of the actual number of overlapping binding sites available for thrombin on the polysaccharide. This is a consequence of the counting of these sites being based on the structural unit (i.e., the disaccharide) rather than the functional unit (i.e., the negative charge) of the polysaccharide.

While a single intrinsic dissociation constant described the binding of thrombin to heparin chains ranging from 3 to ~13 disaccharides in length in these studies, thrombin was observed to bind to larger heparin chains (i.e., from ~22 to ~35 disaccharides) with progressively (up to 4-fold) greater intrinsic affinities (Olson *et al.*, 1991). Sedimentation equilibrium studies of the size of thrombin–heparin complexes indicated the existence of a second, weaker binding site on thrombin for heparin that accounted for both the increased intrinsic affinity of thrombin for larger heparin chains as well as the precipitation of thrombin–heparin complexes at high thrombin binding densities. Other evidence also suggests the existence of two cationic sites on thrombin that can bind the heparin polyanion. Thus, heparin competitively inhibits thrombin clotting of fibrinogen, but at concentrations greatly exceeding those required to saturate the primary thrombin–heparin binding interaction (Fenton *et al.*, 1989). Moreover, the binding of fibrinogen or other competitive ligands to the anion-binding exosite of thrombin does not block the primary interaction of heparin with thrombin or heparin acceleration of the thrombin–antithrombin reaction (Stone and Hofsteenge, 1987; Naski *et al.*, 1990). Together, these observations indicate that the anion-binding exosite of thrombin involved in binding fibrinogen and other macromolecular substrates is not the primary heparin binding site on thrombin involved in accelerating thrombin inactivation by antithrombin, but rather may be a secondary binding site for heparin. The finding that heparin protects lysine residues essential for heparin binding but not for fibrinogen binding from chemical modification (Church *et al.*, 1989b) supports the idea that the fibrinogen binding site on thrombin is distinct from the primary heparin binding site involved in the acceleration of the thrombin–antithrombin reaction. A recent report of the x-ray crystallographic structure of the thrombin–hirudin complex (Rydel *et al.*, 1990) has suggested that a cationic-rich region on the surface of the thrombin molecule distinct from the substrate recognition exosite may represent the primary heparin binding site.

Little information is available on the binding of heparin to other coagulation proteinases. Factor IXa, Xa, and plasmin binding to a fluorescamine-labeled heparin of M_r ~6500, monitored by fluorescence polarization changes, has indicated dissociation constants of 0.25, 8, and ~100 μM, respectively, for these interactions, with a stoichiometry of 1:1 determinable only for the factor IXa interaction (Jordan *et al.*, 1980a). Consideration of the potential nonspecific ionic mode of interaction of these proteinases with heparin suggests that the dissociation constants measured in this study may overestimate the true intrinsic affinity of the interactions (Olson *et al.*, 1991). Nevertheless, such analyses still indicate that these proteinases bind less tightly to heparin than antithrombin. Proteinase–heparin interactions also appear to differ from the antithrombin–heparin interaction in that there is no evidence to suggest that these interactions are accompanied by protein conformational changes.

2.2.3. Mechanism of Heparin Acceleration of Proteinase Inhibition

2.2.3a. Role of Surface Approximation. The correlation between the binding affinity for antithrombin of heparin molecules possessing a specific pentasaccharide sequence recognized by the inhibitor and the ability of these heparin species to accelerate the reaction between antithrombin and thrombin provided an early indication of the importance of inhibitor binding to heparin in heparin rate enhancement (see Section 2.2.1). A direct involvement of this interaction in the reaction pathway was later shown from kinetic studies, in which the heparin rate enhancement for four different target proteinases was found to increase as a function of heparin concentration in parallel with the formation of an antithrombin–heparin complex (Jordan *et al.*, 1979, 1980b). Subsequent kinetic and equilibrium binding studies confirmed that under conditions where thrombin–heparin binary complexes were not formed to any significant extent, the increase in heparin rate enhancement with increasing heparin or antithrombin concentrations directly paralleled antithrombin–heparin complex formation (Peterson and Jørgensen, 1983; Olson, 1988). Moreover, binding of antithrombin to heparin prior to its reaction with thrombin has been directly observed by continuous monitoring of protein fluorescence changes during the reaction by stopped-flow fluorometry

(Olson and Shore, 1986). Together, these studies indicate that the predominant pathway for the heparin-accelerated inhibition of thrombin by antithrombin involves a reaction between an antithrombin–heparin binary complex and free thrombin. The importance of an antithrombin–heparin interaction in the mechanism of heparin rate enhancement is supported by the inability of heparin to accelerate the reaction of thrombin with chemically modified or naturally modified antithrombins which have selectively lost the ability to bind heparin (see Section 2.2.1).

Evidence that thrombin binding to heparin in addition to antithrombin binding is necessary for heparin to accelerate the thrombin–antithrombin reaction was first suggested from studies of the dependence of heparin rate enhancement on heparin chain length. Thus, despite the requirement for only a pentasaccharide to bind antithrombin, heparin chains of at least ~18 saccharides in length are required to accelerate the reaction of antithrombin with thrombin, as well as with factors IXa and XIa (Laurent et al., 1978; Oosta et al., 1981; Holmer et al., 1981; Lane et al., 1984; Danielsson et al., 1986). This contrasts sharply with the ability of heparin oligosaccharide chains as small as the pentasaccharide to accelerate the inactivation of other target enzymes, namely factor Xa, plasma kallikrein, and possibly factor XIIa, to an extent approaching that of full-length heparin chains (Thunberg et al., 1979; Holmer et al., 1980, 1981; Oosta et al., 1981; Choay et al., 1983; Lane et al., 1984; Danielsson et al., 1986; Ellis et al., 1986; Olson and Choay, 1989). These early observations thus suggested that heparin chains have to be long enough to accommodate both antithrombin and the proteinase to accelerate the inactivation of some target enzymes, such as thrombin. For other target enzymes, however, binding of antithrombin to heparin appears to be sufficient for rate enhancement to occur.

Other evidence has supported the importance of a thrombin–heparin interaction in mediating heparin acceleration of the thrombin–antithrombin reaction. Thus, the minimum heparin chain length necessary to accelerate the thrombin–antithrombin reaction also corresponds to the smallest heparin chain able to bind both thrombin and antithrombin simultaneously (Danielsson et al., 1986). Moreover, heparin's accelerating effect can be abolished by chemical modification of lysine or arginine residues of thrombin without affecting the unaccelerated reaction rate, suggesting that an interaction of heparin with basic residues of the proteinase is involved in the rate enhancement (Pomerantz and Owen, 1978; Machovich et al., 1978). Kinetic studies have further suggested a role for a thrombin–heparin interaction in the heparin effect. Heparin acceleration of the reaction of antithrombin with thrombin (but not factor Xa) is thus diminished at high concentrations of polysaccharide in a manner consistent with the binding of thrombin and antithrombin to separate heparin chains (Jordan *et al.*, 1979, 1980b; Oosta *et al.*, 1981; Griffith, 1982b; Petersen and Jørgensen, 1983; Nesheim, 1983; Hoylaerts *et al.*, 1984; Olson, 1988). This observation implies that the binding of the proteinase and inhibitor to the same heparin molecule in a ternary complex is required for heparin rate enhancement. In keeping with this interpretation, active-site-blocked thrombin reduces heparin's rate-enhancing effect in a manner consistent with the inactive thrombin competing with the active enzyme for binding to heparin in a ternary thrombin–antithrombin–heparin complex (Griffith, 1982b; Pletcher and Nelsestuen, 1983; Hoylaerts *et al.*, 1984; Olson, 1988). The importance of thrombin binding to heparin in a ternary complex intermediate has also been deduced from the observation that the initial velocity of heparin-catalyzed thrombin inactivation by antithrombin is saturable with respect to the concentrations of both thrombin and antithrombin (Griffith, 1982c; Nesheim, 1983; Evington *et al.*, 1986a). Moreover, the K_m value for thrombin saturation decreases with increasing heparin chain length for heparin chains having a similar affinity for antithrombin (Hoylaerts *et al.*, 1984; Evington *et al.*, 1986a). Such a dependence of the thrombin K_m on heparin chain length can be explained by an increasing number of thrombin binding sites on larger heparin chains promoting the binding of thrombin first to heparin and then to specifically bound antithrombin by diffusion of the enzyme along the polysaccharide surface (Richter and Eigen, 1974; Winter *et al.*, 1981; Hoylaerts *et al.*, 1984). Rapid-kinetic studies have provided direct evidence for intermediate ternary thrombin–antithrombin–heparin complexes, in which thrombin and antithrombin are bound to heparin both at distant sites on the polysaccharide, where the proteins cannot interact, and at adjacent sites, where the inhibitor interacts with the active site of the enzyme (Olson and Björk, 1991). This observation is consistent with thrombin binding to antithrombin–heparin complex at heparin sites remote from antithrombin, followed by diffusion of the enzyme to a unique heparin site adjacent to the inhibitor. Heparin accelerating activity also increases as the charge density of the polysaccharide increases, despite a similar high affinity of the chains for antithrombin (Hurst *et al.*, 1983), consistent with an ionic thrombin–heparin interaction contributing to heparin rate enhancement.

The evidence that binding of both thrombin and antithrombin to heparin is essential for heparin acceleration of the thrombin–antithrombin reaction has suggested that the rate enhancement is due to the polysaccharide acting as a surface to approximate the proteinase and inhibitor (Olson and Shore, 1982; Griffith, 1982b; Nesheim, 1983; Hoylaerts *et al.*, 1984; Olson, 1988; Olson and Björk, 1991). Consistent with this mechanism, rapid-kinetic studies have shown that heparin accelerates the thrombin–antithrombin reaction by promoting the initial encounter between the two proteins in a ternary complex with heparin, rather than affecting the rate at which this intermediate complex is converted to the stable thrombin–antithrombin complex (Olson and Shore, 1982; Olson and Björk, 1991). An ~10,000 to ~1000-fold enhancement (at I 0.15 and 0.3, respectively) in the affinity of thrombin for antithrombin to form the encounter complex was thus observed when antithrombin was bound to heparin, as compared with when the inhibitor was free. In contrast, heparin had little effect on the 5 s^{-1} rate of conversion of the encounter complex to the stable complex. Other studies have shown that the interactions between thrombin, antithrombin, and heparin involved in assembling the ternary complex intermediate are linked (Olson, 1988). The binding of thrombin to the binary antithrombin–heparin complex to form the ternary complex thus was found to be much tighter than the binding of thrombin to either antithrombin or heparin alone in binary complexes, consistent with, but not proof of, thrombin binding to both antithrombin and heparin in the ternary complex.

　　2.2.3b. *Role of the Antithrombin Conformational Change.* Despite the evidence supporting an essential role for thrombin binding to heparin in the mechanism of the accelerating effect, several investigators have maintained that this binding contributes minimally to heparin rate enhancement. Instead, the conformational change induced in antithrombin by heparin binding has been claimed to be principally responsible for heparin's rate-enhancing effect (Rosenberg and Damus, 1973; Jordan *et al.*, 1980b; Oosta *et al.*, 1981; Carrell *et al.*, 1987). According to this view, the conformational change activates antithrombin to become a better inhibitor of its target proteinases by increasing the accessibility of the reactive-site bond to these enzymes. The requirement for a heparin chain length larger than that needed to bind antithrombin to accelerate thrombin inactivation has been rationalized by suggesting that the larger heparins induce an additional conformational change in the inhibitor that is needed for activation (Stone *et al.*, 1982; Gettins and Choay, 1989). It has also been proposed that the larger heparins simply act to neutralize the cationic change of thrombin and antithrombin to eliminate electrostatic repulsion between the activated inhibitor and the proteinase (Owen *et al.*, 1989a).

　　Spectroscopic evidence has been presented for an additional conformational change in antithrombin induced by heparins having the minimum chain length required for accelerating activity, but not by smaller heparins (Stone *et al.*, 1982; Gettins and Choay, 1989). A similar study, however, suggested that these small spectroscopic differences were not necessarily due to an additional conformational change, but rather might reflect a change in the electrostatic environment of those surface aromatic amino acids whose spectroscopic changes were being monitored (Lindahl *et al.*, 1984). The observation that high heparin concentrations, at which antithrombin and thrombin should be bound to separate heparin chains, still produce a substantial rate enhancement has also suggested a predominant contribution of the antithrombin conformational change to heparin rate enhancement (Jordan *et al.*, 1980b; Oosta *et al.*, 1981). However, the 700-fold rate enhancement observed at high heparin concentrations in one study (Jordan *et al.*, 1980b) was reduced to only 30-fold in a subsequent study when 10-fold higher heparin concentrations were examined (Hoylaerts *et al.*, 1984). Rather than reflecting the contribution of the antithrombin conformational change to heparin rate enhancement, this residual rate enhancement may instead be due to an approximation of thrombin and antithrombin on the same heparin chain that is mediated by the secondary binding site on thrombin for heparin demonstrated in thrombin–heparin binding studies (Naski *et al.*, 1990; Olson *et al.*, 1991). This secondary enzyme binding site would thus permit thrombin to bind to the heparin chain to which antithrombin is bound while being simultaneously bound to a separate heparin molecule.

Other studies which have considered the contribution of the antithrombin conformational change to heparin's accelerating effect have indicated that the conformational change plays little if any role in activating antithrombin for reaction with thrombin. Thus, chemical modification of tryptophan residues of antithrombin resulted in a series of derivatives whose heparin binding affinity and ability to undergo the conformational change, as inferred from spectroscopic changes accompanying the binding, were reduced to varying extents (Peterson and Blackburn, 1987b). Nevertheless, all derivatives were found to accelerate thrombin inactivation to the same extent when saturated with heparin, suggesting that the conformational change was not required for rate enhancement. However, the reduced spectroscopic changes accompanying heparin binding to the chemically modified inhibitors could be due to modification of tryptophan residues which report the conformational change (Shah *et al.*, 1990; Sun and Chang, 1989b), without affecting the conformational change itself. In another study, the contributions of thrombin binding to heparin and of the antithrombin conformational change to heparin promotion of the thrombin–antithrombin ternary complex interaction were resolved by quantitating the ionic and nonionic contributions to ternary complex formation (Olson and Björk, 1991). Rapid-kinetic studies demonstrated that formation of the ternary complex from thrombin and antithrombin–heparin complex was markedly dependent on salt in a manner which paralleled that of thrombin binding to heparin. Moreover, additive contributions of nonionic thrombin–antithrombin and ionic thrombin–heparin binary complex interactions accounted entirely for the binding energy of the thrombin ternary complex interaction. This implied that neither the antithrombin conformational change nor neutralization of the positive charge of thrombin by the binding of the enzyme to heparin had any significant effect on the interaction between thrombin and antithrombin in the ternary complex. Together, these results therefore indicate that heparin promotes the ternary complex encounter of antithrombin and thrombin primarily by approximating the inhibitor and proteinase on the polysaccharide surface.

Another study, in which the heparin pentasaccharide that specifically binds antithrombin and an 8000 M_r full-length heparin containing this sequence were compared with regard to their ability to induce the antithrombin conformational change and to accelerate the reactions of antithrombin with thrombin and factor Xa (Olson *et al.*, 1992), supported and extended these conclusions. The pentasaccharide thus was found to bind antithrombin and induce the conformational change in a manner comparable with that of the full-length heparin. Nevertheless, at physiological ionic strength and pH, the pentasaccharide maximally accelerated antithrombin inhibition of thrombin only 1.7-fold, as compared with the 4000-fold acceleration produced by the full-length heparin. Doubling the salt concentration had no effect on the pentasaccharide acceleration, but markedly decreased the acceleration by the full-length heparin to 200-fold when the polysaccharides were saturated with inhibitor. This confirmed that the rate enhancement by the full-length heparin predominantly was due to an ionic thrombin–heparin interaction promoting the binding of thrombin to antithrombin–heparin complex, with a minimal contribution of the antithrombin conformational change. In sharp contrast to the results obtained with thrombin, the pentasaccharide accelerated factor Xa inhibition by antithrombin a substantial 300-fold, with only a 2-fold greater acceleration (i.e., 600-fold) produced by the full-length heparin. Moreover, doubling the salt concentration had no effect on the pentasaccharide acceleration and decreased just the small additional rate enhancement of the full-length heparin (i.e., to 450-fold). These results are consistent with the heparin rate enhancement of the antithrombin–factor Xa reaction being primarily due to the antithrombin conformational change, with factor Xa binding to heparin making a small additional contribution with full-length heparins. In support of this conclusion, chemical modification of lysine residues of factor Xa eliminates the ability of the enzyme to bind to heparin, but minimally affects the extent of heparin rate enhancement of the antithrombin–factor Xa reaction (Owen and Owen, 1990). A contribution to heparin rate enhancement of factor Xa binding to heparin may, however, be significant for larger heparin chains (Danielsson et al., 1986; Ellis et al., 1986).

2.2.3c. Mechanism of Heparin Catalysis. Heparin is known to act catalytically in accelerating the reactions of antithrombin with its target enzymes (Björk and Nordenman, 1976; Jordan et al., 1979, 1980b; Griffith, 1982c; Pletcher and Nelsestuen, 1983; Evington et al., 1986a; Olson and Shore, 1986). This implies that heparin must dissociate once the stable complex between antithrombin and the proteinase has formed. Heparin indeed binds to the purified antithrombin–thrombin complex with an affinity that is substantially lower than that of antithrombin (Carlström et al., 1977; Jordan et al., 1979, 1980b), but comparable with that of thrombin (Nordenman and Björk, 1978a) or reactive-site-cleaved antithrombin (Björk and Fish, 1982), i.e., ~1000-fold lower than for the binding to antithrombin. These findings imply that a conformational change in the inhibitor and possibly also the proteinase has occurred to reduce their affinity for heparin. Quantitative release of heparin has been demonstrated to accompany the reaction of antithrombin–heparin complex with several target enzymes, including thrombin (Jordan et al., 1979, 1980b). Moreover, product inhibition by preformed antithrombin–thrombin complex was not observed during the heparin-accelerated thrombin–antithrombin reaction (Jordan et al., 1979; Olson and Shore, 1986), consistent with the weak affinity of the complex for heparin. Rapid-kinetic studies have shown that the enhanced protein fluorescence of the antithrombin–heparin complex that results from the antithrombin conformational change is quenched concomitant with the reaction of antithrombin with thrombin to form the stable complex (Olson and Shore, 1986). This suggested that heparin dissociation is either concerted with product formation or occurs in a subsequent more rapid step. Formation of the antithrombin–thrombin complex at a rate constant of 5 s^{-1} is therefore rate-limiting in a single heparin catalytic cycle, in agreement with k_{cat} values of 2–13 s^{-1} reported for heparin catalysis (Griffith, 1982c; Nesheim, 1983; Evington et al., 1986a). Moreover, these rapid-kinetic experiments suggested that the loss in heparin affinity of the reaction product is due to loss of the high-affinity heparin binding conformation of the inhibitor (Olson and Shore, 1986). This conclusion was supported by the observation that the affinity of heparin for the product antithrombin–thrombin complex is similar to that for antithrombin before the conformational change (Olson et al., 1981), as well as for reactive-site-cleaved antithrombin, which does not undergo the conformational change when heparin binds (Björk and Fish, 1982). A change in the heparin-induced conformation of antithrombin following its reaction with thrombin has also been demonstrated immunologically (Peterson and Blackburn, 1987a). Together, these results suggest that antithrombin conformational changes may serve an important function by modulating the heparin binding affinity during the reaction cycle.

2.2.3d. Effect of Heparin on the Substrate Pathway of the Antithrombin–Proteinase Reaction. The small amount of antithrombin that reacts as a normal substrate of thrombin, rather than as an inhibitor, to produce a free reactive-site-cleaved form of the protein (see Section 2.1.2) is increased in the presence of heparin (Fish *et al.*, 1979; Marciniak, 1981; Björk and Fish, 1982; Olson, 1985; Danielsson *et al.*, 1986). At physiological ionic strength and pH, up to 30% of the antithrombin reacting with thrombin in the presence of heparin is thus converted to free proteolytically modified inhibitor (Björk and Fish, 1982; Olson, 1985). The rate of this substrate reaction is accelerated by heparin to the same extent as that of the inhibition reaction, consistent with the products of the substrate and inhibition pathways arising from a common intermediate (Björk and Fish, 1982; Olson, 1985). The enhanced formation of proteolytically modified inhibitor by heparin is specific for heparin species that contain the pentasaccharide binding site for antithrombin and that have the minimum chain length required to accelerate the reaction of antithrombin with thrombin (Björk and Fish, 1982; Olson, 1985; Danielsson *et al.*, 1986). Moreover, the amount of reactive-site-cleaved antithrombin formed during the heparin-accelerated reaction increases with decreasing ionic strength, from 5% at I 0.3 to as much as 95% at I 0.025 (Olson, 1985). These results indicate that heparin binding to the intermediate where partitioning of the reaction occurs, which may be the acyl or tetrahedral intermediate, alters the partitioning in favor of the substrate pathway. The ionic strength dependence of the partitioning is consistent with the branchpoint intermediate having a weak heparin affinity comparable with that of the products of both reaction pathways. If the intermediate is converted to products when heparin is bound, as would be favored at low ionic strength, this conversion preferably occurs via the substrate pathway. However, if heparin dissociates from the intermediate prior to the conversion of the latter to products, which would be favored at physiological ionic strength, the intermediate breaks down preferably via the inhibition pathway, which dominates in the absence of heparin. The observation that the 5 s^{-1} rate constant for conversion of the ternary encounter complex to reaction products is independent of the partitioning between the two reaction pathways (Olson and Björk, 1991) also suggests that the 5 s^{-1} rate constant represents the rate-limiting formation of the branchpoint intermediate. The similarity between the effect of heparin on the partitioning of the reaction pathway and that of a monoclonal antibody directed against the epitope comprising residues 382 to 386 (Asakura *et al.*, 1990), as well as of mutations in this same peptide region (see Section 2.1.2), suggests that heparin may affect this region of the inhibitor. Indeed, as discussed previously, the main heparin binding site of the inhibitor which accommodates the pentasaccharide may extend toward the reactive bond region, so that longer heparin chains may be able to approach the proteinase binding domain (Huber and Carrell, 1989).

2.2.3e. General Mechanism of Heparin Action. Together, the available evidence suggests the following mechanism for heparin acceleration of the reaction of antithrombin with thrombin (Fig. 5). Thrombin, antithrombin, and heparin are first assembled into a productive ternary complex, in which both thrombin and antithrombin are bound to heparin, and an active-site-dependent interaction between the proteinase and inhibitor is established. This assembly occurs primarily by antithrombin binding to heparin before thrombin, due to the much greater affinity of antithrombin for heparin, although a pathway to the ternary complex through an intermediate thrombin–heparin binary complex is possible. The greater affinity of antithrombin for heparin results from the inhibitor binding to a specific pentasaccharide binding site, which causes antithrombin to adopt a high-affinity heparin-binding conformation. Thrombin initially binds to the antithrombin–heparin complex intermediate at any of a number of nonspecific sites on the polysaccharide, which are not contiguous with the bound inhibitor, and subsequently diffuses along the polysaccharide surface to encounter the specifically bound inhibitor. Alternatively, direct binding of thrombin to the unique heparin site adjacent to the bound inhibitor is possible, although statistically less likely. The net result of this process is the promotion by heparin of an intermediate thrombin–antithrombin interaction, mainly through an approximation of the two proteins on the heparin surface. At physiological ionic strength and pH, this approximation effect is evidenced by a nearly 10,000-fold higher affinity of thrombin for heparin-bound antithrombin than for the free inhibitor. The enhanced affinity is primarily due to the additional binding energy contributed by an ionic thrombin–heparin interaction, with the antithrombin conformational change making a relatively insignificant contribution to this enhanced affinity.

Following the encounter of thrombin and antithrombin in the ternary complex, the proteinase reacts with antithrombin, as described earlier for the reaction in the absence of heparin. Thus, an acyl or tetrahedral intermediate analogous to that formed with normal proteinase substrates presumably is formed at a rate of ~ 5 s^{-1}, similar to the rate in the absence of heparin. This step is accompanied by a return of antithrombin in this intermediate to a conformation with low affinity for heparin. Dissociation of heparin from the intermediate then favors a subsequent conformational change in antithrombin that traps the intermediate as a stable complex. If

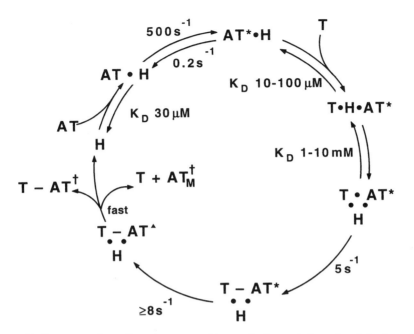

Figure 5. Sequence of reaction steps comprising a single catalytic cycle of the heparin-accelerated reaction between antithrombin and thrombin. Binding and kinetic parameters given for each step refer to values measured at I 0.15, pH 7.4, and 25°C. Noncovalent interactions are represented by dots and the putative covalent antithrombin–thrombin interaction by a line. The reaction cycle starts on the middle left-hand side of the figure with heparin (H) first binding to antithrombin (AT) and changing the inhibitor conformation from a low to a high heparin affinity state (denoted by *). Thrombin (T) subsequently binds to the heparin chain to form a ternary complex and then diffuses along the polysaccharide surface to encounter bound antithrombin. The enzyme–inhibitor encounter complex is next converted to an acyl–enzyme complex by attack of the active-site serine of the enzyme on the reactive bond of the inhibitor. The acylation step causes the inhibitor conformation to revert back from a high- to a low-affinity heparin binding state (denoted by ▲), which facilitates the rapid release of heparin from the acyl–enzyme intermediate. This promotes a further inhibitor conformational change (denoted by †) that traps the enzyme as a stable acyl–enzyme complex (T--AT†). However, a fraction of the acyl intermediate escapes this trapping by deacylating before the conformational change has occurred, thereby resulting in the release of free active enzyme and inactive inhibitor cleaved at the reactive bond ($AT_M^†$).

heparin remains bound to the intermediate, however, deacylation becomes the favored mode of breakdown to release free thrombin and proteolytically modified antithrombin. Heparin dissociation from the intermediate thus not only ensures that antithrombin will react as an inhibitor rather than a substrate of thrombin but also promotes the catalytic recycling of heparin.

The reaction pathway for the heparin-accelerated inhibition by anti-thrombin of other clotting proteinases which require the binding of both inhibitor and enzyme to the polysaccharide, i.e., factors IXa and XIa, is presumably similar to that of thrombin. For clotting proteinases whose heparin rate enhancement does not depend on enzyme binding to the polysaccharide, such as factors Xa, XIIa, and plasma kallikrein, the reaction pathway appears to be the same (Craig *et al.*, 1989), except that the antithrombin conformational change rather than a proteinase–heparin interaction predominantly promotes the binding of the enzyme to anti-thrombin in the ternary complex intermediate. The antithrombin conformational change may additionally accelerate the inhibition of these latter enzymes by increasing the rate of conversion of the ternary encounter complex to the stable antithrombin–proteinase complex.

2.3. Physiological Role of Antithrombin and Heparinlike Polysaccharides

Antithrombin is the major inhibitor of thrombin, factor IXa, and probably also factor Xa when these enzymes are added to plasma *in vitro* or injected *in vivo*. However, α_1-proteinase inhibitor, α_2-macroglobulin, and, in the case of thrombin, heparin cofactor II also contribute to the inhibition of these enzymes, although the contribution of the former two inhibitors is small or absent in the presence of heparin (J. L. Lane *et al.*, 1975; Vogel *et al.*, 1979; Tollefsen and Blank, 1981; Fuchs and Pizzo, 1983; Gitel *et al.*, 1984; Fuchs *et al.*, 1984; Jesty, 1986). In contrast, α_1-proteinase inhibitor and C1-inhibitor are predominantly responsible for the inhibition of factors XIIa, XIa, and kallikrein in plasma (Scott *et al.*, 1982; Schapira *et al.*, 1982; Pixley *et al.*, 1985). The enzymes of the later stages of the blood clotting cascade thus appear to be the primary targets of anti-thrombin, in agreement with the faster rates of inactivation of these enzymes in purified systems. However, antithrombin may also contribute to the regulation of factor XIa and kallikrein in the presence of heparinlike glycosaminoglycans, since high-molecular-weight kininogen increases the heparin rate enhancement of these reactions from 10- to 50-fold to $\sim$1000-fold (Olson, 1989; Olson and Shore, 1989). The observation that factor Xa is protected from antithrombin in the absence or presence of heparin when bound to the surface of activated platelets (Marciniak, 1973; Miletich *et al.*, 1978) has suggested that clotting enzymes may be resistant to inhibition at the site of vascular damage, where aggregation of platelets occurs. Antithrombin may thus act principally as a scavenger of clotting enzymes that have escaped into the circulation and may thereby contribute to the localization of clotting.

Major evidence for an important role of antithrombin in the *in vivo* regulation of blood clotting is provided by the clinical correlation between inherited or acquired deficiencies of the inhibitor and an increased risk of the affected individuals to develop thrombotic disease (Egeberg, 1965; Abildgaard, 1981). Inherited deficiencies can be due either to a low concentration of a normal inhibitor or to the presence of an abnormal protein. The molecular defects in several deficiencies of the former type have been elucidated. Substitution of Pro-407 by Leu (Bock *et al.*, 1988) or of Ala-404 by Thr (Bock *et al.*, 1989) in heterozygous individuals thus causes very low circulating levels of the gene product of the affected allele, resulting in about half the normal concentration of the inhibitor. It thus appears that the structure of the region of antithrombin around residue 405 (Fig. 2) is of importance for maintaining normal circulating levels of antithrombin. This region of the inhibitor has been shown to contain a recognition sequence which mediates the clearance of serpin–enzyme complexes by binding to a specific cell surface receptor (Perlmutter *et al.*, 1990). The recognition sequence, which is located in α_1-proteinase inhibitor within residues 359–374 (equivalent to residues 394–412 in antithrombin), i.e., in a region beginning with the P_1' residue and terminating 16–19 residues toward the C-terminal end of the inhibitor, does not appear to be accessible in the native inhibitor structure but becomes exposed during complex formation with a target enzyme; i.e., presumably following the cleavage of the P_1 and P_1' reactive bond residues of the inhibitor (see Section 2.1.2). Mutations in this C-terminal region of the inhibitor may therefore result in the exposure of the recognition site for the serpin–enzyme complex receptor already in the native inhibitor, thereby increasing the rate of inhibitor clearance and reducing the levels of circulating inhibitor. A reduced antithrombin level has also been shown to result from a frameshift mutation in codon 119, leading to premature termination of the protein synthesized from this allele (Olds *et al.*, 1990). Abnormal antithrombins can be defective either in the ability to inhibit proteinases or in heparin binding; several examples of both of these types have been given above in conjunction with the discussion of antithrombin function. Only heterozygous individuals with antithrombin defective in proteinase binding, who generally have thrombotic tendency, are known, indicating that complete loss of antithrombin activity may not be compatible with life. In contrast, individuals whose antithrombin is defective in heparin binding can be either heterozygous or homozygous. Although homozygous individuals with this defect usually develop thrombotic disorders, heterozygotes generally are asymptomatic (Sakuragawa *et al.*, 1983; Duchange *et al.*, 1987; Brunel *et al.*, 1987; Caso *et al.*, 1990), indicating that the heparin effect may only be of limited physiological importance.

The extravascular location of heparin and the failure to detect the polysaccharide in blood (Jacobsson and Lindahl, 1979) have suggested that heparin may not play a role in regulating thrombin or other coagulation proteinases during normal hemostasis. However, heparan sulfate, a related glycosaminoglycan on the surface of endothelial cells lining the blood vessel wall, binds antithrombin with high affinity and is active in accelerating antithrombin inactivation of its target enzymes, although the maximum rate enhancement *in vivo* has been reported to be only 10- to 16-fold (Marcum and Rosenberg, 1984; Marcum *et al.*, 1984; Stern *et al.*, 1985). About 1% of the vessel wall glycosaminoglycan chains possess the unique 3-*O*-sulfated glucosamine that is a marker of the pentasaccharide binding site for antithrombin and account for the anticoagulant activity of these glycosaminoglycans (Marcum *et al.*, 1986), implying a mechanism of action similar to that of heparin. These glycosaminoglycans have been suggested to be responsible for the nonthrombogenic properties of the blood vessel. However, contrasting studies have suggested that the observed acceleration of thrombin inactivation *in vivo* does not involve a heparinlike component but results from thrombin binding to its endothelial cell receptor, thrombomodulin (Lollar *et al.*, 1984; MacIntosh and Owen, 1987). The reason for the discrepant findings of these studies is not completely clear, but may reflect a considerably higher concentration of thrombomodulin in the region of the vasculature examined in the latter study. This could arise from the two regions of the circulation examined containing different contents of microcirculation, whose larger ratio of surface area to volume results in a much higher concentration of the thrombin receptor (Esmon, 1987). A likely conclusion from these studies is therefore that heparan sulfate chains on the vessel wall may act to accelerate the inhibition of free thrombin but not thrombomodulin-bound thrombin by antithrombin in the circulation.

While heparan sulfate may play a role in normal hemostasis, extravascular heparin may contribute to controlling the hemostatic process at the site of a vascular injury by promoting the scavenging by antithrombin of clotting enzymes that have escaped from their site of action into the extravascular space. Alternatively, the primary function of heparin may be to modulate cellular immune reactions by neutralizing an extravascular procoagulant system of macrophages that is activated by inflammatory stimuli (Lindahl *et al.*, 1982). These roles may be interrelated, given the link between the inflammatory response and the activation of blood coagulation (Esmon, 1987).

2.4. Modulators of the Reaction of Antithrombin and Heparin with Thrombin and Other Clotting Proteinases

2.4.1. Heparin Binding Proteins

Several proteins which are present in plasma or which are released by platelets activated at the site of a damaged vessel are known to bind heparin and thereby to antagonize heparin's accelerating effect on the reaction of antithrombin with its target enzymes. The most important of these, based on their heparin affinity and plasma concentration, are histidine-rich glycoprotein (Lijnen et al., 1983; Lane et al., 1986; Peterson et al., 1987a), platelet factor 4 (Handin and Cohen, 1976; Jordan et al., 1982), vitronectin or S protein (Preissner and Müller-Berghaus, 1986; Lane et al., 1987), fibronectin (Hayashi and Yamada, 1982), and high-molecular-weight kininogen (Björk et al., 1989). These proteins share the ability to completely neutralize the accelerating effect of heparin with high affinity for antithrombin. Moreover, this neutralization can be reversed by heparin having low affinity for the inhibitor, since the proteins all bind low- and high-affinity forms of heparin with comparable affinities. This effect presumably explains the observation that low-affinity heparin potentiates the anticoagulant effect of high-affinity heparin *in vivo* or in plasma, without having any significant anticoagulant activity by itself (Barrowcliffe et al., 1984). Low-affinity heparin thus can buffer the action of these proteins and contribute to the effectiveness of heparin or heparan sulfate *in vivo*. The neutralizing capacity of heparin-binding proteins is diminished or lost with smaller heparin oligosaccharides that are still able to bind antithrombin (Lane et al., 1986, 1987), suggesting that the primary mode of action of these proteins may be to compete for proteinase binding to heparin. However, antithrombin binding to heparin may also be blocked in the case of those proteins having an affinity comparable to that of antithrombin (Peterson et al., 1987a; Lane et al., 1987, Björk et al., 1989). The accelerating effect of heparan sulfate glycosaminoglycans with high affinity for antithrombin can also be neutralized by heparin binding proteins, platelet factor 4 and vitronectin apparently being much more effective than histidine-rich glycoprotein in this respect, of those proteins examined (Lane et al., 1987). Vitronectin may have a more complex mode of regulation of the thrombin–antithrombin reaction, since it binds thrombin as well as the thrombin–antithrombin complex (Ill and Ruoslahti, 1985; Podack et al., 1986) and may require proteolytic or effector activation to express its heparin neutralizing function (Preissner and Müller-Berghaus, 1987; Tomasini and Mosher, 1988).

2.4.2. Components of the Prothrombin Activation Complex and Products of Prothrombin Activation

Antithrombin inactivates thrombin generated from prothrombin and its enzyme activator complex, consisting of factors Xa, Va, Ca^{2+} and phospholipid, 2- to 4-fold slower than purified thrombin (Lindhout *et al.*, 1986; Schoen and Lindhout, 1987). This effect presumably results from the noncovalent association of the products of prothrombin activation, fragment 1.2 or fragment 2, with thrombin, since binding of the latter fragment to thrombin also reduces the rate of antithrombin inhibition by ~3-fold (Walker and Esmon, 1979). Covalent association of fragment 2 with thrombin in the catalytically active intermediate, meizothrombin (des fragment 1), results in an identical reduced rate of antithrombin inactivation (Walker and Esmon, 1979; Schoen and Lindhout, 1987). The heparin-accelerated reaction rate is also reduced when thrombin is generated from prothrombin and the factor Xa activator complex, but in this case, Ca^{2+} ions in addition to the prothrombin activation fragments appear to be involved in the effect (Schoen and Lindhout, 1987). Factor Va or phospholipid alone thus have no effect on the rate of inhibition, although platelets, which presumably provide the membrane surface for the *in vivo* reaction, have been reported to stimulate the inhibition (Jesty, 1986). The effect of Ca^{2+} was suggested to be due to the cation competing with thrombin for binding to heparin (Speight and Griffith, 1983; Schoen and Lindhout, 1987), although it is likely that Ca^{2+} could also affect antithrombin binding to the polysaccharide. Fragment 1 produced during the reaction may additionally bind heparin and neutralize its accelerating effect (Pieters *et al.*, 1987). Heparin does not accelerate the inactivation of meizothrombin (des fragment 1) by antithrombin (Schoen and Lindhout, 1987), presumably due to the absence of a heparin binding site in this thrombin precursor, suggesting that noncovalent association of fragment 2 with thrombin may block heparin's accelerating effect in a similar manner.

2.4.3. Thrombomodulin

The endothelial cell receptor for thrombin, thrombomodulin, has been reported to both stimulate and have no effect on the rate of inhibition of thrombin by antithrombin, depending on the species from which the receptor was isolated. Thus, rabbit lung thrombomodulin produces a 4- to 9-fold accelerating effect (Bourin *et al.*, 1986, 1988; Hofsteenge *et al.*, 1986; Preissner *et al.*, 1987, 1990), while thrombomodulin isolated from bovine lung or human placenta has little or no accelerating effect (Jakubowski *et al.*, 1986; Preissner *et al.*, 1990). The accelerating effect of rabbit thrombomodulin on thrombin inactivation by antithrombin, as well as its blocking of thrombin action on fibrinogen, requires a covalently linked glycosaminoglycan component (Bourin *et al.*, 1986, 1988; Preissner *et al.*, 1990) that has been identified as chondroitin sulfate (Bourin *et al.*, 1990). These thrombomodulin activities, but not the protein C-activating activity of the receptor, are strongly inhibited by the heparin-neutralizing proteins, platelet factor 4 and vitronectin, as well as by the polycations, protamine and polybrene, and weakly inhibited by histidine-rich glycoprotein (Bourin *et al.*, 1986, 1988; Preissner *et al.*, 1987, 1990). These observations have suggested that the glycosaminoglycan moiety of thrombomodulin exerts a heparinlike effect by approximation of thrombin and antithrombin on the polysaccharide surface (Bourin *et al.*, 1988, 1990). However, the comparatively weak accelerating effect of the isolated chondroitin sulfate component of thrombomodulin on antithrombin inhibition of thrombin (Bourin *et al.*, 1990) indicates that an alternative mode of action may be possible, in which thrombomodulin induces a conformational change in thrombin that alters the specificity of the proteinase toward its substrates and inhibitors (Bourin *et al.*, 1990; Preissner *et al.*, 1990). Tight association of vitronectin with human and bovine thrombomodulin may explain the weak or nonexistent acceleration of the thrombin–antithrombin reaction and the weaker inhibition of thrombin cleavage of fibrinogen shown by thrombomodulin isolated from these species (Preissner *et al.*, 1990). Binding of thrombin to thrombomodulin also decreases heparin's accelerating effect on the inactivation of the bound thrombin by antithrombin, possibly by decreasing the affinity of thrombin for heparin (Jakubowski *et al.*, 1986; Bourin, 1989). Alternatively, heparin may bind normally to thrombin when the latter is bound to thrombomodulin, but may no longer be able to bridge antithrombin and thrombin, similar to the effect of fibrin II on the heparin-accelerated reaction (Hogg and Jackson, 1989), as discussed below.

2.4.4. Fibrinogen, Fibrin I, and Fibrin II

The effects of fibrinogen and the products of the sequential cleavage of fibrinopeptides A and B from this protein substrate by thrombin, i.e., fibrin I and fibrin II, on the reaction of antithrombin and heparin with thrombin have also been studied. Plasma concentrations of fibrinogen reduce the rate of antithrombin inactivation of thrombin in the absence of heparin by about 2-fold in a manner consistent with fibrinogen acting as a pure competitive inhibitor of the reaction (Jesty, 1986). In contrast, fibrin I only partially competes with antithrombin or small peptide substrates in the reaction with thrombin, consistent with thrombin binding to fibrin I in two alternative modes, one in which the substrate recognition exosite and active site of the enzyme are both occupied by the fibrin substrate and the other in which only the exosite is occupied and the active site is free (Naski and Shafer, 1990). Antithrombin inhibits thrombin bound in the latter mode with at least a 3-fold greater rate than the free enzyme, suggesting that thrombin inhibition by antithrombin may be accelerated during the action of the enzyme on fibrinogen. Contrary to the stimulatory effect of fibrin I on the thrombin-antithrombin reaction, fibrin II binding to thrombin, which involves only the exosite of the enzyme, decreases the rate of thrombin inhibition, but only by 1.6-fold (Hogg and Jackson, 1989).

In contrast, the heparin-accelerated antithrombin–thrombin reaction is dramatically reduced 300- to 500-fold by fibrin II at a concentration corresponding to that of fibrinogen in plasma (Hogg and Jackson, 1989; Weitz *et al.*, 1990). This effect has been shown to be due to the formation of a thrombin–heparin–fibrin II ternary complex that presumably blocks the bridging or approximation of antithrombin and thrombin by the heparin molecule (Hogg and Jackson, 1990a). The effect of fibrin I on the heparin-accelerated reaction has not been examined. The formation of a thrombin–heparin–fibrin II ternary complex appears to require heparin chains substantially larger than those sufficient to accelerate the thrombin–antithrombin reaction, i.e., molecular weights of at least 11,000 or ~36 saccharides in length, and is independent of the affinity of these chains for antithrombin (Hogg and Jackson, 1990a). Heparan sulfate chains on the vessel wall were proposed to similarly form a ternary complex with fibrin II and thrombin, which would protect the enzyme from antithrombin inactivation at the site of action as well as prevent the diffusion of free thrombin from the clot (Hogg and Jackson, 1990b). A heparin–fibrin–thrombin ternary complex was also suggested to be responsible for the inability of clinically administered heparin to prevent rethrombosis following thrombolytic therapy.

3. HEPARIN COFACTOR II

3.1. Structure and Function

Briginshaw and Shanberge (1974) were the first to describe a second heparin-dependent inhibitor of thrombin in plasma. Tollefsen and Blank (1981) confirmed this observation by demonstrating that another thrombin–inhibitor complex, in addition to the thrombin–antithrombin complex, was formed when I^{125}-labeled thrombin was added to plasma containing heparin. Further evidence for this additional heparin-dependent inhibitor of thrombin came from observations that the level of heparin cofactor activity exceeded the level of antithrombin antigen in individuals with antithrombin deficiency (Friberger et al., 1982; Griffith et al., 1983). Subsequent purification and characterization of this new thrombin inhibitor, termed heparin cofactor II, showed that it was distinct from antithrombin and several other known proteinase inhibitors (Tollefsen et al., 1982). Other names that have been given to this inhibitor include heparin cofactor A (Briginshaw and Shanberge, 1974), antithrombin BM (Wunderwald et al., 1982), dermatan sulfate cofactor (Abildgaard and Larsen, 1984), and human leuserpin 2 (Ragg, 1986).

Heparin cofactor II is a single-chain glycoprotein with a molecular weight of 62,000–66,000 by sedimentation equilibrium analysis and with an isoelectric point of about 5.2 (Tollefsen et al., 1982; Tran et al., 1986). Its concentration in plasma is 0.1 mg/ml or 1.2 μM (Tollefsen and Pestka, 1985). The sequencing of cDNA clones has established that the mature protein consists of 480 amino acids and is synthesized with a 19-amino-acid signal peptide (Ragg, 1986; Inhorn and Tollefsen, 1986; Blinder et al., 1988). The protein shows 25–30% sequence homology with members of the serpin family of proteinase inhibitors with the greatest similarity occurring in the C-terminal two-thirds of the protein (Blinder et al., 1988). It contains three cysteine residues not involved in disulfide bonds (Church et al., 1987a) and three potential N-glycosylation sites (Blinder et al., 1988). A unique structural feature is the presence of two sulfated tyrosine residues, which are part of a short internal repeat rich in acidic residues in the nonhomologous N-terminal region of the protein (Hortin et al., 1986). Removal of the amino-terminal 47 residues of the inhibitor and possibly as much as 66 residues of the N-terminus appears to not affect the rate of thrombin inhibition in the absence or presence of glycosaminoglycans, suggesting that this region of the molecule is not required for inhibitor function (Griffith et al., 1985a; Pratt et al., 1990).

Like antithrombin, heparin cofactor II inhibits thrombin by forming a stable equimolar complex with the enzyme, in which the active site of the proteinase is blocked (Tollefsen et al., 1982). Moreover, this complex, like that of antithrombin with thrombin, is not dissociated by reduction or denaturation, but is dissociated at high pH, consistent with the complex containing a covalent ester linkage between the active-site serine of the enzyme and the P_1 residue of the inhibitor reactive bond (Tollefsen et al., 1982; Griffith et al., 1985a). Thrombin is inhibited by heparin cofactor II with a second-order rate constant at pH 7.4, 25°C of about $8 \times 10^3 \ M^{-1} \ s^{-1}$ in the absence of heparin and $7 \times 10^6 \ M^{-1} \ s^{-1}$ in the presence of optimal concentrations of the polysaccharide, i.e., the reaction is accelerated nearly 1000-fold by heparin (Tollefsen *et al.*, 1982). These rate constants are similar to those of antithrombin. The specificity of heparin cofactor II for serine proteinases is much more restricted than that of antithrombin. Thrombin is thus the only known clotting enzyme that is inhibited, whereas factors VIIa, IXa, Xa, XIa, XIIa, kallikrein, activated protein C, plasmin, urokinase, or tissue plasminogen activator are not significantly affected (Parker and Tollefsen, 1985). Heparin cofactor II additionally inhibits two enzymes, chymotrypsin and cathepsin G, that are not inhibited by antithrombin (Church *et al.*, 1985; Pratt *et al.*, 1990). However, inhibition of the latter enzymes is not accelerated by glycosaminoglycans.

The reactive bond of heparin cofactor II has been identified as that between Leu-444 and Ser-445 from the sequences of the identical C-terminal peptides released on reaction of the inhibitor with thrombin or chymotrypsin and of the one-residue-longer C-terminal peptide released by a snake venom proteinase (Griffith *et al.*, 1985a,b). This assignment was subsequently confirmed from the amino acid sequence derived from the cDNA sequence (Ragg, 1986; Inhorn and Tollefsen, 1986; Blinder *et al.*, 1988). A Leu-Ser reactive bond is also found in the serpin, α_1-antichymotrypsin (Morii and Travis, 1983), which explains the ability of heparin cofactor II to inhibit chymotrypsin and the chymotrypsinlike enzyme, cathepsin G. The inhibition of thrombin by heparin cofactor II, however, is paradoxical, given that thrombin only recognizes substrates with P_1 Arg bonds. It has been suggested that this ability to inactivate thrombin could be related to the P_2 Pro residue of the inhibitor, which is typically found in physiological substrates of thrombin, as well as in the Pro-Arg-Ser reactive-site sequence of a mutant α_1-antitrypsin which inhibits thrombin better than does antithrombin (Inhorn and Tollefsen, 1986). Synthetic substrates with a P_2 Pro residue are also better substrates of thrombin that those containing other P_2 residues (Lottenberg *et al.*, 1983). That the P_1 residue does contribute to the specificity of heparin cofactor II for its target enzymes, however, was shown by studies of a recombinant heparin cofactor II in which the P_1 Leu was mutated to Arg (Derechin *et al.*, 1990). This mutant inhibitor reacted with thrombin 100-fold faster than the wild-type inhibitor and also was able to inhibit factor Xa, kallikrein, and plasmin. Reaction of the mutant inhibitor with thrombin, however, was minimally accelerated by glycosaminoglycans.

Other evidence suggests that a secondary interaction between heparin cofactor II and thrombin, involving the macromolecular substrate recognition exosite of the proteinase, accounts for the ability of heparin cofactor II to inhibit thrombin, but not other clotting enzymes. The homology between an anionic peptide sequence in heparin cofactor II, containing the sulfated tyrosines, and the C-terminal end of the specific thrombin inhibitor, hirudin, which binds to the thrombin exosite, thus indicates that heparin cofactor II may similarly interact with this exosite (Hortin *et al.*, 1989; Ragg *et al.*, 1990). Consistent with this hypothesis, a synthetic peptide corresponding to the anionic sequence in heparin cofactor II was shown to inhibit thrombin cleavage of fibrinogen, as well as the binding of the C-terminal peptide of hirudin to thrombin, but not thrombin hydrolysis of small synthetic substrates (Hortin *et al.*, 1989). Moreover, the hirudin peptide inhibited the reaction of heparin cofactor II with thrombin. However, the heparin cofactor II peptide was found to stimulate the inhibitor–proteinase reaction probably because of its weaker thrombin affinity and heparinlike, polyanionic character.

3.2. Effect of Glycosaminoglycans on the Reaction with Thrombin

Heparin cofactor II differs from antithrombin in that its reaction with thrombin is accelerated both by heparin and by dermatan sulfate, whereas dermatan sulfate minimally accelerates the thrombin–antithrombin reaction (Tollefsen *et al.*, 1983). Heparan sulfate also accelerates the reaction of heparin cofactor II with thrombin, as it does the antithrombin–thrombin reaction, but chondroitin 4-sulfate, chondroitin 6-sulfate, keratan sulfate, and hyaluronic acid show little acceleration of both heparin cofactor II and antithrombin reactions with the enzyme (Tollefsen *et al.*, 1983). The dependence of the acceleration on the concentration of polysaccharide is bell-shaped with an optimum rate enhancement for both heparin and dermatan sulfate of ~1000-fold (Tollefsen *et al.*, 1983). The reduction in rate enhancement at higher glycosaminoglycan concentrations is consistent with a surface approximation mechanism of glycosaminoglycan acceleration, involving the binding of the inhibitor and proteinase to the same polysaccharide chain, similar to the mechanism of heparin's accelerating effect on the antithrombin–thrombin reaction. Consistent with this conclusion, heparin oligosaccharides at least 18–20 units long, i.e., of chain lengths similar to those necessary to accelerate the thrombin–antithrombin reaction (see Section 2.2.3), are required for the acceleration of the heparin cofactor II–thrombin reaction (Maimone and Tollefsen, 1988). Also, dermatan sulfate oligosaccharides 12–14 units long are necessary to accelerate the reaction with thrombin, even though only a hexasaccharide is required for the binding to heparin cofactor II (Tollefsen *et al.*, 1986; Maimone and Tollefsen, 1990). An approximation mechanism is also consistent with the observation that the initial velocity of the heparin cofactor II–thrombin reaction at catalytic levels of heparin is saturable with respect to both proteinase and inhibitor (Griffith, 1983).

Dermatan sulfate chains with low and high affinities for heparin cofactor II and corresponding low and high accelerating activities have been isolated by fractionation of oligosaccharide and full-length dermatan sulfate chains on immobilized heparin cofactor II (Griffith and Marbet, 1983; Tollefsen *et al.*, 1986). Chains with higher affinity and activity were found to have a higher negative charge (Tollefsen *et al.*, 1986). A specific dermatan sulfate hexasaccharide that contains six sulfate groups and binds heparin cofactor II with high affinity has been isolated (Fig. 3; Maimone and Tollefsen, 1990). Hexasaccharides containing five or four sulfate groups also bind to the inhibitor, albeit with weaker affinity, whereas those with three sulfate groups do not bind. While charge density thus appears to be an important factor for the binding of dermatan sulfate chains to heparin cofactor II, the positions of sulfation may also be important (Scully *et al.*, 1988). An analogous correlation between the charge density of heparin chains and their activity in accelerating heparin cofactor II inhibition of thrombin has been shown (Hurst *et al.*, 1983). Other natural and synthetic polyanions are also able to produce substantial accelerating effects on the reaction between heparin cofactor II and thrombin comparable to that of glycosaminoglycans, whereas such polyanions show only small accelerating effects on the antithrombin–thrombin reaction (Scully and Kakkar, 1984; Yamagishi *et al.*, 1984; Church *et al.*, 1987b, 1989a). Like dermatan sulfate, the anticoagulant activity of these polyanions is thus mediated primarily by heparin cofactor II (Tollefsen *et al.*, 1983; Scully and Kakkar, 1984; Church *et al.*, 1987b, 1989a). The ability of a polyanion to accelerate the heparin cofactor II–thrombin reaction may therefore depend more on its negative charge rather than on any specific structural features it may contain. This contrasts with the unique pentasaccharide structure in heparin that is required to significantly accelerate the antithrombin–thrombin reaction.

Heparin cofactor II binds heparin with a lower affinity than that of antithrombin (Griffith, 1983), and higher polysaccharide concentrations are therefore required to achieve a maximum rate of inhibition of thrombin (Tollefsen *et al.*, 1983). The dissociation constant for the interaction has been estimated to be > 30 μM by competitive binding studies (Verhamme *et al.*, 1989). The glycosaminoglycan binding domain of the inhibitor has been mapped by chemical modification and site-directed mutagenesis studies, as well as by identification of the conserved basic residues in heparin cofactor II and antithrombin. Chemical modification studies have indicated that lysine and arginine residues are involved in heparin binding, although the modified residues have not been identified (Church and Griffith, 1984; Church *et al.*, 1986). Site-directed mutagenesis studies have provided evidence that the serpin A and D helixes, which are implicated in heparin binding to antithrombin, may not both be involved in the binding of glycosaminoglycans to heparin cofactor II. Mutagenesis of Arg-103 in heparin cofactor II, the position homologous to Arg-47 in antithrombin, to Leu, Gln, or Trp thus did not affect the ability of the mutant proteins to bind to heparin–agarose, nor the ability of the Leu mutant to be accelerated by heparin or dermatan sulfate (Blinder and Tollefsen, 1990). These results indicated that Arg-103 is not involved in binding either glycosaminoglycan, suggesting that the A helix in heparin cofactor II may not be part of the glycosaminoglycan binding site, as it has been proposed to be in antithrombin. In contrast, the D helix of heparin cofactor II does appear to be involved in glycosaminoglycan binding. Thus, 8 of 15 residues in this helix, beginning with residue number 179, which is homologous to Ile-119 of antithrombin, are identical to those in antithrombin, including residues corresponding to Lys-125, Arg-129, and Arg-132 in the latter inhibitor (Carrell et al., 1987; Blinder et al., 1988). Moreover, Lys-133 in antithrombin is an Arg in heparin cofactor II. Mutagenesis of Lys-185 of heparin cofactor II, which is homologous to Lys-125 of antithrombin, to Met, Asn, or Thr resulted in a decreased affinity of the mutant inhibitors for heparin and a requirement of > 10-fold higher concentrations of the polysaccharide to accelerate the inhibition of thrombin (Blinder and Tollefsen, 1990). Moreover, the mutants were not accelerated by dermatan sulfate, but did exhibit a normal uncatalyzed rate of thrombin inhibition, indicating that Lys-185 is important for binding both glycosaminoglycans. Surprisingly, mutation of Arg-189 of heparin cofactor II to His, the residue which is homologous to Arg-129 of antithrombin, reduces the accelerating effect of dermatan sulfate, but not of heparin, on thrombin inhibition (Blinder et al., 1989), suggesting that different basic residues in the D helix may be involved in binding the two glycosaminoglycans. This conclusion is supported by other mutagenesis studies in which differential effects of mutation of Arg-184, Lys-185, Arg-192, and Arg-193 in the D helix on heparin and dermatan sulfate binding to heparin cofactor II, as well as on the acceleration by these glycosaminoglycans of the inhibitor reaction with thrombin, were observed (Ragg et al., 1990). Since Arg-129 of antithrombin is involved in binding heparin (see Section 2.2.1) while the homologous Arg-189 of heparin cofactor II is not, it also follows from these studies that heparin interacts differently with the D helix of the two inhibitors. This is likely to be a consequence of the sequence-specific interaction of heparin with antithrombin but not with heparin cofactor II.

Other site-directed mutagenesis studies have provided evidence that an N-terminal region of the inhibitor other than the one involved in binding glycosaminoglycans, is also important for glycosaminoglycan rate enhancement of the reaction with thrombin (Ragg et al., 1990). This other region includes the acidic peptide sequence that is homologous to the C-terminal sequence of the thrombin inhibitor, hirudin (Hortin *et al.,* 1989). Mutation in either N-terminal region this results in a reduction or loss of the glycosaminoglycan rate enhancement, while mutation in the basic glycosaminoglycan binding region additionally diminishes glycosaminoglycan–inhibitor interactions as discussed above (Ragg *et al.,* 1990). Based on these results, it was proposed that the basic region is involved in an intramolecular interaction with the acidic region and that glycosaminoglycan binding to the basic region disrupts this intramolecular interaction through a conformational change. As a result of this conformational change, the acidic region becomes exposed so that it can interact with the substrate recognition exosite of thrombin. However, the observation that the acidic peptide region contributes to the inhibition of thrombin in the absence of glycosaminoglycans (Hortin *et al.,* 1989) suggests that this region is available for interaction with the exosite of thrombin also in the native inhibitor. One possible rationalization of these findings is that a weak secondary interaction between the acidic region of heparin cofactor II and the thrombin exosite is greatly enhanced by an inhibitor conformational change triggered by glycosaminoglycan binding. The large increase in affinity of heparin cofactor II for thrombin in the presence of heparin would be consistent with this idea (Verhamme *et al.,* 1989). Thus, both surface approximation as well as an inhibitor conformational change may contribute to the accelerating effects of glycosaminoglycans as well as other polyanions on the inhibition of thrombin by heparin cofactor II.

3.3. Physiological Role

The physiological role of heparin cofactor II is not clear. Heparin cofactor II cannot substitute for antithrombin in individuals deficient in the latter inhibitor (Griffith *et al.,* 1983). An association between heparin cofactor II deficiency and thrombosis has been observed in a few cases (Sie *et al.,* 1985; Tran *et al.,* 1985). However, a determination of the importance of this inhibitor in regulating the activity of thrombin *in vivo* must await larger-scale studies in which the incidence of heparin cofactor II deficiency in healthy individuals and patients with thrombosis is compared. Although heparan sulfate molecules on the endothelial surface have been implicated in accelerating the reactions of antithrombin with its target enzymes, similar intravascular acceleration of the reaction of heparin cofactor II with thrombin does not appear to be of importance (MacIntosh and Owen, 1987). However, dermatan sulfate glycosaminoglycans on fibroblasts have been shown to accelerate the inactivation of thrombin by heparin cofactor II (McGuire and Tollefsen, 1987), suggesting a possible role for the inhibitor in regulating thrombin activity in the extravascular space at the site of blood vessel damage. Platelet factor 4 decreases heparin and dermatan sulfate acceleration of the reaction of heparin cofactor II with thrombin (Tollefsen and Pestka, 1985), suggesting that thrombin may be protected from inhibition at sites of platelet activation. Thrombin similarly is protected   from heparin cofactor II inhibition when bound to thrombomodulin (Jakubowski et al., 1986; Preissner et al., 1987). The recent observation that neutrophil proteinases, which participate in tissue remodeling at sites of vessel injury, release N-terminal peptides from heparin cofactor II with potent chemotactic activity for neutrophils, suggests that heparin cofactor II may play a role in mediating the inflammatory response to injury (Hoffman et al., 1989; Pratt et al., 1990).

4. REFERENCES

Abildgaard, U., 1969, Binding of thrombin to antithrombin III, Scand. J. Clin. Lab. Invest. 24:23–27.
Abildgaard, U., 1981, Antithrombin and related inhibitors of blood coagulation, Recent Adv. Blood Coag. 3:151–173.
Abildgaard, U., and Larsen, M. L., 1984, Assay of dermatan sulfate cofactor (heparin cofactor II) activity in human plasma, Thromb. Res. 35:257–266.
Aiach, M., Nora, M., Feissinger, J. N., Roncata, M., Francois, D., and Alhenc-Gelas, M., 1985, A functional abnormal antithrombin III (AT III) deficiency: AT III Charleville, Thromb. Res. 39:559–570.
Andersson, L. O., Barrowcliffe, T. W., Holmer, E., Johnson, E. A., and Sims, G. E. C., 1976, Anticoagulant properties of heparin fractionated by affinity chromatography on matrix-bound antithrombin III and by gel filtration, Thromb. Res. 9:575–583.
Asakura, S., Hirata, H., Okazaki, H., Hashimoto-Gotoh, T., and Matsuda, M., 1990, Hydrophobic residues 382–386 of antithrombin III, Ala-Ala-Ala-Ser-Thr, serve as the epitope for an antibody which facilitates hydrolysis of the inhibitor by thrombin, J. Biol. Chem. 265:5135–5138.
Atha, D. H., Stevens, A. W., and Rosenberg, R. D., 1984, Evaluation of critical groups required for the binding of heparin to antithrombin, Proc. Natl. Acad. Sci. USA 81:1030–1034.
Atha, D. H., Lormeau, J. C., Petitou, M., Rosenberg, R. D., and Choay, J., 1985, Contribution of monosaccharide residues in heparin binding to antithrombin III, Biochemistry 24:6723–6729.
Barrowcliffe, T. W., Merton, R. E., Havercroft, S. J., Thunberg, L., Lindahl, U., and Thomas, D. P., 1984, Low-affinity heparin potentiates the action of high-affinity heparin oligosaccharides, Thromb. Res. 34:125–133.
Beeler, D. L., Marcum, J. A., Schiffman, S., and Rosenberg, R. D., 1986, Interaction of factor XIa and antithrombin in the presence and absence of heparin, Blood 67:1488–1492.
Björk, I., and Fish, W. W., 1982, Production in vitro and properties of a modified form of bovine antithrombin, cleaved at the active site by thrombin, J. Biol. Chem. 257:9487–9493.
Björk, I., and Lindahl, U., 1982, Mechanism of the anticoagulant action of heparin, Mol. Cell. Biochem. 48:161–182.
Björk, I., and Nordenman, B., 1976, Acceleration of the reaction between thrombin and

antithrombin III by non-stoichiometric amounts of heparin, *Eur. J. Biochem.* **68:**507–511.

Björk, I., Danielsson, Å., Fenton, J. W., II, and Jörnvall, H., 1981, The site in human antithrombin for functional proteolytic cleavage by human thrombin, *FEBS Lett.* **126:**257–260.

Björk, I., Jackson, C. M., Jörnvall, H., Lavine, K. K., Nordling, K., and Salsgiver, W. J., 1982, The active site of antithrombin. Release of the same proteolytically cleaved form of the inhibitor from complexes with factor IXa, factor Xa and thrombin, *J. Biol. Chem.* **257:**2406–2411.

Björk, I., Olson, S. T., Sheffer, R. G., and Shore, J. D., 1989, Binding of heparin to human high molecular weight kininogen, *Biochemistry* **28:**1213–1221.

Blackburn, M. N., Smith, R. L., Carson, J., and Sibley, C. C., 1984, The heparin-binding site of antithrombin III. Identification of a critical tryptophan in the amino acid sequence, *J. Biol. Chem.* **259:**939–941.

Blinder, M. A., and Tollefsen, D. M., 1990, Site-directed mutagenesis of arginine 103 and lysine 185 in the proposed glycosaminoglycan-binding site of heparin cofactor II, *J. Biol. Chem.* **265:**286–291.

Blinder, M. A., Marasa, J. C., Reynolds, C. H., Deavan, L. L., and Tollefsen, D. M., 1988, Heparin cofactor II: cDNA sequence, chromosome localization, restriction fragment length polymorphism, and expression in *Escherichia coli, Biochemistry* **27:**752–759.

Blinder, M. A., Andersson, T. R., Abildgaard, U., and Tollefsen, D. M., 1989, Heparin cofactor II$_{Oslo}$. Mutation of Arg-189 to His decreases the affinity for dermatan sulfate, *J. Biol. Chem.* **264:**5128–5133.

Bock, S. C., Wion, K. L., Vehar, G. A., and Lawn, R. M., 1982, Cloning and expression of the cDNA for human antithrombin III, *Nucleic Acids Res.* **10:**8113–8125.

Bock, S. C., Marrinan, J. A., and Radziejewska, E., 1988, Antithrombin III Utah: Proline-407 to leucine mutation in a highly conserved region near the inhibitor reactive site, *Biochemistry* **27:**6171–6178.

Bock, S. C., Silbermann, J. A., Wikoff, W., Abildgaard, U., and Hultin, M. B., 1989, Identification of a threonine for alanine substitution at residue 404 of antithrombin III Oslo suggests integrity of the 404–407 region is important for maintaining normal plasma inhibitor levels, *Thromb. Haemostas.* **62:**494A.

Bode, W., Mayr, I., Baumann, U., Huber, R., Stone, S. R., and Hofsteenge, J., 1989, The refined 1.9 Å crystal structure of human α-thrombin: Interaction with D-Phe-Pro-Arg chloromethylketone and significance of the Tyr-Pro-Pro-Trp insertion segment, *EMBO J.* **8:**3467–3475.

Borg, J. Y., Owen, M. C., Soria, C., Soria, J., Caen, J., and Carrell, R. W., 1988, Proposed heparin binding site in antithrombin based on arginine 47. A new variant Rouen-II, 47 Arg to Ser, *J. Clin. Invest.* **81:**1292–1296.

Borg, J. Y., Brennan, S. O., Carrell, R. W., George, P., Perry, D. J., and Shaw, J., 1990, Antithrombin Rouen-IV, 24 Arg → Cys. The amino-terminal contribution to heparin binding, *FEBS Lett.* **266:**163–166.

Borsodi, A. D., and Bradshaw, R. A., 1977, Isolation of antithrombin III from normal and α$_1$-antitrypsin-deficient human plasma, *Thromb. Haemostas.* **38:**475–485.

Bourin, M. C., 1989, Effect of rabbit thrombomodulin on thrombin inhibition by antithrombin in the presence of heparin, *Thromb. Res.* **54:**27–39.

Bourin, M. C., Boffa, M. C., Björk, I., and Lindahl, U., 1986, Functional domains of rabbit thrombomodulin, *Proc. Natl. Acad. Sci. USA* **83:**5924–5928.

Bourin, M. C., Ohlin, A. K., Lane, D. A., Stenflo, J., and Lindahl, U., 1988, Relationship between anticoagulant activities and polyanionic properties of rabbit thrombomodulin, *J. Biol. Chem.* **263**:8044–8052.

Bourin, M. C., Lundgren-Åkerlund, E., and Lindahl, U., 1990, Isolation and characterization of the glycosaminoglycan component of rabbit thrombomodulin proteoglycan, *J. Biol. Chem.* **265**:15424–15431.

Brennan, S. O., George, P. M., and Jordan, R. E., 1987, Physiological variant of antithrombin-III lacks carbohydrate sidechain at Asn 135, *FEBS Lett.* **219**:431–436.

Brennan, S. O., Borg, J. Y., George, P. M., Soria, C., Soria, J., Caen, J., and Carrell, R. W., 1988, New carbohydrate site in mutant antithrombin (7 Ile → Asn) with decreased heparin affinity, *FEBS Lett.* **237**:118–122.

Briginshaw, G. F., and Shanberge, J. N., 1974, Identification of two distinct heparin cofactors in human plasma. Separation and partial purification, *Arch. Biochem. Biophys.* **161**:683–690.

Brunel, F., Duchange, N., Fischer, A. M., Cohen, G. N., and Zakin, M. M., 1987, Antithrombin III Alger: A new case of Arg 47 → Cys mutation, *Am. J. Hematol.* **25**:223–224.

Carlson, T. H., and Atencio, A. C., 1982, Isolation and partial characterization of two distinct types of antithrombin III from rabbit, *Thromb. Res.* **27**:23–34.

Carlström, A. S., Liedén, K., and Björk, I., 1977, Decreased binding of heparin to antithrombin following the interaction between antithrombin and thrombin, *Thromb. Res.* **11**:785–797.

Carrell, R. W., and Boswell, D. R., 1986, Serpins: The superfamily of plasma serine proteinase inhibitors, in: *Proteinase Inhibitors* (A. J. Barrett and G. Salvesen, eds.), Elsevier, Amsterdam, pp. 403–420.

Carrell, R. W., and Owen, M. C., 1985, Plakalbumin, α_1-antitrypsin, antithrombin and the mechanism of inflammatory thrombosis, *Nature* **317**:730–732.

Carrell, R. W., Boswell, D. R., Brennan, S. O., and Owen, M. C., 1980, Active site of α_1-antitrypsin: Homologous site in antithrombin III, *Biochem. Biophys. Res. Commun.* **91**:399–402.

Carrell, R. W., Christey, P. B., and Boswell, D. R., 1987, Serpins: Antithrombin and other inhibitors of coagulation and fibrinolysis. Evidence from amino acid sequences, in: *Thrombosis and Haemostasis* (M. Verstraete, J. Vermylen, H. R. Lijnen, and J. Arnout, eds.), Leuven University Press, Leuven, Belgium, pp. 1–15.

Caso, R., Lane, D. A., Thompson, E., Zangouras, D., Panico, M., Morris, H., Olds, R. J., Thein, S. L., and Girolami, A., 1990, Antithrombin Padua I: Impaired heparin binding caused by an Arg47 to His (CGT to CAT) substitution, *Thromb. Res.* **58**:185–190.

Caso, R., Lane, D. A., Thompson, E. A., Olds, R. J., Thein, S. L., Panico, M., Blench, I., Morris, H. R., Freyssinet, J. M., Aiach, M., Rodeghiero, F., and Finazzi, G., 1991, Antithrombin Vicenza, Ala 384 to Pro (GCA to CCA) mutation, transforming the inhibitor into a substrate, *Br. J. Haematol.* **77**:87–92.

Casu, B., Oreste, P., Torri, G., Zopetti, G., Choay, J., Lormeau, J. C., Petitou, M., and Sinaÿ, P., 1981, The structure of heparin oligosaccharide fragments with high anti-(factor Xa) activity containing the minimal antithrombin III binding sequence, *Biochem. J.* **197**:599–609.

Chandra, T., Stackhouse, R., Kidd, V. J., and Woo, S. L. C. 1983, Isolation and sequence characterization of a cDNA clone of human antithrombin, *Proc. Natl. Acad. Sci. USA* **80**:1845–1848.

Chang, J. Y., 1989, Binding of heparin to human antithrombin III activates selective chemical modification at lysine 236. Lys-107, lys-125, and lys-136 are situated within the heparin-binding site of antithrombin III, *J. Biol. Chem.* **264**:3111–3115.

Chang, J. Y., and Tran, T. H., 1986, Antithrombin III Basel. Identification of a Pro-Leu substitution in a hereditary abnormal antithrombin with impaired heparin cofactor activity, *J. Biol. Chem.* **261**:1174–1176.

Choay, J., Petitou, M., Lormeau, J.C., Sinaÿ, P., Casu, B., and Gatti, G., 1983, Structure–activity relationship in heparin: A synthetic pentasaccharide with high affinity for antithrombin III and eliciting high anti-factor Xa activity, *Biochem. Biophys. Res. Commun.* **116**:492–499.

Church, F. C., and Griffith, M. J., 1984, Evidence for essential lysines in heparin cofactor II, *Biochem. Biophys. Res. Commun.* **124**:745–751.

Church, F. C., Noyes, C. M., and Griffith, M. J., 1985, Inhibition of chymotrypsin by heparin cofactor II, *Proc. Natl. Acad. Sci. USA* **82**:6431–6434.

Church, F. C., Villanueva, G. B., and Griffith, M. J., 1986, Structure–function relationships in heparin cofactor II: Chemical modification of arginine and tryptophan and demonstration of a two-domain structure, *Arch. Biochem. Biophys.* **246**:175–184.

Church, F. C., Meade, J. B., and Pratt, C. W., 1987a, Structure–function relationships in heparin cofactor II: Spectral analysis of aromatic residues and absence of a role for sulfhydryl groups in thrombin inhibition, *Arch. Biochem. Biophys.* **259**:331–340.

Church, F. C., Treanor, R. E., Sherrill, G. B., and Whinna, H. C., 1987b, Carboxylate polyanions accelerate inhibition of thrombin by heparin cofactor II, *Biochem. Biophys. Res. Commun.* **148**:362–368.

Church, F. C., Meade, J. B., Treanor, R. E., and Whinna, H. C., 1989a, Antithrombin activity of fucoidan. The interaction of fucoidan with heparin cofactor II, antithrombin III, and thrombin, *J. Biol. Chem.* **264**:3618–3623.

Church, F. C., Pratt, C. W., Noyes, C. M., Kalayanamit, T., Sherrill, G. B., Tobin, R. B., and Meade, J. B., 1989b, Structure and functional properties of human α-thrombin, phosphopyridoxylated α-thrombin and γ_T-thrombin. Identification of lysyl residues in α-thrombin that are critical for heparin and fibrin(ogen) interactions, *J. Biol. Chem.* **264**:18419–18425.

Collen, D., and DeCock, F., 1978, Neoantigenic expression in enzyme–inhibitor complexes. A means to demonstrate activation of enzyme systems, *Biochim. Biophys. Acta* **525**:287–290.

Collen, D., Schetz, J., DeCock, F., Holmer, E., and Verstraete, M., 1977, Metabolism of antithrombin III (heparin cofactor) in man: Effects of venous thrombosis and of heparin administration, *Eur. J. Clin. Invest.* **7**:27–35.

Conard, J., Brosstad, F., Larsen, M. L., Samama, M., and Abildgaard, U., 1983, Molar antithrombin concentration in normal human plasma, *Haemostasis* **13**:363–368.

Craig, P. A., Olson, S. T., and Shore, J. D., 1989, Transient kinetics of heparin-catalyzed protease inactivation by antithrombin III. Characterization of assembly, product formation, and heparin dissociation steps in the factor Xa reaction, *J. Biol. Chem.* **264**:5452–5461.

Daly, M., Ball, R., O'Meara, A., and Hallinan, F. M., 1989, Identification and characterization of an antithrombin III mutant (AT Dublin 2) with marginally decreased heparin reactivity, *Thromb. Res.* **56**:503–513.

Danielsson, Å., and Björk, I., 1978, The binding of low-affinity and high-affinity heparin to antithrombin. Competition for the same binding site on the protein, *Eur. J. Biochem.* **90**:7–12.

Danielsson, Å., and Björk, I., 1980, Slow, spontaneous dissociation of the antithrombin–thrombin complex produces a proteolytically modified form of the inhibitor, *FEBS Lett.* **119**:241–244.

Danielsson, Å., and Björk, I., 1981, Binding to antithrombin of heparin fractions with different molecular weights, *Biochem. J.* **193**:427–433.

Danielsson, Å., and Björk, I., 1982, Mechanism of inactivation of trypsin by antithrombin, *Biochem. J.* **207**:21–28.

Danielsson, Å., and Björk, I., 1983, Properties of antithrombin–thrombin complex formed in the presence and in the absence of heparin, *Biochem. J.* **213**:345–353.

Danielsson, Å., Raub, E., Lindahl, U., and Björk, I., 1986, Role of ternary complexes, in which heparin binds both antithrombin and proteinase, in the acceleration of the reactions between antithrombin and thrombin or factor Xa, *J. Biol Chem.* **261**:15467–15473.

Derechin, V. M., Blinder, M. A., and Tollefsen, D. M., 1990, Substitution of arginine for Leu[444] in the reactive site of heparin cofactor II enhances the rate of thrombin inhibition, *J. Biol. Chem.* **265**:5623–5628.

Devraj-Kizuk, R., Chui, D. H. K., Prochownik, E. V., Carter, C. J., Ofosu, F. A., and Blajchman, M. A., 1988, Antithrombin-III-Hamilton: A gene with a point mutation (guanine to adenine) in codon 382 causing impaired serine protease reactivity, *Blood* **72**:1518–1523.

Duchange, N., Chassé, J. F., Cohen, G. N., and Zakin, M. N., 1987, Molecular characterization of the antithrombin III Tours deficiency, *Thromb. Res.* **45**:115–121.

Egeberg, O., 1965, Inherited antithrombin deficiency causing thrombophilia, *Thromb. Diath. Haemorrh.* **13**:516–530.

Einarsson, R., and Andersson, L. O., 1977, Binding of heparin to human antithrombin III as studied by measurements of tryptophan fluorescence, *Biochim. Biophys. Acta* **490**:104–111.

Ellis, V., Scully, M. F., and Kakkar, V. V. 1986, The relative molecular mass dependence of the anti-factor Xa properties of heparin, *Biochem. J.* **238**:329–333.

Engh, R. A., Wright, H. T., and Huber, R., 1990, Modeling the intact form of the α_1-proteinase inhibitor, *Protein Eng.* **3**:469–477.

Epstein, I. R., 1978, Cooperative and noncooperative binding of large ligands to a finite one-dimensional lattice. A model for ligand–oligonucleotide interactions, *Biophys. Chem.* **8**:327–339.

Erdjument, H., Lane, D. A., Ireland, H., Panico, M., Di Marzo, V., Blench, I., and Morris, H. R. 1987, Formation of a covalent disulfide-linked antithrombin–albumin complex by an antithrombin variant, antithrombin "Northwick Park," *J. Biol. Chem.* **262**:13381–13384.

Erdjument, H., Lane, D. A., Panico, M., Di Marzo, V., and Morris, H. R., 1988a, Single amino acid substitutions in the reactive site of antithrombin leading to thrombosis. Congenital substitution of arginine 393 to cysteine in antithrombin Northwick Park and to histidine in antithrombin Glasgow, *J. Biol. Chem.* **263**:5589–5593.

Erdjument, H., Lane, D. A., Ireland, H., Di Marzo, V., Panico, M., Morris, H. R., Tripodi, A., and Manucci, P. M., 1988b, Antithrombin Milano, single amino acid substitution at the reactive site, Arg393 to Cys, *Thromb. Haemostas.* **60**:471–475.

Erdjument, H., Lane, D. A., Panico, M., Di Marzo, V., Morris, H. R., Bauer, K., and Rosenberg, R. D., 1989, Antithrombin Chicago, amino acid substitution of arginine 393 to histidine, *Thromb. Res.* **54**:613–619.

Esmon, C. T., 1987, The regulation of natural anticoagulant pathways, *Science* **235**:1348–1352.

Evington, J. R. N., Feldman, P. A., Luscombe, M., and Holbrook, J. J., 1986a, The catalysis by heparin of the reaction between thrombin and antithrombin, *Biochim. Biophys. Acta* **870**:92–101.

Evington, J. R. N., Feldman, P. A., Luscombe, M., and Holbrook, J. J., 1986b, Multiple complexes of thrombin and heparin, *Biochim. Biophys. Acta* **871**:85–92.

Fenton, J. W., II, Witting, J. I., Pouliott, C., and Fareed, J., 1989, Thrombin anion-binding exosite interactions with heparin and various polyanions, *Ann. N.Y. Acad. Sci.* **556**:158–165.

Ferguson, W. S., and Finlay, T. H., 1983a, Formation and stability of the complex formed between human antithrombin III and thrombin, *Arch. Biochem. Biophys.* **220**:301–308.

Ferguson, W. S., and Finlay, T. H., 1983b, Localization of the disulfide bond in human antithrombin III required for heparin-accelerated thrombin inactivation, *Arch. Biochem. Biophys.* **221**:304–307.

Fish, W. W., and Björk, I., 1979, Release of a two-chain form of antithrombin from the antithrombin–thrombin complex, *Eur. J. Biochem.* **101**:31–38.

Fish, W. W., Orre, K., and Björk, I., 1979, The production of an inactive form of antithrombin through limited proteolysis by thrombin, *FEBS Lett.* **98**:103–106.

Fish, W. W., Danielsson, Å., Nordling, K., Miller, S. H., Lam, C. F., and Björk, I., 1985, The denaturation behaviour of antithrombin in guanidinium chloride. Irreversibility of unfolding caused by aggregation, *Biochemistry* **24**:1510–1517.

Franzén, L. E., Svensson, S., and Larm, O., 1980, Structural studies on the carbohydrate portion of human antithrombin III, *J. Biol. Chem.* **255**:5090–5093.

Friberger, P., Egberg, N., Holmer, E., Hellgren, M., and Blombäck, M., 1982, Antithrombin assay—The use of human or bovine thrombin and the observation of a 'second' heparin cofactor, *Thromb. Res.* **25**:433–436.

Fuchs, H. E., and Pizzo, S. V., 1983, Regulation of factor Xa in vitro in human and mouse plasma and in vivo in the mouse. Role of the endothelium and plasma proteinase inhibitors, *J. Clin. Invest.* **72**:2041–2049.

Fuchs, H. E., Trapp, H. G., Griffith, M. J., Roberts, H. R., and Pizzo, S. V., 1984, Regulation of factor IXa in vitro in human and mouse plasma and in vivo in the mouse. Role of the endothelium and the plasma proteinase inhibitors, *J. Clin. Invest.* **73**:1696–1703.

Furugren, B., Andersson, L. O., and Einarsson, R., 1977, Small-angle X-ray scattering studies on human antithrombin III and its complex with heparin, *Arch. Biochem. Biophys.* **178**:419–424.

Gandrille, S., Aiach, M., Lane, D. A., Vidaud, D., Molho-Sabatier, P., Caso, R., de Moerloose, P., Fiessinger, J. N., and Clauser, E., 1990, Important role of Arginine 129 in heparin binding site of antithrombin III: Identification of a novel mutation Arginine 129 to Glutamine, *J. Biol. Chem.* **267**:18997–19001.

Gettins, P., 1987, Antithrombin III and its interaction with heparin. Comparison of the human, bovine and porcine proteins by 1H NMR spectroscopy, *Biochemistry* **26**:1391–1398.

Gettins, P., and Choay, J., 1989, Examination, by ^{1}H-N.M.R. spectroscopy, of the binding of a synthetic, high-affinity heparin pentasaccharide to human antithrombin III, *Carbohydr. Res.* **185**:69–76.

Gettins, P., and Harten, B., 1988, Properties of thrombin- and elastase modified human antithrombin III, *Biochemistry* **27**:3634–3639.

Gettins, P., and Wooten, E. S., 1987, On the domain structure of antithrombin III. Localization of the heparin-binding region using ^{1}H NMR spectroscopy, *Biochemistry* **26**:4403–4408.

Gitel, S. N., Medina, V. M., and Wessler, S., 1984, Inhibition of human activated factor X by antithrombin and α_1-proteinase inhibitor in human plasma, *J. Biol. Chem.* **259:**6890–6895.

Griffith, M. J., 1982a, Measurement of the heparin-enhanced antithrombin III/thrombin reaction rate in the presence of synthetic substrate, *Thromb Res.* **25:**245–253.

Griffith, M. J., 1982b, Kinetics of the heparin-enhanced antithrombin III/thrombin reaction. Evidence for a template model for the mechanism of action of heparin, *J. Biol. Chem.* **257:**7360–7365.

Griffith, M. J., 1982c, The heparin-enhanced antithrombin III/thrombin reaction is saturable with respect to both thrombin and antithrombin III, *J. Biol. Chem.* **257:**13899–13902.

Griffith, M. J., 1983, Heparin-catalyzed inhibitor–protease reactions: Kinetic evidence for a common mechanism of action of heparin, *Proc. Natl. Acad. Sci. USA* **80:**5460–5464.

Griffith, M. J., and Marbet, G. A., 1983, Dermatan sulfate and heparin can be fractionated by affinity for heparin cofactor II, *Biochem. Biophys. Res. Commun.* **112:**663–670.

Griffith, M. J., Carraway, T., White, G. C., and Dombrose, F. A., 1983, Heparin cofactor activities in a family with hereditary antithrombin III deficiency: Evidence for a second heparin cofactor in plasma, *Blood* **61:**111–118.

Griffith, M. J., Noyes, C. M., and Church, F. C., 1985a, Reactive site peptide structural similarity between heparin cofactor II and antithrombin III, *J. Biol. Chem* **260:**2218–2225.

Griffith, M. J., Noyes, C. M., Tyndall, J. A., and Church, F. C., 1985b, Structural evidence for leucine at the reactive site of heparin cofactor II, *Biochemistry* **24:**6777–6782.

Handin, R. I., and Cohen, H. J., 1976, Purification and binding properties of human platelet factor four, *J. Biol. Chem.* **251:**4273–4282.

Hayashi, M., and Yamada, K. M., 1982, Divalent cation modulation of fibronectin binding to heparin and to DNA, *J. Biol. Chem.* **257:**5263–5267.

Hoffman, M., Pratt, C. W., Brown, R. L., and Church, F. C., 1989, Heparin cofactor II-proteinase reaction products exhibit neutrophil chemoattractant activity, *Blood* **73:**1682–1685.

Hofsteenge, J., Taguchi, H., and Stone, S. R., 1986, Effect of thrombomodulin on the kinetics of the interaction of thrombin with substrates and inhibitors, *Biochem. J.* **237:**243–251.

Hogg, P. J., and Jackson, C. M., 1989, Fibrin monomer protects thrombin from inactivation by heparin–antithrombin III: Implications for heparin efficacy, *Proc. Natl. Acad. Sci. USA* **86:**3619–3623.

Hogg, P. J., and Jackson, C. M., 1990a, Heparin promotes the binding of thrombin to fibrin polymer. Quantitative characterization of a thrombin–fibrin polymer–heparin ternary complex, *J. Biol. Chem.* **265:**241–247.

Hogg, P. J., and Jackson, C. M., 1990b, Formation of a ternary complex between thrombin, fibrin monomer, and heparin influences the action of thrombin on its substrates, *J. Biol Chem.* **265:**248–255.

Holmer, E., Söderström, G., and Andersson, L. O., 1979, Studies on the mechanism of the rate-enhancing effect of heparin on the thrombin–antithrombin III reaction. *Eur. J. Biochem.* **93:**1–5.

Holmer, E., Lindahl, U., Bäckström, G., Thunberg, L., Sandberg, H., Söderström, G., and Andersson, L. O., 1980, Anticoagulant activities and effects on platelets of a heparin fragment with high-affinity for antithrombin, *Thromb. Res.* **18:**861–869.

Holmer, E., Kurachi, K., and Söderström, G., 1981, The molecular-weight dependence of the rate-enhancing effect of heparin on the inhibition of thrombin, factor Xa, factor IXa, factor XIa, factor XIIa and kallikrein by antithrombin, *Biochem. J.* **193:**395–400.

Höök, M., Björk, I., Hopwood, J., and Lindahl, U., 1976, Anticoagulant activity of heparin:

Separation of high-activity and low-activity heparin species by affinity chromatography on immobilized antithrombin, *FEBS Lett.* **66**:90–93.

Hortin, G., Tollefsen, D. M., and Strauss, A. W., 1986, Identification of two sites of sulfation of human heparin cofactor II, *J. Biol. Chem.* **261**:15827–15830.

Hortin, G. L., Tollefsen, D. M., and Benutto, B. M., 1989, Antithrombin activity of a peptide corresponding to residues 54–75 of heparin cofactor II, *J. Biol. Chem.* **264**:13979–13982.

Hoylaerts, M., Owen, W. G., and Collen, D., 1984, Involvement of heparin chain-length in the heparin-catalyzed inhibition of thrombin by antithrombin III, *J. Biol. Chem.* **259**:5670–5677.

Huber, R., and Carrell, R. W., 1989, Implications of the three-dimensional structure of α_1-antitrypsin for structure and function of serpins, *Biochemistry* **28**:8951–8966.

Hunt, L. T., and Dayhoff, M. O., 1980, A surprising new protein superfamily containing ovalbumin, antithrombin III, and alpha$_1$-proteinase inhibitor, *Biochem. Biophys. Res. Commun.* **95**:864–871.

Hurst, R. E., Poon, M.-C., and Griffith, M. J., 1983, Structure–activity relationships of heparin. Independence of heparin charge density and antithrombin-binding domains in thrombin inhibition by antithrombin and heparin cofactor II, *J. Clin. Invest.* **72**:1042–1045.

Ill, C. R., and Ruoslahti, E., 1985, Association of thrombin–antithrombin III complex with vitronectin in serum, *J. Biol. Chem.* **260**:15610–15615.

Inhorn, R. C., and Tollefsen, D. M., 1986, Isolation and characterization of a partial cDNA clone for heparin cofactor II, *Biochem. Biophys. Res. Commun.* **137**:431–436.

Jacobsson, K. G., and Lindahl, U., 1979, Attempted determination of endogenous heparin in blood, *Thromb. Haemostas.* **42**:84.

Jakubowski, H. V., Kline, M. D., and Owen, W. G., 1986, The effect of bovine thrombomodulin on the specificity of bovine thrombin, *J. Biol. Chem.* **261**:3876–3882.

Jesty, J., 1978, The inhibition of activated bovine coagulation factors X and VII by antithrombin III. *Arch. Biochem. Biophys.* **185**:165–173.

Jesty, J., 1979a, Dissociation of complexes and their derivatives formed during inhibition of bovine thrombin and activated factor X by antithrombin III, *J. Biol. Chem.* **254**:1044–1049.

Jesty, J., 1979b, The kinetics of formation and dissociation of the bovine thrombin–antithrombin III complex, *J. Biol. Chem.* **254**:10044–10050.

Jesty, J., 1986, The kinetics of inhibition of α-thrombin in human plasma, *J. Biol. Chem.* **261**:10313–10318.

Jordan, R. E., 1983, Antithrombin in vertebrate species: Conservation of the heparin-dependent anticoagulant mechanism, *Arch. Biochem. Biophys.* **227**:587–595.

Jordan, R., Beeler, D., and Rosenberg, R., 1979, Fractionation of low molecular weight heparin species and their interaction with antithrombin, *J. Biol. Chem.* **254**:2902–2913.

Jordan, R. E., Oosta, G. M., Gardner, W. T., and Rosenberg, R. D., 1980a, The binding of low molecular weight heparin to hemostatic enzymes, *J. Biol. Chem.* **255**:10073–10080.

Jordan, R. E., Oosta, G. M., Gardner, W. T., and Rosenberg, R. D., 1980b, The kinetics of haemostatic enzyme–antithrombin interactions in the presence of low molecular weight heparin, *J. Biol. Chem.* **255**:10081–10090.

Jordan, R. E., Favreau, L. V., Braswell, E. H., and Rosenberg, R. D., 1982, Heparin with two binding sites for antithrombin or platelet factor 4, *J. Biol. Chem.* **257**:400–406.

Jörnvall, H., Fish, W. W., and Björk, I., 1979, The thrombin cleavage site in bovine antithrombin, *FEBS Lett.* **106**:358–362.

Koide, T., Odani, S., Takahashi, K., Ono, T., and Sakuragawa, N., 1984, Antithrombin III Toyama: Replacement of arginine-47 by cysteine in hereditary abnormal antithrombin III that lacks heparin-binding ability, *Proc. Natl. Acad. Sci. USA* **81**:289–293.

Kowalczykowski, S. C., Paul, L. S., Lonberg, N., Newport, J. W., McSwiggen, J. A., and von Hippel, P., 1986, Cooperative and noncooperative binding of protein ligands to nucleic acid lattices: Experimental approaches to the determination of thermodynamic parameters, *Biochemistry* **25**:1226–1240.

Kress, L. F., and Catanese, J. J., 1981, Identification of the cleavage sites resulting from enzymatic inactivation of human antithrombin III by *Crotalus adamanteus* proteinase II in the presence and absence of heparin, *Biochemistry* **20**:7432–7438.

Lam, L. H., Silbert, J. E., and Rosenberg, R. D., 1976, The separation of active and inactive forms of heparin, *Biochem. Biophys. Res. Commun.* **69**:570–577.

Lane, D. A., and Lindahl, U. (eds.), 1989, *Heparin: Chemical and Biological Properties, Clinical Applications,* Edward Arnold, London.

Lane, D. A., Denton, J. Flynn, A. M., Thunberg, L., and Lindahl, U., 1984, Anticoagulant activities of heparin oligosaccharides and their neutralization by platelet factor 4, *Biochem. J.* **218**:725–732.

Lane, D. A., Pejler, G., Flynn, A. M., Thompson, E. A., and Lindahl, U., 1986, Neutralization of heparin-related saccharides by histidine-rich glycoprotein and platelet factor 4, *J. Biol. Chem.* **261**:3980–3986.

Lane, D. A., Flynn, A. M., Pejler, G., Lindahl, U., Choay, J., and Preissner, K., 1987, Structural requirements for the neutralization of heparin-like saccharides by complement S protein/vitronectin, *J. Biol. Chem.* **262**:16343–16348.

Lane, D. A., Erdjument, H., Flynn, A., Di Marzo, V., Panico, M., Morris, H. R., Greaves, M., Dolan, G., and Preston, F. E., 1989a, Antithrombin Sheffield: Amino acid substitution at the reactive site (Arg 393 to His) causing thrombosis, *Br. J. Haematol.* **71**:91–96.

Lane, D. A., Erdjument, H., Thompson, E., Panico, M., Di Marzo, V., Morris, H. R., Leone, G., De Stefano, V., and Thein, S. L., 1989b, A novel amino acid substitution in the reactive site of a congenital variant antithrombin. Antithrombin Pescara, Arg^{393} to Pro, caused by a CGT to CCT mutation, *J. Biol. Chem.* **264**:10200–10204.

Lane, J. L., Bird, P., and Rizza, C. R., 1975, A new assay for the measurement of total progressive antithrombin, *Br. J. Haematol.* **30**:103–115.

Laskowski, M., Jr., and Kato, I., 1980, Protein inhibitors of proteinases, *Annu. Rev. Biochem.* **49**:593–626.

Latallo, Z. S., and Jackson, C. M., 1986, Reaction of thrombins with human antithrombin III: II. Dependence of rate of inhibition on molecular form and origin of thrombin, *Thromb. Res.* **43**:523–537.

Laurent, T. C., Tengblad, A., Thunberg, L., Höök, M., and Lindahl, U., 1978, The molecular-weight-dependence of the anti-coagulant activity of heparin, *Biochem. J.* **175**:691–701.

Léon, M., Aiach, M., Coezy, E., Guennec, J. Y., and Feissinger, J. N., 1983, Antithrombin III synthesis in rat liver parenchymal cells, *Thromb. Res.* **30**:369–375.

Lijnen, H. R., Hoylaerts, M., and Collen, D., 1983, Heparin binding properties of human histidine-rich glycoprotein. Mechanism and role in the neutralization of heparin in plasma, *J. Biol. Chem.* **258**:3803–3808.

Lindahl, U., Bäckström, G., Thunberg, L., and Leder, I. G., 1980, Evidence for a 3-O-sulfated D-glucosamine residue in the antithrombin-binding sequence of heparin, *Proc. Natl. Acad. Sci. USA* **77**:6551–6555.

Lindahl, U., Kolset, S. O., Bogwald, J., Østerud, B., and Seljelid, R., 1982, Studies, with a luminogenic peptide substrate, on blood coagulation factor X/Xa produced by mouse peritoneal macrophages, *Biochem. J.* **206**:231–237.

Lindahl, U., Thunberg, L., Bäckström, G., Riesenfeld, J., Nordling, K., and Björk, I., 1984, Extension and structural variability of the antithrombin-binding sequence in heparin, *J. Biol. Chem.* **259**:12368–12376.

Lindhout, T., Baruch, D., Schoen, P., Franssen, J., and Hemker, H. C., 1986, Thrombin generation and inactivation in the presence of antithrombin III and heparin, *Biochemistry* **25**:5962–5969.

Liu, C. S., and Chang, J. Y., 1987a, Probing the heparin-binding domain of human antithrombin III with V8 protease, *Eur. J. Biochem.* **167**:247–252.

Liu, C. S., and Chang, J. Y., 1987b, The heparin binding site of human antithrombin III. Selective chemical modification at Lys^{114}, Lys^{125} and Lys^{287} impairs its heparin cofactor activity, *J. Biol. Chem.* **262**:17356–17361.

Loebermann, H., Tukuoka, R., Deisenhofer, J., and Huber, R., 1984, Human α_1-proteinase inhibitor. Crystal structure analysis of two crystal modifications, molecular model and preliminary analysis of the implications for function, *J. Mol. Biol.* **177**:531–556.

Lollar, P., MacIntosh, S. C., and Owen, W. G., 1984, Reaction of antithrombin III with thrombin bound to the vascular endothelium. Analysis in a recirculating perfused rabbit heart preparation, *J. Biol. Chem.* **259**:4335–4338.

Lottenberg, R., Hall, J. A., Blinder, M., Binder, E. P., and Jackson, C. M., 1983, The action of thrombin on peptide *p*-nitroanilide substrates. Substrate selectivity and examination of hydrolysis under different reaction conditions, *Biochim. Biophys. Acta* **742**:539–557.

McGhee, J. D., and von Hippel, P. H., 1974, Theoretical aspects of DNA–protein interactions: Cooperative and noncooperative binding of large ligands to a one-dimensional homogenous lattice, *J. Mol. Biol.* **86**:469–489.

McGuire, E. A., and Tollefsen, D. M., 1987, Activation of heparin cofactor II by fibroblasts and vascular smooth muscle cells, *J. Biol. Chem.* **262**:169–175.

Machovich, R., and Aranyi, P., 1977, Effect of calcium ion on the interaction between thrombin and heparin: Thermal denaturation, *Thromb. Haemostas.* **38**:677–684.

Machovich, R., Staub, M., and Patthy, L., 1978, Decreased heparin sensitivity of cyclohexanedione-modified thrombin, *Eur. J. Biochem.* **83**:473–477.

MacIntosh, S., and Owen, W. G., 1987, Regulation of the clearance and inhibition of intravascular thrombin, *Bull. Sanofi Thromb. Res. Found.* **1**:8–18.

Maimone, M. M., and Tollefsen, D. M., 1988, Activation of heparin cofactor II by heparin oligosaccharides, *Biochem. Biophys. Res. Commun.* **152**:1056–1061.

Maimone, M. M., and Tollefsen, D. M., 1990, Structure of a dermatan sulfate hexasaccharide that binds to heparin cofactor II with high affinity, *J. Biol. Chem.* **265**:18263–18271.

Manning, G. S., 1978, The molecular theory of polyelectrolyte solutions with applications to the electrostatic properties of polynucleotides, *Q. Rev. Biophys.* **11**:179–246.

Marciniak, E., 1973, Factor-Xa inactivation by antithrombin III: Evidence for biological stabilization of factor Xa by factor V–phospholipid complex, *Br. J. Haematol.* **24**:391–400.

Marciniak, E., 1981, Thrombin-induced proteolysis of human antithrombin III: An outstanding contribution of heparin, *Br. J. Haematol.* **48**:325–336.

Marcum, J. A., and Rosenberg, R. D., 1984, Anticoagulantly active heparinlike molecules from vascular tissue. *Biochemistry* **23**:1730–1737.

Marcum, J. A., McKenney, J. B., and Rosenberg, R. D., 1984, Acceleration of thrombin–antithrombin complex formation in rat hind-quarters via heparin-like molecules bound to the endothelium, *J. Clin. Invest.* **74**:341–350.

Marcum, J. A., Atha, D. H., Fritze, L. M. S., Nawroth, P., Stern, D., and Rosenberg, R. D., 1986, Cloned bovine aortic endothelial cells synthesize anticoagulantly active heparan sulfate proteoglycan, *J. Biol. Chem.* **261**:7507–7517.

Matheson, N. R., van Halbeek, H., and Travis, J., 1991, Evidence for a tetrahedral intermediate complex during serpin-proteinase interactions, *J. Biol. Chem.* **266**: 13489–13491.

Miletich, J. P., Jackson, C. M., and Majerus, P. W., 1978, Properties of the factor Xa binding site on human platelets, *J. Biol. Chem.* **253**:6908–6916.

Miller-Andersson, M., Borg, H., and Andersson, L. O., 1974, Purification of antithrombin III by affinity chromatography, *Thromb. Res.* **5**:439–452.

Mizuochi, T., Fujii, J., Kurachi, K., and Kobata, A., 1980, Structural studies of the carbohydrate moiety of human antithrombin III, *Arch. Biochem. Biophys.* **203**:458–465.

Molho-Sabatier, P., Aiach, M., Gaillard, I., Feissinger, J. N., Fischer, A. M., Chadeuf, G., and Clauser, E., 1989, Molecular characterization of antithrombin III (AT III) variants using polymerase chain reaction. Identification of the AT III Charleville as an Ala 384 Pro mutation, *J. Clin. Invest.* **84**:1236–1242.

Morii, M., and Travis, J., 1983, Amino acid sequence at the reactive site of human α_1-antichymotrypsin, *J. Biol. Chem.* **258**:12749–12752.

Mourey, L., Samama, J. P., Delarue, M., Choay, J., Lormeau, J. C., Petitou, M., and Moras, D., 1990, Antithrombin III: Structural and functional aspects, *Biochimie* **72**: 599–608.

Naski, M. C., and Shafer, J. A., 1990, α-Thrombin-catalyzed hydrolysis of fibrin I. Alternative binding modes and the accessibility of the active site in fibrin I-bound α-thrombin, *J. Biol. Chem.* **265**:1401–1407.

Naski, M. C., Fenton, J. W., II, Maraganore, J. M., Olson, S. T., and Shafer, J. A., 1990, The COOH-terminal domain of hirudin: An exosite-directed competitive inhibitor of the action of α-thrombin on fibrinogen, *J. Biol. Chem.* **265**:13484–13489.

Nesheim, M. E., 1983, A simple rate law that describes the kinetics of the heparin-catalyzed reaction between antithrombin III and thrombin, *J. Biol. Chem.* **258**:14708–14717.

Nesheim, M. E., Blackburn, M. N., Lawler, C. M., and Mann, K. G., 1986, Dependence of antithrombin III and thrombin binding stoichiometries and catalytic activity on the molecular weight of affinity-purified heparin, *J. Biol. Chem.* **261**:3214–3221.

Nordenman, B., and Björk, I., 1978a, Studies on the binding of heparin to prothrombin and thrombin and the effect of heparin binding on thrombin activity, *Thromb. Res.* **12**:755–765.

Nordenman, B., and Björk, I., 1978b, Binding of low-affinity and high-affinity heparin to antithrombin. Ultraviolet difference spectroscopy and circular dichroism studies, *Biochemistry* **17**:3339–3344.

Nordenman, B., and Björk, I., 1980, Fractionation of heparin by chromatography on immobilized thrombin. Correlation between the anticoagulant activity of the fractions and their content of heparin with high-affinity for antithrombin, *Thromb. Res.* **19**:711–718.

Nordenman, B., and Björk, I., 1981, Influence of ionic strength and pH on the interaction between high-affinity heparin and antithrombin, *Biochim. Biophys. Acta* **672**:227–238.

Nordenman, B., Nyström, C., and Björk, I., 1977, The size and shape of human and bovine antithrombin III, *Eur. J. Biochem.* **78**:195–203.

Nordenman, B., Danielsson, Å., and Björk, I., 1978, The binding of low-affinity and high-affinity heparin to antithrombin. Fluorescence studies, *Eur. J. Biochem.* **90**:1–6.

Ogston, D., Murray, J., and Crawford, G. P. M., 1976, Inhibition of the activated C1 s̄ subunit of the first component of complement by antithrombin III in the presence of heparin, *Thromb. Res.* **9**:217–222.

Olds, R. J., Lane, D. A., Finazzi, G., Barbui, T., and Thein, S. L., 1990, A frameshift mutation leading to type 1 antithrombin deficiency and thrombosis, *Blood* **76**:2182–2186.

Olson, S. T., 1985, Heparin and ionic strength-dependent conversion of antithrombin III from an inhibitor to a substrate of α-thrombin, *J. Biol. Chem.* **260**:10153–10160.

Olson, S. T., 1988, Transient kinetics of heparin-catalyzed protease inactivation by antithrombin III. Linkage of protease-inhibitor–heparin interactions in the reaction with thrombin, *J. Biol. Chem.* **263**:1698–1708.

Olson, S. T., 1989, High molecular weight-kininogen enhancement of the heparin-accelerated rate of plasma kallikrein inactivation by antithrombin III, *J. Cell Biol.* **107**:827a.

Olson, S. T., and Björk, I., 1991, Predominant contribution of surface approximation to the mechanism of heparin acceleration of the antithrombin/thrombin reaction. Elucidation from salt concentration effects, *J. Biol. Chem.* **266**:6353–6364.

Olson, S. T., and Choay, J., 1989, Mechanism of high molecular weight-kininogen stimulation of the heparin-accelerated antithrombin/kallikrein reaction, *Thromb. Haemostas.* **62**:326.

Olson, S. T., and Shore, J. D., 1981, Binding of high affinity heparin to antithrombin III. Characterization of the protein fluorescence enhancement, *J. Biol. Chem.* **256**:11065–11072.

Olson, S. T., and Shore, J. D., 1982, Demonstration of a two-step reaction mechanism for inhibition of α-thrombin by antithrombin III and identification of the step affected by heparin, *J. Biol. Chem.* **257**:14891–14895.

Olson, S. T., and Shore, J. D., 1986, Transient kinetics of heparin-catalyzed protease inactivation by antithrombin III. The reaction step limiting heparin turnover in thrombin neutralization, *J. Biol. Chem.* **261**:13151–13159.

Olson, S. T., and Shore, J. D., 1989, High molecular weight-kininogen and heparin acceleration of factor XIa inactivation by plasma proteinase inhibitors, *Thromb. Haemostas.* **62**:381.

Olson, S. T., Srinivasan, K. R., Björk, I., and Shore, J. D., 1981, Binding of high-affinity heparin to antithrombin III. Stopped flow kinetic studies of the binding interaction, *J. Biol. Chem.* **256**:11073–11079.

Olson, S. T., Björk, I., Craig, P. A., Shore, J. D., and Choay, J., 1992, Role of the antithrombin-binding pentasaccharide in heparin acceleration of antithrombin-proteinase reactions. Resolution of the antithrombin conformational change contribution to heparin rate enhancement, *J. Biol. Chem.*, in press.

Olson, S. T., Halvorson, H. R., and Björk, I., 1991, Quantitative characterization of the thrombin–heparin interaction: Discrimination between specific and nonspecific binding models, *J. Biol. Chem.* **266**:6342–6352.

Oosta, G. M., Gardner, W. T., Beeler, D. L., and Rosenberg, R. D., 1981, Multiple functional domains of the heparin molecule, *Proc. Natl. Acad. Sci. USA* **78**:829–833.

Oshima, G., Uchiyama, H., and Nagasawa, K., 1986, Effect of NaCl on the association of thrombin with heparin, *Biopolymers* **25**:527–537.

Owen, B. A., and Owen, W. G., 1990, Interaction of factor Xa with heparin does not contribute to the inhibition of factor Xa by antithrombin III–heparin, *Biochemistry* **29**:9412–9417.

Owen, M. C., Brennan, S. O., Lewis, J. H., and Carrell, R. W., 1983, Mutation of antitrypsin to antithrombin, α$_1$-Antitrypsin Pittsburgh (358 Met → Arg), a fatal bleeding disorder, *New Engl. J. Med.* **309**:694–698.

Owen, M. C., Borg, J. Y., Soria, C., Soria, J., Caen, J., and Carrell, R. W., 1987, Heparin binding defect in a new antithrombin III variant: Rouen 47 Arg to His, *Blood* **69**:1275–1279.

Owen, M. C., Beresford, C. H., and Carrell, R. W., 1988, Antithrombin Glasgow, 393 Arg to

His: A P_1 reactive site variant with increased heparin affinity but no thrombin inhibitory activity, *FEBS Lett.* **231**:317–320.

Owen, M. C., Borg, J. Y., and Carrell, R. W., 1989a, Activation of antithrombin by heparin, Abstracts, Kyoto Satellite Symposia of XIIIth Congress of ISTH, p. 76.

Owen, M. C., Shaw, G. J., Grau, E., Fontcuberta, J., Carrell, R. W., and Boswell, D. R., 1989b, Molecular characterization of antithrombin Barcelona-2: 47 arginine to cysteine, *Thromb. Res.* **55**:451–457.

Owen, W. G., 1975, Evidence for the formation of an ester between thrombin and heparin cofactor, *Biochim. Biophys. Acta* **405**:380–387.

Parker, K. A., and Tollefsen, D. M., 1985, The protease specificity of heparin cofactor II. Inhibition of thrombin generated during coagulation, *J. Biol. Chem.* **260**:3501–3505.

Perlmutter, D. H., Glover, G. I., Rivetna, M., Schasteen, C. S., and Fallon, R. J., 1990, Identification of a serpin–enzyme complex receptor on human hepatoma cells and human monocytes, *Proc. Natl. Acad. Sci. USA* **87**:3753–3757.

Petersen, L. C., and Jørgensen, M., 1983, Electrostatic interactions in the heparin-enhanced reaction between human thrombin and antithrombin, *Biochem. J.* **211**:91–97.

Petersen, T. E., Dudek-Wojciechowska, G., Sottrup-Jensen, L., and Magnusson, S., 1979, Primary structure of antithrombin III (heparin cofactor). Partial homology between α_1-antitrypsin and antithrombin III, in: *The Physiological Inhibitors of Blood Coagulation and Fibrinolysis* (D. Collen, B. Wiman, and M. Verstraete, eds.), Elsevier/North-Holland, Amsterdam, pp. 43–54.

Peterson, C. B., and Blackburn, M. N., 1985, Isolation and characterization of an antithrombin III variant with reduced carbohydrate content and enhanced heparin binding, *J. Biol. Chem.* **260**:610–615.

Peterson, C. B., and Blackburn, M. N., 1987a, Antithrombin conformation and the catalytic role of heparin. I. Does cleavage by thrombin induce structural changes in the heparin-binding region of antithrombin? *J. Biol. Chem.* **262**:7552–7558.

Peterson, C. B., and Blackburn, M. N., 1987b, Antithrombin conformation and the catalytic role of heparin. II. Is the heparin-induced conformational change in antithrombin required for rapid inactivation of thrombin? *J. Biol. Chem.* **262**:7559–7566.

Peterson, C. B., Morgan, W. T., and Blackburn, M. N., 1987a, Histidine-rich glycoprotein modulation of the anticoagulant activity of heparin. Evidence for a mechanism involving competition with both antithrombin and thrombin for heparin binding, *J. Biol. Chem.* **262**:7567–7574.

Peterson, C. B., Noyes, C. M., Pecon, J. M., Church, F. C., and Blackburn, M. N., 1987b, Identification of a lysyl residue in antithrombin which is essential for heparin binding, *J. Biol. Chem.* **262**:8061–8065.

Pieters, J. Franssen, J., Visch, C., and Lindhout, T., 1987, Neutralization of heparin by prothrombin activation products, *Thromb. Res.* **45**:573–580.

Pixley, R. A., Schapira, M., and Colman, R. W., 1985, The regulation of human factor XIIa by plasma proteinase inhibitors, *J. Biol. Chem.* **260**:1723–1729.

Pletcher, C. H., and Nelsestuen, G. L., 1983, Two-substrate reaction model for the heparin-catalyzed bovine antithrombin/protease reaction, *J. Biol. Chem.* **258**:1086–1091.

Podack, E. R., Dahlbäck, B., and Griffin, J. H., 1986, Interaction of S-protein of complement with thrombin and antithrombin III during coagulation. Protection of thrombin by S-protein from antithrombin III inactivation, *J. Biol. Chem.* **261**:7387–7392.

Pomerantz, M. W., and Owen, W. G., 1978, A catalytic role for heparin. Evidence for a ternary complex of heparin cofactor, thrombin, and heparin, *Biochim. Biophys. Acta* **535**:66–77.

Pratt, C. W., Tobin, R. B., and Church, F. C., 1990, Interaction of heparin cofactor II with neutrophil elastase and cathepsin G, *J. Biol. Chem.* **265:**6092–6097.

Preissner, K. T., and Müller-Berghaus, G., 1986, S protein modulates the heparin-catalyzed inhibition of thrombin by antithrombin III. Evidence for a direct interaction of S protein with heparin, *Eur. J. Biochem.* **156:**645–650.

Preissner, K. T., and Müller-Berghaus, G., 1987, Neutralization and binding of heparin by S protein/vitronectin in the inhibition of factor Xa by antithrombin III. Involvement of an inducible heparin-binding domain of S protein/vitronectin, *J. Biol. Chem.* **262:**12247–12253.

Preissner, K. T., Delvos, U., and Müller-Berghaus, G., 1987, Binding of thrombin to thrombomodulin accelerates inhibition of the enzyme by antithrombin III. Evidence for a heparin-independent mechanism, *Biochemistry* **26:**2521–2528.

Preissner, K. T., Koyama, T., Müller, D., Tschopp, J., and Müller-Berghaus, G., 1990, Domain structure of the endothelial cell receptor thrombomodulin as deduced from modulation of its anticoagulant functions. Evidence for a glycosaminoglycan-dependent secondary binding site for thrombin, *J. Biol. Chem.* **265:**4915–4922.

Prochownik, E. V., Markham, A. F., and Orkin, S. H., 1983, Isolation of a cDNA clone for human antithrombin III, *J. Biol. Chem.* **258:**8389–8394.

Ragg, H., 1986, A new member of the plasma protease inhibitor gene family. *Nucleic Acids Res.* **14:**1073–1088.

Ragg, H., Ulshöfer, T., and Gerewitz, J., 1990, On the activation of human leuserpin-2, a thrombin inhibitor, by glycosaminoglycans, *J. Biol. Chem.* **265:**5211–5218.

Record, M. T., Jr., Lohman, T. M., and DeHaseth, P., 1976, Ion effects on ligand–nucleic acid interactions, *J. Mol. Biol.* **107:**145–158.

Record, M. T., Jr., Anderson, C. F., and Lohman, T. M., 1978, Thermodynamic analysis of ion effects on the binding and conformational equilibria of proteins and nucleic acids: The roles of ion association or release, screening and ion effects on water activity, *Q. Rev. Biophys.* **11:**103–178.

Richter, P. H., and Eigen, M., 1974, Diffusion controlled reaction rates in spheroidal geometry. Application to repressor–operator association and membrane-bound enzymes, *Biophys. Chem.* **2:**255–263.

Rosenberg, R. D., 1977, Chemistry of the hemostatic mechanism and its relationship to the action of heparin, *Fed. Proc.* **36:**10–18.

Rosenberg, R. D., and Damus, P. S., 1973, The purification and mechanism of action of human antithrombin–heparin cofactor, *J. Biol. Chem.* **248:**6490–6505.

Rosenfeld, L., and Danishefsky, I., 1986, A fragment of antithrombin that binds both heparin and thrombin, *Biochem. J.* **237:**639–646.

Rydel, T. J., Ravichandran, K. G., Tulinsky, A., Bode, W., Huber, R., Roitsch, C., and Fenton, J. W., II, 1990, The structure of a complex of recombinant hirudin and human α-thrombin, *Science* **249:**277–280.

Sakuragawa, N., Takahashi, K., Kondo, S. I., and Koide, T., 1983, Antithrombin III Toyama: A hereditary abnormal antithrombin III of a patient with recurrent thrombophlebitis, *Thromb. Res.* **31:**305–317.

Schapira, M., Scott, C. F., James, A., Silver, L. D., Kueppers, F., James, H. L., and Colman, R. W., 1982, High molecular weight kininogen or its light chain protects human plasma kallikrein from inactivation by plasma protease inhibitors, *Biochemistry* **21:**567–572.

Schechter, I., and Berger, A., 1967, On the size of the active site in proteases. I. Papain, *Biochem. Biophys. Res. Commun.* **27:**157–162.

Schoen, P., and Lindhout, T., 1987, The *in situ* inhibition of prothrombinase-formed human α-thrombin and meizothrombin(desF1) by antithrombin III and heparin, *J. Biol. Chem.* **262:**11268–11274.

Scott, C. F., and Colman, R. W., 1989, Factors influencing the acceleration of human factor XIa inactivation by antithrombin III, *Blood* **73**:1873–1879.

Scott, C. F., Schapira, J., James, H. L., Cohen, A. B., and Colman, R. W., 1982, Inactivation of factor XIa by plasma protease inhibitors. Predominant role of α_1-protease inhibitor and protective effect of high molecular weight kininogen, *J. Clin. Invest.* **69**:844–852.

Scully, M. F., and Kakkar, V. V., 1984, Identification of heparin cofactor II as the principal plasma cofactor for the antithrombin activity of pentosan polysulfate (SP 54), *Thromb. Res.* **36**:187–194.

Scully, M. F., Ellis, V., Seno, N., and Kakkar, V. V., 1988, Effect of oversulfated chondroitin and dermatan sulfate upon thrombin and factor Xa inactivation by antithrombin III or heparin cofactor II, *Biochem. J.* **254**:547–551.

Shah, N., Scully, M. F., Ellis, V., and Kakkar, V. V., 1990, Influence of chemical modification of tryptophan residues on the properties of human antithrombin III, *Thromb. Res.* **57**:343–352.

Sie, P., Dupouy, D., Pichon, J., and Boneu, B., 1985, Constitutional heparin cofactor II deficiency associated with recurrent thrombosis, *Lancet* **2**:414–416.

Sinaÿ, P., Jacquinet, J. C., Petitou, M., Duchaussoy, P., Lederman, I., Choay, J., and Torri, G., 1984, Total synthesis of a heparin pentasaccharide fragment having high affinity for antithrombin III, *Carbohydr. Res.* **132**:C5–C9.

Smith, J. W., and Knauer, D. J., 1987, A heparin binding site in antithrombin III. Identification, purification, and amino acid sequence, *J. Biol. Chem.* **262**:11964–11972.

Smith, J. W., Dey, N., and Knauer, D. J., 1990, Heparin binding domain of antithrombin III: Characterization using a synthetic peptide directed polyclonal antibody, *Biochemistry* **29**:8950–8957.

Soons, H., Janssen-Claessen, T., Tans, G., and Hemker, H. C., 1987, Inhibition of factor XI$_a$ by antithrombin III, *Biochemistry* **26**:4624–4629.

Speight, M. O., and Griffith, M. J., 1983, Calcium inhibits the heparin-catalyzed antithrombin III/thrombin reaction by decreasing the apparent binding affinity of heparin for thrombin, *Arch. Biochem. Biophys.* **225**:958–963.

Stein, P. E., Leslie, A. G. W., Finch, J. T., Turnell, W. G., McLaughlin, P. J., and Carrell, R. W., 1990, Crystal structure of ovalbumin as a model for the reactive centre of serpins, *Nature* **347**:99–102.

Stephens, A W., Thalley, B. S., and Hirs, C. H. W., 1987a, Antithrombin-III Denver, a reactive site variant, *J. Biol. Chem.* **262**:1044–1048.

Stephens, A. W., Siddiqui, A., and Hirs, C. H. W., 1987b, Expression of functionally active human antithrombin III, *Proc. Natl. Acad. Sci. USA* **84**:3886–3890.

Stephens, A. W., Siddiqui, A., and Hirs, C. H. W., 1988, Site-directed mutagenesis of the reactive center (serine 394) of antithrombin III, *J. Biol. Chem.* **263**:15849–15852.

Stern, D., Nawroth, P., Marcum, J., Handley, D., Kisiel, W., Rosenberg, R., and Stern, K., 1985, Interaction of antithrombin III with bovine aortic segments. Role of heparin in binding and enhanced anticoagulant activity, *J. Clin. Invest.* **75**:272–279.

Stone, A. L., Beeler, D., Oosta, G., and Rosenberg, R. D., 1982, Circular dichroism spectroscopy of heparin–antithrombin interactions, *Proc. Natl. Acad. Sci. USA* **79**:7190–7194.

Stone, S. R., and Hofsteenge, J., 1987, Effect of heparin on the interaction between thrombin and hirudin, *Eur. J. Biochem.* **169**:373–376.

Sun, X. J., and Chang, J. Y., 1989a, Heparin binding domain of human antithrombin III inferred from the sequential reduction of its three disulfide linkages. An efficient method for structural analysis of partially reduced proteins, *J. Biol. Chem.* **264**:11288–11293.

Sun, X. J., and Chang, J. Y., 1989b, The heparin and pentosan polysulfate binding sites of human antithrombin overlap but are not identical, *Eur. J. Biochem.* **185:**225–230.

Sun, X. J., and Chang, J. Y., 1990, Evidence that arginine-129 and arginine-145 are located within the heparin binding site of human antithrombin III, *Biochemistry* **29:**8957–8962.

Thunberg, L., Lindahl, U., Tengblad, A., Laurent, T. C., and Jackson, C. M., 1979, On the molecular-weight dependence of the anticoagulant activity of heparin, *Biochem. J.* **181:**241–243.

Thunberg, L., Bäckström, G., and Lindahl, U., 1982, Further characterization of the anti-thrombin-binding sequence in heparin, *Carbohydr. Res.* **100:**393–410.

Tollefsen, D. M., and Blank, M. K., 1981, Detection of a new heparin-dependent inhibitor of thrombin in human plasma, *J. Clin. Invest.* **68:**589–596.

Tollefsen, D. M., and Pestka, C. A., 1985, Modulation of heparin cofactor II activity by histidine-rich glycoprotein and platelet factor 4, *J. Clin. Invest.* **75:**496–501.

Tollefsen, D. M., Majerus, P. W., and Blank, M. K., 1982, Heparin cofactor II. Purification and properties of a heparin-dependent inhibitor of thrombin in human plasma, *J. Biol. Chem.* **257:**2162–2169.

Tollefsen, D. M., Pestka, C. A., and Monafo, W. J., 1983, Activation of heparin cofactor II by dermatan sulfate, *J. Biol. Chem.* **258:**6713–6716.

Tollefsen, D. M., Peacock, M. E., and Monafo, W. J. 1986, Molecular size of dermatan sulfate oligosaccharides required to bind and activate heparin cofactor II, *J. Biol. Chem.* **261:**8854–8858.

Tomasini, B. R., and Mosher, D. F., 1988, Conformational states of vitronectin: Preferential expression of an antigenic epitope when vitronectin is covalently and noncovalently complexed with thrombin–antithrombin III or treated with urea, *Blood* **72:**903–912.

Tran, T. H., Marbet, G. A., and Duckert, F., 1985, Association of hereditary heparin cofactor II deficiency with thrombosis, *Lancet* **2:**413–414.

Tran, T. H., Lammle, B., Zbinden, B., and Duckert, F., 1986, Heparin cofactor II: Purification and antibody production, *Thromb. Haemostas.* **55:**19–23.

Verhamme, I. M., Hogg, P. J., and Jackson, C. M., 1989, Kinetic investigation of the heparin-catalyzed inactivation of thrombin by heparin cofactor II, *Thromb. Haemostas.* **62:**33.

Villanueva, G. B., and Allen, N., 1983a, Demonstration of a two-domain structure of anti-thrombin III during its denaturation in guanidinium chloride, *J. Biol. Chem.* **258:**11010–11013.

Villanueva, G. B., and Allen, N., 1983b, Refolding properties of antithrombin III. Mechanism of binding to heparin, *J. Biol. Chem.* **258:** 14048–14053.

Villanueva, G. B., and Danishefsky, I., 1977, Evidence for a heparin-induced conformational change on antithrombin III, *Biochem. Biophys. Res. Commun.* **74:**803–809.

Villanueva, G. B., and Danishefsky, I., 1979, Conformational changes accompanying the binding of antithrombin III to thrombin, *Biochemistry* **18:**810–817.

Vogel, C. N., Kingdon, H. S., and Lundblad, R. L., 1979, Correlation of in vivo and in vitro inhibition of thrombin by plasma inhibitors, *J. Lab. Clin. Med.* **93:**661–673.

Walker, F. J., and Esmon, C. T., 1979, The effect of prothrombin fragment 2 on the inhibition of thrombin by antithrombin III, *J. Biol. Chem.* **254:**5618–5622.

Wallgren, P., Nordling, K., and Björk, I., 1981, Immunological evidence for a proteolytic cleavage at the active site of antithrombin in the mechanism of inhibition of coagulation serine proteases, *Eur. J. Biochem.* **116:**493–496.

Wasley, L. C., Atha, D. H., Bauer, K. A., and Kaufman, R. J., 1987, Expression and char-acterization of human antithrombin III synthesized in mammalian cells, *J. Biol. Chem.* **262:**14766–14772.

Weitz, J. I., Hudoba, M., Massel, D., Maraganore, J., and Hirsh, J., 1990, Clot-bound thrombin is protected from inhibition by heparin–antithrombin III but is susceptible to inactivation by antithrombin III-independent inhibitors, *J. Clin. Invest.* **86:**385–391.

Wiman, B., and Collen, D., 1979, On the mechanism of the reaction between human α_2-antiplasmin and plasmin, *J. Biol. Chem.* **254:**9291–9297.

Winter, R. B., Berg, O. G., and von Hippel, P. H., 1981, Diffusion-driven mechanisms of protein translocation on nucleic acids. 3. The *Escherischia coli lac* repressor-operator interaction: Kinetic measurements and conclusions, *Biochemistry* **20:**6961–6977.

Wong, R. F., Chang, T., and Feinman, R. D., 1982, Reaction of antithrombin with proteases. Nature of the reaction with trypsin, *Biochemistry* **21:**6–12.

Wong, R. F., Windwer, S. R., and Feinman, R. D., 1983, Interaction of thrombin and antithrombin. Reaction observed by intrinsic fluorescence measurements, *Biochemistry* **22:**3994–3999.

Wright, H. T., Qian, H. X., and Huber, R., 1990, Crystal structure of plakalbumin, a proteolytically nicked form of ovalbumin. Its relationship to the structure of cleaved α_1-proteinase inhibitor, *J. Mol. Biol.* **213:**513–528.

Wunderwald, P., Schrenk, W. J., and Port, H., 1982, Antithrombin BM from human plasma: An antithrombin binding moderately to heparin, *Thromb. Res.* **25:**177–191.

Yamagishi, R., Niwa, M., Kondo, S., Sakuragawa, N., and Koide, T., 1984, Purification and biological property of heparin cofactor II: Activation of heparin cofactor II and antithrombin III by dextran sulfate and various glycosaminoglycans, *Thromb. Res.* **36:**633–642.

Zettlmeissl, G., Wirth, M., Hauser, H., and Küpper, H. A., 1988, Efficient expression system for human antithrombin III in baby hamster kidney cells, *Behring Inst. Mitt.* **82:**26–34.

Zettlmeissl, G., Conradt, H. S., Nimtz, M., and Karges, H. E., 1989, Characterization of recombinant human antithrombin III synthesized in Chinese hamster ovary cells, *J. Biol. Chem.* **264:**21153–21159.

Chapter 6

HIRUDIN INTERACTIONS WITH THROMBIN

Stuart R. Stone and John M. Maraganore

1. PROPERTIES OF HIRUDIN

Hirudin was originally isolated from the salivary glands of the medicinal leech *Hirudo medicinalis*. The anticoagulant activity of leech saliva was first described over 100 years ago by Haycraft (1884) and a polypeptide with antithrombin activity was first isolated over 30 years ago by Markwardt (1957). The amino acid sequence of the major form of hirudin was determined by Bagdy *et al.* (1976) and Dodt *et al.* (1984) (see also Mao *et al.*, 1987) and is given in Fig. 1. The most striking feature of the amino acid sequence is the high proportion of acidic residues (12 out of 65); seven of these occur in the C-terminal region from residues 53 to 65, including the sulfated tyrosine residue at position 63 that is formed by posttranslational modification. Hirudin contains six cysteines that form three disulfide bridges (h-Cys-6–Cys-14, h-Cys-16–Cys-28, and h-Cys-22–Cys-39[1]) that

[1]The numbering of the sequence of thrombin is that of Bode *et al.*, (1989) which is based on chymotrypsin numbering. Residues in hirudin are distinguished from those in thrombin by the prefix "h."

Stuart R. Stone • MRC Centre, University of Cambridge, Cambridge CBZ 2QH, United Kingdom. **John M. Maraganore** • Biogen, Inc., Cambridge, Massachusetts 02142.

Thrombin: Structure and Function, edited by Lawrence J. Berliner. Plenum Press, New York, 1992.

were elucidated by Dodt *et al.* (1985) and are also shown in Fig. 1. Subsequently, more than 20 different isoforms of hirudin have been isolated from *H. medicinalis* and sequenced (Dodt *et al.*, 1986; Harvey *et al.*, 1986; Tripier, 1988; Scharf *et al.*, 1989) and all of these forms seem to have antithrombin activity. All of the isoforms are highly homologous with levels of identity usually greater than 80%; the conserved residues are indicated in Fig. 1. The disulfide bridges and the acidic nature of the molecule, including the sulfated tyrosine, are invariant.

The small size of hirudin has allowed the determination of its tertiary structure by two-dimensional NMR spectroscopy (Clore *et al.*, 1987; Sukumaran *et al.*, 1987; Folkers *et al.*, 1989; Haruyama and Wüthrich, 1989; Haruyama *et al.*, 1989). These studies show that hirudin is composed of a compact N-terminal domain (residues 3–49) held together by three disulfide bonds, and a disordered C-terminal tail (residues 50–65). The N-terminal domain structure is characterized by well-defined turns and a protruding "finger" domain consisting of two antiparallel β-sheets (shown in Fig. 2). The orientation of this "finger" domain with respect to the core could not be precisely determined in the NMR studies. The structure of the N-terminal core of hirudin bound to thrombin as elucidated by x-ray crystallography corresponds to that determined by two-dimensional NMR

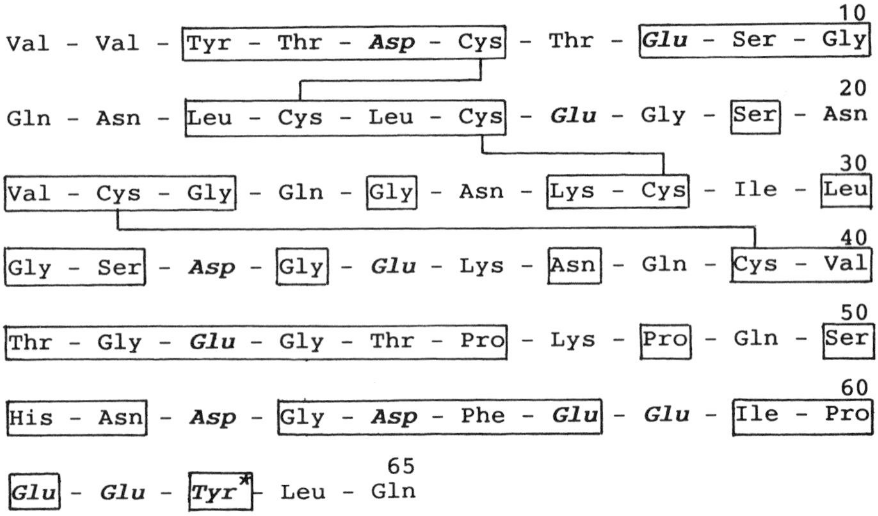

Figure 1. Sequence of hirudin variant 1. The sequence given in the three-letter code is that determined by Bagdy *et al.* (1976) and Dodt *et al.* (1984). The disulfide bridges determined by Dodt *et al.* (1985) are also given. Acidic residues are shown in bold italics and residues invariant in other hirudin sequences (Scharf *et al.*, 1989) are blocked.

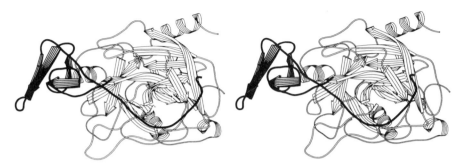

Figure 2. Ribbon drawing of the thrombin–hirudin complex. β-Sheets and α-helices are represented as arrows and coils, respectively. The hirudin structure is darker and is found in front of thrombin. It is composed of four short β-sheets linked by loops and a long C-terminal tail that wraps around the thrombin molecule and binds to the anion-binding exosite. The N-terminus of hirudin binds in the active site of thrombin. The thrombin molecule consists of an A- and B-chain. The A-chain is composed of two helical portions and is found at the back of the B-chain while the B-chain consists mainly of β-sheets. The structure represented is that of Grütter *et al.* (1990) and the plot was made using the program of Priestle (1988).

(see Fig. 2). In the crystal structures, however, the floppy C-terminal tail has a defined structure due to contacts with thrombin (Rydel *et al.*, 1990; Grütter *et al.*, 1990). The conformation of C-terminal fragments of hirudin bound to thrombin has also been studied by NMR spectroscopy using transferred nuclear Overhauser effects (Ni *et al.*, 1990; see also the chapter by Ni et al.) and the observed conformation is similar to that seen in the crystal structure of Rydel *et al.* (1990).

2. MOLECULAR MECHANISM OF THE INTERACTION OF HIRUDIN WITH THROMBIN

Hirudin exhibits a unique specificity toward thrombin. Its potent inhibitory activity is confined to thrombin which it inhibits at picomolar concentrations. It has not been found to inhibit any other proteases even at micromolar concentrations (reviewed by Walsmann and Markwardt, 1981; Sawyer, 1986; Walsmann, 1988; Wallis, 1988). In view of the contamination of earlier preparations of hirudin by other protease inhibitors (Fritz and Krejci, 1976), previous reports of inhibition of coagulation proteases (Kiesel and Hanahan, 1974; Brown *et al.*, 1980) should be reexamined. Hirudin's unique specificity is achieved by a novel mechanism of inhibition involving numerous interactions with thrombin. A number of interaction areas on thrombin and hirudin were identified by studies in solution and the results of these studies have been confirmed and extended by the recently determined crystal structures of thrombin–hirudin complexes (Rydel *et al.*, 1990; Grütter *et al.*, 1990).

The structure presented in Fig. 2 illustrates the molecular basis for hirudin's specificity. Studies on other inhibitors of serine proteases have indicated that they make contacts with their target enzyme predominantly in the region of the active site (reviewed by Read and James, 1986). In contrast, hirudin interacts with thrombin over an extended area including regions far removed from the active site. Hirudin forms an extremely tight complex with thrombin [the values of the dissociation constant (K_d) are 10^{-14} M and 10^{-13} M for native and recombinant hirudin, respectively; Braun *et al.*, 1988a] and this avidity is the result of the large number of favorable contacts between hirudin and thrombin observed in the crystal structures.

The mode of interaction of hirudin with the active site of thrombin represents a completely novel mechanism of inhibition. Hirudin binds with its N-terminal three residues in the active-site cleft of thrombin, with the hirudin polypeptide chain running in a direction opposite to that expected for a substrate and to that observed for the inhibitor D-Phe-Pro-ArgCH$_2$Cl [as shown in Fig. 3 for recombinant hirudin variant 2 (Asn-

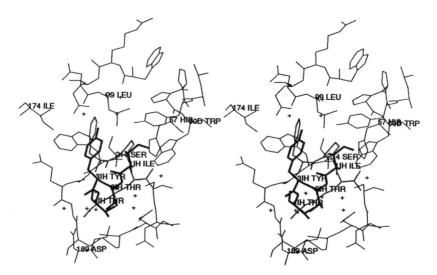

Figure 3. Comparison of the binding of rHV2-Lys[47] and D-Phe-Pro-ArgCH$_2$ to the active site of thrombin. The binding of rHV2-Lys[47] (thick lines; from Rydel *et al.*, 1990) to the active site of human α-thrombin is compared with that of D-Phe-Pro-ArgCH$_2$ (dotted lines; from Bode *et al.*, 1989). rHV2-Lys[47] residues are shown as numbers with an "IH" suffix.

47→Lys), abbreviated as rHV2-Lys[47]]. The residues h-Ile-1 and h-Tyr-3 occupy roughly the binding sites S_1 and S_3 (nomenclature of Schechter and Berger, 1967), respectively (Rydel et al., 1990). The recombinant hirudin variant 1 (rHV1) residues h-Val-1 and h-Tyr-3 occupy similar positions (Grütter et al., 1990). The primary specificity pocket (S_1 subsite; Schechter and Berger, 1967) of thrombin is not occupied (see Fig. 3). While the specificity of other inhibitors of serine proteases is determined by interactions with the S_1 subsite of their target enzymes (see, e.g., Laskowski et al., 1987), that of hirudin is not due to such an interaction but to numerous other interactions both within the active-site cleft and at other distinct binding sites (Fig. 2). The absence of a reactive-site residue in hirudin is presumably why it inhibits no serine proteases other than thrombin. To allow the peptide chain of hirudin to leave the active site, the 140–150 loop of thrombin (also called the γ-loop because it contains a cleavage site for the conversion of α- to γ-thrombin) assumes a conformation different from that observed in the structure of thrombin inactivated by D-Phe-Pro-ArgCH$_2$Cl. The core region of hirudin closes off the active site of thrombin. The C-terminal tail of hirudin lies in a long groove on the surface of thrombin that forms the anion-binding exosite (Figs. 2 and 6). In addition to the numerous electrostatic interactions made between thrombin and hirudin in this area, important hydrophobic contacts are formed between residues in the C-terminal tail of hirudin and the thrombin exosite. Particularly noteworthy in this respect are the last five residues of hirudin that form a 3_{10} helical loop (Rydel et al., 1990).

3. KINETIC MECHANISM FOR THE INHIBITION OF THROMBIN BY HIRUDIN

Studies on the kinetic mechanism of hirudin inhibition have indicated that hirudin is a competitive inhibitor of thrombin with respect to tripeptidyl p-nitroanilide substrates (Stone and Hofsteenge, 1986, 1991a; Stone et al., 1987). The substrate appears to compete with hirudin by binding to the active site and to a second, lower-affinity site (Stone and Hofsteenge, 1986; Stone et al., 1987). The identity of the second, lower-affinity site has not, however, been established. The crystal structures of thrombin–hirudin complexes confirm the proposed competitive mechanism for thrombin inhibition (Rydel et al., 1990; Grütter et al., 1990). The binding of the N-terminus of rHV2-Lys[47] (bold structure) and the substrate-like inactivator D-Phe-Pro-Arg-CH$_2$ (dotted structure) to the active site of thrombin are contrasted in Fig. 3. It can be seen that h-Ile-1 and h-Tyr-3 occupy the sites of D-phenylalanine and proline and, thus, the binding of a tripeptidyl substrate and hirudin to the active site will be mutually exclusive.

While studies by Degryse *et al.* (1989) suggest that rHV2-Lys[47] is a noncompetitive inhibitor with respect to a tripeptidyl chromogenic substrate, a recent study by Stone and Hofsteenge (1991a) indicates that rHV2-Lys[47] is a competitive inhibitor of human α-thrombin. An explanation for the discrepancy between the two studies is the presence of a contamination of γ-thrombin in the thrombin preparation used by Degryse *et al.* (1989). γ-Thrombin is a degraded form of α-thrombin that can arise through autolysis (Fenton *et al.*, 1977a,b). This form of thrombin has a much reduced activity with fibrinogen (Lewis *et al.*, 1987) and its affinity for hirudin is likewise greatly diminished (Landis *et al.*, 1978; Stone *et al.*, 1987). Simulations indicated that a contamination with γ-thrombin as small as 2% could yield data that suggest a noncompetitive mechanism.

On the time scale accessible using a conventional spectrophotometer, it was found that the formation of the thrombin–hirudin complex involves a two-step mechanism as illustrated in the following scheme:

$$E + I \underset{k_2}{\overset{k_1}{\rightleftharpoons}} EI \underset{k_4}{\overset{k_3}{\rightleftharpoons}} EI^*$$

where E and I represent thrombin and hirudin, respectively. The first step is ionic strength-dependent and does not involve the active site of thrombin; the rate of the association of hirudin with thrombin was found to be independent of the substrate concentration (Stone and Hofsteenge, 1986, 1991a; Stone *et al.*, 1987). Under conditions of low hirudin concentrations (picomolar), the first step is rate-limiting. In a second step, hirudin binds to the active site. The ionic strength-dependent interactions between hirudin and thrombin occur predominantly with the C-terminal region of hirudin and, thus, it can be assumed that the first step involves this region of hirudin. Removal of negatively charged residues from the C-terminal region of hirudin by site-directed mutagenesis both decreases the magnitude of the association rate constant for hirudin and reduces the ionic strength dependence of this constant (Braun *et al.*, 1988a; Stone *et al.*, 1989). The extent to which the interaction with the C-terminal region in the first step facilitates the binding of the N-terminal region of hirudin to the active site in the second step has been investigated by using fragments of hirudin. Studies by Konno *et al.* (1988) using circular dichroism spectroscopy indicate that the binding of hirudin induces a conformational change in thrombin. Chromatographic and spectroscopic studies by Mao *et al.* (1988) demonstrate that the binding of a C-terminal fragment of the hirudin molecule induces a similar effect. Evidence that this conformational change affects the active site of thrombin comes from studies of the kinetics of cleavage of tripeptidyl substrate in the presence of C-terminal peptides. In the presence of these peptides, the Michaelis constants of thrombin for tripeptidyl chromogenic and fluorogenic substrate are about 60% lower (Dennis *et al.*, 1990; Naski *et al.*, 1990). This change in the active-site region also favors the combination of an N-terminal fragment of hirudin consisting of residues 1 to 52; its K_d value is also 60% lower in the presence of a C-terminal peptide consisting of residues 53 to 65 (10nM compared with 24 nM without the peptide; Dennis *et al.*, 1990). Thus, the binding of the C-terminal region causes a conformational change that affects the active site of thrombin and slightly facilitates the binding of the rest of the hirudin molecule; the amount of cooperativity observed for the binding of the two regions is, however, small.

Comparison of the crystal structures of D-Phe-Pro-ArgCH$_2$–thrombin and thrombin–hirudin complexes indicates the possible nature of the conformational change induced by the binding of hirudin to thrombin. A large conformational change in the γ-loop of thrombin (140–150 loop) occurs upon the binding of hirudin. The main chain of this loop moves to open the active site of thrombin and to allow the binding of the hirudin N-terminal region. In D-Phe-Pro-ArgCH$_2$–thrombin, Trp-148 is positioned close to the active site and occupies a space that is filled by the main chain of h-Thr-4 and h-Asp-5 in the thrombin–hirudin complex (Rydel *et al.*, 1990; Grütter *et al.*, 1990). Thus, a conformational change in thrombin is necessary for the binding of the inhibitor to the active site. It is not clear which interactions made by the C-terminal region of hirudin trigger this change. Rydel *et al.* (1990) have speculated that it could arise from electrostatic interactions of h-Asp-55 with Lys-149e. The observed conformational changes in the γ-loop in the crystal structures, however, may result from differences in the packing of thrombin molecules in the unit cells of PPACK and hirudin structures (Skrzypczak-Jankun *et al.*, 1991).

Using a conventional recording spectrophotometer, it has been possible to estimate an apparent rate constant for the first step in the combination of hirudin with thrombin. Values of 1.4×10^8 and 4.7×10^8 M^{-1} S^{-1} were obtained for this parameter for rHV1 and native hirudin, respectively (Braun *et al.*, 1988a). The values obtained depend on ionic strength and those given are for an ionic strength of 0.125 M. More recently, estimates for the rate constants of the first and second steps (k_1 and k_3) have been obtained for rHV1 by using a fluorescence stopped flow spectrophotometer (M. Jackman, J. Hofsteenge, and S.R. Stone, unpublished results). The rate of interaction of hirudin with thrombin was followed in two ways: the decrease in fluorescence caused by the displacement of the fluorescent probe *p*-aminobenzamidine from the active site of thrombin (Evans *et al.*, 1982; Olson and Shore, 1982) and the enhancement in the fluorescence of thrombin that was found to occur during the course of complex formation. Although the cause of this fluorescence enhancement is not certain, the movement of Trp-148e caused by the combination of hirudin with thrombin could be at least partially responsible. Using nanomolar concentrations of hirudin and pseudo-first-order conditions, the observed rate constant for the fluorescence enhancement was dependent on the hirudin concentration and decreased with increasing ionic strength. Thus, it can be assumed that the rate of formation of the first complex was being followed. The value of the rate constant (k_1) obtained was in the range of that previously determined using a conventional spectrophotometer (Braun et al., 1988a). Using micromolar concentrations of hirudin, the first step in the formation of the complex was too fast to measure. An estimate of the rate of the second step could, however, be obtained. The rate observed was independent of the concentration of hirudin and did not depend on ionic strength. In contrast, it was influenced by the binding of p-aminobenzamidine to the active site. Thus, this step can be assigned to the isomerization step involved in the binding of hirudin to the active site. The rate constant for this step (k_3) is about 100 s^{-1} (M. Jackman, J. Hofsteenge, and S.R. Stone, unpublished results). At present, there are no estimates for the individual dissociation rate constants $(k_2$ and $k_4)$.

Studies on the pH-dependence of the dissociation constant for hirudin with thrombin between pH 6 and 10 indicated that the formation of the complex is dependent on the ionization states of three groups with pK_a values of about 7.3, 8.4, and 9.0 (A. Betz, J. Hofsteenge, and S. R. Stone, unpublished results). For optimal complex formation, the group with a pK_a of 7.3 must be ionized while the other two groups must be protonated. Two of these pK_a values (7.3 and 9.0) were also seen in the pH profiles for the interaction of thrombin with substrate and an active-site-directed inhibitor. The pK_a value of 7.3 can be assigned to the active site His-57 while that of 9.0 can be assigned to the α-amino group of the N-terminus of the B-chain of thrombin (Ile-16); an intramolecular salt bridge involving this group is essential for the activity of serine proteases (Fersht, 1972, 1985). A pK_a value of 8.4 was determined for the α-amino group of hirudin by titration with 2,4,6-trinitrobenzenesulfonic acid (Wallace et al., 1989). Moreover, when a hirudin variant with an acetylated α-amino group was used, the pK_a value of 8.4 was not observed (A. Betz, J. Hofsteenge, and S.R. Stone, unpublished results). Thus, the pK_a value of 8.4 can be assigned to the α-amino group of hirudin (A. Betz, J. Hofsteenge, and S.R. Stone, unpublished results). This assignment is also consistent with results obtained with chemically modified hirudins that indicate that a positively charged N-terminus is essential for the formation of a tight complex (see below and Wallace et al., 1989). The crystal structures of thrombin–hirudin complexes also indicate that the α-amino group of hirudin is involved in electrostatic interactions (Rydel *et al.*, 1990; Grütter *et al.*, 1990). In contrast to the observed pH-dependence of the dissociation constant, the association rate constant for hirudin is independent of pH over the range 6 to 10. The lack of interacting groups that ionize within this pH range in either the C-terminal tail of hirudin or the exosite of thrombin is consistent with this result. In addition, rhir(1–52) which lacks the C-terminal tail exhibits a pH-dependence identical to that observed with a rHV1 (A. Betz, J. Hofsteenge, and S.R. Stone, unpublished results), confirming that the groups responsible for the observed pK_a values do not interact with the C-terminal tail.

4. INTERACTION AREAS ON THROMBIN AND HIRUDIN

The results of studies aimed at identifying interaction areas on thrombin and hirudin and assessing their importance will be outlined below and discussed in relation to the crystal structures. For a more detailed discussion of the crystal structure of the thrombin–hirudin complex, the reader is referred to the chapter by Bode *et al.* In allowing the quantitative importance of particular interactions to be evaluated, solution studies provide information that cannot be obtained by x-ray crystallography. Thus, this aspect will be discussed wherever possible.

Protein-chemical, site-directed mutagenesis, and x-ray crystallographic studies have identified two regions of contact that seem to be particularly important for complex formation: the binding of the N-terminal residues of hirudin to the active-site cleft of thrombin; and the interaction between the C-terminal tail of hirudin and the anion-binding exosite. Each of these regions will be considered separately.

4.1. Interactions between the N-Terminal Region of Hirudin and the Active-Site Cleft of Thrombin

In the crystal structure determined by Rydel *et al.* (1990), contacts between the three N-terminal amino acids of hirudin and thrombin's active-site cleft accounted for 22% of the intermolecular contacts less than 4 Å. The N-terminal three residues of hirudin form a parallel β-sheet with residues 214–217 of thrombin (Fig. 3). This structure contrasts to the antiparallel β-sheet that has been seen in all previously determined structures of serine protease–inhibitor complexes (Read and James, 1986) including the D-Phe-Pro-ArgCH$_2$–thrombin complex (Bode *et al.*, 1989) and the hirulog 1–thrombin complex (Skrzypczak-Jankun *et al.*, 1991). Possible hydrogen bonds between the α-amino group of hirudin and Ser-195 (Rydel *et al.*, 1990) and the main chain carbonyl of Ser-214 (Grütter *et al.*, 1990) have been identified. Residues 1 and 3 of hirudin make numerous hydrophobic contacts. h-Val-1 (rHV1) and h-Ile-1 (rHV2-Lys⁴⁷) have hydrophobic interactions with His-57, Tyr-60a, Trp-60d, Leu-99, and Trp-215 while h-Tyr-3 binds to a cleft formed by Leu-9, Ile-174, and Trp-215. The side chains of h-Val-2 (rHV1) and h-Thr-2 (rHV2-Lys⁴⁷) make fewer hydrophobic contacts and are bound at the entrance to the primary specificity pocket (Fig. 3).

The importance of interactions between hirudin and the active-site cleft was originally identified by using forms of thrombin with their modified active sites. Diisopropylfluorophosphate inactivates thrombin by covalently modifying the active-site serine to form diisopropylphosphoryl (DIP)-thrombin. Access to the primary specificity pocket appears also to be blocked in DIP-thrombin (Stone *et al.*, 1987). Contact with the active-site histidine (His-57) is also expected to be hindered. Hirudin is still able to bind to DIP-thrombin but with a reduced affinity (Fenton *et al.*, 1979; Stone *et al.*, 1987). The complex between native hirudin and human α-thrombin has a dissociation constant of 20 fM (2×10^{-14} M; Stone & Hofsteenge, 1986) which corresponds to a binding energy of 81 kJ mol^{-1}. The affinity of native hirudin for DIP-thrombin is reduced by about 10^3-fold and the binding energy is 18 kJ mol^{-1} lower (Stone *et al.*, 1987). These results suggest that interactions involving the region of thrombin around the active-site serine contribute a maximum of 18 kJ mol^{-1} to binding energy. This value should be regarded with some caution as modification of this region of thrombin may force hirudin to bind in a manner such that interactions with regions distant from the active-site serine may also be disturbed. D-Phe-Pro-ArgCH$_2$Cl is a specific inactivator of α-thrombin that alkylates the active-site histidine (Kettner and Shaw, 1981; Glover and Shaw, 1971; Bode *et al.*, 1989). The peptide portion of this inactivator binds to the primary specificity pocket of thrombin as well as to the apolar binding site (Kettner and Shaw, 1981; Sonder and Fenton, 1984; Walker *et al.*, 1985; Bode *et al.*, 1989). A tetrahedral hemiketal with Ser-195 O$^\gamma$ and a covalent bond to His-57 N$^\varepsilon$ are also formed (Bode *et al.*, 1989). The affinity of native hirudin for D-Phe-Pro-ArgCH$_2$–thrombin is 10^6-fold lower than its affinity for α-thrombin which corresponds to a decrease in binding energy of 35 kJ mol^{-1} (Stone *et al.*, 1987). This decrease is binding energy can be considered to be the maximum contribution of interactions with thrombin in the region of the active-site serine and histidine and with the apolar binding site.

Results of studies in which the α-amino group of hirudin was modified indicate that a salt bridge or hydrogen bond involving this group makes an important contribution to binding energy. Attempts to express recombinant hirudin indicated that any extension of the N-terminus causes a marked decrease in its inhibitory activity (Bergmann *et al.*, 1986; Loison *et al.*, 1988). Extension of the N-terminus by a single glycyl residue results in a 20 kJ mol^{-1} loss in binding energy (Wallace *et al.*, 1989). A large part of this decrease is due to disruption of electrostatic interactions involving the positively charged α-amino group as was shown by experiments in which this group was specifically modified (Wallace *et al.*, 1989). Removal of the positive charge of the α-amino group by acetylation reduces the binding energy by 27 kJ mol^{-1}. Acetimidation of the α-amino group adds a moiety of similar size to the acetyl moiety but maintains a positive charge and the resultant decrease in binding energy upon this modification is only 12 kJ mol^{-1}. The difference in binding energies between the acetylated and acetimidated hirudin (15 kJ mol^{-1}) can be naively attributed to bonds made by the positively charged α-amino group of hirudin. This value corresponds to that of about 12 kJ mol^{-1} suggested by Fersht (1987) for a charged hydrogen bond. This observation is consistent with the electrostatic interactions observed between the α-amino group and Ser-195 and other residues in its vicinity (Rydel *et al.*, 1990; Grütter *et al.*, 1990). Studies using shortened forms of hirudin have also suggested the importance of interactions with a full-length N-terminal region of hirudin (Dodt *et al.*, 1987).

The importance of the hydrophobic contacts made by the three N-terminal residues of hirudin has been demonstrated by site-directed mutagenesis (Wallace *et al.*, 1989; A. Betz, J. Hofsteenge, and S.R. Stone, unpublished results). Wallace and co-workers substituted both N-terminal valyl residues of rHV1 by a series of amino acids and the results are shown in Fig. 4. Replacement of these residues by polar amino acids (Lys, Glu, and Ser) causes a marked decrease in the binding energy; for example, replacement of h-Val-1 and h-Val-2 by seryl residues causes a decrease in binding energy of 17 kJ mol^{-1}. In contrast, conservative replacements by other hydrophobic amino acids (Ile, Phe, Leu) result in moderate or negligible changes in the strength of binding; substitution with isoleucine actually causes a small increase in binding energy (Fig. 4). The relative importance of the first and second positions of hirudin has been investigated by substitution of h-Val-1 and h-Val-2 separately (Wallace *et al.*, 1989; A. Betz, J. Hofsteenge, and S.R. Stone, unpublished results). The results, illustrated in Fig. 5, indicate that binding sites for h-Val-1 and h-Val-2 vary in their ability to accommodate other amino acid residues. Whereas leucine and glycine are better tolerated in position 1 than in position 2, the converse is true for serine and lysine, and glutamate was equally poor in both positions. Modeling of these mutants based on the

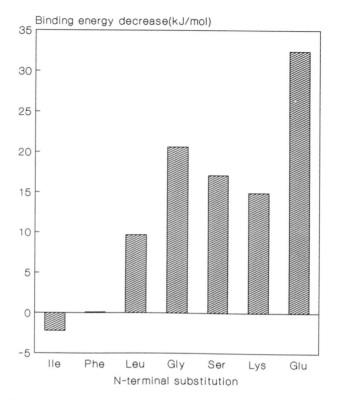

Figure 4. Effect of substitution of h-Val-1 and h-Val-2. The decrease in binding energy caused by the replacement of both h-Val-1 and h-Val-2 by the indicated amino acid was calculated from the data of Wallace *et al.* (1989).

crystal structures provides some insight into the reasons for the observed differences in binding energy (A. Betz, J. Hofsteenge, and S.R. Stone, unpublished results). In both crystal structures, the side chains of the N-terminal residue make a greater number of close contacts with thrombin than the second residue (Rydel *et al.*, 1990; Grütter *et al.*, 1990). Thus, the greater loss of binding energy with h-Val-1→Ser compared with h-Val-2→Ser can be rationalized on the basis of the crystal structure. The effects of other mutations are more difficult to explain. Modeling indicates that, whereas leucine would be able to bind in the h-Val-1 site without any problems, it is not possible to bind leucine in the h-Val-2 site without unfavorable contacts. The decrease in binding energy observed with the h-Val-1→Glu is presumably due to disruption of the electrostatic interactions of the positively charged α-amino group. In the h-Val-2→Glu mu-

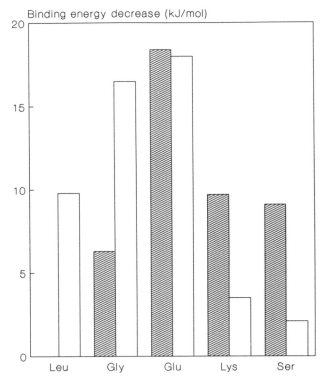

Figure 5. Effect of substitution of h-Val-1 or h-Val-2. The decrease in binding energy caused by the substitution of h-Val-1 and h-Val-2 by the indicated amino acids are represented by the hatched and open bars, respectively (Wallace *et al.*, 1989; A. Betz, J. Hofsteenge, and S. R. Stone, unpublished results).

tant, the negatively charged carboxylate is positioned unfavorably with respect to Glu-192. In contrast, it is possible that a lysyl residue in the second position could bind in the primary specificity pocket with the resultant formation of ionic interaction with Asp-189. The greater effect on binding energy seen with h-Val-2→Gly in comparison with h-Val-1→Gly cannot be explained by the loss of hydrophobic contacts. Perhaps the substitution of glycine in position 2 leads to a great increase in entropy of the free molecule. Crystallographic studies on complexes of thrombin with these mutant hirudins are required to resolve fully these apparent inconsistencies.

Examination of the binding energies for mutants in which h-Val-1 and h-Val-2 have been replaced either separately or together indicates that there is little cooperativity in the binding of these two residues to the active site. In all cases, the sum of the loss of binding energy for the single mutations was within 20% of the observed loss for the double mutation. This lack of cooperativity in the binding of different residues of hirudin is also seen for negatively charged residues in the C-terminal region (discussed below) and in the binding of N-terminal and C-terminal fragments (discussed above).

In the context of interactions with the active-site cleft of thrombin, the role of h-Lys-47 should be mentioned. Thrombin has a clear specificity for substrates with an arginyl residue in the P_1 position. However, the sequence of hirudin does not contain any arginines and, therefore, it was assumed by a number of workers that one of the lysyl residues would be bound in the primary specificity pocket. Since an invariant proline precedes h-Lys-47 in the hirudin sequence and good substrates often have proline in the P_2 position (Chang, 1985; Claeson et al., 1977), it was predicted that h-Lys-47 would be bound in the primary specificity pocket (Walsmann and Markwardt, 1981; Fenton, 1989; Johnson et al., 1989). Site-directed mutagenesis experiments, however, failed to detect a crucial role for h-Lys-47 or, for that matter, any of the other basic residues (Dodt et al., 1988; Braun et al., 1988a; Degryse et al., 1989). All studies indicated a small, but significant, decrease in inhibitory potency when h-Lys-47 was replaced by another amino acid. The role of this residue has subsequently been elucidated by the crystal structures. The N^ε of h-Lys-47 helps to position the N-terminal region by forming hydrogen bonds with the backbone carbonyl of h-Asp-5 and the O^γ of h-Th-4 (Rydel et al., 1990).

4.2. Interactions between the C-Terminal Region of Hirudin and the Anion-Binding Exosite of Thrombin

The crystal structure of Rydel et al. (1990) indicates that the C-terminal tail region of hirudin from residue 49 to 65 makes numerous electrostatic and hydrophobic interactions. The C-terminal tail of hirudin between residues 55 and 65 make 38% of the intermolecular contacts less that 4 Å and our discussion will be centered on these interactions (shown in Fig. 6). Within this region, salt bridges are made by h-Asp-55, h-Glu-57, and the carboxylate of h-Gln-65 with Arg-73/Lys-149e, Arg-77a (and, possibly, Arg-75), and Lys-37, respectively (Rydel et al., 1990). Important hydrophobic contacts are made by h-Phe-56 which binds into a pocket formed by Phe-34, Arg-73, and Thr-74. The residues h-Ile-59, h-Pro-60, and h-Tyr-63 make contacts with hydrophobic residues on the surface of thrombin (Leu-65, Tyr-76, and Ile-82; Rydel et al., 1990).

Before the publication of the crystal structure of the thrombin–hiru-

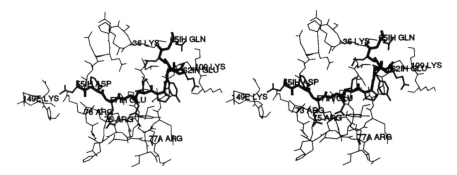

Figure 6. Binding of the C-terminal region of hirudin to human α-thrombin. The main chain of the C-terminal region of hirudin (residues 55 to 65) is shown in thick lines while thrombin residues are in thinner lines. The structure shown is that determined by Rydel *et al.* (1990). Hirudin residues are given as numbers with an "IH" suffix.

din complex, the anion-binding exosite was identified as the binding site for the C-terminal tail of hirudin by using protein chemical methods. Hirudin is able to protect the lysyl residues Lys-60f, Lys-70, Lys-109, Lys-110, and Lys-149e from modification by a lysine-specific reagent (Chang, 1989). All of these residues are located within the vicinity of the anion-binding exosite. Chang *et al.* (1990a) also showed that a hirudin fragment consisting of residues 52 to 65 protects the same residues from modification which suggested that the C-terminal tail of hirudin is bound to the exosite. The studies of Bourdon *et al.* (1990), using an N^{α}-dinitrofluorobenzyl analog of hir(54–64), indicate that Lys-149e forms part of the binding site for the C-terminal tail of hirudin. Hirudin also protects α-thrombin from cleavage at Arg-67, Arg-77a, and Lys-149e by trypsin and from cleavage at Ala-149a by elastase (Dodt *et al.*, 1990; Dennis *et al.*, 1990). C-terminal fragments of hirudin [rHV1(53–65) or rHV1(48–65)] protect thrombin from cleavage by trypsin while the corresponding N-terminal fragments afford little protection. In contrast, the N-terminal fragments protect thrombin from cleavage by elastase and only partial protection is observed with the C-terminal fragments (Dodt *et al.*, 1990; Dennis *et al.*, 1990). These results suggest that the C-terminal tail of hirudin is bound to an exosite that includes residues Arg-67 to Arg-77a. In addition, the accessibility of the Ala-149a–Asn-149b bond to elastase is blocked by the binding of the N-terminal fragment and altered by the binding of the C-terminal fragment.

The importance of interactions between the C-terminal tail of hirudin and the anion-binding exosite was originally demonstrated by the effect of partial proteolysis of hirudin and thrombin on the affinity of the two molecules for each other. Chang (1983) showed that progressive removal of amino acids from the C-terminal region of hirudin results in a progressive decrease in its inhibitory activity. Similar results were obtained by Dodt *et al.* (1987). Recent studies have shown that the removal of 13–22 residues from recombinant hirudin results in a decrease in affinity up to 10^6-fold which corresponds to a 35 kJ mol^{-1} decrease in binding energy (Degryse *et al.*, 1989; Chang *et al.*, 1990b; Schmitz *et al.*, 1991; Dennis *et al.*, 1990). Thus, interactions of the C-terminal tail of hirudin seem to contribute about 40–50% of the total binding energy.

The involvement of the anion-binding exosite of thrombin in the binding of hirudin was demonstrated using proteolytic derivatives of thrombin (Landis *et al.*, 1978; Stone *et al.*, 1987). Cleavage of human α-thrombin between Arg-77a and Asn-78 by trypsin produces β_T-thrombin (Braun *et al.*, 1988b; Bezeaud and Guillin, 1988). Further clevage by trypsin of the bonds Arg-67–Ile-68 and Lys-149e–Gly-150 leads to the formation of γ_T-thrombin (Braun *et al.*, 1988b; Bing *et al.*, 1977; Fenton *et al.*, 1977a,b). β_T-Thrombin has a 140-fold reduced affinity for native hirudin while γ_T-thrombin shows a 10^6-fold reduction in affinity (Stone *et al.*, 1987). The reduced affinity of β_T-thrombin was originally interpreted as being due to a decrease in the strength of ionic interactions between hirudin and the anion-binding exosite (Stone *et al.*, 1987). More recent studies have shown, however, that the strength of ionic interactions is maintained in the β_T-thrombin–hirudin complex but that the hydrophobic interactions with the C-terminal tail of hirudin are disrupted (Stone and Hosfsteenge, 1991b). The importance of the anion-binding exosite was also demonstrated by using antibodies against the sequence Arg-67–Arg-77a. These antibodies were able to block the inhibitory activity of hirudin (Noé *et al.*, 1988).

The important role of acidic residues in the C-terminal tail of hirudin has been defined by studies using site-directed mutagenesis. First, it was found that recombinant hirudin, which lacks the negatively charged sulfate group on h-Tyr-63, has a 10-fold lower affinity for thrombin than the native molecule (Dodt *et al.*, 1988, 1990; Braun *et al.*, 1988a). The full activity is restored by enzymatic phosphorylation (Hofsteenge *et al.*, 1990) or sulfation of h-Tyr-63 (Niehrs *et al.*, 1990). Sequential removal of four more negative charges by mutation of h-Glu-57, h-Glu-58, h-Glu-61, and h-Glu-62 to glutamine results in a progressive decrease in binding energy. The contribution of ionic and nonionic interactions to the binding energy of each of these mutants was assessed by examining the effect of ionic strength on complex formation (Stone *et al.*, 1989). The nonionic contribution to binding energy is about the same for all mutants and for recombinant and native hirudin. In contrast, the ionic contribution decreases in a linear way as negative charges are removed. At zero ionic strength, each charge contributes about 4 kJ mol^{-1} to the overall binding energy. Ionic interactions contribute about one-third of the binding energy at zero ionic strength. This value is ionic strength-dependent and decreases to about 20% at physiological ionic strength (Stone *et al.*, 1989).

The results of the above studies, which suggest that each of the negatively charged residues between residues 57 and 65 contributes equally to binding energy, seem at variance with the crystal structure that indicates that only h-Glu-57 makes a direct ionic interaction. However, calculations of electrostatics field strengths based on the structure of Rydel *et al.* (1990) indicate that the C-terminal tail of hirudin and the anion-binding exosite of thrombin form complementary electrostatic fields and that the interaction of these fields makes a significant contribution to binding energy (A. Karshikov, W. Bode, and S.R. Stone, unpublished results). In this case, removal of negative charges from the C-terminal region of hirudin would reduce its electrostatic field strength and consequently lead to a reduction in binding energy. The contribution of ionic interactions to the overall binding energy may be secondary to other functions that they perform. The ionic interactions involved in the first, ionic strength-dependent step of complex formation could serve to orientate the two molecules as a prelude to the formation of a tighter complex in the subsequent step. Thus, the ionic interactions with the C-terminal tail could be seen as also fulfilling a docking function; that is, they could function to align the two molecules and reduce nonproductive complex formation. In this case, the clustering of negatively charged residues in the C-terminal region of hirudin would create a more effective electrical field for a long-range orientation. A similar docking function for electrostatic interactions has been proposed for other molecules (Matthew, 1985).

Site-directed mutagenesis has also been used to investigate the importance of interactions with hydrophobic residues in the C-terminal tail. These studies have concentrated on three residues: h-Phe-56, h-Pro-60, and h-Tyr-63 (A. Betz, J. Hofsteenge, and S. R. Stone, unpublished results). The results obtained for h-Phe56 are shown in Fig. 7. Studies with synthetic peptides based on the C-terminal sequence of hirudin that are discussed below suggested that h-Phe-56 is critical for efficient binding of these peptides. The data presented in Fig. 7, however, indicate that the decrease in binding energy for rHV1 caused by the loss of the aromatic ring (h-Phe-56→Ala) is small (1.9 kJ mol^{-1}); thus, the contribution of h-Phe-56 to the binding energy of hirudin appears to be small. A similar effect is observed when the h-Phe-56 is replaced by leucine (Köhler *et al.*, 1989). The conservative replacement of h-Phe-56 with tyrosine actually results in an increase in binding energy. Larger decreases in binding

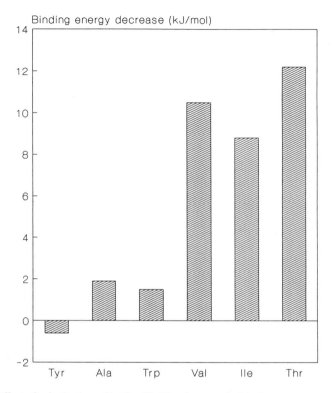

Figure 7. Effect of substitution of h-Phe-56. The decrease in binding energy caused by the replacement of h-Phe-56 by the indicated amino acids was calculated from the data of A. Betz, J. Hofsteenge, and S. R. Stone (unpublished results).

energy occur when h-Phe-56 is replaced by a residue with a side chain branched at the C^β atom (Val, Ile, or Thr). Molecular modeling of these mutants does not indicate an obvious reason for the marked decrease in binding energy observed with these replacements in contrast to h-Phe-56→Ala (A. Betz, J. Hofsteenge, and S.R. Stone, unpublished results). A possible explanation is that, because of unfavorable contacts between one of the C^γ atoms and backbone atoms, the main chain moves and interactions with h-Asp-55, h-Glu-57, or other residues are disturbed. The contribution to binding energy made by h-Pro-60 is somewhat larger; mutation of this residue to an alanine or glycine results in a decrease in binding energy of about 6 kJ mol^{-1} (A. Betz, J. Hofsteenge, and S.R. Stone, unpublished results). The extent to which this decrease is due to loss of hydrophobic interactions with h-Pro-60 or to an increase in entropy of the unbound hirudin is not known. The aromatic ring of h-Tyr-63 appears to make about the same contribution to binding energy as that of h-Phe-56; a 2.7 kJ mol^{-1} decrease was observed when the substitution h-Tyr-63→Ala was made. In contrast to the binding site for h-Phe-56, however, that for h-Tyr-63 seems to be more able to accommodate residues with C$^\beta$ branched side chains. The decrease in binding energy for the h-Tyr-63→Val substitution was only 5.1 kJ mol^{-1} compared with 10.5 kJ mol^{-1} for the h-Phe-56→Val mutant (A. Betz, J. Hofsteenge, and S.R. Stone, unpublished results).

Although the individual contributions of interactions with h-Phe-56, h-Pro-60, and h-Tyr-63 to binding energy are small, the sum of their contributions amounts to more than 10 kj mol^{-1} which is 13–14% of the overall binding energy. Thus, even without considering the contributions of h-Ile-59 and h-Leu-64, which have been shown to be necessary for efficient binding of C-terminal peptides (Krstenansky et al., 1987), the importance of hydrophobic contacts between the C-terminal tail and the exosite is apparent.

5. STRUCTURE–FUNCTION RELATIONSHIPS IN SYNTHETIC HIRUDIN PEPTIDES

As described above, the findings of Chang (1983) showed that the COOH-terminal tail is required for efficient inhibition of thrombin by hirudin. However, these results did not suggest that synthetic peptides of hirudin would themselves exhibit antithrombin activities. This observation was first established by Bajusz et al. (1984) who found that synthetic N-Boc peptide fragments of hirudin corresponding to residues 61–65 inhibited thrombin cleavages of fibrinogen, albeit with IC$_{50}$ values in the millimolar range. The poor activity of these fragments is consistent with subsequent data (Krstenansky and Mao, 1987; Mao et al., 1988; Maraganore et al., 1989) that have shown that longer synthetic peptides are required for more effective antithrombin activity.

Following the report by Bajusz et al., Krstenansky and Mao (1987) documented that antithrombin activities of unsulfated N^α-acetyl-hirudin$_{45–65}$, a 21-residue peptide. This analysis showed clearly that synthetic hirudin fragments as long as 21 amino acids inhibit thrombin's cleavage of fibrinogen. However, the peptides did not inhibit thrombin catalytic activity with small chromogenic substrates. Further, as the fragment studied included the h-Pro$_{46}$-Lys$_{47}$-Pro$_{48}$ sequence, that was presumed to comprise the hirudin domain for active-site inhibition (Walsmann and Markwardt, 1981; Chang, 1985; Fenton, 1989; Johnson et al., 1989; Degryse et al., 1989), the studies with hirudin$_{45–65}$ showed further that inhibition of the thrombin active center by hirudin occurred by a unique mechanism. The dissociation constant of hirudin$_{45-65}$ for thrombin was calculated at ~10 μM and the functional studies showed an IC$_{50}$ value ~10,000-fold greater than that determined for native hirudin (Krstenansky and Mao, 1987).

Together, the findings of Bajusz *et al.* (1984) and Krstenansky and Mao (1987) showed that synthetic hirudin COOH-terminal peptides inhibit thrombin coagulant activity without modifying thrombin amidolytic function. As the characteristics of thrombin inhibited by such peptides were essentially the same as was described for the limit proteolytic forms of thrombin, $β_T$- and $γ_T$-thrombin (Bing *et al.*, 1977; Fenton *et al.*, 1977a; Hofsteenge *et al.*, 1988), it was appropriate to propose that the hirudin fragments were binding to the thrombin anion-binding exosite (Berliner *et al.*, 1985; Fenton *et al.*, 1988). As noted previously, the work of Bourdon *et al.* (1990) and Chang *et al.* (1990a) showed by chemical modification experiments that amino acids of both the β- and γ-thrombin cleavage sites participate in the interactions of hirudin COOH-terminal peptides with thrombin.

Studies on a diverse set of synthetic hirudin peptides have contributed to delineation of structure–function relationships in hirudin. First, studies on the minimal length of such fragments required for efficient antithrombin activity showed that amino acids from residues 53 to 64 are critical for high-affinity interactions with thrombin (Krstenansky *et al.*, 1987; Mao *et al.*, 1988; Maraganore *et al.*, 1989). There was no significant loss of anticoagulant activity when the 20-residue fragment hirudin $_{45-64}$ was truncated to the 12-residue fragment hirudin$_{53-64}$ and further truncation to fragments corresponding to residues 55–64 reduced the activity by only 3-fold. Shortening of the fragment to 8 residues (57–64) abolished activity altogether (Fig. 8) (Maraganore *et al.*, 1989). Thus, antithrombin activity of synthetic hirudin fragments required a minimal peptide segment of 10–12 amino acids.

Studies of synthetic hirudin peptides have also allowed an evaluation of essential residues for interactions of the hirudin COOH-terminus with the thrombin anion-binding exosite, as well as the identification of natural and synthetic amino acid substituents which can improve antithrombin activity (Krstenansky *et al.*, 1988, 1989, 1990). These studies have complemented the site-directed mutagenesis experiments detailed above. The comprehensive work of Krstenansky and co-workers should be briefly reviewed in light of the x-ray crystallographic structure of the hirudin:thrombin complex (Rydel *et al.*, 1990). First, it was found that h-Glu-57, h-Ile-59, h-Pro-60, h-Glu-61, and h-Glu-62 are critical for efficient antithrombin activity. Based on the crystal structure, h-Glu-57, h-Ile-59,

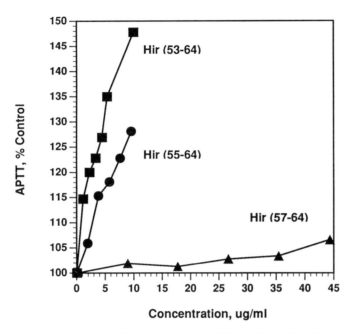

Figure 8. Anticoagulant activites of hirudin peptides of 12 (53–64), 10 (55–64), and 8 (57–64) amino acids in length. Studies on synthetic hirudin peptides have shown that the minimal peptide length with maximal antithrombin activity is 11–12 amino acid residues. APTT assays were performed using pooled, normal human plasma as described by Maraganore *et al.* (1989).

h-Pro-60 make important electrostatic and hydrophobic interactions with thrombin. However, the importance of h-Glu-61 and h-Glu-62 cannot be explained based on crystallographic studies alone. As noted above, these anionic residues may contribute principally to the establishment of complementary electrostatic fields required for efficient hirudin–thrombin interaction. Another key amino acid identified in peptide studies is h-Phe-56. This amino acid could only be replaced by tyrosine or beta-(2-thienyl)alanine, while other changes resulted in significant diminution of activity. The role of h-Phe-56 is clear from the crystallographic studies, where this residue was found to make important van der Waals contacts with Phe-34 and Leu-62 (Rydel *et al.*, 1990).

It should be noted that several amino acid substitutions in hirudin$_{55-65}$ resulted in increased activity. h-Asp-55 can be removed altogether when an NH_2-terminal, anionic alkyl group is employed as a substitution. Such a modification would preserve the electrostatic interactions with Lys-149e and Arg-73 observed in the crystal structure of hirudin–thrombin (Rydel *et al.*, 1990). When h-Glu-58 was replaced by a proline residue, as found in the hirudin isoform, hirudin-PA (Dodt *et al.*, 1986), activity was increased 2-fold. Modeling studies show that this modification could stabilize positioning of the 3_{10} helix (which includes residues 61–65). Likewise, several amino acid substitutions at positions 63-65, including incorporation of a D-amino acid at position 65, were evaluated and some derivatives showed marked increases in inhibitory activities (Krstenansky *et al.*, 1990). Modeling studies would indicate that the consequence of these changes is to stabilize formation of the 3_{10} helix (E. Skrzypczak-Jankun, V. Carperos, P. Bourdon, J. M. Maraganore, and A. Tulinsky, unpublished results).

In addition to defining the minimal hirudin sequence for exosite interactions and allowing a systematic evaluation of important amino acids within this hirudin domain, studies on hirudin COOH-terminal peptides have confirmed a clear role for the O-sulfate ester of h-Tyr-63 (Maraganore *et al.*, 1989). This posttranslational modification in hirudin is absent

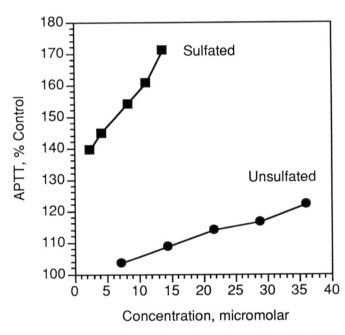

Figure 9. Effects of tyrosine sulfation on the anticoagulant activity of the COOH-terminal dodecapeptide, hirugen. As a result of tyrosine sulfation, the peptide "hirugen," *N*-acetyl-Hir(53–64), shows a marked increase in anticoagulant activity as compared to its unsulfated precursor. X-ray crystallographic analysis of the hirugen–thrombin complex has shown that the sulfate group elaborates an intricate hydrogen-bonding network with thrombin.

in recombinant forms of the protein, and, thus, evaluation of its role in hirudin–thrombin interactions has been difficult until most recently (Hofsteenge et al., 1990; Niehrs et al., 1990). Studies by Maraganore and co-workers showed that sulfation of h-Tyr-63 in a dodecapeptide hirudin fragment increases anticoagulant activity by over one order of magnitude (Fig. 9). The Tyr-sulfated hirudin dodecapeptide, also called "hirugen," inhibited thrombin procoagulant activity with a $K_i \sim 150$ nM. In order to determine the interactions of hirudin with thrombin, Tulinsky's group (E. Skrzypczak-Jankun, V. Carperos, P. Bourdon, J.M. Maraganore, and A. Tulinsky, unpublished results) recently solved the x-ray crystallographic structure of the thrombin–peptide complex (Skrzypczak-Jankun et al., 1991). This analysis showed that sulfation of h-Tyr-63 allows formation of a hydrogen-bonding network including the phenolic hydroxyl group of Tyr-76 and the main-chain amide nitrogen of Ile-82. Of course, in addition to these contacts, Tyr-sulfation may further stabilize thrombin–hirudin interactions by contributing to an anionic electrostatic field. The role of posttranslational modification of h-Tyr-63 sulfation in intact hirudin has been evaluated further by Hofsteenge et al. (1990) and Niehrs et al. (1990), who showed that sulfation or phosphorylation of h-Tyr-63 results in increases in binding affinity.

Results described above show that synthetic fragments of the hirudin COOH-terminal domain interact with the thrombin anion-binding exosite and establish that many interactions are essential for efficient inhibitory activity. Since these fragments inhibit thrombin cleavages in fibrinogen without inhibiting catalytic-site function, the mechanism of action is of particular interest. One explanation could derive from data showing that hirudin peptides induce a conformational change in thrombin (Mao et al., 1988). Hirudin peptides are also known to increase thrombin amidolytic activities toward chromogenic or fluorogenic substrate in a saturable, concentration-dependent fashion (Naski et al., 1990, Dennis et al., 1990), perhaps as a result of the conformational change. While conformational changes in thrombin can explain inhibitory actions of hirudin peptides, the studies of Naski et al. (1990) showed that the fragment of hirudin is a pure competitive inhibitor with respect to fibrinogen. These findings established clearly that hirudin binding to the anion-binding exosite prevents formation of a Michaelis complex of thrombin and fibrinogen. Further, these data confirmed that the anion-binding exosite defines a critical determinant for the binding of thrombin to fibrinogen. Interestingly, similar differential effects on thrombin amidolytic activity versus cleavages in fibrinogen have been observed with thrombomodulin (Jakubowski et el., 1986; Hofsteenge et al., 1986) and antipeptide antibodies toward the thrombin exosite domain (Noé et al., 1988). As with hirugen and other synthetic peptides, inhibition of thrombin by thrombomodulin and the exosite-directed antibodies occurs without inhibition of active-site function. Thus, it would appear that the exosite comprises a discrete functional domain relevant to a broad set of physiologic thrombin activities.

6. HIRUDIN PEPTIDES AS PROBES OF THROMBIN STRUCTURE–FUNCTION

Since hirugen and other hirudin fragments bind to the anion-binding exosite, they represent excellent tools for the evaluation of thrombin activities which are dependent on the exosite.

In addition to cleaving fibrinogen to form fibrin monomers, thrombin also acts to stabilize the polymerized fibrin clot by activation of factor XIII. Recent data indicate that following clot formation, thrombin can bind to fibrin and promote clot accretion. Thrombin's binding to the fibrin clot results in a 30-fold rate enhancement of its activation of factor XIII (Naski *et al.*, 1991) and leads to protection from neutralization by heparin–antithrombin III (ATIII) (Hogg and Jackson, 1989; Weitz *et al.*, 1990). Hirugen was found to inhibit thrombin-catalyzed activation of factor XIII (Naski *et al.*, 1991) and to inhibit thrombin binding to fibrin II (Naski *et al.*, 1990). In both cases, the results were consistent with an inhibition by hirugen of thrombin binding to fibrin, implying that the thrombin exosite interacts with fibrin in addition to the substrate fibrinogen. In the studies of Weitz *et al.* (1990), hirugen was found to inhibit fibrinogen hydrolysis catalyzed by clot-bound thrombin in plasma as measured by immunoassays for FPA. Hirugen and r-hirudin, however, failed to disrupt the binding of [125I] thrombin to the fibrin clot at concentrations significantly greater than those required to inhibit cleavage of fibrinogen. Therefore, it seems that thrombin can also bind to fibrin in a fashion which allows formation of a fibrinogen–thrombin Michaelis complex, which in turn can be inhibited by hirugen. Such binding to fibrin would thus occur in an anion-binding exosite-independent manner. Additional studies are required to resolve the observations that thrombin may bind to fibrin in both anion-binding exosite-dependent and -independent fashions.

The possibility that thrombin's activation of the prothrombinase cofactor, factor V, is dependent on the anion-binding exosite was examined by Ofosu *et al.* (1991). These studies showed that cleavage by thrombin of factor V is inhibited only partially by hirugen. Examination of the pattern of factor V proteolysis by thrombin in the presence of hirugen showed formation of the major heavy chain fragment, but inhibition of light chain fragment generation. This would imply that cleavage by thrombin of the light chain scissile bond is anion-binding exosite-dependent while generation of the heavy chain is exosite-independent. Interestingly, as a result of its failure to inhibit heavy chain formation, hirugen could not completely prevent prothrombin activation induced via the extrinsic or intrinsic pathways of coagulation. These reactions are only delayed, thus explaining the saturable effect of synthetic hirugen peptides on prolongation of blood coagulation (Jakubowski and Maraganore, 1990).

Another macromolecular substrate of thrombin is protein C, whose activation by limit proteolysis results in negative feedback control of coagulation. Examination of the inhibition by hirugen of thrombin-catalyzed activation of protein C showed partial, noncompetitive inhibition of activated protein C formation (Bourdon and Maraganore, 1989). The inhibition of protein C activation is reversed in the presence of either solubilized thrombomodulin (TM) or A549 cells, a human lung carcinoma cell line which expresses large quantities of surface TM. The ability of TM to reverse the inhibitory effects of hirugen on protein C activation would imply that TM binds to the thrombin exosite.

Studies on the effects of TM on thrombin interactions show that TM competes with hirudin for thrombin binding and, most recently, this finding has been found to apply with hirudin COOH-terminal peptides (Tsiang et al., 1990). Furthermore as with the binding of hirudin fragments (Naski et al., 1990; Dennis et el., 1990), TM binding enhances thrombin hydrolysis of tripeptidyl substrates (Hofsteenge et al., 1986). However, unlike hirudin peptide binding (Naski et al., 1990; Dennis et al., 1990), TM binding accelerates the inactivation of thrombin by ATIII as well as the rate of protein C activation (Hofsteenge et el., 1986; Bourin et al., 1986; Preissner et al., 1987). These latter properties of TM would indicate additional interactions with thrombin, possibly the interactions of glycosidic moieties of TM with a separate thrombin exosite (Bourin and Lindahl, 1990; Bourin et al., 1986, 1988; Parkinson et al., 1990). Nevertheless, principal interactions of TM with thrombin are mediated through the anion-binding exosite as defined by the limit proteolytic or catalytic-site-inactivated derivatives of thrombin (Hofsteenge et al., 1986; Noé et al., 1988; Suzuki et al., 1990). Using chemical and proteolytic fragments or recombinant mutants of TM, the anion-binding exosite recognition domain has been mapped within the fifth and sixth epidermal growth factor (EGF)-like domains (Kurosawa et al., 1988; Zushi et al., 1989).

The anion-binding exosite-dependence of thrombin inactivation by ATIII was studied by Naski et al. and Dennis et al. (1990). In addition, since heparin is known to catalyze ATIII inactivation of thrombin by binding to both thrombin and ATIII in the ternary complex (Pomerantz and Owen, 1978; Griffith, 1982; Nesheim, 1983; Olson, 1988), the exosite-dependence of the heparin-catalyzed reaction was also investigated by Naski *et al.* (1990). Using hirugen as a probe of the exosite, the rate of thrombin–ATIII complex formation was found to be inhibited by no more than 2-fold. A similar effect of hirugen was observed on the rate of thrombin–ATIII complex formation in the presence of heparin. These results indicate that the anion-binding exosite is not required for inactivation of thrombin by ATIII. Since complex formation was similarly modified in the presence of heparin, the site on thrombin for the binding of heparin to catalyze ATIII inactivation appears to be distinct from the anion-binding exosite. The evidence of a distinct site on thrombin for heparin binding is supported by the finding that thrombin can bind to heparin–agarose in the presence of saturating hirugen concentrations (J.M. Maraganore and P. Bourdon, unpublished results) and that heparin has only a minor effect on the inhibition of thrombin by hirudin (Stone and Hofsteenge, 1987). Further, chemical modification studies (Church *et al.*, 1989) showed that heparin binding protects modification of Lys-169 and Lys-224 in thrombin. These amino acids of a putative heparin-binding site are quite distant from the anion-binding exosite as defined by the locus for hirudin/hirugen binding, indicating that thrombin may contain an additional site for binding anionic ligands.

While interaction of ATIII and the ATIII–heparin complex appear to be independent of the exosite, the inactivation of thrombin by heparin cofactor II (HCII) may include interactions at this locus (Hortin *et al.*, 1989). A synthetic, anionic peptide segment of HCII, corresponding to residues 54–75, was found to inhibit thrombin cleavages of fibrinogen without inhibiting amidolytic activity. Further, HCII (54–75) inhibited the binding of thrombin to a hirudin C-terminal fragment corresponding to residues 54–66 of hirudin PA. Both hirudin PA and HCII fragments affected a slight enhancement of thrombin-catalyzed hydrolysis of a chromogenic substrate, but HCII (54–74) alone enhanced the rate of HCII inactivation of thrombin. The fact that HCII inhibition of thrombin includes interactions at the anion-binding exosite could explain the exquisite specificity of HCII for thrombin. In contrast, ATIII, which does not inhibit thrombin in an exosite-dependent fashion, is known to inhibit a number of coagulation proteins including factors IXa, Xa, and XIa. Site-directed mutagenesis studies on human leuserpin-2 (LS2) indicate that this proteinase inhibitor, which is highly homologous to HCII, inhibits thrombin in an exosite-dependent manner (Ragg *et al.*, 1990).

Thrombin is a potent, physiologically significant agonist, secretagogue, and mitogen with various cells and platelets (for review, see Fenton, 1981; Fenton and Bing, 1986). With regard to these cellular activities, the anion-binding exosite appears to be essential for thrombin's actions in many systems as demonstrated by studies using C-terminal hirudin fragments. Many of the results obtained with hirudin peptides parallel those observed with γ-thrombin (Fenton, 1981; Fenton and Bing, 1986). Jakubowski and Maraganore (1990) showed that hirugen inhibits thrombin-induced platelet aggregation, dense granule secretion, and eicosanoid generation. Thus, although thrombin's catalytic-site function is critical for platelet activation (Harmon and Jamieson, 1988; Knopp, 1988), studies with hirugen show that thrombin-induced platelet activation is likewise dependent on the anion-binding exosite. Recently, Vu et al. (1991) have shown that hirugen inhibits thrombin-induced signal transduction in *Xenopus* oocytes expressing a functional platelet thrombin receptor. As with platelets, it appears that thrombin activities toward endothelial cells are anion-binding exosite-dependent. Hirugen was found to inhibit thrombin-induced prostacyclin secretion and platelet-activating factor (PAF) production (Prescott et al., 1990). In other studies, hirugen inhibited the secretion of von Willebrand's factor and tissue-type plasminogen activator by endothelial cells treated with thrombin (B. Ewenstein, J. Pober, and J.M. Maraganore, unpublished results). Together, these studies demonstrate that the anion-binding exosite is a critical structural determinant for many of the cellular activities of thrombin.

7. DESIGN OF NOVEL, HIRUDIN-BASED THROMBIN INHIBITORS

Structure–function relationship studies of hirudin C-terminal fragments have provided the framework for design of novel inhibitory peptides capable of binding to thrombin in a bivalent fashion (Maraganore et el., 1990). Prior to determination of the X-ray crystallographic structure of the hirudin–thrombin complex (Rydel et al., 1990; Grütter et al., 1990), Bourdon et al. (1990) established that the NH$_2$-terminus of an undecapeptide corresponding to residues 54–64 of hirudin is proximal to Lys-149e of human α-thrombin. Using a model for the three-dimensional structure of thrombin (Furie et al., 1982), the minimal distance separating the NH$_2$-terminus of h-Gly-54 and the hydroxyl group of the active-site serine was determined to be ~18 Å. Using the crystallographic structures, this distance is, in fact, 18.7 Å. Such data allowed the design of hirudin peptide derivatives capable of binding to the exosite as well as the enzyme catalytic site (Fig. 10). These peptides, called "hirulogs," included: (1) the sequence D-Phe-Pro-Arg-Pro, capable of inhibiting the catalytic site; (2) a segment of hirudin corresponding to residues 53–64, which binds to the exosite; and

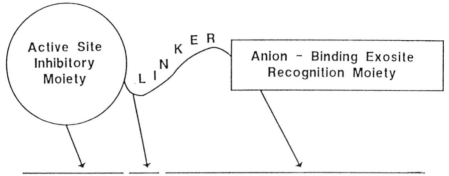

D-Phe-Pro-Arg-Pro-(Gly)₄-Asn-Gly-Asp-Phe-Glu-Glu-Ile-Pro-Glu-Glu-Tyr-Leu

Figure 10. Functional moieties of hirulog peptides and amino acid sequence of hirulog-1.

(3) a linker segment of glycyl residues which serves to bridge the active site and exosite binding moieties of the hirulog peptides.

Unlike the hirudin C-terminal peptide hirugen, hirulog-1 inhibited thrombin-catalyzed hydrolysis of synthetic tripeptidyl substrates with $K_i =$ 2.3 nM. Hirulog-1 inhibited thrombin amidolytic function at namomolar concentrations, while a peptide corresponding to the catalytic-site recognition moiety (D-Phe-Pro-Arg-Pro-Gly) alone failed to significantly inhibit the thrombin rate of reaction at concentrations as high as 10 μM. In order to prove a bivalent mode of interaction with thrombin, the specificity of hirulog action for catalytic-site binding was demonstrated further by the ability of hirulog, but not hirugen, to inhibit the incorporation of [¹⁴C]-diisopropylfluorophosphate in the active center of thrombin. The specificity of hirulog for exosite binding was shown by the ability of saturating concentrations of hirugen to reverse hirulog inhibition of thrombin amidolytic activity. Also, hirulog exhibited a micromolar affinity for γ_T-thrombin, which lacks an intact exosite.

Bivalent interactions in hirulog peptides are achieved in a single oligopeptide via a linker segment of glycine residues. Hirulog fragments containing a short linker segment of two amino acids showed reduced inhibitory activity, while linkers of four to eight glycine residues exhibited maximal effect. This result would suggest that bivalent activity is dependent on a minimal but not maximal linker length, and is consistent with the possibility that the linker segment is as disordered in solution and when bound to thrombin. Also, the absence of a strict dependence on the maximal linker length would indicate the lack of cooperativity in the interactions of hirulog at exo and catalytic sites. In this manner, hirulogs are similar to hirudin, which appear to lack significant cooperativity in the binding of NH_2- and COOH-terminal domains (Dennis *et al.*, 1990). As revealed recently in the x-ray crystallographic structure of the hirulog–thrombin complex (Skrzypczak-Jankun *et al.*, 1991), hirulog binding at the catalytic site of thrombin differs clearly from that of hirudin. Indeed, the D-Phe-Pro-Arg-Pro moiety, like most proteinase inhibitors, binds in a substrate-like manner.

Perhaps as a consequence of its high affinity, and substrate-like interactions with the active center of thrombin, hirulog peptides are cleaved by thrombin at the Arg–Pro bond at a slow rate (Bourdon *et al.*, 1991). Cleavage by thrombin, or any serine proteinase, at an X–Pro bond is unusual and, not surprisingly, k_{cat} for cleavage of hirulog-1 is only 0.21 min^{-1}. In contrast, when the $P_1{}'$) proline is replaced by glycine, the rate of the thrombin-catalyzed hydrolysis is accelerated to 206 min^{-1}. The Gly-4-substituted derivative is an excellent substrate for thrombin with k_{cat}/K_m for the reaction of ~3 × 10^9 M^{-1} min^{-1}, which exceeds that for thrombin cleavage of the fibrinogen Aα-chain. Of course, it has also been possible to prepare derivatives which are resistant to thrombin cleavage while maintaining high-affinity interactions with thrombin (Kline *et al.*, 1991; DiMaoi *et al.*, 1991).

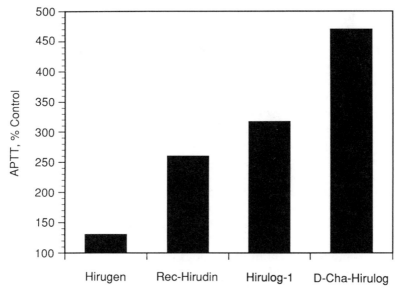

Figure 11. Anticoagulant activities of rec-hirudin and hirulog derivatives. APTT assays were performed using a fixed concentration (1.0 μg/ml) of rec-hirudin and hirulog peptides. D-Cha-hirulog contains an NH_2-terminal D-cyclohexylalanine residue in place of the D-Phe in hirulog-1.

As with hirudin and other simple hirudin tail peptides, the anti-thrombin activity of hirulog-1 is increased upon tyrosine sulfation accounting for a 3.1 kJ mol^{-1} increase in binding energy (Bourdon *et al.*, 1991). However, the most significant change in antithrombin activity was found to be substitution of the NH$_2$-terminal D-Phe residue by D-cyclohexylalanine. This derivative inhibited thrombin-catalyzed hydrolysis of a chromogenic substrate with $K_i = 0.12$ nM (J. M. Maraganore and P. Bourdon unpublished results). Further, as shown in Fig. 11, such a derivative illustrates the successes in peptide approaches to the development of hirudin-based anticoagulants where the activity of hirudin itself has been greatly surpassed. By using hirudin as a model for peptide design, hirulog may emerge as an effective modality for the prevention and treatment of thromboembolism in man.

8. CONCLUSION

Like its interactions with certain physiological substrates, cofactors, receptors, and inhibitors, thrombin interactions with hirudin involve multiple contacts which, while independently weak, together contribute to high affinity and specific complex formation. Kinetic studies on the interactions of hirudin and thrombin show a two-step binding mechanism where: (1) the electrostatic interactions of the hirudin COOH-terminus and the thrombin anion-binding exosite allow rapid and productive association of inhibitor and enzyme; and (2) subsequent binding of the NH$_2$-terminal domain with the active center leads to inhibition of catalytic reactivity. Studies on hirudin mutants and COOH-terminal peptides demonstrate the importance of electrostatic interactions for exosite binding, and x-ray crystallographic analyses of the complex of thrombin with these inhibitors indicate an important role for hydrophobic interactions with the anion-binding exosite. Mutagenesis, chemical modification, and kinetic studies indicate an essential role for the hirudin α-NH$_2$ group for active-site inhibition; this is confirmed in the crystal structures where this group interacts directly with Ser-195 in the catalytic center. The unusual interactions of hirudin with the active center of thrombin certainly explain the high specificity for hirudin action. Another remarkable feature for inhibition of thrombin by hirudin or hirulogs (synthetic, bivalent inhibitors of thrombin based on hirudin) is the absence of significant cooperativity in the binding of exosite and active-site recognition domains. Accordingly, it has been possible to accurately dissect the contributions of specific interactions to the formation of enzyme–inhibitor complex.

As noted above, interactions for hirudin with thrombin are not unlike those of thrombin with its physiologic targets. Hirudin and its peptide fragments have thus allowed a delineation of structural determinants in thrombin for biological function. Clearly, many but not all activities of thrombin are dependent on the anion-binding exosite. Further, studies with hirudin peptides suggest the presence of additional thrombin "exosites" which participate in physiological activities. The delineation of structure–function relationships in these additional exosite domains will certainly be an area of focused investigations in the years to come.

Finally, it is particularly exciting that hirudin and potent hirudin-based peptides, most notably hirulogs, show great promise as antithrombotic drugs. In this regard, the pioneering work of Markwardt (Walsmann and Markwardt, 1981) has established a strong and lasting foundation. Ongoing clinical investigations of these agents may lead to significant, beneficial changes in therapy in the years to come.

Acknowledgments

We thank Drs. C. Roitsch, W. Bode, J. Hofsteenge, J. Dodt, J. W. Fenton II, and M. Jackman for their comments on the manuscript. We also acknowledge Drs. A. Tulinsky, W. Bode, and J. Priestle for providing figures of the thrombin–hirudin structure.

9. REFERENCES

Bagdy, D., Barabas, E., Graf, L., Peterson, T. E., and Magnusson, S., 1976, Hirudin, *Methods Enzymol.* **45**:669–678.

Bajusz, S., Favszt, I., Barabas, E., Dioszegi, M., and Bagdy, D., 1984, Thrombin inhibition by hirudin fragments: Possible mechanism of hirudin–thrombin interactions, in: *Peptides 1984* (U. Ragnarson, ed.), Almquist & Wiksells, Stockholm, pp. 473–476.

Bergmann, C., Dodt, J., Köhler, S., Fink., E., and Gassen, H. G., 1986, Chemical synthesis and expression of a gene coding for hirudin, the thrombin-specific inhibitor from the leech Hirudo medicinalis, *Biol. Chem. Hoppe-Seyler* **367**:731–740

Berliner, L. J., Sugawara, Y., and Fenton, J. W., II, 1985, Human alpha-thrombin binding to non-polymerized fibrin–Sepharose: Evidence for an anionic binding region, *Biochemistry* **24**:7005–7009.

Bezeaud, A., and Guillin, M.-C., 1988, Enzymic and nonenzymic properties of human α–thrombin, *J. Biol. Chem.* **263**:3576–3581.

Bing, D. H., Cory, M., and Fenton, J. W., II, 1977, Exosite affinity labeling of human thrombins, *J. Biol. Chem.* **252**:8027–8034.

Bode, W., Mayr, I., Baumann, U., Huber, R., Stone, S. R., and Hofsteenge, J., 1989, The refined 1.9 Å crystal structure of human α-thrombin: Interaction with D-Phe-Pro-Arg chloromethylketone and significance of the Tyr-Pro-Pro-Trp insertion segment, *EMBO J* **8**:3467–3475.

Bourdon, P., and Maraganore, J. M., 1989, Effects of BG8865 on thrombin-catalyzed protein C activation *in vitro*, *Thromb. Haemostas.* **62**:534 (abstract).

Bourdon, P., Fenton, J. W., II, and Maraganore, J. M., 1990, Affinity labeling of Lys-149 in the anion-binding exosite of human α-thrombin with an Nα-(dinitrofluorobenzyl)hirudin C-terminal peptide, *Biochemistry* **29**:6379–6384.

Bourdon, P., Witting, J., Jablonski, J., Fenton J. W., II, and Maraganore, J. M., 1992, Hirulogs: Interactions with the catalytic site and adjacent regions of human thrombin, submitted for publication.

Bourin, M.-C., and Lindahl, U., 1990, Functional role of the polysaccharide component of rabbit thrombomodulin proteoglycan. Effects on inactivation of thrombin by antithrombin III, cleavage of fibrinogen by thrombin and thrombin-catalyzed activation of factor V, *Biochem. J.* **270**:419–426.

Bourin, M.-C., Boffa, M.-C., Bjork, I., and Lindahl, U., 1986, Functional domains of rabbit thrombomodulin, *Proc. Natl. Acad. Sci. USA* **83**: 5924–5928.

Bourin, M.-C., Ohlin, A.-K., Lane, D. A., Stenflo, J., and Lindahl, U., 1988, Relationship between anticoagulant activities and polyanionic properties of rabbit thrombomodulin, *J. Biol. Chem.* **263**:8044–8052.

Braun, P. J., Dennis, S., Hofsteenge, J., and Stone, S. R., 1988a, Use of site-directed mutagenesis to investigate the basis for the specificity of hirudin, *Biochemistry* **27**:6517–6522.

Braun, P. J., Hofsteenge, J., Chang, J.-Y., and Stone, S. R., 1988b, Preparation and characterization of proteolyzed forms of human α-thrombin, *Thromb. Res.* **50**: 273–283.

Brown, J. E., Baugh, R. F., and Hougie, C., 1980, The inhibition of intrinsic generation of activated factor X by heparin and hirudin, *Thromb. Res.* **17**:267–270.

Chang, J.-Y., 1983, The functional domain of hirudin, a thrombin-specific inhibitor, *FEBS Lett.* **164**:307–313.

Chang, J.-Y., 1985, Thrombin specificity. Requirement for apolar amino acid adjacent to the thrombin cleavage site of polypetide substrate, *Eur. J. Biochem.* **151**:217–224.

Chang, J.-Y, 1989, The hirudin-binding site of human α-thrombin, *J. Biol Chem.* **264**:7141–7146.

Chang, J.-Y., Ngai, P. K., Rink, H., Dennis, S., and Schlaeppi, J.-M., 1990a, The structural elements of hirudin which bind to the fibrinogen recognition site of thrombin are exclusively located within its acidic C-terminal tail, *FEBS Lett.* **261**:287–290.

Chang, J.-Y., Schlaeppi, J.-M, and Stone, S. R., 1990b, Antithrombin activity of the hirudin N-terminal core domain residues 1–43, *FEBS Lett.* **260**:209–212.

Church, F. C., Pratt, C. W., Noyes, C. M., Kalayanamit, T., Sherrill, G. B., Tobin, R. B., and Meade, J. B., 1989, Structural and functional properties of human α-thrombin, phosphopyridoxylated α-thrombin, and γt-thrombin, *J. Biol. Chem.* **264**:18419–18425.

Claeson, G., Aurell, L., Karlsson, G., and Friberger, P., 1977, Substrate structure and activity relationship, in: *New Methods for Analysis of Coagulation using Chromogenic Substrate* (I. Will, ed.), De Gruyter, Berlin.

Clore, G. M., Sukumaran, D. K., Nilges, M., Zarbock, J., and Gronenborn, A. M., 1987, The conformations of hirudin in solution: A study using nuclear magnetic resonance, distance geometry and restrained molecular dynamics, *EMBO J.* **6**:529–537.

Degryse, E., Acker, M., Defreyn, G., Bernat, A., Maffrand, J. P., Roitsch, C., and Courtney, M., 1989, Point mutations modifying the thrombin inhibition kinetics and antithrombin activity in vivo of recombinant hirudin, *Protein Eng.* **2**:459–465.

Dennis, S., Wallace, A., Hofsteenge, J., and Stone, S.R., 1990, Use of fragments of hirudin to investigate the thrombin–hirudin interaction, *Eur. J. Biochem.* **188**:61–66.

DiMaio, J., Ni, F., Gibbs, B., and Konishi., Y., 1991, A new class of potent thrombin inhibitors that incorporates a scissile pseudopeptide bond, *FEBS Lett.* **282**:47–52.

Dodt, J., Müller., A., Seemüller, U., and Chang, J.-Y., 1984, The complete amino acid sequence of hirudin, a thrombin specific inhibitor, *FEBS Lett.* **165**:180–183.

Dodt, J., Seemüller, U., Maschler, R., and Fritz, H., 1985, The complete covalent structure of hirudin. Localisation of the disulfide bonds, Biol. Chem. Hoppe-Seyler **366**:379–385.

Dodt, J., Machleidt, N., Seemüller, U., Maschler, R., and Fritz, H., 1986, Isolation and characterization of hirudin isoinhibitors and sequence analysis of hirudin PA, *Biol. Chem. Hoppe-Seyler* **367**:803–811.

Dodt, J., Seemüller, U., and Fritz, H., 1987, Influence of chain shortening on the inhibitor properties of hirudin and eglin c, *Biol. Chem. Hoppe-Seyler* **368**:1447–1453.

Dodt, J., Köhler, S., and Baici, A., 1988, Interaction of site specific hirudin variants with α-thrombin, *FEBS Lett.* **229**:87–90.

Dodt., J., Köhler, S., Schmitz, T., and Wilhelm, B., 1990, Distinct binding sites of Ala[48]-hirudin[1-47] and Ala[48]-hirudin[48-65] on α-thrombin, *J. Biol. Chem.* **265**:713–718.

Evans, S. A., Olson, S. T., and Shore, J. D., 1982, p-Aminobenzamidine as a fluorescent probe for the active site of serine proteases, *J. Biol. Chem.* **257**:3014–3017.

Fenton, J. W., II, 1981, Thrombin specificity, *Ann. N.Y. Acad. Sci.* **370**:468–495.

Fenton, J. W., II, 1989, Thrombin interactions with hirudin, *Semin. Thromb. Hemostas.* **15**:265–268.

Fenton, J. W., II, and Bing, D. H., 1986, Thrombin active-site regions, *Semin. Thromb. Hemostas.* **12**:200–208.

Fenton, J. W., II, Fasco, M. J., Stackrow, A. B., Aronson, D. L., Young, A. M., and Finlayson, J. S., 1977a, Human thrombins, *J. Biol. Chem.* **252**:3587–3598.

Fenton, J. W., II, Landis, B. H., Walz, D. A., and Finlayson, J. S., 1977b, Human thrombins, in: *Chemistry and Biology of Thrombin* (R. L. Lundblad, J. W. Fenton II, and K. G. Mann, eds.), Ann Arbor Science, Ann Arbor, pp. 43–70.

Fenton, J. W., II, Landis, B. H., Walz, D. A., Bing, D. H., Feinman, R. D., Zabinski, M. P., Sonder, S. A., Berliner, L. J., and Finlayson, J. S., 1979, Human thrombin: Preparative evaluation, structural properties, and enzymic specificity, in: *The Chemistry and physiology of Human Plasma Proteins* (D. H. Bing, ed.), Pergamon Press, Elmsford, N.Y., pp. 151–183.

Fenton, J. W., II, Olson, T. A., Zabinski, M. P., and Wilner, G. D., 1988, Anion-binding exosite of human alpha-thrombin and fibrin(ogen) recognition, *Biochemistry* **27**:7106–7112.

Fersht, A. R., 1972, Conformational equilibria in α-and δ-chymotrypsin. The energetics and importance of the salt bridge, *J. Mol. Biol.* **64**:497–509.

Fersht, A. R., 1985, *Enzyme Structure and Mechanism*, 2nd ed., Freeman, San Francisco.

Fersht, A. R., 1987, The hydrogen bond in molecular recognition, *Trends Biochem. Sci.* **12** 301–304.

Folkers, P. J. M., Clore, G. M., Driscoll, P. C., Dodt, J., Köhler, S., and Gronenborn, A. M., 1989, Solution structure of recombinant hirudin and the Lys-47–Glu mutant: A nuclear magnetic resonance and hybrid distance geometry–dynamical simulated annealing study, *Biochemistry* **28**:2601–2617.

Fritz, H., and Krejci, K., 1976, Trypsin–plasmin inhibitors (Bdellins) from leeches, *Methods Enzymol.* **45**:797–806.

Furie, B., Bing, D. H., Feldmann, R. J., Robinson, D. J., Burnier, J. P., and Furie, B. C., 1982, Computer-generated models of blood coagulation factor Xa, factor IXa, and thrombin based upon structural homology with other serine proteases, *J. Biol. Chem.* **257**:3875–3882.

Glover, G., and Shaw, E., 1971, The purification of thrombin and isolation of a peptide containing the active center histidine, *J. Biol Chem.* **246**:4594–4601.

Griffith, M. J., 1982, The heparin-enhanced antithrombin/thrombin reaction is saturable with respect to both thrombin and antithrombin III, *J. Biol Chem.* **257**:13899–13902.

Grütter, M. G., Priestle, J. P., Rahuel, J., Grossenbacher, H., Bode, W., Hofsteenge, J., and Stone, S. R., 1990, Crystal structure of the thrombin–hirudin complex: A novel mode of serine protease inhibition, *EMBO J.* **9**:2361–2365

Harmon, J. T., and Jamieson, G. A., 1988, Platelet activation by thrombin in the absence of the high affinity thrombin receptor, *Biochemistry* **27**:2151–2156

Haruyama, H., and Wuthrich, K., 1989, The conformation of recombinant desulfatohirudin in aqueous solution determined by nuclear magnetic resonance, *Biochemistry* **28**:4301–4312

Haruyama, H., Quian, Y.-Q., and Wüthrich, K., 1989, Static and transient hydrogen bonding interactions in recombinant desulfatohirudin studied by ^{1}H nuclear magnetic resonance measurements of amide proton exchange rates and pH dependent chemical shifts, *Biochemistry* **28**:4312–4317

Harvey, R. P., Degryse, E., Stefani, L., Schamber, F., Cazenave, J.-P., Courtney, M., Tolstoshev, P., and Lecocq, J.-P., 1986, Cloning and expression of a cDNA coding for the anticoagulant hirudin from the bloodsucking leech, Hirudo medicinalis, *Proc. Natl. Acad. Sci. USA* **83**:1084–1088.

Haycraft, J. B., 1884, Secretion obtained from the medicinal leech, *Proc. R. Soc. London.* **36**:478–487.

Hofsteenge, J., Taguchi, H., and Stone, S. R., 1986, Effect of thrombomodulin on the kinetics of the interaction of thrombin with substrates and inhibitors, *Biochem. J.* **237**:243–251.

Hofsteenge, J., Braun, P. J., and Stone, S. R., 1988, Enzymatic properties of proteolytic derivatives of human α-thrombin, *Biochemistry* **27**:2144–2151.

Hofsteenge, J., Stone, S. R., Donella Deana, A., and Pinna, L. A., 1990, The effect of substituting phosphotyrosine for sulphotyrosine on the activity of hirudin, *Eur. J. Biochem.* **188**:55–59.

Hogg, P. J., and Jackson, C. M., 1989, Fibrin monomer protects thrombin from inactivation by heparin–antithrombin III: Implications for heparin efficacy, *Proc. Natl. Acad. Sci. USA* **86**:3619–3623.

Hortin, G. L., Tollersen, D. M., and Benutto, B. M., 1989, Antithrombin activity of a peptide corresponding to residues 54–75 of heparin cofactor II, *J. Biol. Chem.* **264**:13979–13982.

Jakubowski, H. V., Kline, M. D., and Owen, W. G., 1986, The effect of bovine thrombomodulin on the specificity of bovine thrombin, *J. Biol. Chem* **261**:3876–3882.

Jakubowski, J. A., and Maraganore, J. M., 1990, Inhibition of coagulation and thrombin-induced platelet activities by a synthetic dodecapeptide modeled on the carboxy-terminus of hirudin, *Blood* **75**:399–406.

Johnson, P. H., Sze, P., Winant, R., Payne, P. W., and Lazar, J. B., 1989, Biochemistry and genetic engineering of hirudin, *Semin. Thromb. Hemostas.* **15**:302–315.

Kettner, C., and Shaw, E., 1981, Inactivation of trypsin-like enzymes with peptides of arginine chloromethyl ketone, *Methods Enzymol.* **80**: 826–842.

Kiesel, W., and Hanahan, D. J., 1974, Proteolysis of human factor II by factor Xa in the presence of hirudin, *Biochem. Biophys. Res. Commun.* **59**:570–577.

Kline, T., Hammond, C., Bourdon, P., and Maraganore, J. M., 1991, Hirulog peptides with scissile bond replacements resistant to thrombin cleavage, *Biochem. Biopohys. Res. Comm.* **177**:1049–1055.

Knopp, C. L., 1988, Effect of thrombin inhibitors on thrombin-induced release and aggregation of human platelets, *Thromb. Res.* **49**:231–234.

Köhler, S., Schmitz, T., and Dodt, J., 1989, The interaction of Glu(63)-, Ser(22,39)- and Leu(56)-hirudin with human and bovine α-thrombin, *DECHEMA–Biotechnology Conferences* **3**:405–409.

Konno, S., Fenton, J. W., II, and Villanueva, G. B., 1988, Analysis of the secondary structure of hirudin and the mechanism of its interaction with thrombin, *Arch. Biochem. Biophys.* **267**:158–166.

Krstenansky, J. L., and Mao, S. J. T., 1987, Antithrombin properties of C-terminus of hirudin using synthetic unsulfated N-α-acetyl-hirudin 45–64, *FEBS Lett.* **211**:10–16.

Krstenansky, J. L., Owen, T. J., Yates, M. T., and Mao, S. J. T., 1987, Anticoagulant peptides: Nature of the interaction of the C-terminal region of hirudin with a non-catalytic binding site on thrombin, *J. Med. Chem.* **30**:1688–1691.

Krstenansky, J. L., Owen, T. J., Yates, M. T., and Mao, S. J. T., 1988, Comparison of hirudin and hirudin PA C-terminal fragments and related analogs as antithrombin agents, *Thromb. Res.* **52**:137–141.

Krstenansky, J. L., Payne, M. H., Owen, T. J., Yates, M. T., and Mao, S. J. T., 1989, C-terminal peptide alcohol, acid, and amide analogs of desulfato hirudin as antithrombin agents, *Thromb. Res.* **54**:319–325.

Krstenansky, J. L., Broersma, R. J., Owen, T. J., Payne, M. H., Yates, M. T., and Mao, S. J. T., 1990, Development of MDL 28,050, a small stable antithrombin agent based on a functional domain of the leech protein hirudin, *Thromb. Haemostas.* **63**:208–214.

Kurosawa, S., Stearns, D. J., Jackson, K. W., and Esmon, C. T., 1988, A 10-kDa cyanogen-bromide fragment from the epidermal growth factor homology domain of rabbit thrombomodulin contains the primary thrombin binding site, *J. Biol. Chem.* **263**:5993–5996.

Landis, B. H., Zabinski, M. P., Lafleur, G. J. M., Bing, D. H., and Fenton, J. W., II, 1978, Human α-thrombin and γ-thrombin differential inhibition with hirudin *Fed. Proc.* **37**:1445.

Laskowski, M., Jr., Kato, I., Ardelt, W., Cook, J., Denton, A., Empie, M. W., Kohr, W. J., Park, S. J., Parks, K., Schatzley, B. L., Schoenberger, O. L., Tashiro, M., Vichot, G., Whatley, H. E., Wieczorek, A., and Wieczorek, M., 1987, Ovomucoid third domains from 100 avian species: Isolation, sequences, and hypervariability of enzyme-inhibitor contact residues, *Biochemistry* **26**:202–221.

Lewis, S. D., Lorand, L., Fenton, J. W., II, and Shafer, J. A., 1987, Catalytic competence of human α- and γ-thrombin in the activation of fibrinogen and factor XIII, *Biochemistry* **26**:7597–7603.

Loison, G., Findeli, A., Bernard, S., Nguyen-Juilleret, M., Marquet, M., Riehl-Bellon, N., Cavallo, D., Guerra-Santos, L., Brown, S. W., Courtney, M., Roitsch, C., and Lemoine, Y., 1988, Expression and secretion in S. cerevisiae of biologically active leech hirudin, *Biotechnology* **6**: 72–77.

Mao, S. J. T., Yates, M. T., Blankenship, D. T., Cardin, A. D., Krstenansky, J. L., Lovenberg, W., and Jackson, R. L., 1987, Rapid purification and revised N-terminal sequence of hirudin: A specific thrombin inhibitor of the bloodsucking leech, *Anal. Biochem.* **161**:514–518.

Mao, S. J. T., Yates, M. T., Owen, T. J., and Krstenansky, J. L., 1988, Interaction of hirudin with thrombin: Identification of a minimal binding domain of hirudin that inhibits clotting activity, *Biochemistry* **27**:8170–8173.

Maraganore, J. M., Chao, B., Joseph, M. L., Jablonski, J., and Ramachandran, K. L., 1989, Anticoagulant activity of synthetic hirudin peptides, *J. Biol. Chem.* **264**:8692–8698.

Maraganore, J. M., Bourdon, P., Jablonski, J., Ramachandran, K. L., and Fenton, J. W., II, 1990, Design and characterization of hirulogs: A novel class of bivalent peptide inhibitors of thrombin, *Biochemistry* **29**:7095–7101.

Markwardt, F., 1957, Die Isolierung and chemische Characterisierung des Hirudins, *Hoppe-Seylers Z. Physiol Chem.* **308**:147–156.

Matthew, J. B., 1985, Electrostatic effects in proteins, *Annu. Rev. Biophys. Biophys. Chem.* **14**:387–417.

Naski, M. C., Fenton, J. W., II, Maraganore, J. M., Olson, S. T., and Shafer, J. A., 1990, The COOH-terminal domain of hirudin. An exosite-directed competitive inhibitor of the action of α-thrombin on fibrinogen, *J. Biol. Chem.* **265**:13484–13489.

Naski, M. C., Lorand, L., and Shafer, J. A., 1991, Characterization of the kinetic pathway for fibrin promotion of γ-thrombin-catalyzed activation of factor XIII, *Biochemistry* **30;** 934–941.

Nesheim, M., 1983, A simple rate law that describes the kinetics of the heparin-catalyzed reaction between antithrombin III and thrombin, *J. Biol. Chem.* **258**:14708–14717.

Ni, F., Konishi, Y., and Scheraga, H. A., 1990, Thrombin-bound conformation of the C-terminal fragments of hirudin determined by transferred nuclear Overhauser effects, *Biochemistry* **29**:4479–4489.

Niehrs, C., Huttner, W. B., Carvallo, D., and Degryse, E., 1990, Conversion of recombinant hirudin to the natural form by *in vitro* tyrosine sulfation. Differential substrate specificities of leech and bovine tyrosylprotein sulfotransferases, *J. Biol. Chem.* **265**:9314–9318.

Noé, Hofsteenge, J., Rovelli, G., and Stone, S. R., 1988, The use of sequence specific antibodies to identify a secondary binding site in thrombin, *J. Biol. Chem.* 263:11729–11735.

Ofosu, F. A., Maraganore, J. M., Blajchman, M. A., Fenton, J. W., II, Yang, X., Smith, L., Anvari, N., Buchanan, M. R., and Hirsh, J., 1992, The anticoagulant effectiveness of hirudin, and two synthetic peptides modelled on the carboxyl-terminal dodecapeptide residues of hirudin, *Biochem. J.* **283:** 893–897.

Olson, S. T., 1988, Transient kinetics of heparin-catalyzed protease inactivation by antithrombin III, *J. Biol. Chem.* **263**:1698–1708.

Olson, S. T., and Shore, J. D., 1982, Demonstration of a two-step reaction mechanism for inhibition of α-thrombin by antithrombin III and identification of the step affected by heparin, *J. Biol. Chem.* **257**:14891–14895.

Parkinson, J. F., Grinnel, B. W., Moore, R. E., Hoskins, J., Vlahos, C. J., and Bang, N. U., 1990, Stable expression of a secretable deletion mutant of recombinant human thrombomodulin in mammalian cells, *J. Biol. Chem.* **265**:12602–12610.

Pomerantz, M. W., and Owen, W. G., 1978, A catalytic role for heparin. Evidence for ternary complex of heparin cofactor, thrombin and heparin, *Biochim. Biphys. Acta* **535**: 66–77.

Preissner, K. T., Delvos, U., and Muller-Berghaus, G., 1987, Binding of thrombin to thrombomodulin accelerates inhibition of the enzyme by antithrombin III. Evidence for a heparin-independent mechanism, *Biochemistry* **26**:2521–2528.

Prescott, S. M., Seeger, A. R., Zimmerman, G. A., McIntyre, T. M., and Maraganore, J. M., 1990, Hirudin-based peptides block the inflammatory effects of thrombin on endothelial cells, *J. Biol. Chem.* **265**:9614–9616.

Priestle, J. P., 1988, RIBBON: A stereo cartoon drawing program for proteins, *J. Appl. Crystallogr.* **21**:572–576.

Ragg, H., Ulshofer, T., and Gerewitz, J., 1990, On the activation of human leuserpin-2, a thrombin inhibitor, by glycosaminoglycans, *J. Biol. Chem.* **265**:5211–5218.

Read, R. J., and James, M. N. G., 1986, Introduction to the proteinase inhibitors: X-ray crystallography, in: *Proteinase Inhibitors* (A. J. Barret and G. Salvesen, eds.), Elsevier, Amsterdam, pp. 301–336.

Rydel, T. J., Ravichandran, K. G., Tulinksy, A., Bode, W., Huber, R., Roitsch, C., and Fenton, J. W., II, 1990, The structure of a complex of recombinant hirudin and human α-thrombin, *Science* **249**:277–280.

Sawyer, R. T., 1986, *Leech Biology and Behaviour*, Vol. 2, Oxford University Press (Clarendon), London.

Scharf, M., Engels, M., and Tripier, D., 1989, Primary structures of new 'iso-hirudins,' *FEBS Lett.* **255**:105–110.

Schechter, T., and Berger, A., 1967, On the size of the active site in proteases. I. Papain, *Biochem. Biophys. Res. Commun.* **27**:157–162.

Schmitz, T., Rothe, M., and Dodt, J., 1991, The mechanism of inhibition of α-thrombin by hirudin-derived fragments hirudin (1–47) and hirudin (45–65), *Eur. J. Biochem.* **195**:246–257.

Skrzypczak-Jankun, E., Carperos, V., Ravichandran, K G., Tulinsky, A., Westbrook, M., and Maraganore, J. M., 1991, The structure of hirugen and hirulog1 complexes of α-thrombin, *J. Mol. Biol.* **221**:1379–1393.

Sonder, S. A., and Fenton, J. W., II, 1984, Proflavin binding within the fibrinopeptide groove adjacent to the catalytic site of human α-thrombin, *Biochemistry* **23**:1818–1823.

Stone, S. R., and Hofsteenge, J., 1986, Kinetics of the inhibition of thrombin by hirudin, *Biochemistry* **25**:4622–4628.

Stone, S. R., and Hofsteenge, J., 1987, Effect of heparin on the interaction between thrombin and hirudin, *Eur. J. Biochem.* **169**:373–376.

Stone, S. R., and Hofsteenge, J., 1991a, Recombinant hirudin. Kinetic mechanism for the inhibition of human thrombin, *Protein Eng.* **4**:295–301.

Stone, S. R., and Hofsteenge, J., 1991b, The basis for the reduced affinity of β_T- and γ_T-thrombin for hirudin, *Biochemistry* **30**:3950–3955.

Stone, S. R., Braun, P. J., and Hofsteenge, J., 1987, Identification of regions of α-thrombin involved in its interaction with hirudin, *Biochemistry* **26**:4617–4624.

Stone, S. R., Dennis, S., and Hofsteenge, J., 1989, Quantitative evaluation of the contribution of ionic interactions to the formation of the thrombin–hirudin complex, *Biochemistry* **28**:6857–6863.

Sukumaran, D. K., Clore, G. M., Preuss, A., Zarbock, J., and Gronenborn, A. M., 1987, Proton nuclear magnetic resonance study of hirudin: Resonance assignment and secondary structure, *Biochemistry* **26**:333–338.

Suzuki, K., Nishioka, J., and Hayashi, T., 1990, Localization of thrombomodulin-binding site within human thrombin, *J. Biol. Chem.* **265**:13263–13267.

Tripier, D., 1988, Hirudin: A family of isoproteins. Isolation and sequence determination of new hirudins, *Folia Haematol. (Leipzig)* **115**:30–35.

Tsiang, M., Lentz, S. R., Dittman, W. A., Wen, D., Scarpati, E. M., and Sadler, J. E., 1990, Equilibrium binding of thrombin to recombinant human thrombomodulin: Effect of hirudin, fibrinogen, factor Va, and peptide analogues, *Biochemistry* **29**:10602–10612.

Vu, T.-V. H., Hung, D. T., Wheaton, V. I., and Coughlin, S. R., 1991, Molecular cloning of a functional thrombin receptor reveals a novel proteolytic mechanism of receptor activation, *Cell* **64**:1057–1068.

Walker, B., Wikstöm, P., and Shaw, E., 1985, Evaluation of inhibitor constants and alkylation rates for a series of thrombin affinity labels, *Biochem. J.* **230**:645–650.

Wallace, A., Dennis S., Hofsteenge, J., and Stone, S. R., 1989, Contribution of the N-terminal region of hirudin to its interaction with thrombin, *Biochemistry* **28**:10079–10084.

Wallis, R. B., 1988, Hirudins and the role of thrombin: Lessons from leeches, *Trends Pharm. Sci.* **9**:425–427.

Walsmann, P., 1988, Über den Einsatz des spezifischen Thrombininhibitors Hirudin für diagnostische and biochemische Untersuchungen, *Pharmazie* **43**:737–744.

Walsmann, P., and Markwardt, F., 1981, Biochemische and pharmakologische Aspekte des Thrombininhibitors Hirudin, *Pharmazie* **36**:653–660.

Weitz, J. I., Hudoba, M., Massel, D., Maraganore, J. M., and Hirsh, J., 1990, Clotbound thrombin is protected form inhibition by heparin–antithrombin III but is susceptible to inactivation by antithrombin III-independent inhibitors, *J. Clin. Invest.* **86**:385–391.

Zushi, M. Gomi, K., Yamamoto, S., Maruyama, I., Hayashi, T., and Suzuki, K., 1989, The last three consecutive epidermal growth factor-like structures of human thrombomodulin comprise the minimum functional domain for protein C-activating cofactor activity and anticoagulant activity, *J. Biol. Chem.* **264**:10351–10353.

Chapter 7

ACTIVATION OF HUMAN PLASMA FACTOR XIII BY THROMBIN

L. Lorand and J. T. Radek

1. DUAL ROLE OF THROMBIN IN BLOOD CLOTTING: DISSECTING THE PHYSIOLOGICAL PATHWAY FOR ACTIVATION OF FACTOR XIII

A few years after the discovery that thrombin effected the fibrinogen–fibrin conversion by a process of limited proteolysis (Bailey *et al.*, 1951; Lorand, 1951, 1952; Lorand and Middlebrook, 1952), it was reported from this laboratory (Lorand, 1961; Lorand and Konishi, 1964) that thrombin played another entirely similar and parallel role by activating the fibrin-stabilizing factor, now known as factor XIII. Actually, both thrombin and Ca^{2+} ions were found necessary to bring about the activation of this *plasma precursor* to generate the enzyme (factor XIII_a) which cross-links

L. Lorand and J. T. Radek • Department of Biochemistry, Molecular Biology and Cell Biology, Northwestern University, Evanston, Illinois 60208.

Thrombin: Structure and Function, edited by Lawrence J. Berliner. Plenum Press, New York, 1992.

fibrin.[1] The activation pathway could be resolved into the consecutive steps of a Ca^{2+}-independent hydrolysis by thrombin followed by some Ca^{2+}-specific, but thrombin-independent, changes on the proteolytically clipped factor XIII' intermediate:

$$\text{Factor XIII} \xrightarrow{\text{Thrombin}} \text{Factor XIII'} \xrightarrow{Ca^{2+}} \text{Factor XIII}_a$$

Thus, in the context of blood clotting, thrombin could be shown to perform a dual function by catalyzing not only the conversion of fibrinogen to fibrin but also regulating the rate of production of factor XIII_a, a transamidating enzyme which serves the purpose of strengthening the blood clot by cross-linking (XL) fibrin by N^{ε}-(γ-glutamyl)lysine bridges within the required physiological time frame (Bruner-Lorand et al., 1966; Roberts et al., 1973; Mockros et al., 1974; Shen and Lorand, 1983; for a review see Lorand and Conrad, 1984). The hereditary absence of factor XIII or the sudden appearance of inhibitors either against activation of the zymogen or the functioning of the activated cross-linking enzyme in the circulation may give rise to life-threatening hemorrhage (for a review see Lorand et al., 1980). The following scheme outlines this important coordinating role of thrombin:

$$n \text{ Fibrinogen} \longrightarrow n \text{ Fibrin} \rightleftharpoons (\text{Fibrin})_n \text{ clot}$$
$$\text{Factor XIII} \xrightarrow{\text{THROMBIN}} \text{XIII'} \xrightarrow{Ca^{2+}} \text{XIII}_a \downarrow Ca^{2+}$$
$$\text{XL(Fibrin)}_n \text{ clot}$$

Factor XIII is made up of two different subunits in an ab protomeric structure which can associate into an a_2b_2 tetramer (Schwartz et al., 1971, 1973; Carrell et al., 1989).[2] Only the a subunits ($M_r \sim 75,000$ for each a and $\sim 80,000$ for b) are modified by thrombin and—as with the release of fibrinopeptides from fibrinogen (Lorand, 1951, 1952)—in this case, too, an N-terminal activation peptide (AP) is liberated (Mikuni et al., 1973; Nakamura et al., 1974; Takagi and Doolittle, 1974). In keeping with the known preference of thrombin for the breaking of arginyl bonds (first demonstrated by Sherry and Troll, 1954), the primary cleavage site in the a subunit is at such a residue in position 37. Secondary cleavage sites for

---

[1] This chapter deals with the activation of the zymogen occurring in plasma and not with the platelet factor which was also shown to be activated by thrombin (Buluk et al., 1961). The physiological role, if any, of the activated platelet factor is unknown.

[2] In contrast to plasma factor XIII, the platelet factor or that obtained from placenta (including the recombinant form of the latter; Bishop et al., 1990) comprises only the a subunits as the homodimer, a_2 (Schwartz et al., 1971, 1973). Activation patterns for the plasma and the platelet or placental zymogens are quite different.

thrombin were reported at Lys-513 and Ser-514 (Takahashi *et al.*, 1986). Both α- and γ-thrombin can be used for activating factor XIII (Lorand and Credo, 1977), but the specificity constant obtained for the release of AP with γ-thrombin is about five times less than with α-thrombin (Lewis *et al.*, 1987).

It is the hydrolytically clipped a subunit, denoted as a', which contains the active site cysteine necessary for the expression of transamidase activity. However, in the heterologous $a'b$ combination, the important cysteine residue is still buried and unmasking requires Ca^{2+} ions (Curtis *et al.*, 1973, 1974; Lorand *et al.*, 1974). The effect of the Ca^{2+} is twofold: it brings about the dissociation of

$$a'b \xrightarrow{\text{Ca}^{2+}} a' + b \quad \text{or} \quad a_2'b_2 \xrightarrow{\text{Ca}^{2+}} a_2' + b_2$$

and promotes a conformational change for inducing the enzymatically active configuration:

$$a' \xrightarrow{\text{Ca}^{2+}} a^* \quad \text{or} \quad a_2' \xrightarrow{\text{Ca}^{2+}} a_2^*$$

(where the free a^* or the a_2^* structure represents factor XIII_a). The

$$\text{Factor XIII} \xrightarrow[\text{Ca}^{2+}]{\text{Thrombin}} XIII_a$$

conversion may thus be described by the protomeric formulation as

$$ab \xrightarrow{\text{Thrombin}} a'b \xrightarrow{\text{Ca}^{2+}} a' + b$$
$$\downarrow \qquad\qquad\qquad \downarrow \text{Ca}^{2+}$$
$$\text{AP} \qquad\qquad\qquad a^*$$

or by the tetrameric formulation as

$$a_2b_2 \xrightarrow{\text{Thrombin}} a_2'b_2 \xrightarrow{\text{Ca}^{2+}} a_2' + b_2$$
$$\downarrow \qquad\qquad\qquad \downarrow \text{Ca}^{2+}$$
$$2\text{AP} \qquad\qquad\qquad a_2^*$$

2. FIBRIN PROMOTES THE PROTEOLYTIC ACTIVATION OF FACTOR XIII BY THROMBIN, BUT CROSS-LINKED FIBRIN BLOCKS THIS PROMOTING EFFECT

Interesting regulatory aspects relating to the limited proteolytic attack by thrombin on factor XIII became evident when the reaction was examined in the presence of fibrin (Lewis *et al.*, 1985). High-performance liquid chromatography was used to monitor the kinetics of release of the N-terminal activation peptide, AP. Polymeric fibrin I and II (polymerized des-A and des-AB fibrinogens), the physiological substrates of factor XIII$_a$, were shown to be potent promoters of the thrombin-catalyzed release of AP from the zymogen. Moreover, fibrins I and II were found to be far more efficient in this regard than fibrinogen, even though fibrinogen itself is known to bind factor XIII very tightly (Greenberg and Shuman, 1982). This suggests that factor XIII interacts differently with polymeric fibrins I and II than with fibrinogen.

Recent kinetic and thermodynamic studies by Naski et al. (1991), showed that the cofactor activity of fibrin I in the thrombin-catalyzed proteolysis of AP from factor XIII could, indeed, be attributed to the formation of a tight factor XIII–fibrin I complex ($K_D \cong 65$nM). The release of AP by thrombin from factor XIII in the complex was 80-fold more efficient ($k_{cat}/K_M = 1.2 \times 10^7$ M^{-1} s^{-1}) than with uncomplexed factor XIII ($k_{cat}/K_M = 1.4 \times 10^5$ M^{-1} s^{-1}). This increase in the k_{cat}/K_M specificity constant is largely due to an increase in the apparent affinity of thrombin for the factor XIII–fibrin I complex, as reflected by a 30-fold decrease in the Michaelis constant observed for the interaction of thrombin with the factor XIII–fibrin I complex relative to that with uncomplexed factor XIII in solution.

Naski et al. (1991) also addressed the question whether, in the ternary complex of thrombin-factor XIII–fibrin I, thrombin would still be able to release fibrinopeptide from the N-terminus of the Bβ chain of fibrin I and, in fact, this further cleavage of the fibrin I substrate could be readily demonstrated. The observation shows that in the above complex, thrombin is competent to catalyze both the release of fibrinopeptide B from fibrin I and the removal of AP from factor XIII. It seems that thrombin is anchored to fibrin I through some site (exosite) distinct from its serine residue-containing active site and that the latter can react alternatively either with the AP moiety of factor XIII or with the fibrinopeptide B moiety of the Bβ chain of fibrin I. The conclusion was supported by the finding that a 12-residue peptide (hirugen) which binds to an exosite of thrombin could competitively block the thrombin-catalyzed release of AP and also the release of fibrinopeptide B from the factor XIII–fibrin I complex.

From the point of view of physiological controls in blood clotting, the regulatory significance of promotion of the thrombin-catalyzed activation of factor XIII by fibrin should be obvious. Promotion of the initial hydrolytic step in the activation of factor XIII by polymeric fibrin ensures that significant amounts of factor XIII are not activated until its physiological substrate, polymeric fibrin, is present. This method of control would also minimize the wasteful and premature activation of factor XIII_a and thus the possibly dangerous cross-linking of other plasma proteins by the transamidase.

It should be noted that the discussed mechanism of promotion of factor XIII activation by fibrin functions only with α-thrombin as activator and does not occur in the γ-thrombin-catalyzed reaction (Lewis et al., 1987).

With α-thrombin as the activator of factor XIII, yet another interesting physiological regulation was observed (Lewis et al., 1985) in that the promoting activity of polymeric fibrin was rapidly lost when the catalytically competent factor XIII_a was allowed to form. This finding is consistent with the view that the factor XIII_a-mediated cross-linking of fibrin inactivates fibrin as a promoter for the release of AP from the factor XIII zymogen by thrombin. Clearly, such a feedback shut-off regulation would serve the purpose of a safeguard against the continued generation of factor XIII_a after its fibrin substrates have been cross-linked, thereby avoiding the overproduction of the catalytically active factor XIII_a species.

3. FACTOR XIII CAN BE ACTIVATED WITHOUT PRIOR PROTEOLYTIC CLEAVAGE

It is actually possible to activate the purified factor XIII zymogen to express full transamidating enzyme activity without the prior removal of AP by thrombin. Addition of a high enough concentration of Ca^{2+} (0.1 M) alone is sufficient to bring about this conversion (Credo et al., 1978; Lorand et al., 1981)

The Ca^{2+} requirement is substantially reduced and the thrombin-independent conversion of the zymogen can be greatly accelerated by the addition of chaotropic ions with the relative efficacies of p-toluenesulfonate ≥, thiocyanate > iodide > bromide. At 37°C, pH 7.5, and with 0.05 M Ca^{2+} and 0.2 M p-toluenesulfonate (μ = 0.4), for example, complete conversion was achieved in about 10 min.

Since the AP moiety still remains attached in such an in vitro transformation of the a subunit, the process should be distinguished from the physiological production of the a* enzyme, and the active species will be referred to in this case by the symbol a°. Curtis, Credo, Janus, Haggroth and Lorand (unpublished results) studied the thrombin-independent conversion of factor XIII in some detail with the following conclusions. Specificity for Ca^{2+} was demonstrated by the considerably greater efficacy of Ca^{2+} over Ba^{2+}, Sr^{2+}, and Mg^{2+}. It was shown by gel filtration chromatography that Ca^{2+} ions caused a dissociation of the heterologous protomer. Moreover, upon removal of Ca^{2+}, reassociation could take place:

$$ab \underset{-Ca^{2+}}{\overset{+Ca^{2+}}{\rightleftarrows}} a^{\circ} + b \quad \text{or} \quad a_2 b_2 \underset{-Ca^{2+}}{\overset{+Ca^{2+}}{\rightleftarrows}} a_2^{\circ} + b_2$$

The unmasking of the active center cysteine thiol group as well as the generation of catalytic activity in the a° species was also examined. With regard to the titratability of the cysteine sulfhydryl group with iodoacetamide, essentially full unmasking of the active center (ca. 1 mol of thiol per a° subunit) was found. The catalytic nature of a° was proven (Janus *et al.*, 1981) by the incorporation of isotopic putrescine into *N,N*-dimethylcasein and also by rigorous kinetic measurements with the fully synthetic substrate pair of β-phenylpropionylthiocholine and methanol (Parameswaran and Lorand, 1981). The steady-state analysis of the latter system allowed a meaningful comparison between the catalytic properties of the a° species and those of a^* produced on the thrombin-dependent activation pathway. Both enzymes displayed purely Michaelian kinetics, without any evidence of a°-to-a° or a^*-to-a^* cooperativity if, in fact, the homodimeric molecular formulation of a_2° or a_2^* would apply to these enzymes. Importantly, the apparent Michaelis constants for the acyl group-containing substrate (β-phenylpropionylthiocholine) as well as for the nucleophile (methanol) were identical for both enzymes, and the molar turnover numbers (k_{cat}) were also indistinguishable.[1]

Studies on the direct mode of activating factor XIII, without any prior exposure to thrombin, highlight the advantages afforded by the participation of thrombin in the physiological process. The fact that the a° enzyme displays catalytic activities quite commensurate to a^*, with the former still retaining the AP moiety lost by the latter, indicates that the presence of this 37-amino-acid-containing N-terminal region of the protein does not *per se* prevent expression of catalytic activity. Thus, the deeper rationale for the removal of AP as a first step in the physiological method of activation of the zymogen must be sought elsewhere.

4. RATIONALIZING THE INVOLVEMENT OF THROMBIN IN THE ACTIVATION OF FACTOR XIII

Evidence for the Ca^{2+}-induced heterologous dissociation of subunits of the thrombin-clipped zymogen ensemble came from electrophoretic

---

[1] It is, of course, possible that differences between the enzymatic activities of a° and a^* might emerge when tested on protein substrates (e.g., fibrin).

measurements where the release of b subunits could be readily demonstrated under nondenaturing conditions, and the ensuing $a°$-to-$a*$ conformational change in the thrombin-modified separated catalytic subunit was inferred from the unmasking of the active center residue as measured by titration with [14C]iodoacetamide (Curtis *et al.*, 1973, 1974; Lorand *et al.*, 1974). These experiments established that the thrombin-dependent pathway of factor XIII activation carries the advantage that the generation of cross-linking activity can occur at physiological ionic strength and at a greatly reduced concentration of Ca^{2+} when compared to the activation of the zymogen without prior proteolytic cleavage by thrombin (Credo *et al.*, 1978; Lorand *et al.*, 1981). In fact, in the presence of fibrin(ogen), dissociation of the $a'b$ ensemble proceeds to rapid completion well below the concentration of free Ca^{2+} prevailing in plasma (Credo *et al.*, 1978, 1981). Translated into molecular terminology, the thrombin-catalyzed removal of AP from the N-terminus of the a subunit during the ab-to-$a'b$ conversion weakens the heterologous association in the protomeric entity so that the subunits can separate with much greater ease under physiological conditions. Expressed in another way, at the physiological concentration of Ca^{2+} in plasma, and particularly in the presence of fibrin(ogen) the dissociation constant for the

$$ab \xrightleftharpoons{Ca^{2+}} a + b$$

equilibrium is expected to be considerably greater than that for

$$a'b \xrightleftharpoons{Ca^{2+}} a' + b$$

As yet, no data are available to numerically compare these equilibrium constants, but the enormous differences between the Ca^{2+} sensitivities for of the ab and $a'b$ structures to dissociate can be amply illustrated by our recent (as yet unpublished) results using the technique of fluorescence depolarization. Such a study relating to factor XIII activation was initiated by Freyssinet *et al.* (1978), in which the entire ab or a_2b_2 ensemble was labeled with fluorescein isothiocyanate. Addition of thrombin and Ca^{2+} to the labeled protein produced a large drop in polarization value. Since the decrease in the polarization value, as observed by these workers, could have been due partly to the previously described dissociation of $a'b$ to a' and b and partly to the conformational change of the fluorescently labeled a' to $a*$, we thought that a clear interpretation of depolarization data was possible only if the fluorescein label was introduced selectively into the subunits. Labeling only the b subunit, which is released during the dissociation step, would allow one to focus exclusively on the thrombin and

Ca^{2+}-dependent separation of subunits from the reconstituted zymogen assembly. On the other hand, exclusive labeling of a could be employed to study the direct effects of thrombin and Ca^{2+} on this subunit alone.

Our research was greatly helped by the availability of a recombinant form (r) of the a subunit expressed in yeast from the cDNA of the human placental factor. This product was kindly provided to us by Dr. P. Bishop of Zymogenetics, Seattle, Washington, and was reported to exist in solution as the homodimer of a (Bishop et al., 1990). Accordingly, the recombinant material will be referred to as ra_2. Using the homodimeric formulation, labeling by fluorescein isothiocyanate (F) could be carried out either with the native b_2 protein isolated from human plasma (Lorand et al., 1981) or with the yeast recombinant ra_2, yielding the fluorescent products of b_2^F and ra_2^F. These labeled proteins could then be reconstituted with their un-labeled counterparts to generate $ra_2b_2^F$ which upon treatment by thrombin is converted to the $ra_2'b_2^F$ ensemble or, if necessary, $ra_2^Fb_2$, which, by treatment with thrombin, yields $ra_2'^Fb_2$.

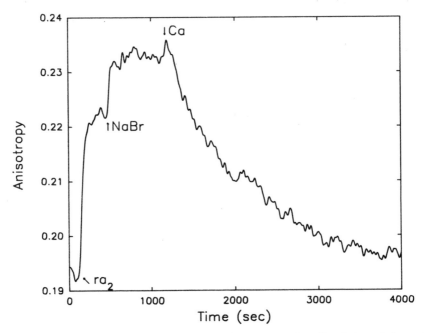

Figure 1. Changes of fluorescence anisotropy demonstrate that dissociation of the $ra_2b_2^F$ ensemble can occur in the absence of treatment by thrombin merely upon the addition of a chaotropic agent plus Ca^{2+}. Fluorescein-labeled native human plasma b subunits ($b_2^F = 33$ nM) were mixed (at 120 s) with the yeast recombinant form of human placental a subunits ($ra_2 = 38$ nM), and NaBr (0.45 M) was added (at 480 s) prior to Ca^{2+} (50 nM at 1140 s). The experiment was performed at 37°C, with 75 mM Tris–HCl buffer of pH 7.5.

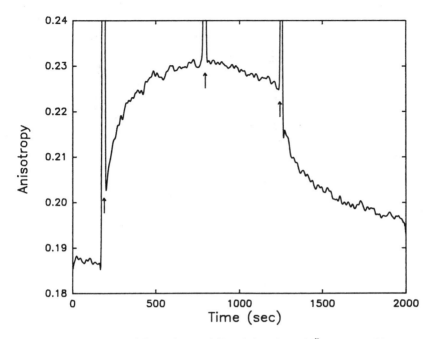

Figure 2. Reconstruction of the pathway of dissociating the $ra_2b_2{}^F$ zymogen with treatment by thrombin and addition of Ca^{2+}. Using 75 mM Tris–HCl buffer with 0.15 M NaCl and 2 mM EDTA at 37°C, $b_2{}^F$ (47 nM) was mixed with ra_2 (48 nM; arrow at 180 s). Human thrombin (5 NIH units) was injected first (at 800 s), followed by Ca^{2+} (30 mM; at 1250 s).

The results presented in Figs. 1–4 pertain to mixtures of fluorescein-labeled $b_2{}^F$ with ra_2, forming the $ra_2b_2{}^F$ ensemble. Dissociation was then examined under various experimental conditions. Adding equimolar ra_2 to $b_2{}^F$ (Figs. 1 and 2) produced an essentially maximal rise in the measured anisotropy of the system (for definition, see Lakowicz, 1983, pp. 145–153), showing that saturation was achieved with regard to forming the largest rotating $ra_2b_2{}^F$ structure. No change in anisotropy would have ensued upon addition of 30 mM Ca^{2+} alone, but if Ca^{2+} was applied after the addition of 0.4 M NaBr there was a drop in anisotropy back to near the value seen with the original free $b_2{}^F$ species (Fig. 1). In contrast to the thrombin-independent dissociation of the $ra_2b_2{}^F$ unit requiring such forced conditions, when Ca^{2+} was added after thrombin injection a similar drop in anisotropy was demonstrable at $\mu \sim 0.15$ (Fig. 2). Clearly thus, treatment of the $ra_2b_2{}^F$ equivalent of factor XIII with thrombin with the attendant conversion of the zymogen to $ra_2'b_2{}^F$ weakened the heterologous interaction of the subunits sufficiently for them to dissociate with Ca^{2+} in the absence of any chaotropic agent. The sensitivity of the thrombin-clipped zymogen, $ra_2'b_2^F$, to dissociation at different concentrations of Ca^{2+} is illustrated in Fig. 3. It is known that fibrin(ogen) can greatly enhance the effect of Ca^{2+} for dissociating the thrombin-modified factor XIII ensemble and this could also be confirmed in the recent experiments where the anisotropy of $ra_2'b_2^F$ was monitored in the presence of fibrinogen, with Gly-Pro-Arg-Pro (Laudano and Doolittle, 1980) added to inhibit clotting. As seen in Fig. 4, injection of thrombin alone was sufficient to elicit the dissociation event which, of course, would likely have been enhanced by traces of Ca^{2+}.

Fluorescence anisotropy measurements are more difficult to interpret when the a subunit is labeled. An experiment with ra_2^F and native b_2 is shown in Fig. 5. Conceptually, this experiment was similar to that presented in Fig. 2, but it is striking that the drop in anisotropy upon adding Ca^{2+} to the thrombin-treated zymogen, $ra_2'^Fb_2$, fell far below the anisotropy value of the original free ra_2^F. Since there is no independent evi-

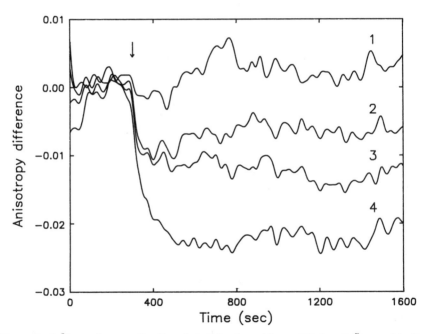

Figure 3. Ca^{2+} requirement for dissociating the thrombin-modified $ra_2'b_2^F$ ensemble. Preformed $ra_2b_2^F$ (25 nM) in 75 mM Tris–HCl, pH 7.5, and 0.15 M NaCl was treated with thrombin (10 NIH units) at 37°C for 5 min prior to adding Ca^{2+} (arrow) to concentrations of 7.5 mM (line 2), 15 mM (line 3), and 30 mM (line 4). Line 1 represents the control without added Ca^{2+}.

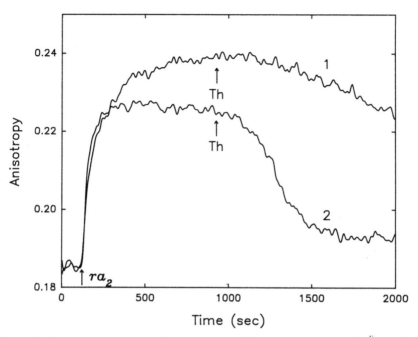

Figure 4. Fibrin(ogen) promotes the dissociation of thrombin-activated $ra_2'b_2^F$ even in the absence of Ca^{2+}. The experiment was carried out at 37°C and the initial buffer contained 75 mM Tris–HCl of pH 7.5, 0.15 M NaCl, 0.5 mM EDTA, 22.5 nM b_2^F, either 2 µM human fibrinogen (line 2) or 2 µM human plasma fibronectin (line 1), and 2.5 mM Gly-Pro-Arg-Pro to inhibit fibrin polymerization (Laudano and Doolittle, 1980). At 120 s ra_2 protein was added (25 nM) and at 900 s 2.4 NIH units of thrombin (Th) was injected. It is seen that the drop in fluorescence anisotropy characteristic for the dissociation of the $ra_2'b_2^F$ structure occurred only in the presence of fibrin(ogen) but not in that of fibronectin.

dence for suggesting that $ra_2'^F$ could dissociate into ra'^F units under the conditions employed, it may be tentatively assumed that conformational changes associated with the $ra_2'^F$-to-ra_2*^F transition contribute greatly to the change in anisotropy. Whatever the explanation may be, this event shows a high degree of Ca^{2+} sensitivity, as illustrated in Fig. 6, where changes in anisotropy values of the thrombin-treated subunit, $ra_2'^F$, were examined at Ca^{2+} concentrations below 1 mM.

5. CONCLUSION

Thrombin has a dual role in the final stages of normal blood coagulation. It catalyzes the limited proteolytic reactions of converting fibrinogen to fibrin and that of another plasma protein (factor XIII or fibrin-stabiliz-ing factor) to factor XIII_a, a transamidase which strengthens the clot by N^ε-(γ-glutamyl))lysine side-chain bridges.

It is significant from the point of view of physiological controls that the hydrolytic attack by thrombin on the factor XIII zymogen is greatly ac-celerated following the removal of the first set of N-terminal segments (fibrinopeptides A) from the Aα chains of fibrinogen. It is equally impor-tant that this promoting effect of fibrin on the activation of factor XIII by thrombin is terminated upon the cross-linking of the fibrin substrate by factor XIII_a.

The thrombin-mediated removal of the N-terminal activation peptide from the a subunit of factor XIII causes labilization of this ab protomeric

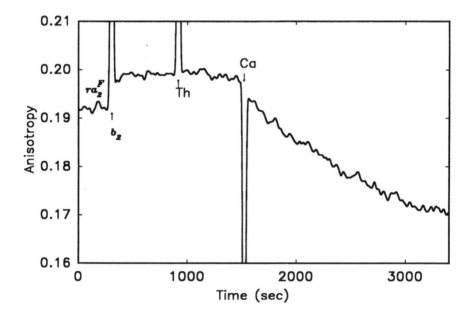

Figure 5. Changes in the anisotropy of fluorescence cannot be solely attributed to subunit dissociation when the $ra_2{}^F b_2$ labeled zymogen structure is treated with thrombin and Ca^{2+}. Experimental conditions were identical to those given for Fig. 2, except that the solution contained 74 nM fluorescein-labeled $ra_2{}^F$, to which 75 nM native b_2 subunits from human plasma were added (300 s). Human thrombin (5 NIH units) was injected at 900 s, followed by Ca^{2+} (to 10 mM at 1500 s). Note that fluorescence anisotropy after Ca^{2+} addition drops well below that of the starting value found with $ra_2{}^F$ alone prior to forming the $ra_2{}^F b_2$ complex.

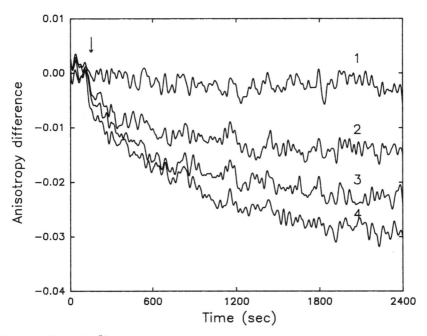

Figure 6. Effect of Ca^{2+} on the fluorecence anisotropy of the thrombin-treated recombinant $ra_2'^F$ protein alone. The experiment was carried out (37°C) with 20 nM ra_2^F in 75 mM Tris–HCl and 0.15 M NaCl. Following treatment with 3.5 NIH units of thrombin for 5 min, Ca^{2+} was added (at the arrow) to concentrations of 0.2 mM (line 2), 0.25 mM (line 3), and 1 mM (line 4). Line 1 represents the control in the absence of Ca^{2+} (or with 10 mM Mg^{2+} *in lieu* of Ca^{2+}). It might be mentioned that, without any prior exposure to thrombin, a drop in anisotropy of ra_2^F could be demonstrated merely by the additon of Ca^{2+}, albeit with about two orders of magnitude higher concentration of Ca^{2+} than was required for the thrombin-treated $ra_2'^F$.

zymogen ensemble, leading to the separation of subunits in the presence of Ca^{2+}. Dissociation, a prerequisite for generating the factor XIII_a enzyme, occurs with the greatest ease in the presence of fibrin.

Ca^{2+} also induces a conformational change in the thrombin-cleaved subunit of factor XIII, necessary for assuming the active configuration of the transamidase.

Ilustrations are provided from recent studies, using the method of fluorescence anisotropy, with either the *a* or the *b* subunit being labeled with fluorescein as a reported group.

Acknowledgments

This work was aided by a USPHS Career Award (HL-03512) and by grants from the National Institutes of Health (HL-02212 and HL-16346).

6. REFERENCES

Bailey, K., Bettelheim, F. R., Lorand, L., and Middlebrook, W. R., 1951, Action of thrombin in the clotting of fibrinogen, *Nature* **167**:233–234.

Bishop, P. D., Teller, D. C., Smith, R. A., Lasser, G. W., Gilbert, T., and Seale, R. L., 1990, Expression, purification, and characterization of human factor XIII in Saccharomyces cerevisiae, *Biochemistry* **29**:1861–1869.

Bruner-Lorand, J., Pilkington, T. R. E., and Lorand, L., 1966, Inhibitors of fibrin cross-linking: Relevance for thrombolysis, *Nature* **210**:1273–1274.

Buluk, K., Januszko, T., and Olbromski, J., 1961, Conversion of fibrin to desmofibrin, *Nature* **191**:1093–1094.

Carrell, N. A., Erickson, H. P., and McDonagh, J., 1989, Electron microscopy and hydrodynamic properties of factor XIII subunits, *J. Biol. Chem.* **264**:551–556.

Credo, R. B., Curtis, C. G., and Lorand, L., 1978, Ca^{2+}-related regulatory function of fibrinogen, *Proc. Natl. Acad. Sci. USA* **75**:4234–4237.

Credo, R. B., Curtis, C. G., and Lorand, L., 1981, α-Chain domain of fibrinogen controls generation of fibrinoligase (coagulation factor XIIIa). Calcium ion regulatory aspects, *Biochemistry* **20**:3770–3778.

Curtis, C. G., Stenberg, P., Chou, C.-H. J., Gray, A., Brown, K. L., and Lorand, L., 1973. Titration and subunit localization of active center cysteine in fibrinoligase (thrombin-activated fibrin stabilizing factor). *Biochem. Biophys. Res. Commun.* **52**:51–56.

Curtis, C. G., Brown, K. L., Credo, R. B., Domanik, R. A., Gray, A., Stenberg, P., and Lorand, L., 1974, Calcium-dependent unmasking of active center cysteine during activation of fibrin stabilizing factor, *Biochemistry* **13**:3774–3780.

Freyssinet, J.-M., Lewis, B. A., Holbrook, J. J., and Shore, J. D., 1978. Protein–protein interactions in blood clotting, *Biochem. J.* **169**:403–410.

Greenberg, C. S., and Shuman, M. A., 1982, The zymogen forms of blood coagulation factor XIII bind specifically to fibrinogen, *J. Biol. Chem.* **257**:6096–6101.

Janus, T. J., Credo, R. B., Curtis, C. G., Haggroth, L., and Lorand, L., 1981, Novel mode of activating human factor XIII zymogen without the hydrolytic cleavage of the *a* subunits, *Fed. Proc.* **40**:1585 (abstract).

Lakowicz, J. R., 1983, *Principles of Fluorescence Spectroscopy*, Plenum Press, New York.

Laudano, A. P., and Doolittle, R. F., 1980, Studies on synthetic peptides that bind to fibrinogen and prevent fibrin polymerization. Structural requirements, number of binding sites, and species differences, *Biochemistry* **19**:1013–1019.

Lewis, S. D., Janus, T. J., Lorand, L., and Shafer, J. A., 1985. Regulation of the formation of factor XIII_a by its fibrin substrates, *Biochemistry* **24**:6772–6777.

Lewis, S. D., Lorand, L., Fenton, J. W., II, and Shafer, J. A., 1987, Catalytic competence of human α- and γ-thrombin in the activation of fibrinogen and factor XIII, *Biochemistry* **26**: 7597–7603.

Lorand, L., 1951, "Fibrino-peptide": New aspects of the fibrinogen–fibrin transformation, *Nature* **167**:992–993.

Lorand, L., 1952, Fibrino-peptide, *Biochem. J.* **52**:200–203.

Lorand, L., 1961, in: *Progress in Coagulation*, Suppl 1. *Thromb. Diath. Haemorrh.* **7**:238–248.

Lorand, L., and Conrad, S. M., 1984, Transglutaminases, *Mol. Cell. Biochem.* **58**:9–35.

Lorand, L., and Credo, R. B., 1977, Thrombin and fibrin stabilization in: *Chemistry and Biology of Thrombin* (L. Lundblad, J. W. Fenton, II, and K. G. Mann, eds.), Ann Arbor Science, Ann Arbor, pp. 311–323.

Lorand, L., and Konishi, K., 1964, Activation of the fibrin stabilizing factor of plasma by thrombin, *Arch. Biochem. Biophys.* **105**:58–67.

Lorand, L., and Middlebrook, W. R., 1952, The action of thrombin on fibrinogen, *Biochem. J.*, **52**:196–199.

Lorand, L., Gray, A. J., Brown, K., Credo, R. B., Curtis, C. G., Domanik, R. A., and Stenberg, P., 1974. Dissociation of the subunit structure of fibrin stabilizing factor during activation of the zymogen, *Biochem. Biophys. Res. Commun.* **56**:914–922.

Lorand, L., Losowsky, M. S., and Miloszewski, K. J. M., 1980, Human factor XIII: Fibrin stabilizing factor, in: *Progress in Hemostasis and Thrombosis*, Vol. 5 (T. H. Spaet, ed.), Grune & Stratton, New York, pp. 245–290.

Lorand, L., Credo, R. B., and Janus, T. J., 1981, Factor XIII (fibrin stabilizing factor), *Methods Enzymol.* **80**:333–341.

Mikuni, Y., Iwanaga, S., and Konishi, K., 1973, A peptide released from plasma fibrin stabilizing factor in the conversion to the active enzyme by thrombin, *Biochem. Biophys. Res. Commun.* **54**:1393–1402.

Mockros, L. F., Roberts, W. W., and Lorand, L., 1974, Viscoelastic properties of ligation-inhibited fibrin clots, *Biophys. Chem.* **2**:164–169.

Nakamura, S., Iwanaga, S., Suzuki, T., Mikuni, Y., and Konishi, K., 1974, Amino acid sequence of the peptide released from bovine factor XIII following activation by thrombin, *Biochem. Biophys. Res. Commun.* **58**:250–256.

Naski, M. G., Lorand, L., and Shafer, J., 1991, Characterization of the kinetic pathway for fibrin promoting of α-thrombin-catalyzed activation of plasma factor XIII, *Biochemistry* **30**:934–941.

Parameswaran, K. N., and Lorand, L., 1981, New thioester substrates for fibrinoligase (coagulation factor XIII$_a$) and for transglutaminase. Transfer of the fluorescently labeled acyl group to amines and alcohols, *Biochemistry* **20**:3703–3711.

Roberts, W. W., Lorand, L., and Mockros, L. F., 1973, Viscoelastic properties of fibrin clots, *Biorheology* **10**:29–42.

Schwartz, M. L., Pizzo, S. V., Hill, R. L., and McKee, P. A., 1971, The subunit structures of the human plasma and platelet factor XIII (fibrin-stabilizing factor), *J. Biol. Chem.* **246**:5851–5854.

Schwartz, M. L., Pizzo, S. V., Hill, R. L., and McKee, P. A. 1973, Human factor XIII from plasma and platelets, *J. Biol. Chem.* **248**:1395–1407.

Shen, L., and Lorand, L., 1983, Contribution of fibrin stabilization to clot strength. Supplementation of factor XIII-deficient plasma with the purified zymogen, *J. Clin. Invest.* **71**:1336–1341.

Sherry, S., and Troll, W., 1954, The action of thrombin on synthetic substrates, *J. Biol. Chem.* **208**:95–105.

Takagi, T., and Doolittle, R. F., 1974, Amino acid sequence studies on factor XIII and the peptide released during its activation by thrombin, *Biochemistry* **13**:750–756.

Takahashi, N., Takahaski, Y., and Putnam, F. W., 1986, Primary structure of blood coagulation factor XIIIa (fibrinoligase, transglutaminase) from human placenta, *Proc. Natl. Acad. Sci. USA* **83**:8109–8023.

Part 3

PHYSIOLOGY

Chapter 8

EFFECT OF CHEMICAL MODIFICATION OF α-THROMBIN ON ITS REACTION WITH PLATELETS AND NUCLEATED CELLS

Nicholas J. Greco

1. INTRODUCTION

Considering the pivotal role of the serine protease thrombin (EC 3.4.21.5) in blood coagulation and cell activation, an intense evaluation of the structure of thrombin is warranted and could allow a molecular evaluation of how thrombin functions. Insight into the biological control mechanisms in which thrombin is involved may be aided by observing how thrombin interacts with its many diverse substrates. Results of experiments observing the effects of thrombin on cells and tissues *in vitro* should be interpreted with caution unless it can be demonstrated that these reactions also occur *in vivo*. Some of the *in vitro* effects of thrombin may mimic physiologic

Nicholas J. Greco • Biomedical Research and Development, The Jerome H. Holland Laboratory, American Red Cross, Rockville, Maryland 20855

Thrombin: Structure and Function, edited by Lawrence J. Berliner. Plenum Press, New York, 1992.

events while others may arise from the ability of thrombin to substitute for other serine proteases or hormones found *in vivo*.

Thrombin is proteolytically generated from its zymogen, prothrombin, which circulates in the blood as a single-chain polypeptide composed of F1, F2, and prothrombin 2 domains (Seegers *et al.*, 1975; Aronson *et al.*,1977; Fenton and Bing, 1986). Thrombin has decisive, important functions at all levels of hemostasis, not only influencing the plasma levels and activation states of other coagulation factors but also affecting blood cells and the vasculature. The regulation of hemostasis and the diverse biological functions of α-thrombin have been extensively reviewed (Fenton, 1981; Berliner, 1984).

In the clotting of whole blood, greater than 80% of the prothrombin is activated, but less than 10% of the expected concentration of thrombin is achieved (i.e. 5 to 15 u/ml) (Aronson *et al.*, 1977; Fenton, 1986) although much lower values (0.006–0.27 u/ml) have been reported (Shuman and Levine, 1978). Thrombin concentrations are probably near these lower levels because of the rapid inclusion of thrombin into fibrin clots (Aronson *et al.*, 1977), inhibition by antithrombin III (see Olson and Bjork, this volume), binding to thrombomodulin (Takahashi *et al.*, 1984; Bezeaud *et al.*, 1985) or other cellular binding sites. Inhibition of thrombin represents a crucial and effective point in the coagulation system and in the diverse bioregulatory effects of the enzyme.

Once generated, α-thrombin can act in many ways: first, thrombin is a potent activation stimulus for a number of diverse cells including platelets (Tam *et al.*, 1980; McGowan and Detwiler, 1986; Harmon and Jamieson, 1986), endothelial cells (Levin *et al.*, 1984; Weksler *et al.*, 1978), fibroblasts (Carney *et al.*, 1984; Chen and Buchanan, 1975; Perdue *et al.*, 1981a,b), smooth muscle cells (Haver and Namm, 1984), nerve cells (Snider *et al.*, 1984), and leukocytes (Bar-Shavit *et al.*, 1983). Thrombin is also intricately involved in the coagulation process by hydrolyzing a limited number of Arg-Gly peptide bonds; at a Gly-Val-Arg-Gly sequence in the A chain of fibrinogen liberating fibrinopeptides A and B in the conversion of fibrinogen to fibrin (Iwanaga *et al.*, 1969; Blombäck and Blombäck, 1972), at an Ile-Pro-Arg-Gly sequence in prothrombin to form thrombin (Magnusson *et al.*, 1975), and at a Val-Pro-Arg-Gly sequence in factor XIII converting it to factor XIIIa (Takagi and Doolittle, 1974). Fibrin clots are stabilized by factor XIIIa which catalyzes covalent bond formation between Lys and Gln residues on adjacent fibrin chains (Lorand, 1975). α-Thrombin rapidly disappears from the blood by binding to fibrin monomers which may protect thrombin from inactivation (Hogg and Jackson, 1989) and the proteolytic activity of thrombin can be inhibited by the covalent coupling to several protease inhibitors including antithrombin III (see

Olson and Bjork, this volume), protease nexin (Low *et al.*, 1981), thrombomodulin (Bezeaud *et al.*, 1985), α_2-macroglobulin, and α_1-antitrypsin (Fenton, 1981). Additionally, thrombin has been implicated in stimulating cellular processes of wound healing since it is generated upon tissue damage (Chen and Buchanan, 1975) and it induces the release of prostacyclin, a potent inhibitor of platelet aggregation and adherence, from endothelial cells (Weksler *et al.*, 1978).

These control mechanisms can be further divided into protease inhibitors present in the plasma (antithrombin III, α_2-macroglobulin, and $\alpha_1$antitrypsin) and those supplied by the cells themselves (protease nexin and thrombomodulin) which provides a mechanism for autoregulation at cell surfaces. The proteolytic activity of thrombin arises from a catalytic triad in the pocket of its active site composed of residues His-43, Asp-99, and Ser-205 which corresponds to residues His-57, Asp-102, and Ser-195 in trypsin, chymotrypsin and elastase. All exhibit secondary and tertiary structural homology regardless of substantial differences in primary sequence and substrate specificity.

Extensive metabolic and morphological changes are associated with the interaction of thrombin with cells. These changes are, at least in part, receptor-mediated since the binding of thrombin can be shown to be specific, saturable, reversible, and to lead to physiological responses. This chapter will primarily focus on the interaction of thrombin with platelets although other cells and substrates affected by thrombin will also be discussed.

2. MODIFIED THROMBINS

Several chemical modifying agents which differ in their affinity for the active site of serine proteases are diisopropylphosphorofluoridate (DIP [Ser-195]), phenylmethylsulfonylfluoride (PMSF [Ser-195]) (Nemerson and Esnouf, 1973), tosyl-lysylchoromethyl ketone (TLCK [His-57]) (Glover and Shaw, 1971; Workman *et al.*, 1977a), and phenylalanyl-prolyl-arginyl chloromethyl ketone (PPACK [His-57]) (Kettner and Shaw, 1978, 1981).The precise amino acid residue within the active site that is modified is indicated in brackets. These compounds act as affinity reagents since they contain the amino acid sequence corresponding to the cleavage site of physiological substrates of thrombin, e.g., fibrinogen. The use of these reagents indicates that both Ser-195 and His-57 are essential for proteolytic activity. DIP-thrombin retains some ($\sim$0.4%) fibrinogen clotting activity (Olson *et al.*, 1986). DIP-thrombin has been found to initiate biochemical changes (e.g., Ca^{2+} mobilization) in platelets (Jones *et al.*, 1989) and chemotaxis of human leukemic-60 cells (Bar-Shavit *et al.*, 1986), but it is unclear whether these DIP-thrombin preparations were completely inert in terms of proteolytic activity at the concentrations used. This consideration is especially important in the treatment of tissues in culture since these treatments are typically conducted for hours or days. PMSF is also incomplete as an inhibitor of serine proteases (Fahrney and Gold, 1963). If the PMSF moiety is removed from the active site, the resulting active-site-inhibited thrombin (anhydro-thrombin) maintains 0.15% amidolytic activity of fibrinogen hydrolysis (Tomono and Sawada, 1986).

Likewise, TLCK-thrombin has been reported to inhibit α-thrombin-induced serotonin release (Harmon and Jamieson, 1986) and does not interact with GPIb as detectable by crossed immunoelectrophoresis (Hagen *et al.*, 1982). A later study (Greco *et al.*, 1990) demonstrated that derivatization of α-thrombin with TLCK is incomplete and the product contains about 4% residual α-thrombin as assayed by hydrolysis of a chro-

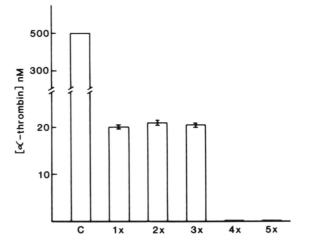

Figure 1. Residual α-thrombin amidolytic activity in preparations of TLCK-thrombin. The ordinate represents the concentration of α-thrombin as measured by its amidolytic activity with the chromogenic substrate S2238. C indicates the concentration of α-thrombin in the original reaction in the absence of TLCK. The abscissa represents three cycles of treatment (1×, 2×, 3×) of 13 μmol/liter α-thrombin with 10,000-fold molar excesses of TLCK each for 1 h. The 3× preparation was further treated with two cycles of derivatization with PPACK (4×, 5×) in 15-fold excess. Mean values (± SE) are given for six experiments carried out in duplicate or triplicate. These conditions had no effect on the activity of α-thrombin (10 nmol/liter) in the absence of TLCK or PPACK. (From Greco *et al.*, 1990, with permission.)

mogenic substrate; sequential addition of solid TLCK (10,000-fold molar excess) to solutions of α-thrombin (13 μM) did not completely inhibit amidolytic activity (Fig. 1). TLCK alone does not hydrolyze the chromogenic substrate. These data imply that these preparations are equilibrium mixtures of TLCK-thrombin and α-thrombin and cannot be used for evaluating competition between these two agents. However, residual α-thrombin amidolytic activity in TLCK-thrombin preparations can be completely inhibited by the PPACK reagent (Fig. 1). This apparent reversibility or lack of complete active-site inhibition must be taken into account in determining the effects of these analogs on cells and their ability to compete with α-thrombin for binding sites and/or receptors. Free, noncomplexed reagents such as TLCK may affect cell function or morphology (Weber *et al.*, 1975).

PPACK is an irreversible inhibitor reacting stoichiometrically to inactivate thrombin. It has been inferred from the recently determined crystal structure of human PPACK-thrombin that, in the structure of α-thrombin, there is restricted access to the active site cleft which might be responsible for the inability of most protease inhibitors to bind to α-thrombin. In contrast, the specific binding of the tripeptide PPACK might be explained by a hydrophobic cage formed by various residues including Ile-174, Trp-215, Leu-99, His-57, Tyr-60A, and Trp-60D (Skrzypczak-Jankun *et al.*, 1989; Bode *et al.*, 1989). Specifically, the D-phenylalanine residue (P3) binds in a hydrophobic pocket near the active site of thrombin (Sonder and Fenton, 1984) and the arginine residue (P1) binds in the proteinase specificity pocket.

Several reagents react with α-thrombin at sites distant from the active center and modify its function. Two such reagents are *N*-bromosuccinimide which modifies tryptophan residue(s) at or near the fibrinogen binding site and tetranitromethane (Alexander *et el.*, 1983) which modifies tyrosine residues. Nitrated α-thrombin is not uniformly modified or uniformly soluble and different kinetic properties are observed from batch to batch. These data indicate that consideration must also be given to kinetic parameters in evaluating differences in the properties of modified thrombins.

γ-Thrombin, a proteolytic derivative of α-thrombin, retains the ability to cleave small substrates but not fibrinogen and has been used extensively in attempts to characterize cell receptors and binding sites (Berliner, 1984; Alexander *et al.*, 1983). Most of the discussion in this chapter will be limited to active-site inhibition of thrombin with PPACK (PPACK-thrombin) because of its high-affinity, irreversible binding to and selectivity for thrombin.

3. EFFECTS OF α-THROMBIN ON PLATELETS

Platelets respond to thrombin by undergoing shape change, aggregation, secretion of granule constituents, Ca^{2+} mobilization, changes in $Na^+$and proton fluxes, and synthesis of prostanoids. These responses are rapid and occur at physiologically relevant thrombin concentrations suggesting that thrombin is an important mediator of platelet function in hemostasis and thrombosis. Thrombin binds to intact platelets (Tollefsen *et al.*, 1974; Okumura and Jamieson, 1976; Martin *et al.*, 1976; Alexander *et al.*, 1983; Harmon and Jamieson, 1988) and to platelet membranes (Ganguly and Sonnichsen, 1976; Tam and Detwiler, 1978; Harmon and Jamieson, 1988; Harmon *et al.*, 1991). Initial studies were taken to indicate 500–1000 high-affinity (K_d 10^{-9} M) and 50,000–75,000 low-affinity (K_d 10^{-7} M) thrombin binding sites per platelet (Tollefsen *et al.*, 1876; Workman *et al.*, 1977b) or, alternatively, were interpreted to indicate a single binding site exhibiting negative cooperativity (Tollefsen and Majerus, 1976). More recent studies using computer-assisted binding analysis (LIGAND; Munson and Rodbard, 1980) have determined the best fit for the binding of thrombin to platelets to be a three-site model (Harmon and Jamieson, 1985) with high-affinity (50 sites/platelet; K_d 0.3 nM), moderate-affinity (1700 sites/platelet; K_d 11 nM), and low-affinity (590,000 sites/platelet; K_d 2900 nM) binding sites but without nonspecific binding.

Since proteolytically active thrombin is required for platelet activation (Davey and Luscher, 1967; Workman *et al.*, 1977a), it has been proposed that activation by thrombin is strictly a proteolytic, not a receptor-mediated event (Phillips and Agin, 1977; Mosher *et el.*, 1979). A simple proteolytic event is inconsistent with the kinetics of activation. Specifically, in a proteolytic reaction, the rate of reaction is proportional to the enzyme concentration whereas the extent of proteolysis should be independent of the enzyme concentration; complete proteolysis would be expected even at low enzyme concentrations. Over a range of thrombin concentrations, from low concentrations causing only shape change to high concentrations giving complete activation, the rate of heat production is constant but the extent is proportional to the thrombin concentration (Ross *et al.*, 1973). Secretion of Ca^{2+} and ATP followed comparable kinetics with various thrombin concentrations (Detwiler and Feinman, 1973; Martin *et al.*, 1975, 1976). These studies suggest that the interaction of thrombin with platelets has properties of both an enzyme-mediated reaction and an agonist–receptor equilibrium since there is not a catalytic turnover of thrombin and both receptor mechanisms appear to be operative in platelets. It is known that thrombin is bound to the platelet surface by a portion of the molecule adjacent to, but not at the active site (Workman *et al.*, 1977a,b) indicating that binding may serve to orient thrombin for the proposed proteolytic event (Okumura and Jamieson, 1976; Harmon and Jamieson, 1986); alternatively, proteolysis and binding may be separate events.

Binding of a protease such as α-thrombin or trypsin would proceed through an acyl bond intermediate which would be formed between Ser-205 (195) and a cleavage site located C-terminal to a lysine or arginine residue of a receptor protein. It has been postulated that the generated amino-terminus could be a biological signal as is observed during the activation of zymogen substrates (Neurath, 1975). This proteolytic cleavage may be on the receptor or on an adjacent molecule. These proteolytic events may generate a transmembrane signal which acts intracellularly. Perhaps the proteolytic event allows the interaction of surface proteins or glycoproteins with cytoskeletal structures initiating redistribution (Crossin and Carney, 1981), aggregation, or oligomeric receptor interaction between catalytic enzymes (e.g. phospholipase A_2 or C) or the formation of ion channels (e.g. Na^+/H^+; Na^+,K^+-ATPase)(Stiernberg et al., 1984). Amiloride, an inhibitor of a Na^+/H^+ antiporter (Baron and Limbird, 1988), inhibits thrombin-initiated DNA synthesis in fibroblasts (Stiernberg et al., 1984).

Interpretation of possible proteolytic mechanisms of thrombin can be complicated if protease inhibitors are used without proper characterization of their site(s) of action. For example, leupeptin (acetyl-leu-leu-arginal), a dual sulfhydryl and serine protease inhibitor, has been used in an attempt to identify proteolytically mediated activation pathways on human platelets (Ruggiero and Lapetina, 1985; Lazarowski and Lapetina, 1990). In these studies, it has been shown that leupeptin (100 µg/ml) inhibits such thrombin-induced responses as platelet aggregation and serotonin release (Ruggiero and Lapetina, 1985), formation of inositol phosphates, activation of protein kinase C and myosin light chain kinase, activation of phospholipase C, and phosphorylation reactions (Lazarowski and Lapetina, 1990). From these findings, it has been suggested that a reactive sulfhydryl and/or serine protease may play a crucial role in the activation of platelets by thrombin. Various other investigators (Ruda and Scrutton, 1987; Greco, unpublished) have demonstrated that 100 µg leupeptin/ml inhibits platelet aggregation, but not shape change, induced by α-thrombin and also inhibits the hydrolysis of S2238 (H-D-phenylalanyl-L-pipecolyl-L-arginine-p-nitroanilide, Kabi Vitrum) (Lottenberg et al., 1982) by 2.5 nM α-thrombin. These inhibitory effects are specific toward thrombin since 100 µg leupeptin/ml does not inhibit aggregation induced by 2.5 µM ADP, 100 µM arachidonate, or 2.5 µg collagen/ml (Ruda and Scrutton, 1987; Greco, unpublished). These data, taken together, suggest a proteolytic effect of thrombin on platelets but interpretation is difficult since leupeptin inhibits the proteolytic activity of thrombin itself.

In support of a receptor-mediated mechanism are reports that the kinetics of the thrombin-induced platelet activation are more consistent with a receptor-mediated mechanism than with an enzyme-mediated mechanism (Ross *et al*., 1973; Detwiler and Feinman, 1973), and that the binding of thrombin and platelet sensitivity are affected in parallel by changes in membrane fluidity (Tandon *et el*., 1983), in Bernard-Soulier platelets which lack glycoprotein Ib (Jamieson and Okumura, 1978), and in myeloid leukemia (Ganguly *et al*., 1978).

Four platelet-associated molecules have been considered as thrombin receptors: glycoprotein V (GPV), protease nexin, glycoprotein Ib (GPIb) and a recently cloned molecule (M_r42,000).

Glycoprotein V (M_r 82,000). Considering the requirement for proteolytically active thrombin to initiate platelet activation, hydrolyzed platelet membrane components might be considered prime candidates as thrombin receptor(s). One component, GPV, appears to be released from the cell surface by the action of thrombin (Phillips and Agin, 1977; Mosher *et al*., 1979). However, independent experimental approaches suggest that the binding of α-thrombin and hydrolysis of GPV are separate and unrelated events (Knupp and White, 1985); (1) Essentially all of the GPV an be removed by chymotrypsin yet full platelet activation can be observed. (2) The kinetics of proteolytic liberation of this proposed mediator do not parallel the kinetic requirements for thrombin activation of platelets (McGowan *et al*., 1983). (3) Hydrolysis is not blocked by the use of TLCK-thrombin (Knupp and White, 1985). Interpretation of this latter result was uncertain since (a) the relative affinities of α-thrombin and TLCK-thrombin for binding sites were not compared and (b) TLCK-thrombin preparations continue to hydrolyze a chromogenic substrate S2238 at concentrations necessary for competition experiments (Greco *et al*., 1990). (4) Polyclonal anti-GPV antibodies, which block the degradation of GPV by α-thrombin, do not alter platelet activation by α-thrombin. Results relating to the rate or extent of GPV hydrolysis and to the extent of platelet activation by thrombin suggest that hydrolysis of GPV is not related to platelet activation by α-thrombin (McGowan *et al*., 1983; Bienz *et al*., 1986).

Protease nexin (M_r 41,000). Protease nexin forms a M_r 77,000 covalent complex with thrombin (Bennett and Glenn, 1980; Yeo and Detwiler, 1985). Since the time required for complex formation and the number of calculated binding sites (~500) approximates the estimated number of high-affinity platelet binding sites, it was suggested that this complex constituted the high-affinity thrombin receptor (Lerea and Glomset, 1987). This suggestion was obviated since complex formation is cell-activation dependent, i.e., protease nexin is secreted only when platelets are activated and therefore cannot be the primary receptor in nonstimulated cells.

Protease nexin may provide a crucial function by acting as an antithrombin and limiting the potent reactivity of thrombin in circulation and limiting extensive thrombus formation.

Glycoprotein Ib (M_r 170,000). GPIb appears to play as a role as a target receptor for thrombin on the platelet surface: (1) glycocalicin, a hydrolytic product of GPIb, acts as a competitive inhibitor of the activation of platelets by thrombin (Okumura and Jamieson, 1976); (2) there is reduced thrombin binding concomitant with reduced GPIb content in Bernard-Soulier platelets (Jamieson and Okumura, 1978); (3) thrombin can be chemically cross-linked to GPIb in intact platelets (Larsen and Simons, 1981; Jung and Moroi, 1983); (4) isolated GPIb binds ^{125}I thrombin with high affinity (Cooper *et al.*, 1981; Harmon and Jamieson, 1986); (5) platelet sensitivity to thrombin is decreased and the pattern of activation is altered after GPIb is removed by action of a specific protease of Serratia marcescens (Cooper *et al.*, 1981; Harmon and Jamieson, 1986); and (6) PPACK-thrombin competes with α-thrombin and inhibits platelet aggregation and serotonin release (Greco *et al.*, 1990). Each of these observations satisfies a criterion for a thrombin receptor. It has been suggested (Alexander *et al.*, 1983) that GPIb is not the essential receptor but a component which accelerates the response of the platelet to thrombin. This issue cannot, at the present time, be unequivocally resolved until the moderate-affinity binding site for thrombin is characterized and shown to be absolutely required for platelet activation.

In support of our observations regarding the existence of an additional receptor besides GPIb for α-thrombin (Yamamoto *et al.*, 1991), it has been reported that a functional thrombin receptor distinct from GPIb was expressed in Xenopus oocytes with mRNA isolated from megakaryocytic cell lines, fibroblasts, or human umbilical vein endothelial cells (Pipili-Synetos *et al.*, 1990; Van Obberghen-Schilling *et al.*, 1990; Vu *et al.*, 1991; Rasmussen *et al.*, in press). The clones (Vu *et al.*, 1991; Rasmussen *et al.*, 1991) contain a putative cleavage site (LDPRS) resembling a cleavage site in protein S. A proposed mechanism of action is that thrombin binds to the hirudin-like region within the receptor, proteolyzes the molecule and thereby reveals a new amino terminus which then serves as the cellular agonist. Synthetic peptides of this amino terminus or tethered ligand peptide activate platelets causing shape change, aggregation, and activation of phospholipase C (Vu *et al.*, 1991; Huang *et al.*, 1991).

At the present time it is uncertain as to whether the cloned receptor represents the high or moderate affinity binding site (Harmon and Jamieson, 1985; 1986; 1988). Because its deduced molecular weight of ≈ 42,000 based on amino acid composition agrees with the functional molecular size (30,000±9,000) determined from the radiation inactivation technique (Harmon and Jamieson, 1985), the cloned receptor acts synergistically with GPIb to induce the mobilization of intracellular Ca^{2+} (Greco et al., unpublished results), and the tethered ligand peptide activates phospholipase C (Huang et al., 1991), it may, at present, be assigned as the moderate affinity receptor.

Within the cloned receptor is an acidic sequence, E_{53}PFWEDE-EKNES, having homology to an acidic region of the C-terminus of hirudin, DFEEIPEE (Jandrot-Perrus et al., 1991; Vu et al., 1991). A similar sequence is also found within GPIb, D_{277}YYDEEDTEGD, and evidence to support that this site directly binds α-thrombin includes the findings that monoclonal antibodies directed within or near this acidic sequence inhibit thrombin binding to platelets and platelet activation (Yamamoto et al., 1985; Mazurov et al., 1991; Yamamoto et al., 1991; De Marco et al., 1991) and peptides derived from the primary sequence of GPIb inhibit thrombin-induced platelet activation by directly binding α-thrombin (Katagiri et al., 1990). A similar inhibitory effect is observed with peptides derived from the hirudin sequence (Jakubowski and Maraganore, 1990). At the present time it is unknown as to the relationship between the two receptors.

Recently, several lines of evidence indicate that α-thrombin activates platelets by two distinct pathways and a model summarizing these has been proposed (Jamieson, 1988). Specifically, the two pathways differ in the binding affinities of thrombin to its putative receptors (Harmon and Jamieson, 1985), in the role of nucleotide regulatory proteins (Houslay et al., 1986; Banga et al., 19882), and in their requirements for receptor occupancy (Holmsen et al., 1981; 1984; Huang and Detwiler, 1987). Those thrombin-induced responses requiring only transient receptor occupancy may occur by a proteolytic mechanism while those requiring the continual presence of thrombin may be caused by a receptor-mediated mechanism. Results obtained using hirudin to inactivate thrombin show that platelet responses requiring only transient exposure to thrombin include aggregation, dense granule secretion, and polyphosphoinositide breakdown. In contrast, arachidonate release, phosphatidic acid formation, and secretion of acid hydrolases require the continual presence of thrombin (Holmsen et al., 1981; 1984). Other evidence for two pathways includes the observation that prior treatment of platelets with chymotrypsin inhibits the thrombin-induced inhibition of adenylate cyclase and activation of phospholipase A_2 while aggregation, secretion, and the activation of phospholipase C and protein kinases are unaffected (McGowan and Detwiler, 1986). Amiloride antagonizes platelet activation induced by low concentrations of α-thrombin while it is ineffective at higher thrombin concentrations (Horne and Simons, 1978; Sweatt et al., 1986). Lastly, a monoclonal antibody, TM60, directed against GPIb (Yamamoto *et al.*, 1985), inhibits platelet activation at ≤ 0.5 nM α-thrombin but not a greater concentrations.

GPIb appears to represent a necessary thrombin receptor on platelets and is involved in at least one of the proposed pathways of platelet activation. However, there are pathways for thrombin-induced platelet activation independent of GPIb, as indicated by the ability of chymotrypsinized platelets (McGowan and Detwiler, 1986), Bernard-Soulier platelets (Jamieson and Okumura, 1978), human leukocyte elastase treated platelets (Wicki and Clemetson, 1985), and Serratia marcescens protease-treated platelets (Harmon and Jamieson, 1986; Yamamoto *et al.*, 1991) to respond to thrombin, although at concentrations 10 to 20 times higher than that effective for intact platelets. Of the proteases described above, only the one derived from Serratia marcescens appears to be specific for the removal of GPIb.

Previous studies evaluated the binding of ^{125}I-thrombin to platelets treated with Serratia marcescens (Harmon and Jamieson, 1988), which appeared to remove > 97% of the surface localized GPIb as analyzed using sodium periodate (^{3}H) labeled platelets followed by sodium dodecylsulfate polyacrylamide gel electrophoresis (SDS-PAGE); this analysis would imply that at least 900 GPIb molecules remain on each platelet after Serratia marcescens treatment. An alternative approach in evaluating the expression and quantity of platelet surface antigens is the use of monoclonal antibodies directed against cell-surface specific antigens in combination with fluorescence-activated flow cytometry (FACS; Michelson 1987). By this technique, it was determined that Serratia marcescens treatment results in a decrease in binding of 99.8 ± 0.1% (Yamamoto *et al.*, 1991) when a specific monoclonal antibody, TM60, is used (Table 1). TM60 binds to the 45 KDa amino-terminal peptide of GPIb (Yamamoto *et al.*, 1986) and, at saturating concentrations necessary to occupy all surface GPIb (~30,000), inhibits ^{125}I-thrombin binding to unactivated platelets by 50%. Using this technique it was shown that Serratia treatment 1) did not result in the exposure of GMP-140 (Table 1; S12), a component of α-granules only expressed on the platelet surface after platelet activation (Stenberg *et al.*, 1985); 2) did not enzymatically alter other cell surface-associated glycoproteins as indicated by the continuing ability of monoclonal antibodies to recognize them (Table 1; 10E5); and 3) eliminated the ability of a low thrombin concentration (0.5 nM) but not a high thrombin concentration (10 nM) to permit detection of GMP-140 when compared to intact platelets where the expression of this activation-dependent antigen was seen at both concentrations. These experiments clearly demonstrate that the effect of Serratia marcescens protease is restricted solely to the hydrolysis of GPIb and, concomitantly, loss of the high-affinity binding site.

Table I. Effect of Serratia marcescens Protease on Platelet Surface Glycoproteins[a]

Antibody	Serratia protease concentration (µg/ml)				
	0.1	0.25	0.5	1.0	2.5
TM60	62[b]	36	11	1	N.D.
	(2)	(1)	(1)	(1)	
10E5	100	104	103	107	103
	(2)	(1)	(4)	(4)	(3)
S12	1	1	2	2	3
	(1)	(1)	(1)	(1)	(1)

[a]Washed platelets were incubated (30 min, 22°C) with or without Serratia protease, fixed, incubated with a saturating concentration of a fluorescent-labeled monoclonal antibody, and analyzed by flow cytometry. For antibodies TM60 and 10E5, the fluorescence intensity of non-Serratia protease-treated platelets was assigned 100 units. For antibody S12, the fluorescence intensity of maximally activated (thrombin 10 nM) non-Serratia protease-treated platelets was assigned 100 units. Data are mean ± S.E.M. (shown in parentheses). $N = 3$. N.D., not detectable.
[b]Values represent the percent of control values observed in the absence of Serratia protease. Monoclonal antibodies are directed against the following platelet antigens: TM60 (GPIb, Yamamoto et al., 1985), 10E5 (GPIIb/IIIa complex, Collen, 1985), and 512 (GMP 140, Stenberg et al., 1985).

Of critical importance in determining the components involved in the mechanism(s) by which thrombin interacts with platelets is the observation that platelet shape change, representing the first measurable interaction of thrombin, and aggregation are observed at very low thrombin concentrations (≤ 1 nM, ≤ 0.01 u/ml) (Fig. 2A). Platelets treated with Serratia marcescens to remove GPIb respond with less sensitivity than control platelets to thrombin and show two apparent differences from intact platelets in platelet aggregation plots (Fig. 2B): (1) aggregation of Serratia-treated platelets required an approximately 10-fold increase in concentration of α-thrombin above those used for intact platelets and (2) the shape change response is attenuated and this was even more marked at lower α-thrombin concentrations. It was noted that the measurement of the extent of activation depends on the parameter measured, i.e. 50% of maximal aggregation with intact platelets is observed at 0.28 nM α-thrombin whereas with Serratia-treated platelets, 50% aggregation requires 4.5 nM, a rightward shift of about 20-fold. In contrast, only a 5-fold rightward shift in sensitivity was detected in Serratia-treated versus intact platelets when serotonin release is used as the marker (Harmon and Jamieson, 1986, 1988). From these figures it may be concluded that Serratia. marcescens-proteolyzed platelets retain the ability to aggregate and release serotonin. Conclusive evidence that platelet activation after Serratia treatment is not

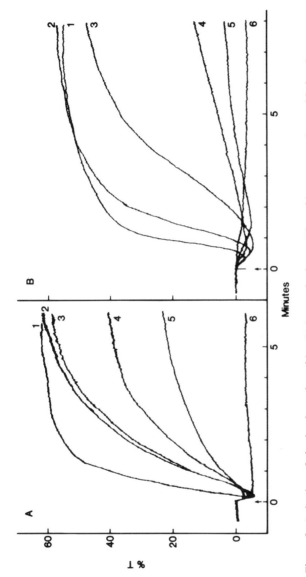

Figure 2. Activation of platelets by α-thrombin. (A) Intact platelets. α-Thrombin concentrations, added at the arrow, were as follows: curve 1, 5 nM; curve 2, 0.64 nM; curve 3, 0.48 nM; curve 4, 0.32 nM; curve 5, 0.24 nM; curve 6, 0.15 nM. (B) Serratia-proteolyzed platelets. α-Thrombin concentrations were as follows: curve 1, 20 nM; curve 2, 10 nM; curve 3, 5 nM; curve 4, 2.5 nM; curve 5, 1.25 nM; curve 6, 0.5 nM. This figure is representative of seven separate determinations. (From Greco et al., 1990, with permission.)

mediated by residual GPIb was obtained by exposing platelets to a large molar excess of TM60 (15–90 µg/ml) which dose-dependently inhibited [³H] serotonin release (Greco et al., 1985, 1990) by 0.5 nM thrombin from intact platelets (~80% inhibition at 90 µg/ml). The antibody (30–1200 µg/ml) was without effect on Serratia marcescens-treated platelets.

Most investigators conduct experiments designed to evaluate time-dependent effects of thrombin at high concentrations rather than concentration-dependent effects. Our data would indicate that differences in receptors might be distinguished based on their level of occupancy and therefore strict attention should be paid to the thrombin concentration used. In our work, we use unmodified human α-thrombin with a defined specific activity (~2,900 u/mg where 0.01 u/ml = 0.1 nM) compared to a well-accepted standard, i.e., NIH thrombin (Fenton et al., 1977a,b), and specific activity is evaluated both for a low-molecular-weight tripeptide chromogenic substrate (e.g., S2238) and for natural substrates (e.g., fibrinogen). Since α-thrombin is generated from prothrombin preparations (Fenton et al., 1977a,b; Greco et al., 1990), purification is assayed by SDS–PAGE of [¹²⁵I]-thrombin preparations to ensure that proteolytically active meizothrombin or the degraded forms of α-thrombin, β or γ-thrombin, are not present. Using a lower than optimal specific activity, where 100% of active sites cannot be titrated with p-nitrophenyl-p-guanidino-benzoate (Shaw, 1975; Greco et al., 1990) or fluorescein mono-p-guanidinobenzoate (Bock et al., 1989), may indicate the presence of partially denatured thrombin which may have undergone tertiary structural changes that may cause it to bind differently to cells and substrates. The loss of amidolytic activity and fibrinogen clotting activity does not occur simultaneously; the ability to clot fibrinogen is lost rapidly whereas the loss of ability to cleave small substrates is much less marked (Elion et al., 1986).

4. EFFECTS OF α-THROMBIN ON NUCLEATED CELLS

Thrombin binds to and initiates diverse effects on a variety of nucleated cells. In fibroblasts, it initiates mitogenesis and DNA synthesis (Chen and Buchanan, 1975; Carney et al., 1979; Glen and Cunningham, 1979; Glenn et al., 1980; Hall and Ganguly, 1980), stimulates arachidonic acid release, inositol phosphate and diacylglycerol generation (Raben et al., 1987), and activates phospholipase C and protein phosphorylation (Chambard and Pouyssegur, 1983). Thrombin causes smooth muscle cell retrac-

tion and associated cytoskeletal changes (Rowland *et al.*, 1984) and affects
Ca^{2+} homeostasis in chick embryonic heart cells (Chien *et al.*, 1990). Functional Ca^{2+}-mobilizing α-thrombin receptors have recently been expressed
in oocytes and appear to be coupled by G-protein-mediated mechanisms
to intracellular Ca^{2+} mobilization and Ca^{2+}-dependent Cl^- channel activation (Van Obberghen-Schilling *et al.*, 1990; Vu *et al.*, 1991).

Thrombin binds to specific sites on endothelial cells (Awbrey *et al.*,
1979; Lollar and Owen, 1980; de Groot *et el.*, 1987) and has several direct
effects on their properties (1) it initiates mitogenesis (Baker *et al.*, 1979); (2)
it generates a substained elevation of plasminogen activator (Loskutoff,
1979); (3) it induces the release of fibronectin (Mosher and Vaheri, 1978)
and release of stored nucleotides (Pearson and Gordon, 1979); and (4) it
stimulates phospholipase A_2 (Lollar and Owen, 1980)

Catalytically active thrombin is required for binding to chick or human fibroblasts whereas it is not required for binding to mouse or hamster
cells (Glenn *et al.*, 1980). Therefore, in chick and human fibroblasts a
mechanism proposed for activation involves proteolysis of an M_r 43,000
cell surface receptor (Glenn and Cunningham, 1979) which has been
identified as protease nexin (Baker *et al.*, 1980; Low *et al.*, 1981; Van
Nostrand *et al.*, 1990). A specific interaction of thrombin with protease
nexin has not suggested how or if this proteolytic event stimulates
fibroblasts and in platelets this complex is formed only after activation.
Binding sites for thrombin or fibroblasts appear to be clustered on the cell
surface prior to thrombin binding (Carney and Bergman, 1982) and may
indicate a mechanism by which thrombin is localized to a site of proteolysis.

It appears in fibroblasts that activation by thrombin involves two
signals, one generated by receptor occupancy of a high-affinity receptor by
α-thrombin or an enzymatically inactive thrombin (i.e., DIP-thrombin)
and the second generated by the proteolytic activity of proteases such as
thrombin, γ-thrombin, or trypsin at a lower-affinity, perhaps nonsaturable
receptor (Glenn *et al.*, 1980). It has been shown in experiments using
DIP-thrombin and γ-thrombin that the initiation of cell proliferation is not
due to residual thrombin since similar results are obtained in the presence
of protease nexin or antithrombin III, both of which bind residual α-
thrombin (Carney *et al.*, 1984, 1986). In further support of this hypothesis
are studies using a monoclonal antibody directed against a high-affinity
thrombin binding site on fibroblasts which initiates DNA synthesis in the
presence of α-thrombin similar to α-thrombin alone (Carey *et al.*, 1986;
Frost *et al.*, 1987). This monoclonal antibody recognizes a molecule with
M_r 150,000 on immunoblots (Herbosa *et al.*, 1984). In addition, α-thrombin has been chemically cross-linked to a component of M_r 150,000 on

fibroblasts and a 40-fold excess of either PPACK-thrombin or PMS-thrombin did not block α-thrombin-induced DNA synthesis (Van Obberghen-Schilling and Pouyssegur, 1985).

5. COMPARISONS OF THE EFFECTS OF α-THROMBIN BETWEEN CELL TYPES

5.1. α-Thrombin Binding Components

Two components of M_r 150,000–170,000 and 40,000–45,000 can be cross-linked to thrombin in many tissue types. The M_r 40,000–45,000 component, protease nexin, appears to be similar whether derived from platelets or nucleated cells. By a number of criteria, GPIb appears to be the high-affinity receptor for thrombin on platelets. Monoclonal antibodies developed against GPIb (Yamamoto *et al.*, 1985; McGregor *et al.*, 1983) inhibit thrombin-induced platelet activation (Yamamoto *et al.*, 1985; De Marco *et al.*, 1991) but only when the concentration of thrombin is approximately equal to the K_d (0.3 nM) of the high-affinity receptor (Harmon and Jamieson, 1986). At higher thrombin concentrations, inhibition is not observed presumably because GPIb-independent pathways are activated. Similar data have not been reported for the effects of antibodies developed against binding sites in nucleated cells.

5.2. Responses to Nonenzymatically Active Thrombin

A blocked thrombin (e.g., DIP-thrombin) may function nonenzymatically in a hormone like manner initiating proliferative responses of fibroblasts (Carney *et al.*, 1984). The chemotactic activity of thrombin, which resides in an anionic binding site composed of arginyl and/or lysyl residues (Bar-Shavit *et al.*, 1984), also appears to be nonenzymatic (Bar-Shavit *et al.*, 1983). Similar effects of enzymatically inactive thrombin on platelets have not been reported. PPACK-thrombin appears to retain a heparin binding site since it binds to and is eluted from a heparin column (Olson *et al.*, 1986). In contrast, we have found that the antithrombin III binding site on PPACK-thrombin is physically blocked or absent due to active-site modification since there is no apparent interaction between PPACK-thrombin and antithrombin III. It should also be noted that DIP-thrombin does not bind antithrombin III (Carney *et el.*, 1986), in contrast to anhydro-thrombin (Tomono and Sawada, 1986). These results may imply that occupancy of the active site by substituents based on size may determine whether antithrombin III binds to enzymatically inactive thrombin (Rosenberg and Damus, 1973; Fenton, 1981) but it also may imply that bulky substituents distort the active site sufficiently to affect tertiary structure. If distortion is observed, it may affect the interpretation of structures determined by x-ray crystallography of PMSF-thrombin (McKay *et al.*, 1977) or PPACK-thrombin (Bode *et al.*, 1989).

5.3. Receptor Occupancy

In fibroblasts, occupancy of high-affinity receptors is required for signal generation and increased cAMP levels (Carney *et al.*, 1984; 1986; Gordon and Carney, 1986). Also it has been suggested that γ-thrombin reacts at a second site, other than the high-affinity site, on fibroblasts and activates phosphoinositide turnover and Ca^{2+} mobilization (Carney *et al.*, 1986). Similar observations have been made in platelets using γ-thrombin or in the absence of GPIb (McGowan and Detwiler, 1986; Yamamoto *et al.*, 1991).

6. MODIFICATION OF CELL ACTIVATION BY ACTIVE-SITE-INHIBITED α-THROMBIN

Initial studies examined the effect of specific active-site-directed inhibitors to obtain information concerning the active center residues. Affinity labels, synthesized to correspond to physiological cleavage sites in natural substrates, modify the proteolytic effects of enzymes. Inactivation of a specific enzyme by a particular affinity reagent is completely dependent on the structure of the inhibitor. Affinity labeling by chloromethyl ketones has permitted identification and characterization of serine proteases with chymotryptic, tryptic, and elastolytic activities (Shaw, 1975; Bajusz *et al.*, 1978). Since thrombin is a serine protease with kinetic similarities to trypsin and chymotrypsin, a common hydrolytic mechanism may be involved and a similar three-dimensional structure among all three serine proteases may be inferred from the homology within the B-chain sequence (Siegler *et al.*, 1968).

Models for structure–function relationships for thrombin can be addressed knowing the specific determinants involved in the interactions of thrombin with substrates, inhibitors, cofactors, and cellular binding sites. To approach these structure–function relationships for platelet–thrombin or cell –thrombin interactions, degraded and chemically modified forms of thrombin have been used. Results of these studies must be interpreted with the knowledge that extensive modification might significantly alter the tertiary structure and that full characterization of the modified thrombin (e.g., binding to natural substrates) might be lacking. The sites of specific stoichiometric substitution into the primary sequence of modified thrombin may also be lacking. These modified thrombins, devoid of catalytic activity, have been used in an attempt to distinguish between receptor-mediated versus proteolytically mediated mechanisms of thrombin activation.

6.1. Preparation of PPACK-Thrombin

Critical considerations in evaluating the effect of PPACK-thrombin are to ensure that free, unreacted PPACK is completely removed from blocked thrombin preparations and that residual α-thrombin activity is absent. This critical step can be assayed by examining the effect of PPACK-thrombin preparations on thrombin-induced hydrolysis of S2238 which occurs C-terminally to an arginine residue at the P3 position and liberates p-nitroanilide. The method for the determination of activity is based on the difference in absorbance between the p-nitroanilide formed and the original substrate. The rate of p-nitroanilide formation, i.e., the increase in absorbance per time at 405 nm, is proportional to the enzymatic activity of thrombin.

PPACK reacts rapidly with α-thrombin at 37°C: with stoichiometric amounts of PPACK, a 98% inhibition of thrombin-induced S2238 hydrolysis and a 52% inhibition of thrombin-induced platelet aggregation are noted. With a five-fold molar excess of PPACK, respective inhibitions of 100 and 96% are noted (Table II). PPACK selectively alters the platelet aggregation response to α-thrombin without affecting platelet shape change and the aggregation of platelets by 2.5 µg collagen/ml, 2.5 µM ADP,

**Table II. Effect of PPACK on the Hydrolysis of S2238
and the Extent of Washed Platelet Aggregation
Caused by 0.5 nM α-Thrombin[a]**

PPACK (nM)	Percent of maximal absorbance	Percent of maximal aggregation
0.025	102	100
0.05	43	92
0.25	32	71
0.5	2	48
2.5	N.D.	4

[a]Hydrolysis of S2238 (absorbance at 405 nm) and the percent of platelet aggregation were measured at 37°C in the presence of increasing concentrations of PPACK. Values represent the percent of control values observed in the presence of PPACK; maximal aggregation is observed using 10 nM α-thrombin. N.D., not detectable.

and 100 μM arachidonate is unaffected by 5 nM PPACK. This finding suggests that proteolysis of a site on the platelet by α-thrombin elicits an aggregation response or, alternatively, that the inhibition by PPACK may indicate that a reactive extracellular-localized serine is involved in a proteolytic event on the platelet surface.

Unreacted PPACK can be removed by extensive dialysis in high-salt (≥0.3 M) buffers (Fenton *et al.*, 1977a; Landis *et al.*, 1981). To confirm that PPACK-thrombin preparations are devoid of residual PPACK, admixtures of 0.3 nM α-thrombin and micromolar concentrations of active-site-blocked thrombin are mixed and the resulting absorption at 405 nm is measured. If unreacted PPACK is present, this absorption will decrease. This assay is sufficiently sensitive such that < 0.5 nM of unreacted PPACK can be determined. Typical preparations of PPACK-thrombin utilize 50–70 μM α-thrombin and a 15-fold molar excess (750–1050 μM) of PPACK reagent (Greco *et al.*, 1990). Therefore, the chromogenic assay can determine a contamination of 0.00005% unreacted PPACK. The chromogenic assay is approximately 10 to 20 fold more sensitive than the determination of fibrinogen clot formation by α-thrombin/blocked-thrombin mixtures and approximately 5-fold more sensitive than serotonin release measurements. A sensitive assay for residual α-thrombin activity, besides the chromogenic substrate assay, involves platelet shape change which occurs at an α-thrombin concentration as low as 0.1 nM. Shape change is the first measurable indication that thrombin has interacted with the cell. A complication of dialysis in high salt is the need to correct for the effects of the resulting high salt concentrations on platelet aggregation. High salt concentrations are required for maintaining the solubility of derivatized thrombin as previously noted for native thrombin (Fenton *et al.*, 1977a,b). The activity of thrombin is affected by pH and ionic strength (Landis *et al.*, 1981) and by monovalent cations (Orthner and Kosow, 1980). Reports on kinetic data in the literature using either bovine or human thrombin do not always state the conditions. To effect a rapid, initial separation, gel filtration of the dialyzed product can be carried out on columns of Sepharose G-25M before dialysis. Extensively dialyzed PPACK-thrombin (1000 nM) does not (1) convert fibrinogen to fibrin (0.5 nM α-thrombin clots fibrinogen in approximately 3.5 min) (Harmon and Jamieson, 1986, 1988; Greco *et al.*, 1990), (2) hydrolyze S2238 (Greco *et al.*, 1990), or (3) interfere with the ability of 1 nM thrombin to cleave fibrinogen using pooled human plasma. Preparations of PPACK-thrombin appear to remain stable for periods up to 6 months at −70°C.

All preparations of PPACK-thrombin discussed in this chapter have used a 15-fold molar excess of PPACK to thrombin and were extensively dialyzed before use in competitive assays. In addition, preparations of

PPACK-thrombin tested at 1000 nM do not cleave S2238, indicating residual α-thrombin activity is not present. It should be noted that a comparison of autoradiograms derived form 10% SDS–PAGE analysis of $[^{125}I]$-thrombin versus $[^{125}I]$-PPACK-thrombin revealed an identical pattern of a single species (nonreduced gels) of M_r 36,000; autolytic products of α-thrombin, i.e. β- or γ-thrombin, were not detected (< 1%) by this technique.

6.2. Effect of PPACK-Thrombin on the Interaction of α-Thrombin with Platelets

A low concentration of thrombin (0.4 nM) causes washed intact platelets to change shape, release granule constituents (measured by $[^3H]$serotonin release), and aggregate. Published studies on thrombin-induced platelet activation have typically used concentrations of at least 0.1 u/ml (1 nM) and frequently as high as 10u/ml. Since maximal serotonin release is observed at much lower thrombin concentrations (Harmon and Jamieson, 1986, 1988) and 50% of high-affinity receptor occupancy is achieved at 0.3 nM (Jamieson, 1988), PPACK-thrombin was tested for its ability to alter platelet responses at low (0.3–0.5 nM: 0.03–0.05 u/ml) α-thrombin concentrations. The simultaneous addition to or preincubation of PPACK-thrombin with platelets dose-dependently inhibited platelet aggregation and $[^3H]$serotonin release but shape change was unaffected. This inhibition appeared to be competitive with a graphically determined IC_{50} of 110 nM for platelet aggregation when 0.4 nM α-thrombin was used. The affinity of PPACK-thrombin for the high-affinity platelet binding site is only about 1/17 of that for intact α-thrombin (Harmon and Jamieson, 1985).

The effects of PPACK-thrombin were specific for α-thrombin since 500 nM PPACK-thrombin did not affect aggregation of platelets using minimal aggregating concentrations of 1 μM epinephrine, 2.5 μM collagen/ml (in the presence or absence of exogenous Ca^{2+} and fibrinogen), 50 μM arachidonate, 2.5 μM ADP or 1.1 mg ristocetin/ml.

As previously indicated in this chapter, Na^+/H^+ exchange via an antiporter appears to play an important role in stimulus–response coupling, appearing to affect many cellular functions including DNA synthesis, protein synthesis, glycolytic energy production, cell growth, skeletal muscle contractility, and secretory processes (for review see Pouyssegur, 1985). In platelets, Na^+/H^+ exchange has been suggested to initiate specific functions, namely, adhesion to subendothelial structures, shape change, and aggregation and release of platelet stores of ADP (Connolly and Limbird, 1983). In human fibroblasts, the Na^+/H^+ exchange may be regulated by

Ca^{2+}/calmodulin (Paris and Pouyssegur, 1984) or by a direct receptor-linked mechanism.

Activation of Na^+/H^+ antiporter can be measured by determining pH changes when an intracellular pH indicator dye, 2′,7′-bis(carboxyethyl)-5,6-carboxyfluorescein (BCECF), is introduced into platelets (Rink et al., 1982; Rink and Hallam, 1984). Initial experiments demonstrated that at low α-thrombin concentrations (0.3–0.5 nM), a persistent cytoplasmic acidification was observed in intact platelets (Fig. 3). Only at higher thrombin concentrations was there an initial transient acidification followed by a marked alkalinization. Many investigators have noted persistent cytoplasmic alkalinization (Greco et al., 1990, and references therein) and these results perhaps arose since most investigators have used thrombin concentrations two to three times greater than the minimum required to elicit

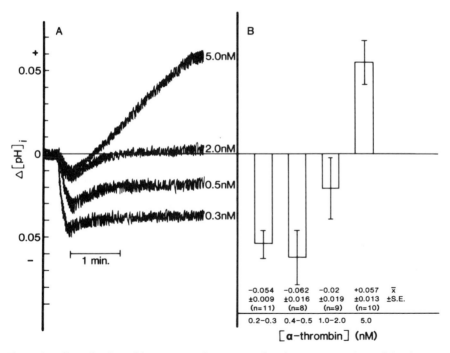

Figure 3. Effect of α-thrombin concentration on cytoplasmic pH. (A) Tracings of the change in pH with time at the indicataed α-thrombin concentrations as measured by BCECF fluorescence in a representative experiment. (B) Mean values (± S.E.) of the relative pH change at the 2-min time point from 8 to 11 separate determinations. (From Greco et al., 1990, with permission.)

full platelet aggregation and have used thrombin of undefined specific activity and purity. Therefore, it appears that alkalinization of the platelet interior is not required for platelet activation and presumably is a result of platelet activation and not the cause of it. This reasoning would be consistent with the observation that cytoplasmic alkalinization is not required for Ca^{2+} mobilization (Sanchez et al., 1988; Zavoico and Cragoe, 1988).

PPACK-thrombin (500 nM) did not affect the cytoplasmic acidification induced by 0.4 nM thrombin (Fig. 4). Only in those experiments which

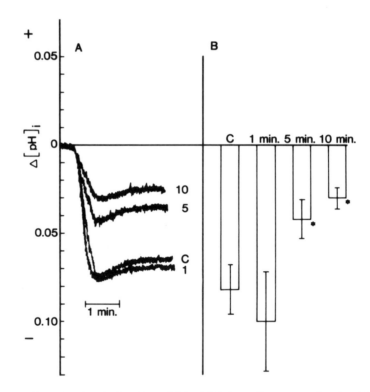

Figure 4. Effect of preincubation time with PPACK-thrombin on cytoplasmic acidification induced by α-thrombin. Platelets were washed using a Hepes-Mg^{2+} buffer method and loaded with BCECF prior to incubation with PPACK-thrombin (500 nM) for 1, 5, and 10 min as indicated. (A) Tracings of change in pH versus time in a representative experiment in the absence of 500 nM PPACK-thrombin (c) or with increasing times of incubation (1, 5, and 10 min) prior to addition of α-thrombin (0.4 nM). (B) Summation of pH changes measured after 2 min. Mean values (± S.E.) are given for five experiments. The asterisks indicate values that differ significantly from control ($p < 0.05$). (From Greco et al., 1990, with permission.)

allowed a 5 to 10 min preincubation of platelets with PPACK-thrombin was there a progressive reduction in the degree of acidification (Fig. 4). PPACK-thrombin alone had no effect on cytoplasmic pH in the absence of α-thrombin.

Mobilization of Ca^{2+} from intracellular stores appears to play a pivotal event in cell activation (Rink and Hallam, 1984; Sage and Rink, 1987). To determine intracellular Ca^{2+} mobilization, fluorescence measurements from an intracellularly trapped indicator dye (Fura-2) were used (Sage and Fink, 1987). In the absence of extracellular Ca^{2+}, α-thrombin (0.4 nM) causes a rapid increase in cytoplasmic Ca^{2+} concentration in intact platelets from baseline values of 79 ± 17 nM to 459 ± 16 (S.D.) nM within 5 to 10 sec (Fig. 5). In the presence of 500 nM PPACK-thrombin, this level decreased to 402 ± 33 nM and 249 ± 28 nM with a 1 or 5 min preincubation of PPACK-thrombin, respectively; again, PPACK-thrombin itself had no effect on Ca^{2+}mobilization at this concentration.

After removal of GPIb by Serratia-protease treatment, PPACK-throm-

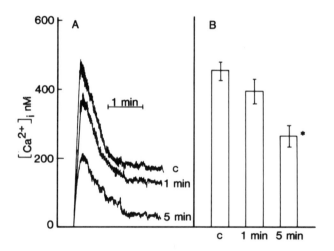

Figure 5. Effect of preincubation time with PPACK-thrombin on cytoplasmic Ca^{2+} levels induced by α-thrombin. (A) Tracings of the change in Ca^{2+} mobilization versus time in the presence (1 or 5 min preincubation) or the absence (c) of PPACK-thrombin as measured by Fura-2 fluorescence in a representative experiment. (B) Mean values (± S.E.) of the maximal Ca^{2+} mobilization in five experiments under the conditions described above (A). The asterisk indicates that values determined at 5 min differ significantly from control ($p < 0.05$). Note that PPACK-thrombin (5 min preincubation) affects only the maximal Ca^{2+} mobilization and does not affect the reuptake of Ca^{2+}. These experiments were carried out in the absence of exogenous Ca^{2+}. (From Greco et al., 1990, with permission.)

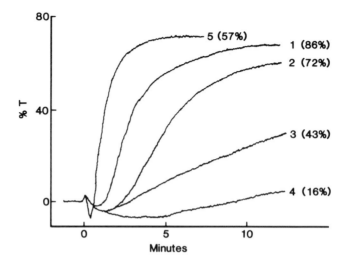

Figure 6. Aggregation of *Serratia*-treated platelets and inhibition by PPACK-thrombin. *S. marcescens*-treated platelets (1.1 × 10⁸/ml) showed a full response with 10 nM α-thrombin (curve 1) and an almost full response with 5 nM (curve 2): serotonin secretion data for each curve are shown in parentheses. When PPACK-thrombin was incubated for 1 min with the *Serratia*-treated platelets prior to the addition of 5 nM α-thrombin, there was about 50% inhibition at μM PPACK-thrombin (curve 3) and almost complete inhibition at 2 μM PPACK-thrombin (curve 4). Note that platelet shape change is observed in each case. For comparison, curve 5 shows the aggregation tracing for intact platelets in the presence of 0.5 nM α-thrombin. (From Greco *et al.*, 1990, with permission.)

bin continues to cause an inhibition of thrombin-induced aggregation and serotonin release (Fig. 6) This finding is consistent with previous studies (Harmon and Jamieson, 1988) showing that PPACK-thrombin is able to complete with α-thrombin for all platelet binding sites and further suggests that platelet activation by α-thrombin and inhibition by PPACK-thrombin can occur by GPIb-independent pathways.

Several conclusions can be drawn from these studies using PPACK-thrombin to antagonize platelet activation by α-thrombin: (1) Since PPACK-thrombin affects platelet shape change, aggregation, and granule release in a dose-dependent fashion, the simplest explanation is that PPACK-thrombin serves as receptor antagonist, albeit weak, of thrombin. (2) Since it is required that PPACK-thrombin be preincubated with platelets 5 min before addition of thrombin and the measurement of intracellular Ca^{2+} mobilization and cytoplasmic acidification, it is again suggested that the effect of PPACK-thrombin may be as an inhibitor of signal trans-duction events. However, since aggregation requires only transient re-ceptor occupancy (Holmsen *et al.*, 1981, 1984; Huang and Detwiler, 1987), and is observed using chymotrypsinized platelets (McGowan and Detwiler, 1986), platelet shape change, Ca^{2+} mobilization, and cytoplasmic pH changes may require single proteolytic events at thrombin receptor(s).

7. STRUCTURE–FUNCTION RELATIONS OF PROTEIN MODIFICATIONS DISTINCT FROM THE CATALYTIC SITE

A complementary approach to using active-site inhibited thrombin to evaluate structure–function relationships is to utilize genetic variants of thrombin which have a single amino acid altered. Two characterized thrombin variants derived from dysfunctional prothrombins are Quick I and Quick II (Quick *et al.*, 1955; Henriksen and Owen, 1980). Thrombin Quick I has an increased K_m (7 μM to 44 μM) and a decreased k_{CAT} (91 s^{-1} to 7 s^{-1}). Endopeptidase digestion of thrombin Quick I showed that Arg-382 in the B chain is replaced by Cys (Henriksen and Owen, 1987; Hen-riksen and Mann, 1988). Thrombin Quick I retains catalytic activity to-ward benzoyl arginine ethyl ester (74% of thrombin level) and nitroanilide substrates (55% of thrombin levels), is complexed with antithrombin III at the same rate as thrombin, and hydrolyzes bovine prothrombin and bo-vine protein C at about 30% of the rate of thrombin. The k_{CAT}/K_m ratio is equal to 0.012 of the value obtained for thrombin with a relative fibrinogen Aα chain clotting activity of 0.013. Likewise, prostacyclin release from human umbilical vein endothelial cells and the initiation of platelet ag-gregation by thrombin Quick I are only 2.4 and 1.7% respectively of the levels elicited by thrombin. These results indicate that Arg-382 is essential in determining specificity toward fibrinogen and cell binding sites whereas different determinants are involved in the binding of other substrates. An extended region of interaction of the thrombin surface is probably in-volved in the reactivity of thrombin Quick with protein substrates and cell binding sites. Similar data have been obtained for thrombin Tokushima where Arg-418 is replaced by Trp as the primary structural alteration (Miyata *et al.*, 1987).

In contrast to Quick I, thrombin Quick II does not cleave fibrinogen or hydrolyze *p*-nitroanilide substrates. Gly-558 is conserved in serine pro-teases such as chymotrypsin and trypsin, forming part of the substrate binding pocket for aromatic and basic side chains. Crystallographic structure evaluation for serine proteases indicates that the structure of the substrate binding pocket is complementary to arginine residues. Since thrombin hydrolyses the peptide chain C-terminal to a limited number of arginine residues, modification of this site could explain altered substrate binding and hydrolysis.

It is not possible to completely determine if these residues (Arg-382, Quick I; Gly-558, Quick II) are directly involved in binding interactions or whether the residues are critical for maintaining tertiary protein structure necessary for binding of substrates at other sites on thrombin. These studies further indicate that the catalytic activity of thrombin is sensitive to structural alterations occurring at sites necessary for substrate binding. Further knowledge of the tertiary structure of thrombin should resolve these questions.

8. UTILITY OF CHEMICALLY MODIFIED THROMBINS: THROMBOSIS INTERVENTION AND TREATMENT

Antithrombin therapy has typically involved the infusion of heparin to prevent stasis-type venous thrombosis wherein thrombin is inactivated by activation of plasma antithrombin III. This approach prevents fibrin formation. The inhibitory effects of heparin are dependent on the plasma levels of antithrombin III, interactions with platelets and fibrinolytic components, and neutralization from platelet-derived antiheparin (e.g., platelet factor 4). Under normal physiological conditions of high-shear circulatory flow, heparin does not prevent arterial thrombosis or hemostatic plug formation, suggesting that these processes may be independent of thrombin formation. This resistance of thrombosis to heparin may be explained by a number of factors including the binding of thrombin to platelet-released protease nexin and platelet factor 4, restricted exposure of the heparin–antithrombin III complex to thrombin sequestered within the thrombus, and meizothrombin-induced platelet activation since meizothrombin is not affected by the heparin–antithrombin III complex. An effective antithrombotic strategy must consider targeting thrombin or inhibiting thrombin–platelet interactions. Inhibitors of the catalytic activity of thrombin impair activation of blood platelets regardless of their effect on thrombin binding.

In contrast, several studies have indicated a critical role for thrombin in activating platelets under high-shear conditions. These studies have shown that infusion of the synthetic peptide derivative PPACK (1) reduced intravascular thrombosis in dogs by blocking the ability of thrombin to cleave fibrinogen to fibrin (Schaeffer *et al.*, 1984; 1986), (2) reduced cardiac thrombus formation and mortality effected by thrombin infusion in rabbits (Collen *et al.*, 1982), and (3) interrupted acute platelet-dependent thrombosis in baboons (Hanson and Harker, 1988). Besides the direct effect of PPACK itself on α-thrombin, the stable active-site-inhibited thrombin, PPACK-thrombin, competes at a high-affinity thrombin receptor on platelets (Harmon and Jamieson, 1986, 1988) but does not initiate activation as evaluated by platelet shape change, [^{3}H]serotonin release, Ca^{2+} mobilization and acidification or alkalinization of the platelet interior (Greco *et al.*, 1990). PPACK-thrombin competitively inhibits thrombin-induced platelet aggregation and serotonin release but Ca^{2+} mobilization and cytoplasmic acidification induced by ≤ 0.5 nM thrombin are unaffected. Platelet shape change induced by 0.1 nM α-thrombin can be inhibited 50% by 500 nM PPACK-thrombin, a 5000-fold excess. These results imply that the transient exposure of platelets to proteolytically-active thrombin may be sufficient to initiate platelet shape, cytoplasmic pH changes, and Ca^{2+}mobilization. Both Ca^{2+} mobilization and cytoplasmic acidification occur within 5 s after exposure to α-thrombin.

A selective affinity reagent such as PPACK may also be effective in reducing thrombin generation by inactivating platelet-bound meizothrombin generated on the platelet surface (Lindout *et al.*, 1986). *In vivo* treatment of thrombosis by a transient infusion of PPACK would (1) inactivate the proteolytic activity of thrombin thereby removing the ability of thrombin to activate cells via a proteolytic mechanism, (2) form a stable derivative PPACK-thrombin which competes for platelet activation at all sites accessible to thrombin, and furthermore (3) allows the interaction of remaining proteolytically active thrombin with plasma antithrombin III since PPACK-thrombin does not retain the ability to bind to antithrombin III. It has been suggested that irreversible inhibitors react with amino or thiol groups in constituents of the blood and tissue by alkylation or acylating reactions and therefore this property makes them unsuitable for use in an antithrombotic therapy (Hauptmann and Markwardt, 1980). The selectivity of PPACK for thrombin (Kettner and Shaw, 1978) and the low levels necessary for efficient antithrombosis (Hanson and Harker, 1988) and its apparent nontoxicity *in vivo* (Collen *et al.*, 1982) however, could allow the use of PPACK for short-term thromboprophylactic treatment.

These studies suggest that synthetic antithrombin peptides may be useful treatment for the interruption of thrombosis plug formation and vascular graft thrombosis. Reocclusion of coronary arteries successfully treated by thrombolytic therapy, i.e., tissue plasminogen activator or preventing occlusion during angioplasty and endarterectomy may also be affected by treatment with antithrombin peptides.

9. UNANSWERED QUESTIONS

Despite intensive efforts, many questions regarding the ability of α-thrombin to interact with and initiate physiological responses in cells are unanswered. These questions can be summarized as follows:

Are binding and proteolysis necessary for cell activation or can receptor-mediated and proteolytically mediated events be differentiated? As shown in this chapter, tertiary alterations in thrombin structure are likely to affect its interaction with its binding sites. Interpretation of changes in cellular functions will be dependent on knowing which structural changes have occurred. Attempts to characterize thrombin receptors should evaluate which, if any, changes in the overall tertiary structure of active-site-inhibited thrombins have occurred by using physical methods (e.g., x-ray crystallography, NMR, circular dichroism) in conjunction with substrate binding experiments. A novel approach evaluating the relationship between the amino-binding exosite and the catalytic site of α-thrombin involves synthetic peptide "hirulogs" (Maraganore *et al.*, 1990) which contain an active-site specificity sequence (PPACK) and an anion-active binding exosite (Berliner *et al.*, 1985) (the C-terminal 12 residues of hirudin) coupled by a polymeric linker of glycyl residues. This C-terminal peptide of hirudin, derivatized with dinitrodifluorobenzene, specifically modifies Lys-149 of α-thrombin (Bourdon and Maraganore, 1989). Studies of the crystallographic structure of the hirudin—thrombin complex (Konno *et al.*, 1988; Rydel *et al.*, 1990) and PPACK-thrombin (Bode *et al.*, 1989) and the proposed three-dimensional model of the B chain of human α-thrombin (Furie *et al.*, 1982) indicate that the NH_2-terminus of hirulog peptides would be 18–20 Å from the active-site β-hydroxyl group of Ser-195 as determined using a minimum spacing arm of four glycyl residues (hirulog-1). Hirulog-1 has an anticoagulant activity comparable to hirudin while sulfated hirudin peptides are approximately 100-fold less active (Maraganore *et al.*, 1990). Hirudin peptides inhibit the fibrinogenolytic activity of thrombin by binding to the exosite and interfering with substrate recognition while affinity reagents such as PPACK modify the active site. Hirulogs retain high affinity and specificity for thrombin and may be useful to evaluate the sites on thrombin necessary for cell binding and activation in combination in studies employing all of these reagents.

Is proteolytically active thrombin required for the activation of all cell types? The simplest explanation for the requirement for proteolytically active thrombin is that a proteolytic step is an absolute requirement for cell activation. Chemical modification of thrombin could sterically alter the conformation of the active site and alter cell binding site(s) on the thrombin molecule. PPACK-thrombin demonstrates altered binding affinities to intact platelets and to antithrombin III. By examining possible structural changes from the physical aspect (x-ray crystallography and tertiary structure analysis) and biological aspect (exosite binding moieties), we should be able to secure an understanding of thrombin-surface regions necessary for activation (Sugawara *et al.*, 1985). An implicit assumption in evaluating the ability of active-site-modified thrombins or modified thrombins (e.g., PPACK-thrombin) to activate or compete for activation, is that the altered thrombins bind to the same site(s) as α-thrombin and would activate platelets through identical mechanisms(s). Peptide bond specificity of various proteases can determine whether an enzyme activates platelets (e.g. thrombin, trypsin) or an enzyme with a different substrate specificity (e.g. chymotrypsin) does not (Martin *et al.*, 1975; Davey and Luscher, 1967). Since some homology exists between the B chains of thrombin and trypsin, correlations between the x-ray structure of thrombin and trypsin, which activate platelets, in comparison to chymotrypsin, which does not, may allow a deduction of common exosites necessary for receptor binding. Correlations between the x-ray structures of PPACK-thrombin and hirudin–thrombin complexes may also allow similar conclusions if taken with the realization that these complexes may not reflect "native" structures.

Perhaps the saturable binding site accessible to α-thrombin and not other serine proteases, i.e., trypsin and γ-thrombin (Alexander *et al.*, 1983), may be the relevant physiological interactive site necessary for accelerating the platelet activation whereas nonsaturable binding sites are nonspecific proteolytic cleavage sites. The nonsaturable binding sites occupied by trypsin and γ-thrombin require concentrations of typically 20–100 nM. Considering that an estimate for the amount of thrombin generated *in vivo* during blood coagulation is approximately 0.06–2.7 nM (Shuman and Levine, 1978), the high-affinity α-thrombin receptor may be the only relevant binding site. The proposal that saturable binding may not be necessary for platelet activation but contributes to its rapidity (Alexander *et al.*, 1983) may be important *in vivo*.

What are the identities, characteristics, and cofactor requirements of cell receptors and/or binding sites for α-thrombin? Only in platelets does there appear to be a characterized high-affinity binding site (GPIb) satisfying a number of criteria for a receptor. A precise primary sequence domain within the GPIb α chain binds thrombin in a relatively strict conformation (Katagiri *et al.*, 1990). Is GPIb on platelets identical to the M_r 150,000 species observed on fibroblasts (Moss *et al.*, 1983; Van Obberghen-Schilling and Pouyssegur, 1985) or are they immunologically related or homologous? Does α-thrombin interact with this M_r 150,000 species and accelerate the mitogenic response of fibroblasts and other cells similar to the proposed role of GPIb in platelets? It has been demonstrated that the molecular weight of the high-affinity receptor complex on platelets is ~900,000 by a radiation inactivation technique (Harmon and Jamieson, 1985). Is there a similar structure of a complex nature existing on other cells? Thrombin receptors appear to be clustered on fibroblasts before cell activation by thrombin (Carney, 1983) and perhaps oligomers of the high-affinity receptor serve to localize α-thrombin where a direct enzymatic event may occur. This high-molecular-weight complex may be composed of subunits including G-proteins, adenylate cyclase, and phospholipase A_2 as suggested (Jamieson, 1988). Perhaps there are common receptor(s) or subunits on various cells which do not differ in their transmembrane signal generation.

Can complex activation processes be separated by classical receptor-occupancy models or can complex stimulus-coupling mechanisms be assigned to specific binding sites?

Can a common activation model be developed for various cell types having diverse responses?

Are secondary binding sites on thrombin, presumed necessary for efficient cell activation, disrupted by active-site inhibition and how would these potential alterations affect the interpretation of studies of competition, activation, and so forth?

Are there alternate approaches for evaluating thrombin receptors or binding sites? Most cross-linking methods used to identify thrombin-binding components are restricted to reagents reacting randomly with protein amino groups (Duncan et al., 1983). Consequently, several "binding sites" may be labeled due to their proximity and not necessarily due to their involvement in the reactions in question. Extensive topographical derivatization of thrombin by photoactivatable or chemically cross-linked groups to provide increased access of the reagents may alter the biological properties of α-thrombin.

An alternate approach to cross-linking would be to use highly selective active-site affinity labeling of thrombin with a reagent such as the tripeptide chloromethyl ketone derivative (PPACK) to couple reactive groups in the active site. This approach would be analogous to the incorporation of spectroscopic probes (Berliner and Wong, 1974; Bock, 1988) into the active site of α-thrombin (i.e., bound to histidine). PPACK is coupled to succinimidyl (acetyl thio) acetate forming a thio ester-reactive group at the amino-terminus of the tripeptide. Since thrombin does not contain thio groups, reactive groups (i.e, azides) can be covalently attached via this probe to the active site. The interaction of this radiolabeled modified α-thrombin with a binding site or receptor could (theoretically) be trapped as an acyl intermediate present in enzyme–substrate complexes.

10. REFERENCES

Alexander, R. J., Fenton, J. W., and Detwiler, T. C., 1983, Thrombin–platelet interactions: An assessment of the roles of saturable and nonsaturable binding in platelet activation, *Arch. Biochem. Biophys.* **222**:266–275.

Aronson, D. L., Stevan, L., Ball, A. P., Franza, R., and Finlayson, J. S., 1977, Generation of the combined prothrombin activation peptide (F1.2) during the clotting of blood and plasma, *J. Clin. Invest.* **60**:1410–1418.

Awbrey, B. J., Hoak, J. C., and Owen, W. G., 1979, Binding of human thrombin to cultured human endothelial cells, *J. Biol. Chem.* **254**:4092–4095.

Bajusz, S., Barabas, E., Tolnay, P., Szell, E., and Bagdy, D., 1978, Inhibition of thrombin and trypsin with tripeptide analogues, *Int. J. Pept. Protein Res.* **12**:217–221.

Baker, J. B., Simmer, R. L., Glen, K. C., and Cunningham, D. D., 1979, Thrombin and epidermal growth factor become linked to cell surface receptors during mitogenic stimulation, *Nature* **278**:743–745.

Baker, J. B., Low, D. A., Simmer, R. L., and Cunningham, D. D., 1980, Protease-nexin: A cellular component that links thrombin and plasminogen activator and mediates their binding to cells, *Cell* **21**:37–45.

Banga, H. S., Walker, R. K., Winberry, L. L., and Rittenhouse, S. E., 1988, Platelet adenylate cyclase and phospholipase C are affected differentially by ADP-ribosylation, *Biochem. J.* **252**:297–300.

Baron, B. M., and Limbird, L. E., 1988, Human platelet phospholipase A_2 activity is responsive in vitro of pH and Ca^{2+} variations which parallel those occurring after platelet activation in vivo, *Biochim. Biophys. Acta* **971**:103–111.

Bar-Shavit, R. A., Kahn, A., Wilner, G. D., and Fenton, J. W., 1983, Monocyte chemotaxis: Stimulation by specific exosite region in thrombin, *Science* **220**:728–731.

Bar-Shavit, R. A., Kahn, A., Mudd, M. S., Wilner, G. D., Mann, K. G., and Fenton, J. W., 1984, Localization of a chemotactic domain in human thrombin, *Biochemistry* **23**:397–400.

Bar-Shavit, R., Hruska, K. A., Kahn, A. J., and Wilner, G. D., 1986, Hormone-like activity of human thrombin in: *Bioregulatory Functions of Thrombin* (D. A. Walz, J. W. Fenton, and M. A. Shuman, eds.), New York Academy of Sciences, New York, pp. 335–348.

Bennett, W. F., and Glenn, K. C., 1980, Hypersensitivity of platelets to thrombin: Formation of stable thrombin–receptor complexes and the role of shape change, *Cell* **22**:621–627.

Berliner, L. J., 1984, Structure–function relationship in human alpha- and gamma-thrombin, *Mol. Cell. Biochem.* **61**:159–172.

Berliner, L. J., Sugawara, Y., and Fenton, J. W., 1985, Human-thrombin binding to non-polymerized fibrin-sepharose: Evidence for an anionic binding region, *Biochem.* **24**:7005–7009.

Berliner, L. J., and Wong, S. S., 1974, Spin-labeled sulfonyl fluorides as active site probes of protease structure, *J. Biol. Chem* **249**:1668–1577.

Bezeaud, A., Denninger, M. -H., and Guillin, M. -C., 1985, Interaction of human α-thrombin and gamma-thrombin with antithrombin III, protein C and thrombomodulin, *Eur. J. Biochem.* **153**:491–496.

Bienz, D., Schnippering, W., and Clemetson, K. J., 1986, Glycoprotein V is not the thrombin activation receptor on human blood platelets, *Blood* **68**:720–725,

Blombäck, B., and Blombäck, M., 1972, The molecular structure of fibrinogen, *Ann. N. Y. Acad. Sci.* **202**:77–97.

Bock, P. E., 1988, Active site selective labeling of serine proteases with spectroscopic probes using thioester peptide chloromethyl ketones, *Biochemistry* **276**633–6639.

Bock, P. E., Craig, P. A., Olson, S. T., and Singh, P., 1989, Isolation of human blood coagulation α-factor Xa by soybean trypsin inhibitor–Sepharose chromatography and its active-site titration with fluorescein mono-p-guanidino benzoate, *Arch. Biochem. Biophys.* **273**:375–388.

Bode, W., Mayr, I., Baumann, U., Huber, R., Stone, S. R., and Hofsteenge, J., 1989, The refined 1.9 Å crystal structure of human alpha-thrombin: Interaction with D-Phe-Pro-Arg chloromethyl ketone and singificance of the Tyr-Pro-Pro-Trp insertion segment, *EMBO J.* **8**:3467–3475.

Boudon, P., and Maraganore, J. M., 1989, On the interaction of BC8865 with thrombin, *Thromb. Hemostas.* **62**:533.

Carney, D. H., and Bergman, J. S., 1982, ^{125}I-thrombin binds to clustered receptors on noncoated regions of mouse embryo cell surfaces, *J. Cell Biol.* **95**:697–703.

Carney, D. H., 1983, Immunofluorescent visualization of specifically bound thrombin reveals cellular heterogeneity in number and density of preclustered receptors, *J. Cell. Physiol* **117**:297–307.

Carney, D. H., Glenn, K. C., Cunningham, D. D., Das, M., Fox, C. F., and Fenton, J. W., 1979, Photoaffinity labeling of a single receptor for α-thrombin on mouse embryo cells, *J. Biol Chem.* **254**:6244–6247.

Carney, D. H., Stiernberg, J., and Fenton, J. W., 1984, Initiation of proliferative events by human thrombin requires both receptor binding and enzymic activity, *J. Cell Biochem.* **26**:181–195.

Carney, D. H., Herbosa, G. J., Stiernberg, J., Bergmann, J. S., Gordon, E. A., Scott, D., and Fenton, J. W., 1986, Double-signal hypothesis for thrombin initiation of cell proliferation, *Semin. Thromb. Hemostas.* **12**:231–240.

Chambard, J -C., and Pouyssegur, J., 1983, Thrombin-induced protein phosphorylation in resting platelets and fibroblasts: Evidence for common post-receptor molecular events, *Biochem. Biphys. Res. Commun.* **111**:1034–1044.

Chen, L. B., and Buchanan, J. M., 1975, Mitogenic activity of blood components. I. Thrombin and prothrombin, *Proc. Natl. Acad. Sci. USA* **72**:131–135.

Chien, W. W., Mohabir, R., and Clusin, W. T., 1990, Effect of thrombin on calcium homeostasis in chick embryonic heart cells, *J. Clin. Invest.* **85**:1436–1443.

Collen, D., Matuso, O., Stassen, J. M., Kettner, C., and Shaw, E., 1982, In vivo studies of a synthetic inhibitor of thrombin, *J. Lab. Clin. Med* **99**76–83.

Coller, B. S., 1985, A new murine monoclonal antibody reports an activation-dependent change in the conformation and/or microenvironment of the platelet glycoprotein IIb/IIIa complex, *J. Clin. Invest.* **76**:101–108.

Connolly, T. M., and Limbird, L. E., 1983, Removal of extraplatelet Na^+ eliminates indomethacin-sensitive secretion from human platelets stimulated by epinephrine, ADP and thrombin, *Proc. Natl. Acad. Sci. USA* **80**:5320–5327.

Cooper, H. A., Bennett, W. P., Kreger, A., Lyerly, P., and Wagner, R. H., 1981, The effect of extracellular protease from gram-negative bacteria on the interaction of von Willebrand factor with human platelets, *J. Lab. Clin. Med.* **97**:379–389.

Crossin, K. L., and Carney, D. H., 1981, Evidence that microtubule depolymerization early in the cell cycle is sufficient to initiate DNA synthesis, *Cell* **23**:61–71.

Davey, M. G., and Luscher, E. F., 1967, Actions of thrombin and other coagulant and proteolytic enzymes on blood platelets, *Nature* **216**857–858.

de Groot, P. G., Reinders, J. H., and Sixma, J. J., 1987, Perturbation of human endothelial cells by thrombin or PMA changes the reactivity of their extracellular matrix towards platelets, *J. Cell Biol.* **104**:697–704.

De Marco, L., Mazzucato, M., Masotti, A., Fenton, J. W., and Ruggeri, Z. M., 1991, Function of glycoprotein Ibα in platelet activation induced by α–thrombin, *J. Biol. Chem* **266:** 23776–23783.

Detwiler, T. C., and Feinman, R. D., 1973, Kinetics of the thrombin-induced release of calcium (II) by platelets, *Biochemistry* **12**:282–289.

Duncan, R. J. S., Weston, P. D., and Wrigglesworth, R., 1983, A new reagent which may be used to introduce sulfhydryl groups into proteins, and its use in the preparations of conjugates for immunoassay, *Anal. Biochem.* **132**:68–73.

Elion, J., Boissel, J. -P., Le Bonniec, B., Bezeaud, A., Jandrot-Perrus, M., Rabiet, M. -J., and Gullin M. -C., 1986, Proteolytic derivatives of thrombin, in: *Bioregulatory Functions of Thrombin* (D. A. Walz, J. W. Fenton, and M. A. Shuman, Eds.), New York Academy of Sciences, New York, pp. 16–26.

Fahrney, D., and Gold, A., 1963, Sulfonyl fluorides as inhibitors of esterase. I. Rates of reaction with acetylcholinesterase, chymotrypsin and trypsin, *J. Am. Chem. Soc.* **85**:997–1002.

Fenton, J. W., 1981, Thrombin specificity, *Ann. N.Y. Acad. Sci.* **370**:468–495.

Fenton, J. W., 1986, Thrombin, *Ann. N.Y. Acad. Sci.* **485**:5–15.

Fenton, J. W., and Bing, D. H., 1986, Thrombin active-site regions, *Sem. Thromb. Hemostas.* *12*:200–208.

Fenton, J. W., Fasco, M. J., and Stackrow, A. B., 1977a, Human thrombins: Production, evaluation and properties of α-thrombin, *J. Biol. Chem.* **252**:3587–3598.

Fenton, J. W., Landis, B. H., Walz, A. D., and Finlayson, J. S., 1977b, Human thrombins, in: *Chemistry and Biology of Thrombin.* (R. L. Lundblad, J. W., Fenton, and K. G. Mann, eds.), Ann Arbor Science, Ann Arbor, pp. 43–70.

Frost, G. H., Thompson, W. C., and Carney, D. H., 1987, Monoclonal antibody to the thrombin receptor stimulates DNA synthesis in combination with gamma-thrombin or phorbol myristate acetate, *J. Cell Biol.* **105**: 2551–2558,

Furie, B., Bing, D. H., Feldman, R. J., Robinson, D. J. Burnier, J. P., and Furie, B. C., 1982, Computer-generated models of blood coagulation factor Xa, factor IXa, and thrombin, based upon structural homology to other serine proteases, *J. Biol. Chem.* **257**:3875–3882.

Ganguly, P., and Sonnichsen, W. J., 1976, Binding of thrombin to human platelets and its possible significance, *Br. J. Haematol,* **34**: 291–301.

Ganguly, P., Sutherland, S. B., and Bradford, H. R., 1978, Defective binding of thrombin to platelets in myeloid leukaemia, *Br. J. Haematol* **39**:599–605.

Glenn, K. C., and Cunningham, D. D., 1979, Thrombin-stimulated cell division involves proteolysis of its cell surface receptor, *Nature* **278**:711–714.

Glenn, K. C., Carney, D. H., Fenton, J. W., II, and Cunningham, D. D., 1980, Thrombin active size regions required for fibroblast receptor binding and initiation of cell division, *J. Biol. Chem.* **255**: 6609–6616.

Glover, G., and Shaw, E., 1971, The purification of thrombin and isolation of a peptide containing the active center histidine, *J. Biol. Chem.* *246*:4594–4601.

Gordon, E. A., and Carney, D. H., 1986, Thrombin receptor occupancy initiates cell proliferation in the presence of phorbol myristic acetate, *Biochem. Biophys. Res. Commun.* **141**:650–656.

Greco, N. J., Arnold, J. H., O'Dorisio, T. M., Cataland, S., and Panganamala, R., 1985, Action of platelet activating factor on type I diabetic human platelets, *J. Lab. Clin. Med.* **105**:410–416.

Greco, N. J., Tenner, T. T., Jr., Tandon, N. N., and Jamieson, G. A., 1990, PPACK-thrombin inhibits thrombin-induced platelet aggregation and cytoplasmic acidification but does not inhibit platelet shape change, *Blood* **75**:1983–1990.

Hagen, I., Bjerrum, O. J., Gogstad, G., Korsomo, R., and Solum, N. O., 1982, Involvement of divalent cations in the complex between the platelet glycoproteins IIb and IIIa, *Biochim. Biophys, Acta* **701**:1–6.

Hall, W. M., and Ganguly, P., 1980, Binding of thrombin to cultured human fibroblasts: Evidence of receptor modulation, *J. Cell Biol.* **87**:601–610.

Hanson, S. R., and Harker, L. A., 1988, Interruption of acute platelet-dependent thrombosis by the synthetic antithrombin D-phenylalanyl-L-prolyl-L-arginal chloromethylketone, *Proc. Natl. Acad. Sci USA* **85**: 3184–3188.

Harmon, J. T., and Jamieson, G. A., 1985, Thrombin binds to a high-affinity ~900,000-dalton site on human platelets, *Biochemistry* **24**: 58–64.

Harmon, J. T., and Jamieson, G. A., 1986, Activation of platelets by α-thrombin is a receptor-mediated event, *J. Biol. Chem.* **261**:15928–15933.

Harmon, J. T., and Jamieson, G. A., 1988, Platelet activation by thrombin in the absence of the high-affinity thrombin receptor, *Biochemistry* **27**:2151–2157.

Harmon, J. T., Greco, N. J., and Jamieson, G. A., 1992, Isolation of human platelet plasma membranes by glycerol lysis, *Methods Enzymol.* **215**: 32–36.

Hauptmann, J., and Markwardt, F., 1980, Studies on the anticoagulant and antithrombotic action of an irreversible thrombin inhibitor, *Thromb. Res.* **20**:347–351.

Haver, V. M., and Namm, D. H., 1984, Characterization of the thrombin-induced contraction of vascular smooth muscle, *Blood Vessels* **21**: 53–63.

Henriksen, R. A., and Mann, K. G., 1988, Identification of the primary structural defect in the dysthrombin thrombin Quick I: Substitution of cysteine for arginine-382, *Biochemistry* **27**:9160–9165.

Henriksen, R. A., and Mann, K. G., 1989, Substitution of valine for glycine-558 in the congenital dysthrombin thrombin Quick II alters primary substrates specificity, *Biochemistry* **28**:2078–2082.

Henriksen, R. A., and Owen, W. G., 1980, Identification of a congenital dysthrombin, thrombin Quick, *J. Clin. Invest.* **66**:934–940.

Henriksen, R. A., and Owen, W. G., 1987, Characterization of the catalytic defect in the dysthrombin, thrombin Quick, *J. Biol. Chem.* **262**:4664–4669.

Herbosa, G., Thompson, W. C., and Carney, D. H., 1984, Thrombin receptor characterization by monoclonal antibodies, *J. Cell Biochem* **8A**:237.

Hogg, P. J., and Jackson, C. M., 1989, Fibrin monomer protects thrombin from inactivation by heparin–antithrombin III: Implications for heparin efficacy, *Proc. Natl. Acad. Sci. USA* **86**:3619–3623.

Holmsen, H., Dangelmaier, C. A., and Holmsen, H. K., 1981, Thrombin-induced platelet responses differ in requirement for receptor occupancy, *J. Biol. Chem.* **256**: 9393–9396.

Holmsen, H., Dangelmaier, C. A., and Rongved, S., 1984, Tight coupling of thrombin-induced acid hydrolase secretion and phosphatidate synthesis to receptor occupancy in human platelets, *Biochem. J.* **222**: 157–167.

Horne, W. C., and Simmons, E. R., 1978, Effects of amiloride on the response of human platelets to bovine α-thrombin, *Thromb. Res.* **13**:599–607.

Houslay, M., Bojanic, D., O'Hagan, S., and Wilson, A., 1986, Thrombin, unlike vasopression, appears to stimulate two distinct guanine nucleotide regulatory proteins in human platelets, *Biochem. J.* **238**:109–113.

Huang, E. M., and Detwiler, T. C., 1987, Thrombin-induced phosphoinositide hydrolysis in platelets, *Biochem J.* **242**: 11–18.

Huang, R. S., Sorisky, A., Church, W. R., Simons, E. R., and Rittenhouse, S. E., 1991, Thrombin receptor-directed ligand accounts for activation by thrombin of platelet phospholipase C and accumulation of 3-phosphorylated phosphoinositides, *J. Biol. Chem.* **266**:18435–18438.

Iwanaga, S., Wallen, P., Grondahl, N. J., Henschen, A., and Blombäck, B., 1969, On the primary structure of human fibrinogen, *Eur. J. Biochem.* **8**:189–199.

Jamieson, G. A., 1988, The activation of platelets by thrombin: A model for activation by high and moderate affinity receptor pathways, in: *Platelet Membrane Receptors Molecular Biology, Immunology, Biochemistry, and Pathology* (G. A. Jamieson, ed.), Liss, New York, pp. 137–158.

Jamieson, G. A., and Okumura, T., 1978, Reduced thrombin binding and aggregation in Bernard-Soulier platelets, *J. Clin. Invest.* **61**:861–864.

Jakubowski, J. A., and Maraganore, J. M, 1990 Inhibition of coagulation and thrombin-induced platelet activities by a synthetic dodecapeptide modeled on the carboxy-terminus of hirudin, *Blood* **75**:399–406.

Jandrot-Perrus, M., Husse, M., G., Kristenansky, J. L., Bezeaud, A., and Guillin, M. -C., 1991, Effect of the hirudin carboxy-terminal peptide 54–65 on the interaction of thrombin with platelets, *Thromb. Haemost.* **66**:300–305.

Jones, G. D., Carty, D. J., Freas, D. L., Spears, J. T., and Gear, A. R. L., 1989, Effects of separate proteolytic and high-affinity binding activities of human thrombin on rapid platelet activation, *Biochem. J.* **262**:611–616.

Jung, S. M., and Moroi, M., 1983, Crosslinking of platelet glycoprotein Ib by N-succinimidyl(4-azidophenylidithio)propionate and 3.3'-dithiobis-(sulfosuccinimidyl propionate), *Biochim. Biophys Acta* **761**:152–162.

Katagiri, Y.L., Hayashi, Y., Yamamoto, K., Tanoue, K., Kosaki, G., and Yamazaki, H., 1990, Localization of von Villebrand factor and thrombin-interactive domains on human platelet glycoprotein Ib. *Thromb. Haemostas.* **63**:122–126.

Kettner, C., and Shaw, E., 1978, Synthesis of peptides of arginine chloromethyl ketone. Selective inactivation of human plasma kallikrein, *Biochemistry* **17**:4778–4784.

Kettner, C., and Shaw, E., 1981, Inactivation of trypsin-like enzymes with peptides of arginine chloromethyl ketone, *Methods Enzymol.* **80**:826–842.

Knupp, C. L., and White G. C., 1985, Effect of active-site modified thrombin on the hydrolysis of platelet-associated glycoprotein V by native thrombin, *Blood* **65**:578–583.

Konno, S., Fenton, J. W., II, and Villanueva, G. B., 1988, Analysis of the secondary structure of hirudin and the mechanism of its interaction with thrombin, *Arch. Biochem. Biophys.* **267**:158–166.

Landis, B. H., Koehler, K. A., and Fenton, J. W., 1981, Human thrombins, group Ia and IIa salt-dependent properties of α-thrombin, *J. Biol. Chem.* **256**: 4604–4610.

Larsen, N. E., and Simons, E. R., 1981, Preparation and application of a photoreactive thrombin analogue: Binding to human platelets, *Biochemistry* **20**:4141–4147.

Lazarowski, E. R., and Lapetina, E. G., 1990, Persistent activation of platelet membrane phospholipase C by proteolytic action of trypsin and thrombin, *Arch. Biochem, Biophys.* **276**:265–269.

Lerea, K. M., and Glomset, J. A., 1987, Agents that elevate the concentration of cAMP in platelets inhibit the formation of NaDodSO$_4$-resistant complex between thrombin and a 40-kDa protein, *Proc. Natl. Acad. Sci. USA* **84**:5620–5624.

Levin, E. G., Marzec, V., Anderson, J., and Harker, L. A., 1984, Thrombin stimulates tissue plasminogen activator release from cultured human endothelial cells, *J. Biol. Chem.* **74**:1988–1995.

Lindout, T., Baruch, D., Schoen, P., Franssen, J., and Hemker, H. C., 1986, Thrombin generation in the presence of antithrombin III and heparin, *Biochemistry* **25**: 5962–5969.

Lollar, P., and Owen, W. G., 1980, Evidence that the effects of thrombin on arachidonate metabolism in cultured human endothelial cells are not mediated by a high affinity reception, *J. Biol. Chem.* **255**: 8031–8034.

Lorand, L., 1975, Controls in the clotting of fibrinogen, in: *Proteases and Biological Control* (E. Reich, D. B., Rifkin, and E. Shaw, eds.), Cold Spring Harbor Press, Cold Spring Harbor, N.Y., pp. 79–84.

Loskutoff, D. J., 1979, Effect of thrombin on the fibrinolytic activity of cultured bovine endothelial cells, *J. Biol. Chem.* **64**:329–332.

Lottenberg, R., Hall, J. A., Fenton, J. W., and Jackson, C. M., 1982, The action of thrombin on peptide p-nitroanilide substrates: Hydrolysis of tosyl-gly-pro-arg-pNa and D-phe-pip-arg-pNA by human alpha- and gamma- and bovine alpha- and beta thrombins, *Thromb. Res.* **28**:313–332.

Low, D. A., Baker, J. B., Koonce, W. C., and Cunningham, D. D., 1981, Released protease-nexin regulates cellular binding, internalization and degradation of serine proteases, *Proc. Natl. Acad. Sci. USA* **78**:2340–2344.

McGowan, E. B., and Detwiler, T. C., 1986, Modified platelet response to thrombin. Evidence of two types of receptors or coupling mechanisms, *J. Biol. Chem.* **261**:739–746.

McGowan, E. B., Ding, A., and Detwiler, T. C., 1983, Correlation of thrombin-induced glycoprotein V hydrolysis and platelet activation, *J. Biol. Chem.* **258**:11243–11248.

McGregor, J. L., Brochier, J., Wild, F., Follea, G., Trzeciak, M. -C., James, E., Dechavanne, M., McGregor, L., and Clemetson, K. J., 1983, Monoclonal antibodies against platelet membrane glycoproteins. Characterization and effect on platelet function, *Eur. J. Biochem.* **131**:427–434.

McKay, D. B., Kay, L. M., and Stround, R. M., 1977, Protein structure of thrombin in: *Chemistry and Biology of Thrombin* (R. L., Lundblad, J. W. Fenton, and K. G., Mann, eds), Ann Arbor Science, Ann Arbor, pp. 113–121.

Magnusson, S., Petersen, T. E., Sottrup-Jensen, L., and Claeys, H., 1975, Complete primary sequence of prothrombin: Isolation, structure and reactivity of ten carboxylated glutamic acid residues and regulation of prothrombin activation by thrombin, in: *Proteases and Biolgical Control* (E. Reish, D. B., Rifkin, and E. Shaw, eds.), Cold Spring Harbor Press, Cold Spring Harbor, N.Y., p. 123.

Maraganore, J. M., Bourdon, P., Jablonski, J., Ramachandran, K. L., and Fenton, J. W., II, 1990, Design and characterization of hirulogs: A novel class of bivalent peptide inhibitors of thrombin, *Biochemistry* **29**:7095–7101.

Martin, B. M., Feinman, R. D., and Detwiler, T. C., 1975, Platelet stimulation by thrombin and other proteases, *Biochemistry* **14**:1308–1314.

Martin, B. M., Wasiewski, W. W., Fenton, J. W., and Detwiler, T. C., 1976, Equilibrium binding of thrombin to platelets, *Biochemistry* **15**:4886–4893.

Mazurova, A. V., Vinegradov, D. V., Vilasik, T. N., Repin, V. S., Earth, W. J., and Berndt, M. C., 1991, Characterization of an antiglycoprotein Ib monoclonal antibody that specifically inhibits platelet-thrombin interaction, *Thromb. Res.* **62**:673–684.

Michelson, A. D., 1987, Flow cytometric analysis of platelet surface glycoproteins: Phenotypically distinct subpopulations of platelets in children with chronic myeloid leukemia, *J. Biol. Chem.* **110**:346–354.

Miyata, T., Morita, T., Inomoto, T., Kawauchi, S., Shirakami, A., and Iwanaga, S., 1987, Prothrombin Tokushima, a replacement of arginine-418 by tryptophan that impairs the fibrinogen clotting activity of derived thrombin Tokushima, *Biochemistry* **26**:1117–1122.

Mosher, D. F., and Vaheri, A., 1978, Thrombin stimulate the production and release of a major surface-associated glycoprotein (fibronectin) in cultures of human fibroblasts, *Exp. Cell Res.* **112**:323–334.

Mosher, D. F., Vaheri, A., Choate, J. J., and Gahmberg, C. G., 1979, Action of thrombin on surface glycoproteins of human platelets, *Blood* **53**:437–445.

Moss, M., Wiley, H. S., Fenton, J. W., and Cunningham, D. D., 1983, Photoaffinity labeling of specific alpha thrombin binding sites on Chinese hamster lung cells, *J. Biol. Chem.* **258**:3996–4002.

Munson, P. J., and Rodbard, D., 1980, LIGAND: A versatile computerized approach for characterization of ligand-binding systems, *Anal. Biochem.* **107**:220–239.

Nemerson, Y., and Esnouf, M. P., 1973, Activation of a proteolytic system by a membrane lipoprotein: The mechanism of action of tissue factor, *Proc. Natl. Acad. Sci. USA* **70**:310–314.

Neurath, H., 1975, Limited proteolysis and zymogen activation, in: *Proteases and Biological Controls* (E. Reich, D. B., Rifkin, and E. Shaw, eds.), Cold Spring Harbor Press, Cold Spring Harbor, N.Y., pp. 51–64.

Okumura, T., and Jamieson, G. A., 1976, Platelet glycocalicin: A single receptor for platelet aggregation induced by thrombin or ristocetin, *Thromb. Res.* **8**:701–706.

Olson, T. A., Sonder, S. A., Wilner, G. D., and Fenton, J. W., 1986, Heparin binding in proximity to the catalytic site of human α-thrombin, in: *Chemistry and Biology of Thrombin* (R. L. Lundblad, J. W. Fenton, and K. G. Mann, eds.), Ann Arbor Science, Ann Arbor, pp. 96–103.

Orthner, C. L., and Kosow, D. P., 1980, Evidence that human α-thrombin is a monovalent cation activated enzyme, *Arch. Biochem. Biophys.* **202**:63–75.

Paris, S., and Pouyssegur, J., 1984, Growth factors activate the Na^+/H^+ antiporter in quiescent fibroblasts by increasing its affinity for intracellular H^+, *J. Biol. Chem.* **259**:10898–10994.

Pearson, J. D., and Gordon, J. L., 1979, Vascular endothelial and smooth muscle cells in culture selectively release adenine nucleotides, *Nature* **281**:384–386.

Perdue, J. F., Lubensky, W., Kivity, E., and Fenton, J. W., 1981a, Characterization of human α-thrombin binding on cultured avian and mammalian cells, in: *Plasma and Cellular Modulatory Proteins* (D. H. Bing and R. A. Rosenbaum, eds.), Center for Blood Research, Boston, pp. 109–123.

Perdue, J. F., Lubensky, W., Kivity, E., Sonder, S. A., and Fenton, J. W., 1981b, Protease mitogenic response to chick embryo fibroblasts and receptor binding/processing of human α-thrombin, *J. Biol. Chem.* **256**:2767 2776,

Phillips, D. R., and Agin, P. P., 1977, Platelet plasma membrane glycoproteins. Identification of a proteolytic substrate for thrombin, *Biochem. Biophys. Res. Commun.* **75**:940–947.

Pipili-Synetos, E., Gershengan, M. C., and Jaffe, E. A., 1990, Expression functional thrombin receptors in Xenopus oocytes injected with human endothelial cell mRNA, *Biochem. Biophys. Res. Commun.* **171**:013–919.

Pouyessegur, J., 1985, The growth factor-activatable Na^+/H^+ exchange system: A genetic approach, *Trends Biochem. Sci.* **10**:453–455.

Quick, A. J., Pisciotta, A. V., and Hussey, C. V., 1955, Congenital hypoprothrombinemic states, *Arch. Intern. Med.* **95**: 2–14.

Raben, D. M., Yasuda, K., and Cunningham, D. D., 1987, Modulation of thrombin-stimulated lipid responses in cultured fibroblasts. Evidence for two coupling mechanisms, *Biochemistry* **26**: 2759–2765.

Rasmussen, U. B., Vouret-Craviari, V., Jallat, S., Schlesinger, Y., Pages, G., Pavirani, A., LeCocq, J. -P., Pouyseegaur, J., and Van Obberhegen-Schilling, E., 1991, cDNA cloning and expression of a hamster-thrombin receptor coupled to Ca^{2+} mobilization, *FEBS Lett.* **288**(1,2): 123–128.

Rink, T. J., and Hallam, T. J., 1984, What turns platelets on? *Trends Biochem. Sci* **9**:215–219.

Rink, T. J., Tsien, R. Y., and Possan, T., 1982, Cytoplasmic pH and free Mg^{2+} in lymphocytes, *J. Biol. Chem.* **95**:189–196.

Rosenberg, R. D., and Damus, P. S., 1973, The purification and mechanism of action of human antithrombin–herapin cofactor, *J. Biol. Chem.* **248**:6490–6505.

Ross, P. D., Fletcher, A. D., and Jamieson, G. A., 1973, Microcalorimetric study of isolated blood platelets in the presence of thrombin and other aggregation agents, *Biochim. Biophys. Acta* **313**:106–118.

Rowland, F. N., Donovan, M. J., Picciano, P. T., Wilner, G. D., and Kreutzer, D. L., 1984, Fibrin-mediated vascular injury. Identification of fibrin peptides that mediate endothelial cell retraction, *Am. J. Pathol.* **117**: 418–428.

Ruda, E. M., and Scrutton, M. C., 1987, Effect of leupeptin on platelet aggregation, fibrin formation and amidolysis induced by thrombin, *Thromb. Res.* **47**:611–619.

Ruggiero, M., and Lapetina, E. G., 1985, Leupeptin selectively inhibits human platelet responses by thrombin and trypsin: A role for proteolytic activation of phospholipase C, *Biochem. Biophys. Res. Commun.* **131**: 1198–1205.

Rydel, T. J., Ravichandran, K. G., Tulinsky, A., Bode, W., Huber, R., Roitsch, C., and Fenton, J. W., II, 1990, The structure of a complex of recombinant hirudin and human α-thrombin, *Science* **249**: 277–280.

Sage, S. O., and Rink, T. J., 1987, The kinetics of changes in intracellular calcium concentration in Fura-2-loaded human platelets, *J. Biol. Chem.* **262**:16364–16369.

Sanchez, A., Alonso, M. T., and Collazos, J. M., 1988, Thrombin-induced changes of intracellular (Ca^{2+}) and pH in human platelets. Cytoplasmic alkalization is not a prerequisite for calcium mobilization, *Biochim. Biopys. Acta* **938**:497–500.

Schaeffer, R. C., Chilton, S.M., Hadden, T. J., and Carlson, R. W., 1984, Pulmonary fibrin microembolism with Echis carinatus venom in dogs: A model of increased vascular permeability, *J. Appl. Physiol.* **57**: 1824–1828.

Schaeffer, R. C., Brisont, C., Chilton, S. M., and Carlson, R. W., 1986, Disseminated intravascular coagulation following Echis carinatus venom in dogs: Effects of a synthetic thrombin inhibitor, *J. Lab. Clin. Med.* **107**:488–497.

Seegers, W. H., Hassouna, H. I., Hewett-Emett, D., Walz, D. A., and Andary, T. J., 1975, Prothrombin and thrombin: Selected aspects of thrombin formation, properties, inhibition, and immunology, *Semin. Thromb. Hemostas.* **1**:211–283.

Shaw, E., 1975, Synthetic protease inhibitors acting by affinity labeling, in: *Proteases and Biological Control* (E. Reich, D. B. Rifkin, and E. Shaw, eds.), Cold Spring Harbor Press, Cold Spring Harbor, N. Y., pp. 455–465.

Shuman, M. A., and Levine, S.P., 1978, Thrombin generation and secretion of platelet factor 4 during blood clotting, *J. Clin. Invest.* **61**:1102–1106.

Siegler, P. B., Blow, D. M., Matthews, B. W., and Henderson, R., 1968, Structure of crystalline α-chymotrypsin, *J. Mol. Biol.* **35**:143–164.

Skryzpczak-Jankun, E., Rydel, T. J., Tulinsky, A., Fenton, J. W., and Mann, K. G., 1989, Human D-phe-pro-arg-CH_2-α-thrombin crystallization and diffraction data, *J. Mol. Biol.* **206**:755–757.

Snider, R. M., McKinney, M., and Fenton, J. W., 1984, Activation of cyclic nucleotide formation in murine neuroblastoma NIE-115 cells by modified human thrombins, *J. Biol. Chem.* **259**:9078–9081.

Sonder, S. A., and Fenton, J. W., 1984, Proflavin binding within the fibrinopeptide groove adjacent to the catalytic site of human α-thrombin, *Biochemistry* **23**:1818–1823.

Stenberg, P. E., McEver, R. P., Shuman, M. A., Jacques, Y. V., and Bainton, D. F., 1985, A platelet alpha-granule membrane protein (GMP140) is expressed on the plasma membrane after activation, *J. Biol. Chem.* **101**: 880–886.

Stiernberg, J., Carney, D. H., Fenton, J. W., II, and LaBelle, E. F., 1984, Initiation of DNA synthesis by human thrombin: Relationships between receptor binding, enzymic activity, and stimulation of ^{86}Rb^{+} influx, *J. Cell, Physiol,* **120**: 289–295.

Sugawara, Y., Birktoft, J. J., and Berliner, L. J., 1985, Human α- and γ-thrombin inhibition by trypsin inhibitors supports predictions from molecular graphics experiments, *Semin. Thromb. Hemostas.* **12**:209–212.

Sweatt, J. D., Blair, I. A., Cragoe, E. J., and Limbird, L. E., 1986, Inhibitors of Na^{+}/H^{+}exchange block epinephrine- and ADP-induced stimulation of human platelet phospholipase C blockade of arachidonic acid release at a prior step, *J. Biol. Chem.* **261**:8660–8666.

Takagi, T., and Doolittle, R. F., 1974, Amino acid sequence studies on factor XIII and the peptide released during its activation by thrombin, *Biochemistry* **13**:750–756.

Tam, S. W., and Detwiler, T.C., 1978, Binding of thrombin to human platelet membranes, *Biochim. Biophys. Acta* **543**:194–201.

Tam, S. W., Fenton, J. W., II, and Detwiler, T. C., 1980, Platelet thrombin receptors: Binding of α-thrombin is coupled to signal generation by a chymotrypsin-sensitive mechanism *J. Biol. Chem.* **255**: 6626–6632.

Tandon, N. N., Harmon, J. T., Rodbard, D., and Jamieson, G. A., 1983, Thrombin receptors define responsiveness of cholesterol-modified platelets, *J. Biol. Chem.* **258**:1184011845.

Tollefsen, D. M., and Majerus, P. W., 1976, Evidence for a single class of thrombin-binding sites on human platelets, *Biochemistry* **15**:2144–2149.

Tollefsen, D. M., Feagler, J. R., and Majerus, P. W., 1974, The binding of thrombin to the surface of human platelets, *J. Biol. Chem.* **249**:2646–2651.

Tomono, T., and Sawada, E., 1986, Preparation of anhydro-thrombin and its interaction with plasma anti-thrombin III, *Acta Haematol JN* **49**:969–979.

Van Nostrand, W. E., Wagner, S. L., Farrow, J. S., and Cunningham, D. D., 1990, Immunopurification and protease inhibitory properties of protease nexin-2/amyloid β-protein precursor, *J. Biol. Chem.* **265**:9591–9594.

Van Obberghen-Schilling, E., and Pouyssegur, J., 1985, Affinity labeling high-affinity alpha-thrombin binding site on the surface of fibroblasts, *Biochim. Biphys, Acta* **847**:335–343.

Van Obberghen-Schilling, E., Chambard, J. C., Lory, P., Nargeot, J., and Pouyssegur, J., 1990, Functional expression of Ca^{2+}-mobilizing-α-thrombin receptors in mRNA-injected Xenopus oocytes, *FEBS Lett.* **262**:330–334.

Vu, T.-K. H., Hung, D. T., Wheaton, V. I., and Loughlin, S. R., 1991, Molecular cloning of a functional thrombin receptor reveals a novel proteolytic mechanism of receptor activation, *Cell* **64:** 1057–1068.

Weber, M. J., Hale, A. H., and Roll, D. E., 1975, Role of protease activity in malignant transformation by Rous sarcoma virus, in: *Proteases and Biological Control* (E. Reich, D. B., Rifkin, and E. Shaw, eds.), Cold Spring Harbor Press, Cold Spring Harbor, N. Y., pp. 915–930.

Weksler, B. B., Ley, D. W., and Jaffe, E. A., 1978, Stimulation of endothelial cell prostacyclin production by thrombin, trypsin, and ionophore A23187, *J. Clin. Invest.* **62**:923–930.

Wicki, A. N., and Clemetson, K. J., 1985, Structure and function of platelet membrane glycoproteins Ib and V. Effects of leukocyte elastase and other proteases on platelet response of von Willebrand factor and thrombin, *Eur. J. Biochem* **153**:1–7.

Workman, E. F., White, G. C., and Lundblad, R. L., 1977a, Structure–function relationships in the interaction of α-thrombin with blood platelets, *J. Biol. Chem.* **252**:7118–7123.

Workman, E. F., White, G. C., and Lundblad, R. L., 1977b, High affinity binding of thrombin to platelets: Inhibition by tetranitromethane and heparin, *Biochem. Biophys. Res. Commun* **75**:925–932.

Yamamoto, K., Yamamoto, N., Kitagawa, H., Tanoue, K., Kosaki, G., and Yamazaki, H., 1986, Localization of thrombin-binding site on human platelet membrane glycoprotein Ib determined by a monoclonal antibody, *Thromb. Haemostas.* **55**:162–167.

Yamamoto, N., Kitagawa, H., Tanoue, K., and Yamazaki, H., 1985, Monoclonal antibody to glycoprotein Ib inhibits both thrombin- and ristocetin-induced platelet aggregations, *Thromb. Res.* **39**:751–759.

Yamamoto, N., Greco, N. J., Barnard, M. R., Zain, B., Tanoue, K., Yamazaki, H., Jamieson, G. A., and Michelson, A. D., 1991, GPIb-dependent and GIPb-independent pathways of thrombin-induced platelet activation, *Blood.* **77**:1740–1748.

Yeo, K. T., and Detwiler, T. C., 1985, Analysis of the fate of platelet-bound thrombin, *Arch. Biochem. Biophys.* **236**:399–410.

Zavoico, G. B., and Cragoe, E. J., 1988, Ca^{2+} mobilization can occur independent of acceleration of Na^{2+}/H^+ exchange in thrombin-stimulated human platelets, *J. Biol. Chem.* **263**:9635–9639.

Chapter 9

FUNCTIONAL DOMAINS IN THROMBIN OUTSIDE THE CATALYTIC SITE

Cellular Interactions

Rachel Bar-Shavit, Miriam Benezra, Valerie Sabbah, Elisabetta Dejana, Israel Vlodavsky, and George D. Wilner

1. THROMBIN INVOLVEMENT IN INFLAMMATION

1.1. Mononuclear Phagocyte Chemotaxis

Thrombin is known to be sequestered within the matrix of the fibrin gel, where it may remain active and intact for extended periods of time (Wilner

Abbreviations used: ATIII, anti-thrombin III; bFGF, basic fibroblast growth factor; BMDMs, bone-marrow-derived macrophages; DIP-α-thrombin, diisopropylfluorophosphate α-thrombin; EC, endothelial cell; ECM, extracellular matrix; FMLP, f-Met-Leu-Phe-OH; MeSO$_2$, methylsulfonyl fluoride; PPACK, D-phenylalanyl-L-prolyl-L-arginyl-chloromethyl ketone; SMC, smooth muscle cell; TLCK, N-α-tosyl-L-lysine chloromethyl ketone.

Rachel Bar-Shavit, Miriam Benezra, Valerie Sabbah, and Israel Vlodavsky • Department of Oncology, Hadassah University Hospital, Jerusalem 91120, Israel. **Elisabetta Dejana** • Instituto di Ricerche Farmacologiche, Milano, Italy. **George D. Wilner** • Department of Medicine, Division of Hematology, The Albany Medical College, Albany, New York 12208.

Thrombin: Structure and Function, edited by Lawrence J. Berliner. Plenum Press, New York, 1992.

et al., 1981; Mann, 1987). Since deposition of fibrin accompanies wound healing, thrombin may participate in modulation of this process. Inflammation is a localized protective response elicited by injury or destruction of tissues, which serves to destroy both the injurious agent and the injured tissue. Most forms of acute and chronic inflammation are amplified and propagated as a result of the recruitment of humoral and cellular components of the immune system. Immunologically mediated elimination of foreign material proceeds through a series of integrated steps. The actual destruction of antigens by immune mechanisms is mediated by cells with phagocytic capability. Such cells may migrate freely or may exist at fixed sites as components of the mononuclear phagocyte system. Since inflammation in its early stages is characterized by an influx of inflammatory cells, we wondered whether thrombin might participate in recruitment of such cells.

1.1.1. Thrombin-Mediated Chemotaxis

Purified human α-thrombin promoted a dose-dependent migration of human peripheral blood monocytes at concentrations ranging from 10^{-10} to 10^{-6} M. In contrast, prothrombin, the zymogen, failed to stimulate significantly monocyte movement (Bar-Shavit *et al.*, 1983a). In addition, thrombin complexed with antithrombin III (ATIII), a potent circulating serine protease inhibitor, and with hirudin, a low-molecular-weight, high-affinity, leech-derived thrombin inhibitor, were similarly inactive. These indicated that the cell movement observed was specifically due to thrombin, and not to minor contaminants in thrombin preparations (Bar-Shavit *et al.*, 1983a).

One of the questions raised by the inability of prothrombin and thrombin–inhibitor complexes to stimulate cell movement was whether chemotaxis, like other thrombin-elicited biological functions, required retention of enzymatic activity. Esterolytically inactive forms of thrombin [i.e., diisopropylfluorophosphate (DIP)-α-thrombin] and noncoagulant, proteolytically degraded forms of thrombin (i.e., γ-thrombin), when tested, were found as effective as the intact α-thrombin in stimulating monocyte movement, thus demonstrating that this function is independent of the enzymatic activity (Bar-Shavit *et al.*, 1983a,b; Bar-Shavit and Wilner, 1986).

1.1.2. Monocyte/Macrophage Interactive Exosite of Thrombin

In order to localize and identify the site on thrombin required for eliciting directed cell migration, the chemotactic activity of thrombin fragments was analyzed. Cyanogen bromide (CNBr) digestion of prethrombin 1 (a single chain precursor of α-thrombin) resulted in the generation of a limited number of relatively large polypeptide fragments which are separable by gel filtration chromatography (Butkowski *et al.*, 1977; Bar-Shavit *et al.*, 1985). Of these fragments, only fragment CB67-129, representing residues 33–84 (by sequence data according to the chymotrypsin numbering system) (Bar-Shavit *et al.*, 1985) of the human thrombin B chain, promotes significant mononuclear cell chemotaxis. This peptide was chemotactic for both human peripheral blood monocytes and murine macrophages. In addition, the CNBr-derived peptide was capable of completely inhibiting α-thrombin and DIP-α-thrombin-mediated chemotaxis, but not that stimulated by f-Met-Leu-Phe-OH (FMLP), thus indicating that peptide CB67-129 and thrombin compete for the same binding sites on the mononuclear cell membrane. Moreover, these findings suggest that the chemotactic activity associated within thrombin resides with the portion of the enzyme represented by the CNBr-derived fragment.

1.1.3. Construction of Computer Models of CB67-129

To assist in delineating the features in the CNBr-derived fragment that may be required for its biological functions, computer-generated models of the exosite region were prepared (Bar-Shavit *et al.*, 1985). For this purpose, the structures of chymotrypsin and trypsin were selected, since they possess a high degree of homology with thrombin (Elion *et al.*, 1977). Comparison of the aligned sequences of thrombin with chymotrypsin and other serine proteases identified well-conserved regions (constant regions), as well as regions showing little preservation of the structure (variable regions) (Furie *et al.*, 1982). These variable regions, characterized by insertions and deletions of amino acids unique to each enzyme (Furie *et al.*, 1982; Butkowski *et al.*, 1977; Bing *et al.*, 1981; Elion *et al.*, 1977). Based on computer modeling studies of this sequence, it was apparent that only those residues carboxy-terminal to the active site histidine (i.e., residues 364–400) are exposed to the surface and therefore accessible to cell interactions. Obviously, the structural requirements for cell interaction must be located in this surface-expressed region of the protein. The thrombin B chain was found to contain a surface-exposed, relatively large insertion sequence adjacent to the active site histidine (His-363). This insertion sequence, termed "loop B," was found to contain all of the carbohydrate moieties known to be present in thrombin. Moreover, "loop B" was found in chymotrypsin, the protease most nearly homologous to thrombin. Since deglycosylated thrombin was fully chemotactic, the carbohydrate residues do not play a role in promoting cell movement. This functional deficiency was not due only to the absence of the "loop B" insertion sequence, since experiments with a synthetic peptide analog encompassing this insertion site showed that it was neither itself chemotactic nor capable of blocking the cell movement stimulated by thrombin or CB67-129. These findings suggested that other elements carboxy-terminal to the insertion sequence are also essential for thrombin-mediated chemotactic activity (Bar-Shavit *et al.*, 1984, 1985).

1.2. Growth-Promoting Activity in Macrophages

It has been recognized for many years that α-thrombin, the enzymatically active form of thrombin (EC 3.4.21.5), was capable of initiating proliferation in quiescent fibroblastic cells (Chen and Buchanan, 1975; Perdue *et al.*, 1981). As with other growth factors, this stimulation resulted in enhanced phosphorylation of the S6 ribosomal protein (Chambard *et al.*, 1983), phosphoinositide turnover releasing inositol phosphate (Carney *et al.*, 1985) and activation of the Na^+/H^+-antiport system leading to alkalinization of the cytosol (Paris and Pouyssegur, 1984). The ability of thrombin to stimulate fibroblast proliferation was intimately linked to its proteolytic activity. Thus, while native α-thrombin was capable of evoking DNA synthesis in G_0/G_1-arrested fibroblasts, neither the enzymatically inactive thrombin DIP-α-thrombin nor partially degraded thrombin γ-thrombin share this capability (Perdue *et al.*, 1981). In contrast, we found that in several macrophagelike cell lines (J774, P388D$_1$, RAW, and PU-5) proteolytically inactive thrombin forms stimulated significantly thymidine ([^{3}H]-TdR) incorporation into quiescent cells (Bar-Shavit *et al.*, 1986a). This stimulation was blocked upon prior complexing with hirudin, a leech-derived thrombin inhibitor.

This observation raised the possibility that the thrombin region responsible for induction of macrophage proliferation may either coincide with or overlap the thrombin chemotactic exosite (Bar-Shavit *et al.*, 1984, 1986a,b). To test this hypothesis, the thrombin-derived chemotactic fragment (Bar-Shavit *et al.*, 1986a) peptide CB67-129 was assessed for its ability to act as a mitogen. Indeed, CB67-129 stimulated [^{3}H]-TdR incorporation in J774 cells over a concentration range of 10^{-11} to 10^{-6} M, with optimal stimulation occurring at 10 nM. In contrast to macrophagelike cells, CB67-129 had no stimulatory effect on CHL fibroblasts. Flow cytometric analysis indicated that CB67-129 was capable of increasing the proportion of cells in S phase from 5.6% to 18.75%. The degree of response elicited with this CNBr-derived fragment was approximately equal to or greater than that elicited with either active or enzymatically inactive thrombins when both were tested at their optimal concentrations.

To determine whether the mitogenic effects of the CNBr peptide might be expressed by smaller peptide fragments, the lysine groups of CB67-129 were reversibly blocked by citrocanylation (Bar-Shavit *et al.*, 1986a), then digested with trypsin. Under these conditions, cleavage of CB67-129 was limited exclusively to sites containing arginyl bonds. Following removal of the lysine side-chain-blocking groups, the resultant digest was tested for the ability to induce a mitogenic response in J774 cells. This CB67-129 digest was found to stimulate mitogenic activity at levels comparable to that of the intact CB67-129 fragment. Furthermore, fractionation of the digest by reversed-phase HPLC demonstrated that activity was present in fractions containing residues 356–382 of the thrombin B chain. This region contains the thrombin unique insertion sequence termed "loop B," absent in homologous serine proteases (Elion *et al.*, 1977). To establish whether the "loop B" sequence itself might account for the biological activity of the parent CB67-129 peptide, a "loop B"-containing homolog of the CNBr fragment, representing residues 367–380 of the human thrombin B chain, was synthesized, and found to be capable of stimulating both $[^3H]$-TdR incorporation and protein synthesis in G_0/G_1-arrested J774 cells (over a concentration range paralleling that of both thrombin and CB67-129, with an optimal stimulation occurring at 10^{-8} M) (see Table I). Moreover, this growth-promoting effect was not limited only to J774 cells; other murine macrophagelike cell lines, including $P388D_1$, RAW, and PU-5, were likewise stimulated by this synthetic peptide. Thus,

Table I. Thrombin-Directed Cell Migration and Mitogenesis on Monocyte/Macrophages[a]

	Migration (cells/hpf[b])	$[^3H]$-TdR incorporation J774 cells (% increase)
α-thrombin	120 ± 10	142 ± 9
iPR$_2$P-α-thrombin	135 ± 9	170 ± 8
Hirudin–α-thrombin complex	9 ± 3	12 ± 2
Prothrombin	10 ± 2	12 ± 3
ATIII–α-thrombin complex	4 ± 0.5	10 ± 3
CB67-129	110 ± 6	138 ± 7
CB67-120 digest	2 ± 0.5	139 ± 11
Synthetic B-loop peptide	3 ± 0.4	140 ± 9

[a]Chemotaxis was determined as described (Bar-Shavit *et al.*, 1983a,b) using modified Boyden chambers and human peripheral blood monocytes. Mitogenesis was established by the degree of [3H]-TdR incorporation into TCA-insoluble fraction of stimulated J774 cells.
[b]hpf-high power field.

the "loop B" region in the thrombin molecule appears responsible, at least in part, for the nonproteolytic macrophage-growth-promoting activity exhibited by the enzymatically inactive thrombin.

1.1.4. Growth-Promoting Effects of Proteolytically Inactive Thrombin on Bone-Marrow-Derived Macrophages (BMDMs)

Our previous studies (Bar-Shavit *et al.*, 1986a,b) had demonstrated that proteolytically inactive thrombin (10^{-8} M) acts as a growth promoter for certain macrophagelike tumor cell lines, such as J774. Since J774 growth was independent of CSF-1 regulation, it was not surprising that other factors (including thrombin) played such a role. By contrast, authentic nontransformed macrophage populations, such as BMDMs, require CSF-1 for both growth and survival (Tushinski *et al.*, 1982). The question arose, therefore, as to whether thrombin can exert a growth-promoting effect on these cells. To address this issue, homogeneous populations of BMDMs were isolated from femoral marrows of normal 6- to 9-week-old male A/J mice. After obtaining a relatively homogeneous cell monolayer, quiescent BMDMs were cultured under serum-free conditions with either DIP-α-thrombin (10^{-8} M) alone, 1000 CSF-1/ml, or both. In contrast to the effects observed with J774 cells, thrombin alone at 10^{-8} M failed to act as a growth promoter. In fact, in the presence of thrombin alone, BMDMs failed to survive (Clohisy *et al.*, 1990). Cells cultured with both thrombin and CSF-1 displayed the same time course regarding the peak of [^{3}H]-TdR incorporation, but the degree of incorporation was increased at least two-fold when compared to controls. These data demonstrated that proteolytically inactive thrombin assists CSF-1 in promoting BMDM cell proliferation, although the enzyme alone is incapable of such influence.

1.3. Thrombin Cell Surface Receptors on Mononuclear Phagocytic Cells

Thrombin-mediated chemotaxis and growth-promoting activity in mononuclear phagocytic cells suggest the existence of specific thrombin binding sites on these cells (Bar-Shavit *et al.*, 1983a,b, 1986a). Chemotactic assays in the presence of putative competitive agents were conducted in order to determine whether thrombin-mediated chemotaxis was (1) receptor-mediated and (2) involved chemotactic receptors that were distinct from those mediating responses with a formylated peptide. Progressive inhibition of cell movement was observed as a function of increasing α-thrombin concentrations in the upper compartment of the Boyden chamber, when contrasted with a fixed concentration of α-thrombin (10^{-8} M) in the lower compartment. Similar results were obtained when varying amounts of α-thrombin and DIP-α-thrombin in the upper compartment competed with 10^{-8} M α-thrombin in the lower. By contrast, no inhibition of cell movement was noted when increased amounts of FMLP were compared with 10^{-8} M α-thrombin in the lower compartment. These findings were interpreted as showing that α-thrombin and its modified forms stimulated chemotaxis via a common receptor on the J774 cell membrane. Additionally, this receptor was found separate and distinct from that which was responsive to formylated peptide. Direct binding studies using [^{125}I]-α-thrombin revealed the presence of approximately 14,000 binding sites on J774 cells with an apparent K_d of 7.5×10^{-9} M (Bar-Shavit et al., 1983c). Binding studies were carried out on BMDMs as well using [^{125}I]-α-thrombin (Clohisy et al., 1990). Scatchard analysis of the data revealed a curvilinear plot, which was mathematically resolvable into two segments with differing slopes. The number of high-affinity sites was estimated at approximately 20,000 binding sites/cell (K_d 7×10^{-9} M) while the number of low-affinity sites was predicted to be 2×10^6 sites/cell (K_d 9×10^{-7} M).

1.4. Early Events Associated with Mononuclear Phagocyte Activation

Most chemotactic agents evoke the directed migration of both granulocytes and monocytes, and relatively few chemically well-defined agents have been described which function as selective chemotaxins for monocytes/macrophages. Thrombin's ability to elicit cell movement in monocytes but not granulocytes has been further explored. Toward this end, the human promyelocytic leukemia cell line HL-60 was used as a model system to assess the development of the chemotactic response to thrombin within the context of cell lineage and differentiation. This cell line can be differentiated along a monocytic pathway in the presence of TPA (Lotem and Sachs, 1979) or 1,25(OH)$_2$D$_3$ (Z. Bar-Shavit et al., 1983) or along a granulocytic pathway by DMSO (Collins et al., 1978) or retinoic acid (Honma et al., 1980). We found that neither undifferentiated HL-60 cells nor HL-60 cells induced along the granulocytic pathway responded chemotactically to thrombin (Bar-Shavit et al., 1987). In contrast, thrombin is chemotactic for monocyte-differentiated HL-60 cells, and elicited early biochemical events associated with cell migration, including the rapid elevation in levels of free cytosolic Ca^{2+} and the concomitant induction of cytosolic actin reorganization (Bar-Shavit *et al.*, 1987).

1.4.1. Thrombin Effect on Cytosolic Ca^{2+}

Exposure of $1,25(OH)_2D_3$-differentiated HL-60 to thrombin elicited an increase in free cytosolic Ca^{2+} levels as detected by the Ca^{2+} fluorescent probe, quin-2. The maximal increase was reached within 15–20 s, which then decreased to control levels by 5 min. The elevation of free cytosolic Ca^{2+} occurred in a dose-dependent manner as a function of thrombin concentration. A thrombin concentration of 10^{-8} M elevated free cytosolic Ca^{2+} levels from a basal value of 142 nM to 170 nM, while the greatest increase (1.42×10^{-7} M to 2.81×10^{-7} M) occurred following stimulation by 10^{-6} M thrombin. Therefore, the thrombin concentration required for maximally increasing cytosolic Ca^{2+} coincided with the concentration of the enzyme required for optimal chemotaxis.

1.4.2. Thrombin Effect on Actin Polymerization

Actin was postulated to be a positive force transducer for cell migration. Since chemotaxis involves changes in cell shape and motility, differential organization of actin pools within the cell may be observed following exposure of cells to chemotactic agents (White *et al.*, 1983). Cytosolic proteins were isolated in Triton-containing buffer, and the insoluble cytoskeletal fraction (which contains polymeric F-actin) was further analyzed by gradient SDS–PAGE run under reducing conditions. Exposure of $1,25(OH)_2D_3$-differentiated HL-60 to thrombin or FMLP for 30 s induced an increase in cytoskeleton-associated actin. By contrast, these agents produced no significant effect on undifferentiated HL-60 cells. Densitometric quantitation of the 42-kDa actin band on gradient gels indicated that the response to thrombin was elicited over a range of 10^{-10} to 10^{-6} M (30 to 75% increase above control), similar to the chemotactic effects on these cells. The increase in Triton-insoluble actin occurred very rapidly following thrombin stimulation, with maximal effects observed within 30 s. Subsequently, the level of actin declined and returned to control levels within 5 min following stimulation. The half-life of reversibility of actin association was approximately 150 s. The time course of cytosolic Ca^{2+} elevation paralleled changes in cytoskeletal actin association, suggesting a relationship not only between these two events, but also between these events and thrombin-mediated chemotaxis (Bar-Shavit *et al.*, 1987).

2. THROMBIN REGULATION OF VESSEL WALL COMPONENTS

Control of cell proliferation and differentiation is determined, to a large extent, by cell interaction with the extracellular matrix (ECM). *In vivo*, epithelial cells are found in contact with a basement membrane, exhibiting a high rate of cell turnover (Gospodarowicz *et al.*, 1980a,b). Likewise, cultured cells, in order to proliferate and express their normal phenotype, require, in addition to nutrients and growth factors, an appropriate substratum upon which they can attach and spread (Reddi and Anderson, 1976; Bernfield *et al.*, 1972; Dodson, 1963), indicating that the substratum where the cells rest is a decisive element in their proliferative and differentiation responses. Based on these and other observations, the ECM is regarded as an insoluble complex of factors that regulate cellular growth, morphogenesis, and differentiation.

Cultured endothelial cells secrete and lay ECM closely resembling the subendothelium *in vivo* in its morphological appearance and molecular composition. It contains collagen (mostly types III and IV, with smaller amounts of types I and V), proteoglycans (mostly heparan sulfate and dermatan sulfate proteoglycans, with smaller amounts of chondroitin sulfate proteoglycans), laminin, fibronectin, and elastin (Gospodarowicz *et al.*, 1980a, 1984; Vlodavsky *et al.*, 1980). Moreover, because the ECM is secreted in a polar fashion, exclusively underneath the endothelial cell monolayer, and because it is firmly attached to the entire area of the tissue culture dish, the cell layer can be removed while the underlying ECM remains intact and free of nuclei, cytoskeletal elements, and other cellular debris (Gospodarowicz *et al.*, 1983). Data from our laboratory have indicated that the pluripotent angiogenic factor, basic fibroblast growth factor (bFGF), is an ECM component required for supporting cell proliferation and differentiation (Vlodavsky *et al.*, 1987; Rogelj *et al.*, 1989). bFGF has been identified in the subendothelial ECM produced *in vitro* and in basement membranes of the cornea and blood vessels *in vivo*. bFGF is bound to heparan sulfate (HS) in the ECM and is released in an active form when ECM-HS is degraded by cellular heparanase (Bashkin *et al.*, 1989). In addition, two recent reports describe a role for heparan sulfate proteoglycans in localizing colony-stimulating factors (GM-CSF) and interleukin 3 (IL-3) to the stromal cell matrix where interaction with stem cells occurs (Gordon *et al.*, 1987; Roberts *et al.*, 1988). The ECM seems to represent a site where proteolytic enzymes, their proforms, and their inhibitors reside and regulate such activations (Laiho and Keski-Oja, 1989). These observations and the ability of certain extracellular components (i.e., heparan sulfate proteoglycans) to regulate the localization and activity of bFGF, IL-3, and GM-CSF indicate that growth factors and enzymes may elicit their effects at a very restricted area and that these activities are modulated by ECM components.

In addition to molecules that are secreted by the ECM-producing cells, the ECM may serve as a storage reservoir for active molecules produced by other cell types. Blood normally circulates through endothelium-lined vessels without coagulation or platelet activation taking place (Colman *et al.*, 1987). With vascular injury, however, rapid activation of the hemostatic process is initiated after exposure of the subendothelium. This would then lead to thrombin generation, platelet activation, and fibrin clot formation to establish the hemostatic plug (Kaplan, 1982). It has also been demonstrated that under certain circumstances, the vascular endothelium can actively bind various coagulation factors, resulting in thrombin production on the endothelial surface (Stern *et al.*, 1985). Furthermore, thrombin has been shown to interact specifically with cell surface receptors on the endothelium (Awbrey *et al.*, 1979; Savion *et al.*, 1981). Thrombin interaction with these or other binding sites on endothelial cells (EC) promotes the production and release of diverse cellular mediators and proteins, such as: prostacyclin (PGI$_2$) (Weksler *et al.*, 1978), adenine nucleotides (Pearson and Gordon, 1979), plasminogen activator (Loskutoff, 1979), plasminogen activator inhibitor (Gelehrber and Sznycer-Laszuk, 1986), vWF (Sporn *et al.*, 1987), fibronectin (Galdal *et al.*, 1985), platelet-activating factor (Camussi *et al.*, 1983; Prescott *et al.*, 1984), and platelet-derived growth factor (PDGF) (Harlan *et al.*, 1986). It has been demonstrated also that under normal conditions, where the integrity of the endothelium is intact, thrombin can induce gap formation between adjacent EC in a rapid, noncytotoxic, and reversible manner (Laposada *et al.*, 1983; Lerner *et al.*, 1979; Garcia *et al.*, 1986). Thus, thrombin can pass through the EC layer and reach subendothelial structures. Indeed, exposure of EC to either thrombin or TPA has been shown to enhance the thrombogenic properties of their underlying ECM (de Groot *et al.*, 1987). Moreover, thrombin may also be accessible to the vascular subendothelium even at postclotting events because upon fibrinolysis, it can be released from a fibrin clot, intact and functionally active (Wilner *et al.*, 1981).

In view of thrombin's accessibility to the subendothelial basement membrane, we have addressed the possibility that thrombin is capable of interacting with the subendothelium and thus may contribute to its thrombogenic properties.

2.1. Thrombin Immobilization to the Subendothelial ECM

Immobilization of proteins to a solid support might improve dramatically their kinetic parameters as compared to a fluid phase. This was demonstrated for plasminogen (Knudsen *et al.*, 1986), enabling activation at minimal circulating concentrations. Such an effect on thrombin might take place in the hemostatic system when immobilized either on a fibrin gel, on cell surfaces, or on the subendothelial ECM. The ECM produced by EC provides a natural barrier to penetration of cells and macromolecules through the vessel wall, and was shown to behave as an actively dynamic regulator of diverse biological processes.

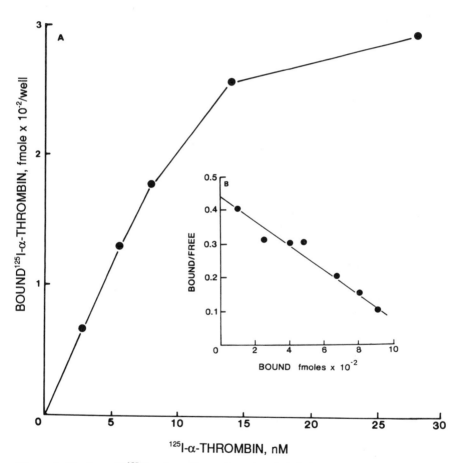

Figure 1. Binding of $[^{125}I]$-α-thrombin to the subendothelial ECM. ECM-coated 96-well plates were incubated with 0.1% BSA for 1 h at 37°C before binding. Increasing amounts of $[^{125}I]$-α-thrombin were then added to the wells and incubated for 4 h at 37°C. Specific binding was determined by subtracting the nonspecific binding obtained in the presence of 100-fold excess unlabeled thrombin. Similar results were obtained using bovine $[^{125}I]$-α-thrombin for binding. (Inset) Scatchard plot analysis. (From Bar-Shavit *et al.*, 1989.)

2.1.1. Binding Properties

Binding of [^{125}I]-α-thrombin to ECM-coated plates was analyzed as a function of thrombin concentration (Fig. 1) (Bar-Shavit *et al.*, 1989). Scatchard analysis of the data indicated a single class of binding sites (Fig. 1, insert). The apparent dissociation constant of thrombin interaction with ECM was 13×10^{-9} M, with an estimated 5.1×10^9 binding sites per square millimeter of ECM, corresponding to the area occupied by 10^3 confluent bovine aortic endothelial cells. Apparent equilibrium of thrombin binding to the ECM was reached after 2 h at 37°C. Up to 70% of the bound thrombin remained associated with the ECM after being rinsed and incubated for 24 h with PBS at 37°C. It should be emphasized that the above binding parameters are given to permit comparison with other binding systems, bearing in mind that Scatchard analysis assumes reversible binding conditions (unlike our data; Scheinberg, 1982). Note also that [^{125}I]-α-thrombin does not bind covalently to the ECM, as > 75% of the ECM-bound thrombin was released upon a mild treatment with acetic acid (0.2 M glacial acetic acid, 0.5 M NaCl, 5 min) (Haigler *et al.*, 1980).

To characterize the ECM binding domain of thrombin, we tested catalytically blocked thrombin (DIP-α-thrombin) and various thrombin fragments (corresponding to the region of "loop B" macrophage mitogenic peptide) (Bar-Shavit *et al.*, 1989) for their ability to compete with thrombin binding to the ECM. DIP-α-thrombin competed with [^{125}I]-α-thrombin binding to ECM with the same efficiency as native α-thrombin, indicating that the enzyme proteolytic site was not involved in thrombin binding to ECM. When competition studies were performed with the synthetic tetradecapeptide corresponding to residues 367–380 of the thrombin B chain, 50% inhibition was obtained at 10^{-5} M, compared with 10^{-9} M of native thrombin. This inhibition was specific, as when only one amino acid was substituted (Trp for Tyr) at position 370, the inhibitory effect on thrombin binding was abolished. There was also no inhibition in the presence of a shorter synthetic peptide of 10 residues and representing residues 371–380 of the thrombin B chain.

2.1.1a. Effect of glycosaminoglycans (GAGs) on Thrombin binding to ECM. Thrombin possesses highly cationic residues that bind effectively to the negatively charged sulfated polysaccharide heparin. Interestingly, such a positively charged cluster is located in the vicinity of the thrombin loop B insertion site (Bing *et al.*, 1986). Therefore, we have analyzed the effect of heparin on thrombin binding to ECM. Heparin inhibited effectively thrombin binding to ECM with 50% inhibition of 0.5 μg/ml heparin. Similar studies indicated that heparan sulfate and dermatan sulfate competed effectively with thrombin binding, whereas chondroitin sulfate, keratin sulfate, and hyaluronic acid at concentrations as high as 100 μg/ml failed to affect thrombin binding to ECM. Two other heparin binding proteins, protamine sulfate (PS) (Taylor and Folkman, 1982) and lipoprotein lipase (LPL) (Olivecrona *et al.*, 1971), had opposing effects on thrombin binding to ECM. Although PS had no effect, LPL effectively inhibited thrombin binding to ECM, suggesting, perhaps, competition over the same binding site in the matrix.

2.1.1b. ECM-degrading enzymes and their effect on thrombin binding to ECM. The possibility that thrombin binds to a specific GAG in the ECM was addressed using different degradative enzymes that cleave certain GAGs while leaving others intact. For this purpose, ECM was treated with either heparitinase, heparinase, chondroitinase AC, or chondroitinase ABC, washed extensively, and tested for its ability to bind thrombin. Treatment with chondroitinase ABC resulted in a significant inhibition of thrombin binding to ECM while treatment with heparitinase (which cleaves heparan sulfate side chains) or chondroitinase AC (which cleaves all GAGs except heparan sulfate and dermatan sulfate) was ineffective. Based on the inhibitory effect of chondroitinase ABC (which degrades all GAGs except heparan sulfate), we concluded that thrombin is bound through a dermatan sulfate moiety in the ECM.

2.1.2. Functional Activities of ECM-Bound Thrombin

The ability of DIP-α-thrombin to compete effectively with the native enzyme for binding to ECM indicated that thrombin binding is not mediated through its catalytic site. Indeed, when citrated platelet-poor plasma (PPP) was added to an ECM plate that was preincubated with thrombin, clot formation was observed within 10–30 s, whereas ECM plates alone did not induce clot formation. Quantitative analysis was further performed using the Chromozyme-TH assay to monitor for the amidolytic activity of the enzyme. The results indicated increased amidolytic activity in thrombin-bound ECM plates in a linear fashion parallel to the amount of bound thrombin, whereas ECM-coated plates alone did not exhibit a detectable amidolytic activity. Analysis of comparable thrombin concentrations in solution indicated similar levels of Chromozyme-TH activity. Thus, ECM-bound thrombin contains fully functional and exposed catalytic sites of the molecule exhibited also by the ability to activate platelets to release serotonin. ECM alone induced only basal levels of serotonin release during the time interval tested, similar to that obtained by incubation of labeled platelets with ECM-bound DIP-α-thrombin (which by itself does not induce

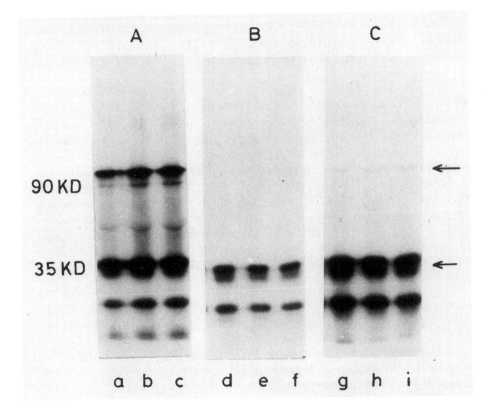

Figure 2. Thrombin–ATIII complex formation by ECM-bound thrombin as compared with thrombin in solution. Time dependence of thrombin–ATIII complex formation was determined by incubating a fixed amount of [^{125}I]-α-thrombin (0.685 pmol) either in solution or ECM-bound with ATIII (13.7 nM) in the presence of heparin (0.3 u/ml). Incubations were carried out at 37°C for 5 s (a, d, g), 20 s (b, e, h), and 5 min (c, f, i) followed by SDS–PAGE. (A) Complex formation of ATIII and thrombin (1.37 nM) in solution. (B) Complex formation of ATIII and ECM-bound thrombin. ATIII and heparin were added to ECM-bound [^{125}I]-α-thrombin (0.685 pmol). At the indicated time intervals, the reaction was stopped by addition of SDS–PAGE sample buffer. (C) Complex formation of ATIII and an increased amount of ECM-bound thrombin. As shown in B, except that excess of [^{125}I]-α-thrombin (2 pmol) was initially bound to the ECM. (From Bar-Shavit *et al.*, 1989.)

platelet activation). These results indicate that thrombin when bound to subendothelial ECM is capable of facilitating platelet activation and fibrin clot formation. ECM-bound thrombin was also functionally active in promoting neighboring cell proliferation (Bar-Shavit *et al.*, 1990) as will be outlined later.

2.1.3. Protection of ECM-Bound Thrombin

Circulating thrombin is rapidly inactivated by ATIII, as well as three other minor inhibitors present in plasma (Rosenberg and Damus, 1973; Jolyon, 1986). We have investigated the ability of ECM-bound thrombin to form complexes with ATIII and compared it with the extent of complex formation with thrombin in solution. ECM-bound [^{125}I]-α-thrombin (0.68 pmol) was allowed to interact with ATIII (13.7 nM) and heparin (0.3 u/ml). Complex formation was stopped at different time intervals by adding SDS–PAGE sample buffer. Similar experiments were carried out with the same amount of thrombin in solution. Samples were subjected to SDS–PAGE for identification of a band at 93 kDa, corresponding to the combined molecular masses of thrombin (35 kDa) and ATIII (58 kDa). In a fluid phase the complex was formed within 5 s, reaching a maximum at 20 s. In contrast, thrombin–ATIII complexes were not detected upon incubation of ATIII with ECM-bound thrombin (Fig. 2B). Moreover, even when a higher amount of thrombin was bound to ECM (2 pmol), only trace levels of complexes could be detected (Fig. 2C). In addition, experiments were carried out to determine the amidolytic activity of equivalent amounts of ECM-bound thrombin and thrombin in solution that were incubated with increasing concentrations of ATIII. The results indicated that inhibition of soluble thrombin was detected at a ratio of 1:5 (thrombin/ATIII), reaching complete inhibition at a ratio of 1:10. On the other hand, when ECM-bound thrombin was tested, no inhibition was obtained even at a thrombin/ATIII ratio of 1:140. In contrast, when hirudin (a high-affinity leech-derived thrombin inhibitor) was used, inhibition of thrombin amidolytic activity was observed with both ECM-bound thrombin and thrombin in solution (data not shown). This result indicates that hirudin binds to a blocking site on thrombin which is not affected by the interaction with ECM. Recently, Hanson and Harker (1988) demonstrated the role of thrombin as an important mediator of hemostatic plug formation and acute high-shear thrombosis. They showed that by continuous infusion of the synthetic antithrombin inhibitor D-phenylalanyl-L-prolyl-L-arginyl-chloromethyl ketone (PPACK), platelet deposition and thrombus formation were abolished. Interestingly, however, sustained treatment with a comparable anticoagulating level of heparin had no such effect. Our finding that thrombin immobilized to a solid support (ECM) becomes less accessible for inhibition by ATIII supports this notion. Experiments using hirudin as a direct inhibitor of thrombin, demonstrated that ECM-bound thrombin could be inhibited through a different blocking site. These observations suggest that a new class of anticoagulant compounds may be therapeutically superior to heparin in cases of acute arterial thrombosis.

Other plasma proteins participating in the hemostatic process were also shown to bind specifically to the ECM. These include vWF (Jaffe *et al.*, 1974) mediating platelet adhesion to the vascular subendothelium, and plasminogen participating in the fibrinolytic system. Moreover, immobilized plasminogen was found to be a better substrate for tissue plasminogen activator than soluble plasminogen and was protected from its inhibitor α_2-plasmin inhibitor (Knudsen *et al.*, 1986). Therefore, a concept emerges wherein the ECM not only binds and localizes different hemostatic proteins, but also modulates their mode of action and provides a protective environment from the circulating plasma inhibitors.

PDGF was shown to bind to collagen while retaining its mitogenic activity (Smith *et al.*, 1982). bFGF was identified in the subendothelial ECM both *in vivo* and *in vitro* (Bashkin *et al.*, 1989) and released upon degradation of the ECM heparan sulfate (Folkman *et al.*, 1988; Bashkin *et al.*, 1989). These observations together with our findings indicate that various biologically active molecules can be sequestered and stabilized by the ECM and participate in the induction of various cellular responses, allowing a more persistent and localized effect, compared with the same molecules in a fluid phase. In view of other biological activities of thrombin (e.g., chemotaxis, mitogenesis), its presence in an active form bound to the ECM may have further implications in processes such as inflammation and localized wound healing.

2.2. Thrombin as a Growth Factor for Smooth Muscle Cells: Nonenzymatic Mode of Action

Defining factors that regulate proliferation of cells within the vessel wall provides a continuous challenge. The vascular system is composed of the intimal endothelial cells lining the lumen and the underlying medial smooth muscle cells (SMCs), both co-existing in a quiescent growth state (Castellot *et al.*, 1982; Ross, 1986). Unregulated proliferation of the medial SMCs may thus have significant implications in the progression of arterial-wall diseases such as atherosclerosis. Accumulating evidence suggests that initiation of atherogenesis *in vivo* does not necessarily require products released external to the artery (Ross, 1986) and may, in fact, occur within the intact vessel wall via intrinsic mediators. Other factors released mainly from mononuclear phagocytes have been described as playing a key role in the progression of vascular diseases. Among these are PDGF-like molecules (DiColerto and Bowen-Pope, 1983; Martinet *et al.*, 1986), FGF (Baird *et al.*, 1985; Klagsbrun and Edelman, 1989), and the multipotent mediator IL-1 (Libby *et al.*, 1988). Because under certain circumstances thrombin may be present within the vessel wall, firmly bound to the subendothelial basement membrane (Bar-Shavit *et al.*, 1989), we investigated whether it promotes vascular SMC proliferation.

2.2.1. Induction of SMC Proliferation by Modified Thrombin Preparations

When either native α-thrombin or the proteolytically inactive form of DIP-α-thrombin was added to quiescent SMCs, stimulation of DNA synthesis was obtained. The mitogenic effect of thrombin on SMCs was observed over a wide range of concentrations (0.1 nM–1 M), with a half-maximal response at 1 nM. Both forms of thrombin elicited a five- to sixfold stimulation of [^{3}H]-TdR incorporation, measured 48 h after addition of the mitogen. In addition, both induced a two-to fourfold increase in cell number over a period of 1–5 days. Furthermore, we have analyzed the extent of cell populations entering the S phase by autoradiography of [^{3}H]-TdR-labeled nuclei. Treatment of G_0/G_1-arrested cells with either α-thrombin, DIP-α-thrombin, or bFGF for 24 h, increased the degree of nuclear label from 5% to 65–80%. These data indicate that, under our experimental conditions, the majority of the cells responded to α-thrombin or DIP-α-thrombin and progressed to the S phase.

Selectively modified thrombin preparations were examined to determine whether alterations in the procoagulant exosite, or in the catalytic site, affect the mitogenic activity of thrombin in SMCs. The nonclotting forms γ-thrombin and NO_2-α-thrombin elicited [^{3}H]-TdR incorporation by SMCs to an extent that was $\sim$ 80% of that induced by native thrombin. Moreover, serine- or histidine-blocked thrombin forms (i.e., $MeSO_2$-α-thrombin and TLCK-α-thrombin, respectively) induced a mitogenic response similar to that exerted by DIP-α-thrombin. The same level of induction was obtained by catalytically inactive PPACK-α-thrombin subjected to fibrinopeptide exosite affinity labeling. These results clearly indicate that thrombin stimulation of DNA synthesis in SMCs is mediated largely through a nonproteolytic domain and cannot be related to a residual proteolytic activity occasionally found in DIP-α-thrombin preparations. However, the thrombin-derived chemotactic peptide CB67-129 and the macrophage mitogenic "loop B" peptide (Bar-Shavit *et al.*, 1990) failed to elicit stimulated incorporation of [^{3}H]-TdR in quiescent SMCs.

The leech-derived thrombin inhibitor, hirudin, inhibited prolifera-

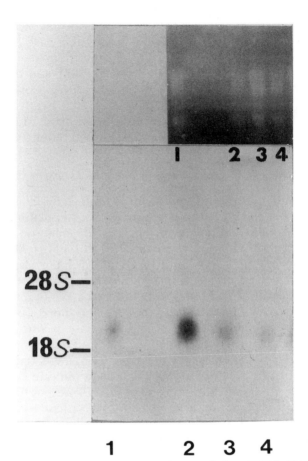

Figure 3. DIP-α-thrombin-mediated transient induction of c-fos mRNA in SMCs. Quiescent SMCs were stimulate with DIP-α-thrombin (10 μM) for the indicated periods of time. At the end of each incubation period, plates were washed with cold PBS and dissolved in guanidinium thiocyanate buffer for RNA isolation. Northern blot analysis and hybridization were performed as described. Lanes: (1) quiescent SMCs at G_0/G_1 phase; (2) 30 min; (3) 2 h; and (4) 4 h; stimulation with DIP-α-thrombin. (Inset) RNA samples corresponding to the above lanes separated on 1% agarose–formaldehyde gel and stained with ethidium bromide. (From Bar-Shavit *et al.*, 1990.)

tion induced by DIP-α-thrombin, but not serum-stimulated proliferation. Thus, the mitogenic domain for SMCs may reside in the vicinity of thrombin's cell interaction exosite, described for macrophages (Bar-Shavit *et al.*, 1986a).

DIP-α-thrombin induces a rapid and transient expression of c-fos protooncogene in arterial SMCs. Addition of growth factors (i.e., serum,

PDGF, IL-1) to fibroblasts or SMCs rapidly induces mRNA that encodes the c-fos protooncogene (Libby *et al.*, 1988). The product of this gene is a nucleus-associated protein that may be involved in signaling the early events in the commitment of cellular division (Greenberg and Ziff, 1984; Kruijer *et al.*, 1984; Muller *et al.*, 1984; Kindy and Sonenshein, 1986). Because thrombin was found to stimulate SMC DNA synthesis in a non-enzymatic fashion, we tested the effect of DIP-α-thrombin on c-fos protooncogene transcript levels. Northern blot analysis revealed that RNA isolated from DIP-α-thrombin-treated SMC cultures contained elevated levels of c-fos transcript compared with nonstimulated cells (Fig. 3). This increase reached maximal levels after 30 min incubation with DIP-α-thrombin, decreased back to control levels by 90 min, and remained unchanged thereafter, up to 4 h. Similar effects on the expression of c-fos mRNA were observed with PDGF and other growth factors, demonstrating that DIP-α-thrombin induces early events typical to growth-committed cells.

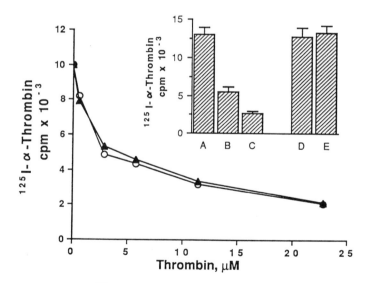

Figure 4. Competition of [^{125}I]-α-thrombin binding to SMCs by unlabeled α-thrombin and DIP-α-thrombin. Confluent cultures of SMCs (5 × 10^5 cells/16-mm well) were incubated (4 h, 4°C) in binding medium with [^{125}I]-α-thrombin (approx. 7 nm, 200,000 cpm/well) in the presence of increasing amounts of unlabeled α-thrombin (O) or DIP-α-thrombin (▲). (Inset) SMC cultures were incubated (4 h, 4°C) with [^{125}I]-α-thrombin (14 nM) in (A) the absence and (B) the presence of (B) 50 μg/ml (6.25 μm) and (C) 100 μg/ml (13 μM) of unlabeled chemotactic peptide CB67-129, or (D) 50 μg/ml (approx. 30 μM) and (E) 100 μg/ml (approx. 60 μM) of unlabeled "loop B" peptide. The cultures were washed (3 times) with PBS and the amount of cell-bound [^{125}I]-α-thrombin measured. (From Bar-Shavit *et al.*, 1990.)

2.2.2. Binding of [125I]-α-Thrombin to SMCs

Scatchard analysis of [125I]-α-thrombin binding to SMCs revealed a single class of binding sites over the concentration range of thrombin studied, with an estimated 540,000 binding sites per cell with an apparent dissociation constant of 6.3×10^{-9} M, similar to that observed with fibroblasts (Glenn *et al.*, 1980; Perdue *et al.*, 1981). As native and DIP-α-thrombin compete with [125I]-α-thrombin binding to the same degree (50% inhibition at 28 nM), both share common cell-surface receptor sites. Moreover, the chemotactic peptide CB67-129 inhibited the binding of [125I]-α-thrombin by 76% at 13.4 μM and by 57.5% at 6.7 μM (Fig. 4). The macrophage mitogenic peptide "loop B" failed to compete with [125I]-α-thrombin binding to SMCs. We conclude that, although the chemotactic peptide CB67-129 is not capable of promoting vascular SMC proliferation, it shares common cell-surface binding sites with α-thrombin. The synthetic "loop B" peptide, however, did not compete for [125I]-α-thrombin binding to the putative thrombin receptor and hence did not promote cell pro-

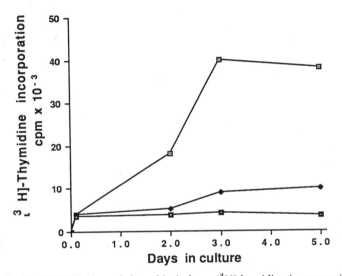

Figure 5. PF-HR-9/ECM-bound thrombin induces [³H]thymidine incorporation in SMCs. SMCs were seeded (5×10^4 cells/dish) into 35-mm dishes (◆) and into dishes coated with ECM produced by PF-HR-9 cells (▣, ▪). Some of the ECM-coated plates were preincubated (4 h, 37°C) with 10^{-6} M α-thrombin and washed free of unbound thrombin before seeding of the cells (▣). SMCs maintained in medium containing 0.2% FCS were pulsed at various times after seeding with [³H]thymidine (1 Ci/plate) for 24 h and measured for DNA synthesis. (From Bar-Shavit *et al.*, 1990.)

liferation. Thrombin binding sites on the surface of SMCs were identified by the affinity cross-linking technique. For this purpose, intact SMCs were incubated with increasing concentrations of the bifunctional cross-linkers [1-ethyl-(3,3-dimethylaminopropyl)carbodiimide]HCL (EDC) (0.5 or 1 mM) or disuccinimidyl substrate (DSS) (0.5 mM). Subsequent analysis by SDS–PAGE under reducing conditions and autoradiography revealed a 90-kDa, iodinated component representing the product of the cross-linking reaction. This result suggests that the apparent molecular mass of the thrombin receptor on SMCs is 55 kDa. Cross-linking of [^{125}I]-α-thrombin to its specific cell-surface receptors was inhibited in the presence of excess unlabeled thrombin.

2.2.3. ECM-Bound Thrombin is Mitogenic for Vascular SMCs

Because thrombin can be immobilized on the subendothelial ECM in a manner that leaves the molecule functionally active while protected from inhibition by circulating ATIII (Bar-Shavit et al., 1989), we investigated whether ECM-bound thrombin is also capable of promoting SMC proliferation. The subendothelial ECM produced by bovine vascular and corneal endothelial cells contains bFGF (Vlodavsky et al., 1987), which is a potent mitogen for SMCs. Therefore, use of these matrices might mask the mitogenic effect of ECM-immobilized thrombin. Thus, ECM produced by PF-HR-9 mouse endodermal carcinoma cells which is devoid of bFGF (Bar-Shavit et al., 1990) was employed. As shown in Fig. 5, PF-HR-9 ECM-bound thrombin was capable of promoting a three- to fourfold stimulation of vascular SMC proliferation, compared with cells maintained on PF-HR-9 ECM alone (which had no effect on SMC growth). Maximal stimulation was observed 3–5 days after seeding the cells in contact with ECM-bound thrombin in medium containing 0.2% fetal calf serum (FCS). The possibility that some of the ECM-immobilized α-thrombin was released during the 5-day assay for SMC proliferation was addressed next and found to be insignificant. These results suggest that thrombin, when immobilized to the subendothelial basement membrane in vivo, may elicit a localized, long-acting stimulation of SMC proliferation.

Our data demonstrated that thrombin-derived chemotactic peptide competed with [^{125}I]-α-thrombin binding to SMCs although it was incapable of promoting cell division. Thus, it appears that the thrombin binding region to SMC receptors resides within the CB67-129 chemotactic peptide, which requires additional sequence(s) to elicit a mitogenic response. The SMC growth-promoting site in thrombin differs, however, from the macrophage mitogenic domain, because the synthetic tetradec-

apeptide, representing residues 367–380 of the thrombin B chain, failed
to elicit any mitogenic response in growth-arrested SMCs but exhibited a
potent growth-promoting activity on macrophagelike cells (Bar-Shavit *et
al.*, 1986b). Delineation of the SMC-specific mitogenic domain in the
thrombin B-chain exosite may have important physiological implications
toward localization/identification of potential sequences in circulating degradation products of thrombin. Such fragments may escape inhibition by
the traditional thrombin inhibitors: protease nexin (Kramer and Vogel,
1984), ATIII (Rosenberg, 1977), or heparin cofactor II (Tollefsen *et al.*,
1982).

3. CELL ADHESION TO THROMBIN

Cell–substratum and cell–cell recognition are critical events in diverse
biological processes such as embryogenesis, tumor cell metastasis, wound
healing, coagulation, and immunological recognition. Recently, a family of
cell adhesion receptors, termed "integrins," has been described that is
functionally implicated in each of these biological events (Hynes, 1987;
Ruoslahti and Pierschbacher, 1987). Integrins exist as heterodimers of
noncovalently associated α and β subunits that are expressed on the cell
surface and promote cell attachment to fibronectin (Pytela *et al.*, 1985a),
vitronectin (Pytela *et al.*, 1985b), laminin (Horowitz *et al.*, 1985), collagen
(Wayner and Carter, 1987), fibrinogen (Languino *et al.*, 1989), and vWF
(Ruggeri *et al.*, 1982; Plow *et al.*, 1985). To date, individual members of the
integrin family of cell adhesion receptors have been divided into sub-
families based on the presence of distinct, but homologous α subunits
termed α_1, α_2, α_3, and α_4 (Hynes, 1987; Ruoslahti and Pierschbacher,
1987; Kajiji *et al.*, 1989). This family is further diversified since each βsu-
bunit can associate with one of a number of α subunits (Hynes, 1987).

The integrity of the vascular system is maintained by factors that
mediate the attachment and growth of EC. Normal hemostasis and pathological development of thrombotic vascular occlusion are events, associated with the cell-adhesion process, involving the induction of platelet
aggregation and EC attachment, contributing to the formation of thrombus that initially seals the vessel to prevent excessive blood loss.

Amino acid analysis of the thrombin B chain reveals the presence of
an Arg-Gly-Asp-Ala sequence at residues 187–190 (Furie *et al.*, 1982).
Because thrombin is ubiquitous to injury sites, we studied the possible
involvement of thrombin in EC adhesion, contributing thus to repair
mechanisms of vascular lesions.

3.1. Endothelial Cell Adhesion to Thrombin

By analyzing thrombin species that were chemically modified to alter the enzyme's procoagulant or proteolytic functions, we found that NO_2-α-thrombin was the most potent adhesive substratum, inducing attachment of about 80% of the cells within 2 h (Fig. 6). The modified thrombin preparations proteolytically inactivated, Exo-α-thrombin and $MeSO_2$-α-thrombin were highly active in promoting bovine aortic endothelial cell (BAEC) attachment. α-thrombin, lacking the procoagulant site of thrombin, was also active. In contrast, the native enzyme (α-thrombin) and the esterolytically inactive forms (TLCK-α-thrombin and DIP-α-thrombin) exhibited low attachment activity.

3.2. Effect of Synthetic Peptides on EC Attachment to NO_2-α-Thrombin

The involvement of the RGD sequence in EC attachment to thrombin was studied using the synthetic hexapeptides GRGDSP or GRGESP. While over 95% inhibition was obtained in the presence of 0.01 mM GRGDSP, no significant inhibition was observed with GRGESP. These results indicate that the attachment activity of various modified thrombin preparations can be attributed to an RGD sequence in the protein. An RGDA sequence in thrombin is located downstream at residues 187–190 of the human thrombin B chain, when residues are aligned with residues 16 through 245 of bovine chymotrypsin (Furie et al., 1982). Other described functional domains in thrombin, outside the proteolytic pocket, are located upstream to the RGDA region in distinct and separate regions. The "loop B" domain is located between residues 60 and 61 within the variable region 2 (VR2) of the thrombin B chain, while the chemotactic peptide CB67-129 is located at residues 33–84 according to this numbering system. When the synthetic peptide corresponding to the "loop B" mitogenic site, as well as various constructs of this region were tested for their ability to compete with EC attachment to NO_2-α-thrombin, only 10–20% inhibition was observed. Similarly, the thrombin-derived chemotactic fragment CB67-129 inhibited EC attachment to thrombin by about 15%. In contrast, soluble α-thrombin inhibited the attachment of EC to NO_2-α-thrombin in a dose-dependent manner (55% inhibition at 20 µg/ml). This may indicate that the RGD domain in thrombin is only partially cryptic, and that additional neighboring sequences may assist in recognition of this RGD sequence by the appropriate cell surface receptors. The low attachment-inducing activity of native α-thrombin (Fig. 6) does not appear to be due to inactiva-

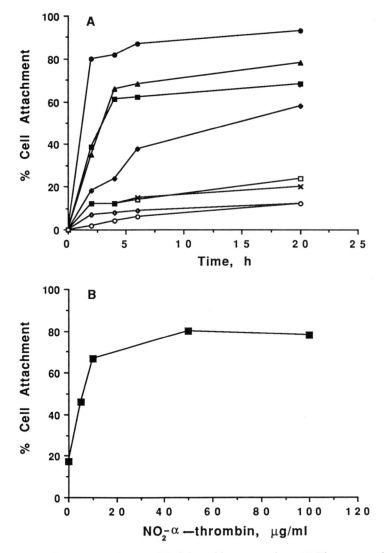

Figure 6. EC adhesion to various modified thrombin preparations. (A) Time course. Dishes were coated with different thrombin analogs (10–50 μg/ml) in PBS containing 0.1% BSA. EC (1.3 × 10⁵ cells/well) were added and unattached cells removed at various time intervals after seeding. Substrates: (●) NO₂-α-thrombin; (▲) exosite affinity-label thrombin; (■) MeSO₂-α-thrombin; (◆) γ-thrombin; (□)TLCK-α-thrombin; (x) DIP-α-thrombin; (◇) α-thrombin; (O) BSA. (B) Dose dependence. Four-well plates were coated with increasing concentrations of NO₂-α-thrombin. EC (1.3 × 10⁵ cells/well) were seeded and unattached cells removed after 2 h incubation at 37°C. The degree of cell attachment, expressed as percentage of attached cells relative to the number of cells seeded, was evaluated by the uptake of methylene blue. (From Bar-Shavit *et al.*, 1991.)

tion as a result of coating on a solid support, since all of the other modified thrombin preparations retained their cell attachment activity both when present in solution (competition experiments) and when absorbed onto plastic or glass surfaces.

3.3. Effect of Specific Thrombin Inhibitors on EC Attachment

ATIII inhibited the EC attachment to NO_2-α-thrombin (up to 93%) without affecting the attachment of another potent adhesive protein— fibronectin. On the other hand, hirudin, the leech-derived high-affinity inhibitor of thrombin, did not inhibit EC attachment to NO_2-α-thrombin even at a thrombin/hirudin ratio of 1:5.

Antiprothrombin antibodies which also recognize thrombin, inhibited in a dose-dependent manner EC attachment to thrombin. Complete inhibition was obtained at a 1:25 dilution of the antiserum, while there was no significant effect in the presence of preimmune serum. This inhibition further demonstrates the specificity of EC attachment to thrombin.

3.4. Effect of NO_2-α-Thrombin on EC Cytoskeletal Organization

As demonstrated in Fig. 7, NO_2-α-thrombin induced spreading of EC in a manner comparable to the subendothelial ECM. The distribution of the microfilamentous cytoskeleton and the localization of vinculin were studied by fluorescence microscopy on fixed and permeabilized EC. F-actin was visualized by fluorochrome-labeled phalloidine (PHD) and vinculin was identified with a specific monoclonal antibody that recognizes the mammalian form of vinculin. The extent of adhesion was evaluated by IRM, a technique that records the distance between the adhesion substratum and the cell ventral membrane: the smaller the distance the darker the signal intensity, black indicating tight adhesion. Upon plating of EC on NO_2-α-thrombin, the cells spread and organized a network of thick microfilamentous bundles. Formation of stress fibers occurred with appearance of vinculin streaks at their endings in correspondence with focal contacts. These cells showed multiple IRM black streaks which corresponded to the endings of stress fibers and distinct arrowhead-shaped vinculin speckles. Cell spreading and cytoskeletal organization on NO_2-α-thrombin was comparable to that observed upon seeding of EC on a vitronectin substratum (Dejana et al., 1988). Immunofluorescence staining using anti-α_v and anti-β_3 antibodies (Fig. 8) gave a peculiar pattern of oval and arrowhead-shaped spots usually located at stress fiber endings. The amount of β_3 clusters was, however, reduced in comparison to EC adhering to vitronectin (Fig. 8). A comparable distribution was observed in

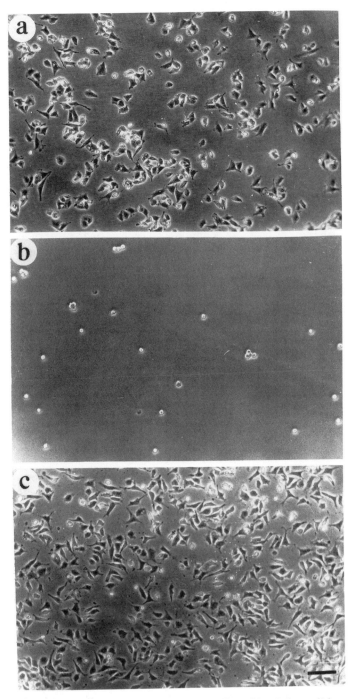

Figure 7. Attachment of endothelial cells to NO_2-α-thrombin. Culture dishes were coated with 20 μg/ml of (a) NO_2-α-thrombin or (b) BSA. Dishes were also coated with (c) naturally produced subendothelial ECM. Phase micrographs were taken after 2 h incubation at 37°C with the cells. Bar = 100 μm. (From Bar-Shavit *et al.*, 1991.)

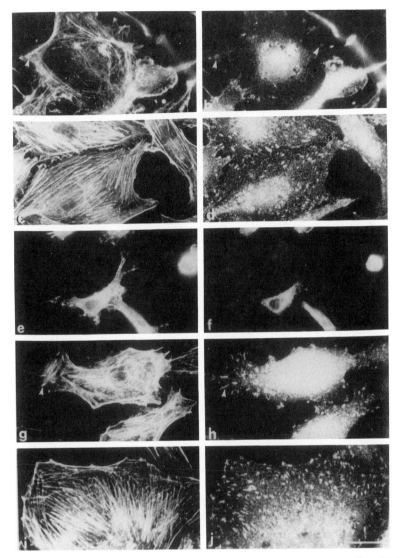

Figure 8. Distribution of α_v receptors, β_3 receptors, and F-actin. HUVECs (a–f) and BAECs (g–j) were incubated (1.3×10^5 cells/well) for 4 h on coverslips coated with vitronectin (c, d, i, j), NO_2-α-thrombin (a, b, g, h), or thrombin (e, f). The cells were permeabilized and stained with anti-α_vmAB (b, d, f), rabbit anti-GpIIIa antiserum (h, j), or F-PHD (a, c, e, g, i). Plating of HUVECs on NO_2-α-thrombin supports spreading and stress fiber formation to a lesser extent than on vitronectin (compare a and c; b and d, respectively); α_v clusters were fewer and more weakly stained, see arrowheads (b, d). In HUVECs the organization of β_3 was comparable to that of α_v (not shown). (From Bar-Shavit *et al.*, 1991.)

studies with human umbilical vascular endothelial cells (HUVECs) using anti-α_v (LM142) mAB (Fig. 8), or an anti-$\alpha_v\beta_3$ complex mAB (LM609), or anti-β_3 monoclonal and polyclonal antibodies (data not shown). On BAEC only the polyclonal anti-β_3 antibodies gave a good staining at adhesion plaques, while staining with the monoclonal antibodies used in this study was less easily detectable, possibly due to lack of cross-reactivity. Altogether these data strongly suggest that $\alpha_v\beta_3$ plays a leading role in mediating EC interaction with NO_2-α-thrombin. The vitronectin receptor is very promiscuous and recognizes, besides vitronectin, other substrata including von Willebrand factor and fibrinogen (Cheresh, 1987; Lawler *et al.*, 1988). The data reported here provide evidence that this receptor binds modified thrombin as well, thus identifying a novel ligand for this integrin.

3.5. Attachment of EC to Native α-Thrombin

To explore the possibility that under certain conditions (i.e., limited autoproteolysis) the RGD region of native α-thrombin can be exposed and function in promoting cell adhesion, the enzyme was preincubated at 37°C prior to coating of plastic surfaces. When α-thrombin was preincubated at 37°C, a potent induction of EC attachment was obtained as compared to lack of such activity in the absence of preincubation. This effect was rather rapid since a significant attachment activity was observed after 15 min preincubation of α-thrombin at 37°C. When the protease inhibitor TLCK was present during the preincubation of thrombin at 37°C, the ability of thrombin to induce EC attachment was reduced by over 65%. These data suggest that preincubation at 37°C and the associated autoproteolysis of the thrombin molecule may result in surface exposure of the RGD domain in thrombin which may hence mediate EC adhesion following absorbance to the tissue culture plastic.

The biological relevance of the adhesive properties of modified thrombin molecules remains to be fully elucidated. The native enzyme has very little adhesive properties, despite having an RGD domain. However, it may still compete for adhesion to modified NO_2-α-thrombin and to other adhesive proteins, for example fibronectin. The fact that the thrombin molecule is composed of various functional domains some of which are exosite exposed regions and others, like the RGD domain, are hindered, is intriguing. We have demonstrated that prior autoproteolysis of the enzyme resulted in induced attachment activity. We postulate that the RGD site can be surface exposed in response to specific signals which will cause the required modifications to induce attachment. This may occur under certain stress or pathological conditions. In wound healing and inflammatory processes, extensive recruitment of leukocytes takes place.

Moreover, it has been demonstrated that thrombin, which is ubiquitous to vascular injury, is a potent chemotaxin for mononuclear phagocytes (Bar-Shavit *et al.*, 1983a,b). Therefore, we suggest that during tissue repair when excess mononuclear phagocytic cells (very rich in proteolytic enzymes) are present, these cells may act on thrombin in a manner that will modify the molecule and expose the RGD domain for direct participation in cellular attachment. The effect of various proteolytic enzymes (e.g., elastase, plasmin, cathepsin G) on thrombin attachment activity is under investigation.

4. CONCLUDING REMARKS: FUNCTIONAL DOMAINS WITHIN THROMBIN AND DIVERSE CELLULAR INTERACTIONS

Proteolytic enzymes are presumed to have arisen in the earliest phases of evolution from a common ancestry gene (Neurath, 1984). During the process of evolution, proteases have evolved from simple digestive functions to highly specialized restricted proteolysis characterized by mutual fitting of the substrate to the configuration of the protease active site (Neurath, 1986). Using synthetic peptides and computer-derived models, we have identified in thrombin—the major procoagulant serine protease—other biologically active domains (e.g., chemotactic domain, "loop B" mitogenic site, SMC growth-promoting exosite, cell attachment domain RGD). Recent data on the gene sequence of the serine protease family suggest that the amino acid insertion sequence in thrombin (such as the one present in CB67-129) arises from gene insertion at the exon/intron boundaries (Craik *et al.*, 1982). It is tempting to postulate that these functional regions, absent in homologous serine proteases, were acquired during evolution, thus giving rise to a multifunctional protein that may participate in diverse cellular processes. The current concept on the structure of the gene predicts that exons code for functional protein structure while introns serve as linkers between exons. It seems that boundaries of these exon/intron regions may represent dynamic regions which could acquire additional sequences derived from diverse sources. Therefore, we can directly link activation of the procoagulant cascade with cellular events such as inflammation, wound healing, metastasis, and progression of the atherosclerotic plaque formation. Unless controlled, proteolytic enzymes may become a potential hazard, since they can degrade essential protein components. Thus, the fine tuning and balance between proteases and their inhibitors may provide a key element in various fundamental physiological processes, such as growth control and tumor metastasis. Recently it has been demonstrated that proteases and their inhibitors may play important complementary roles in the nervous system and that imbalance in proteolysis might be involved in the etiology of Alzheimer disease (Ponte et al., 1988). Indeed, reduced levels of protease nexin-1, a cell-secreted protease inhibitor, particularly of thrombin, were found in this disease (Wagner et al., 1989).

Proliferation of SMCs in the intima of the artery occurs in the early stages of atherogenesis. Therefore, the capability of ECM-bound thrombin to induce proliferation in the neighboring SMCs indicates that thrombin may act as a risk factor, involved in the progression of atherosclerotic plaque formation. As opposed to the above surface-exposed exosites, the RGD sequence is naturally hindered and only upon various modifications becomes surface exposed rendering adhesive properties to thrombin. Such modifications may result from specific signaling due to certain pathological situations. Since ECM-immobilized thrombin is protected from inactivation by the circulating physiological inhibitor ATIII, future work should focus on the design of adjuvant therapeutic drugs as specific antithrombin agents. The potential application of such drugs should follow several criteria such as their half-life, toxicity, and involvement in the pathogenesis of the disease. Thus, a wide spectrum of drugs with minimal, structural requirements that have to be met by thrombin, should be developed for the ultimate choice of a potent, safe, and specific drug used in therapeutic trials.

Acknowledgments

This work was supported by GIF, The German–Israeli Foundation for Scientific Research and Development, the Israel Academy of Sciences (to R.B.), and PHS grant CA32089 (to I.V.).

5. REFERENCES

Awbrey, B. J., Hoak, J. C., and Owen, W. G., 1979, Binding of human thrombin to cultured endothelial cells, J. Biol. Chem. 254:4092–4095.

Baird, A., Mormede, P., and Bohlen, P., 1985, Immunoreactive fibroblast growth factor in cells of peritoneal exudate suggests its identity with macrophage-derived growth factor, Biochem. Biophys. Res. Commun. 126:358–364.

Bar-Shavit, R., and Wilner, G. D., 1986, Mediation of cellular events by thrombin, Int. Rev. Exp. Pathol. 29:213–241.

Bar-Shavit, R., Kahn, A., Fenton, J. W., II, and Wilner, G. D., 1983a, Chemotactic response of monocytes to thrombin, J. Cell Biol. 96:282–285.

Bar-Shavit, R., Kahn, A., Wilner, G. D., and Fenton, J. W., II, 1983b, Monocyte chemotaxis: Stimulation by specific exosite region in thrombin, Science 220:728–731.

Bar-Shavit, R., Kahn, A. J., Fenton, J. W., II, and Wilner, G. D., 1983c, Receptor-mediated chemotactic response of macrophages to thrombin, *Lab. Invest.* **49:**702–707.

Bar-Shavit, Z., Teitelbaum, S. L., Reitsman, P., Hall, A., Pegg, L. E., Trial, J. A., and Kahn, A. J., 1983d, Induction of monocyte differentiation and bone resorption by 1,25-dihydroxyvitamin D$_3$, *Proc. Natl. Acad. Sci. USA* **80:**5907–5911.

Bar-Shavit, R., Kahn, A., Mudd, M. S., Wilner, G. D., Mann, K. G., and Fenton, J. W., II, 1984, Thrombin chemotactic domain is localized within a B-chain CNBr fragment, *Biochemistry* **23:**397–400.

Bar-Shavit, R., Bing, D. H., Kahn, A. J., and Wilner, G. D., 1985, Thrombin-mediated chemotaxis: Relationship of ligand structure to biological activity, in: *UCLA Symposia on Molecular and Cellular Biology* (M. P. Czech and C. R. Kahn, eds.), Liss, New York, Vol. 23, pp. 329–338.

Bar-Shavit, R., Kahn, A. J., Mann, K. G., and Wilner, G. D., 1986a, Identification of a thrombin sequence with growth factor activity on macrophages, *Proc. Natl. Acad. Sci. USA* **83:**976–980.

Bar-Shavit, R., Kahn, A. J., Mann, K. G., and Wilner, G. D., 1986b, Growth promoting effects of esterolytically inactive thrombin, *J. Cell Biochem.* **32:**261–272.

Bar-Shavit, R., Hruska, K. A., Kahn, A. J., and Wilner, G. D., 1987, Thrombin chemotactic stimulation of HL-60 cells. Studies on thrombin responsiveness as a function of differentiation, *J. Cell Physiol.* **131:**255–261.

Bar-Shavit, R., Eldor, A., and Vlodavsky, I., 1989, Binding of thrombin to subendothelial extracellular matrix: Protection and expression of functional properties, *J. Clin. Invest.* **84:**1096–1104.

Bar-Shavit, R., Benezra, M., Eldor, A., Hy-Am, E., Fenton, J. W., II, and Vlodavsky, I., 1990, Thrombin immobilized to extracellular matrix is a potent mitogen for vascular smooth muscle cells: Nonenzymatic mode of action, *Cell Regul.* **1:**453–463.

Bar-Shavit, R., Sabbah, V., Lampugnani, M. G., Marchisio, P. C., Fenton, J. W., II, Vlodavsky, I., and Dejana, E., 1991, An Arg-Gly-Asp sequence within thrombin promotes endothelial cell adhesion, *J. Cell Biol.* **112:**335–344.

Bashkin, P., Klagsbrun, M., Doctrow, S., Svahn, C. M., Folkman, J., and Vlodavsky, I., 1989, Basic fibroblast growth factor binds to subendothelial extracellular matrix and is released by heparanase and heparin-like molecules, *Biochemistry* **28:**1737–1743.

Bernfield, M. R., Banerjee, S. D., and Cohn, R. H., 1972, Dependence of salivary epithelial morphology and branching morphogenesis upon acid mucopolysaccharide protein of the epithelial surface, *J. Cell Biol.* **52:**674–689.

Bing, D. H., Laura, R., Robinson, D. J., Furie, B., Furie, B. C., and Feldman, R. J., 1981, Structure–function relationship of thrombin based on computer generated three dimensional model of the B-chain of bovine thrombin, *Ann. N.Y. Acad. Sci.* **370:**496–501.

Bing, D. H., Feldman, R. J., and Fenton, J. W., II, 1986, A computer-generated three-dimensional model of the B chain of bovine α-thrombin, *Ann. N.Y. Acad. Sci.* **485:**104–119.

Butkowski, R. J., Elion, J., Downing, M. R., and Mann, K. G., 1977, Primary structure of human prethrombin 2 and α-thrombin, *J. Biol. Chem.* **252:**4942–4957.

Camussi, G., Afglietta, M., Malavasi, F., Tetta, C., Piacibello, W., Sanavio, F., and Bussolino, F., 1983, The release of activating factor from human endothelial cells in culture, *J. Immunol.* **131:**2397–2403.

Carney, D. H., Scott, L., Gordon, E. A., and LaBelle, E. F., 1985, Neomycin inhibits thrombin-stimulated phosphoinositide turnover and initiation of DNA synthesis: Role of phosphoinositides in thrombin mitogenesis, *Cell* **42:**479–488.

Castellot, J. J., Jr., Favreau, L. V., Karnovsky, M. J., and Rosenberg, R. D., 1982, Inhibition of vascular smooth muscle cell growth by endothelial cell-derived heparin, *J. Biol. Chem.* **257:**11256–11260.

Chambard, J. C., Franchi, A., Le Cam, A., and Pouyssegur, J., 1983, Growth factor-stimulated protein phosphorylation in G_0/G_1-arrested fibroblasts, *J. Biol. Chem.* **258:**1706–1713.

Chen, L. B., and Buchanan, J. M., 1975, Mitogenic activity of blood components. I. Thrombin and prothrombin, *Proc. Natl. Acad. Sci. USA* **72:**131–135.

Cheresh, D. A., 1987, Human endothelial cells synthesize and express an Arg-Gly-Asp directed adhesion receptor involved in attachment to fibrinogen and von Willebrand factor, *Proc. Natl. Acad. Sci. USA* **84:**6471–6475.

Clohisy, D. R., Erdman, J. M., and Wilner, G. D., 1990, Thrombin binds to murine bone marrow-derived macrophages and entrances colony stimulating factor-1-driven mitogenesis, *J. Biol. Chem.* **265:**7729–7732.

Collins, S. J., Ruscetti, F. W., Gallagher, R. E., and Gallo, R. C., 1978, Terminal differentiation of human promyelocytic leukemia cells induced by dimethyl sulfoxide and other polar compounds, *Proc. Natl. Acad. Sci. USA* **75:**2458–2462.

Colman, R. W., Marder, V. J., and Salzman, E. W., and Hirsh, J., 1987, Introduction, in: *Hemostasis Thrombosis, basic principles and clinical practice*, 2nd ed. (R. W. Colman, J. Hirsh, V. J. Marder, and E. W. Salzman, eds.), Lippincott, Philadelphia, pp. 3–17.

Craik, C. S., Sprang, C., Fletterick, R., and Rutter, R., 1982, Intron–exon splice junctions map at protein surfaces, *Nature* **299:**180–182.

de Groot, P. G., Reinders, J. H., and Sixma, J. J., 1987, Perturbation of human endothelial cells by thrombin or PMA changes the reactivity of their extracellular matrix toward platelets, *J. Cell Biol.* **104:**697–704.

Dejana, E., Colella, S., Conforti, G., Abbadini, M., Gaboli, M., and Marchisio, P. C., 1988, Fibronectin and vitronectin regulate the organization of their respective Arg-Gly-Asp adhesion receptors in cultured human endothelial cells, *J. Cell Biol.* **107:**1215–1223.

DiColerto, P. E., and Bowen-Pope, D. F., 1983, Cultured endothelial cells produce a platelet-derived growth factor-like protein, *Proc. Natl. Acad. Sci. USA* **80:**1919–1923.

Dodson, J. W., 1963, On the nature of tissue interactions in embryonic skin, *Exp. Cell Res.* **31:**233–240.

Elion, J., Downing, M. R., Butkowski, R. J., and Mann, K. G., 1977, Human and bovine thrombin sequence: Comparison to other serine proteases, in: *Chemistry and Biology of Thrombin* (R. L. Lundblad, J. W. Fenton, II, and K. G. Mann, eds.), Ann Arbor Science, Ann Arbor, p. 97.

Folkman, J., Klagsbrun, M., Sasse, J., Vadzinski, M., Ingber, D., and Vlodavsky, I., 1988, A heparin binding angiogenic protein, basic fibroblast growth factor, is stored within basement membrane, *Am. J. Pathol.* **130:**393–399.

Furie, B., Bing, D. H., Feldman, R. J., Robinson, D. J., Burnier, J. P., and Furie, B. C., 1982, Computer-generated models of blood coagulation factor Xa, factor IXa, and thrombin based upon structural homology with other serine proteases, *J. Biol. Chem.* **257:**3875.

Galdal, K. S., Evensen, S. A., and Nilsen, E., 1985, The effect of thrombin on fibronectin in cultured human cells, *Thromb. Res.* **37:**583–593.

Garcia, J. G. N., Siflinger-Birnboim, A., Bixios, R., Del Vecchio, P. J., Fenton, J. W., and Malik, A. B., 1986, Thrombin induced increase in albumin permeability across the endothelium, *J. Cell Physiol.* **128:**96–104.

Gelehrber, T. D., and Sznycer-Laszuk, R., 1986, Thrombin induction of plasminogen activator-inhibitor in cultured human endothelial cells, *J. Clin. Invest.* **77:**165–169.

Glenn, K. C., Carney, D. H., Fenton, J. W., II, and Cunningham, D. D., 1980, Thrombin active site regions required for fibroblast receptor binding and initiation of cell division, *J. Biol. Chem.* **255**:6609–6616.

Gordon, M. Y., Riley, G. P., Watt, S. M., and Greaves, M. F., 1987, Compartmentalization of a haemopoietic growth factor (GM-CSF) by glycosaminoglycans in the bone marrow microenvironment, *Nature* **326**:403–405.

Gospodarowicz, D., Vlodavsky, I., and Savion, N., 1980a, The extracellular matrix and the control of proliferation of vascular endothelial and vascular smooth muscle cells, *J. Supramol. Struct.* **13**:339–372.

Gospodarowicz, D., Delgado, D., and Vlodavsky, I., 1980b, Permissive effect of the extracellular matrix on cell proliferation *in vitro*, *Proc. Natl. Acad. Sci. USA* **77**:4094–4098.

Gospodarowicz, D., Gonzales, R., and Fujii, D. K., 1983, Are factors originating from serum, plasma or cultured cells involved in the growth promoting effect of the extracellular matrix produced by cultured bovine corneal endothelial cells? *J. Cell Physiol.* **114**:191–202.

Gospodarowicz, D., Lepine, J., Massoglia, S., and Wood, I., 1984, Comparison of the ability of basement membranes produced by corneal endothelial and mouse derived endodermal PF-HR-9 cells to support the proliferation and differentiation of bovine kidney tubule epithelial cells in vitro, *J. Cell Biol.* **99**:947–1961.

Greenberg, M. E., and Ziff, E. B., 1984, Stimulation of 3T3 cells induces transcription of the *c-fos* proto-oncogene, *Nature* **311**:433–438.

Haigler, H. T., Maxfield, F. R., Willingham, M. C., and Pastan, I., 1980, Dansylcadaverine inhibits internalization of ^{125}I-epidermal growth factor, *J. Biol. Chem.* **255**:1239–1241.

Hanson, S. R., and Harker, L. A., 1988, Interruption of acute platelet-dependent thrombosis by the synthetic antithrombin D-phenylalanyl-L-prolyl-arginyl-chloromethyl ketone, *Proc. Natl. Acad. Sci. USA* **85**:3184–3188.

Harlan, J. M., Thompson, P. J., Ross, R. R., and Bowen-Pope, D. F., 1986, Thrombin induces release of platelet-derived growth factor like molecule(s) by cultured human endothelial cells, *J. Cell Biol.* **103**:1125–1133.

Honma, Y., Takenaga, K., Kasukabe, T., and Hozumi, M., 1980, Induction of differentiation of cultured human promyelocytic leukemia cells by retinoids, *Biochem. Biophys. Res. Commun.* **95**:507–512.

Horowitz, A., Duggan, K., Greggs, R., Decker, C., and Buck, C., 1985, The cell substrate attachment (CSAT) antigen has properties of a receptor for laminin and fibronectin, *J. Cell Biol.* **101**:2134–2144.

Hynes, R. O., 1987, Integrins: A family of cell surface receptors, *Cell* **48**: 549–555.

Jaffe, E. A., Hoyer, L. W., and Nachman, R. L., 1974, Synthesis of von Willebrand factor by culture of human endothelial cells, *Proc. Natl. Acad. Sci. USA* **71**:1906–1909.

Jolyon, J., 1986, The kinetics of inhibition of α-thrombin in human plasma, *J. Biol. Chem.* **261**:10313–10318.

Kajiji, S., Tamura, R. N., and Quaranta, V., 1989, A novel integrin (αEβ4) from human epithelial cells suggests a fourth family of integrin adhesion receptors, *EMBO J.* **8**:673–680.

Kaplan, K., 1982, Interaction of platelets with endothelial cells, in: *Pathobiology of the Endothelial Cell* (H. L. Nofssel and H. J. Vogel, eds.), Academic Press, New York, pp. 337–349.

Kindy, M. S., and Sonenshein, G. E., 1986, Regulation of oncogene expression in cultured aortic smooth muscle cells: Posttranscriptional control of c-myc mRNA, *J. Biol. Chem.* **261**:12865–12868.

Klagsbrun, M., and Edelman, E. R., 1989, Biological and biochemical properties of fibroblast growth factor. Implications for the pathogenesis of atherosclerosis, *Arteriosclerosis* **9**:269–278.

Knudsen, B. S., Silverstein, R. L., Leung, L. L. K., Harpel, P. C., and Nachman, R. L., 1986, Binding of plasminogen to extracellular matrix, *J. Biol. Chem.* **261**:10765–10771.

Kramer, H. R., and Vogel, K. G., 1984, Selective degradation of basement membrane macromolecules by metastatic cells, *J. Natl. Cancer Inst.* **72**:889–899.

Kruijer, W., Copper, J. A., Hunter, T., and Verma, I. M., 1984, Platelet-derived growth factor induces rapid but transient expression of the c-fos gene and protein, *Nature* **312**:711–716.

Laiho, M., and Keski-Oja, J., 1989, Growth factors in the regulation of pericellular proteolysis: A review, *Cancer Res.* **49**:2533–2553.

Languino, L. R., Collela, S., Zanetti, A., Andrieux, J. J., Pyckewaert, M. H., Charon, P. C., Marchisio, E. F., Plow, M. H., Ginsberg, G., Margurie, G., and Dejana, E., 1989, Fibrinogen–endothelial cell interaction in-vitro: A pathway mediated by an Arg-Gly-Asp recognition specificity, *Blood* **73**:734–742.

Laposada, M., Dovuarsky, D. K., and Solkin, H. S., 1983, Thrombin induced gap formation in confluent endothelial cell monolayers, *Blood* **62**:549–556.

Lawler, J., Weinstein, R., and Hynes, O. R., 1988, Cell attachment to thrombospondin: The role of Arg-Gly-Asp, calcium and integrin receptors, *J. Cell Biol.* **107**:351.

Lerner, R. G., Chenong, L. C., and Nelson, I. C., 1979, Thrombin induced endothelial cell retraction, *Thromb. Haemostas.* **42**:244–247.

Libby, P., Warner, S. J. C., and Friedman, G. B., 1988, Interleukin 1: A mitogen for human vascular smooth muscle cells that induces the release of growth inhibitory prostanoids, *J. Clin. Invest.* **81**:487–498.

Loskutoff, D. J., 1979, Effect of thrombin on the fibrinolytic activity of cultured bovine endothelial cells, *J. Clin. Invest.* **64**:329–340.

Lotem, J., and Sachs, L., 1979, Regulation of normal differentiation in mouse and human myeloid leukemia cells by phorbol esters and the mechanism of tumor protection, *Proc. Natl. Acad. Sci. USA* **76**:5158–5162.

Mann, K. G., 1987, The assembly blood clotting complexes on membranes, *Trends Biochem. Sci.* **12**:229–234.

Martinet, Y., Bitterman, P. B., Mornex, J. F., Grotendorst, G. R., Martin, G. R., and Crystal, R. G., 1986, Activated human monocytes express c-sis proto-oncogene and release a mediator showing PDGF-like activity, *Nature* **319**:158–160.

Muller, R., Bravo, R., Bruckhardt, J., and Curran, T., 1984, Induction of c-fos gene and protein by growth factors precedes activation of c-myc, *Nature* **312**:716–720.

Neurath, H., 1984, Evolution of proteolytic enzymes, *Science* **224**:350–357.

Neurath, H., 1986, The versatility of proteolytic enzymes, *J. Cell Biochem.* **32**:35–49.

Olivecrona, T., Egelrud, T., Iverius, P. H., and Lindahl, V., 1971, Evidence for an anionic binding of lipoprotein lipase to heparin, *Biochem. Biophys. Res. Commun.* **45**:524–529.

Paris, S., and Pouyssegur, J., 1984, Growth factors activate the Na^+/H^+ antiporter in quiescent fibroblasts by increasing its affinity for intracellular H^+, *J. Biol. Chem.* **259**:10989–10994.

Pearson, J. D., and Gordon, J. L., 1979, Vascular endothelial and smooth muscle cells in culture selectively release adenine nucleotides, *Nature* **281**:384–386.

Perdue, J. F., Lubenskyi, W., Kivity, E., Sonder, S. A., and Fenton, J. W., II, 1981, Protease mitogenic response of chick embryo fibroblasts and receptor binding/processing of human thrombin, *J. Biol. Chem.* **256**:2767–2776.

Plow, E. F., McEver, R. P., Coller, B. S., Woods, V. L., Jr., and Marguerie, G. A., 1985,

Related binding mechanisms for fibronectin, von Willebrand factor and thrombin-stimulated human platelets, *Blood* **66**:724–727.

Ponte, P., Gonzalez-DeWhitt, P., Schilling, J., Miller, J., Hsu, D., Greenberg, B., Davis, K., Wallace, W., Lieberburg, I., Fuller, F., and Cordell, B., 1988, A new A4 amyloid in mRNA contains a domain homologous to a serine proteinase inhibitor, *Nature* **331**:525–527.

Prescott, S. M., Zimmerman, G. A., and McIntyre, T. M., 1984, Human endothelial cells in culture produce BAF when stimulated with thrombin, *Proc. Natl. Acad. Sci. USA* **81**:3534–3538.

Pytela, R., Pierschbacher, M. D., and Ruoslahti, E., 1985a, Identification and isolation of 140kd cell surface glycoprotein with properties expected of a fibronectin receptor, *Cell* **40**:191–198.

Pytela, R., Pierschbacher, M. D., and Ruoslahti, E., 1985b, A 125/155 kDa cell surface receptor specific for vitronectin interacts with the arginine-glycine-aspartic acid adhesion sequence derived from fibronection, *Proc. Natl. Acad. Sci. USA* **82**:5766–5770.

Reddi, A. H., and Anderson, W. A., 1976, Collagenous bone matrix induced endochondral ossification and hemopoiesis, *J. Cell Biol.* **68**:557–572.

Roberts, R., Gallagher, J., Spooncer, S., Allen, T. D., Bloomfield, F., and Dexter, T. M., 1988, Heparin-sulphate bound growth factors: A mechanism for stromal cell mediated haemopoiesis, *Nature* **332**:376–378.

Rogelj, S., Klagsbrun, M., Atzmon, R., Kurokawa, M., Haimovitz, A., Fuks, Z., and Vlodavsky, I., 1989, Basic fibroblast growth factor is an extracellular matrix component required for supporting the proliferation of vascular endothelial cells and the differentiation of PC12 cells, *J. Cell Biol.* **109**:824–831.

Rosenberg, R. D., 1977, Biologic actions of heparin, *Semin. Hematol.* **14**:427–440.

Rosenberg, R. D., and Damus, D. S., 1973, The purification and mechanism of action of human anti thrombin–heparin cofactor, *J. Biol. Chem.* **248**:6490–6505.

Ross, R., 1986, The pathogenesis of atherosclerosis—an update, *N. Engl. J. Med.* **314**:488–499.

Ruggeri, Z. M., Bader, R., and DeMarco, L., 1982, Glanzmann's thrombasthenia: Deficient binding of von Willebrand factor to thrombin-stimulated platelets, *Proc. Natl. Acad. Sci. USA* **79**:6038–6041.

Ruoslahti, E., and Pierschbacher, M. D., 1987, New perspectives in cell adhesion: RGD and integrins, *Science* **238**:491–497.

Savion, N., Isaacs, J. D., Gospodarowicz, D., and Shuman, M. A., 1981, Internalization and degradation of thrombin and upregulation of thrombin-binding sites in corneal endothelial cells, *J. Biol. Chem.* **256**:4514–4519.

Scheinberg, I. H., 1982, Scatchard plots, *Science* **215**:312–314.

Smith, J. C., Singh, J. P., Lillquist, J. S., Goon, D. S., and Stiles, C. D., 1982, Growth factors adherent to cell substrate are mitogenically active in situ, *Nature* **296**:154–156.

Sporn, L. A., Marder, V. J., and Wagner, D. D., 1987, von Willebrand factor released from Wiebel–Palade bodies binds more avidly to extracellular matrix than that secreted constitutively, *Blood* **69**:1531–1534.

Stern, D., Nawroth, P., Handley, D., and Kisiel, W., 1985, An endothelial cell-dependent pathway of coagulation, *Proc. Natl. Acad. Sci. USA* **82**:2523–2527.

Taylor, S., and Folkman, J., 1982, Protamine is an inhibitor of angiogenesis, *Nature* **297**:307–312.

Tollefsen, D. M., Majerus, P. W., and Blank, M. K., 1982, Heparin cofactor II, *J. Biol. Chem.* **257**:2162–2169.

Tushinski, R. J., Oliver, I. T., Guilbert, L. J., Tynan, P. W., Warner, J. R., and Stanley, E. R.,

1982, Survival of mononuclear phagocytes depends on a lineage-specific growth factor that the differentiated cells selectively destroy, *Cell* **28:**71–81.

Vlodavsky, I., Liu, G. M., and Gospodarowicz, D., 1980, Morphological appearance, growth behavior and migratory activity of human tumor cells maintained on extracellular matrix vs plastic, *Cell* **19:**607–616.

Vlodavsky, I., Folkman, J., Sullivan, R., Fridman, R., Ishai-Michaeli, R., Sasse, J., and Klagsbrun, M., 1987, Endothelial cell-derived basic fibroblast growth factor: Synthesis and deposition into subendothelial extracellular matrix, *Proc. Natl. Acad. Sci. USA* **84:**2292–2296.

Wagner, S. L., Gededdes, J. W., Cotman, C. W., Lau, A. L., Gurewitz, D. G., Isackson, P. J., and Cunningham, D. D., 1989, Protease nexin-1, an antithrombin with neurite outgrowth activity is reduced in Alzheimer disease, *Proc. Natl. Acad. Sci. USA* **86:**8284–8288.

Wayner, E. A., and Carter, W. G., 1987, Identification of multiple cell adhesion receptors for collagen and fibronectin in human fibrosarcoma cells possessing unique and common subunits, *J. Cell Biol.* **105:**1873–1884.

Weksler, B. B., Ley, C. W., and Jaffe, E. A., 1978, Stimulation of endothelial cell prostacyclin (PGI$_2$) production by thrombin, trypsin and the ionophore A23187, *J. Clin. Invest.* **62:**923–930.

White, J. R., Naccache, P. H., and Sha'afi, R. I., 1983, Stimulation by chemotactic factor of actin association with the cytoskeleton in rabbit neutrophils, *J. Biol. Chem.* **258:**14041–14047.

Wilner, G. B., Danitz, M. P., Mudd, M. S., Hsieh, K. H., and Fenton, J. W., II, 1981, Selective immobilization of α-thrombin by surface bound fibrin, *J. Lab. Clin. Med.* **97:**403–407.

Chapter 10

POSTCLOTTING CELLULAR EFFECTS OF THROMBIN MEDIATED BY INTERACTION WITH HIGH-AFFINITY THROMBIN RECEPTORS

Darrell H. Carney

1. THROMBIN AS A GROWTH FACTOR

1.1. Stimulation of Cell Proliferation by Trypsin and Thrombin

Current knowledge of thrombin involvement in postclotting stimulation of cell proliferation has evolved from initial studies showing that trypsin and other serine proteases stimulated DNA synthesis and proliferation in certain lines of mouse 3T3 cells (Burger, 1970; Noonan, 1976; Noonan and Burger, 1973), and quiescent primary cultures of chick embryo fibroblasts (Blumberg and Robbins, 1975; Cunningham and Ho, 1975; Hovi and Vaheri, 1975; Sefton and Rubin, 1970; Vaheri *et al.*, 1974). The idea that

Darrell H. Carney • Department of Human Biological Chemistry and Genetics, University of Texas Medical Branch, Galveston, Texas 77550.

Thrombin: Structure and Function, edited by Lawrence J. Berliner. Plenum Press, New York, 1992.

cell proliferation could be studied with proteases was intriguing at that time since these enzymes were well characterized and readily available. Interestingly, other proteases which cleaved a number of cell surface molecules were not as effective as trypsin in stimulating these cells (Blumberg and Robbins, 1975). This suggested that the stimulation might involve more that just proteolytic cleavage of cell surface molecules. Hodges *et al.* reported that trypsin accumulated inside of cells, raising the possibility that the apparent specificity was dependent upon receptor-mediated internalization (Hodges *et al.*, 1973). We were able to show, however, that trypsin action at the cell surface was sufficient to initiate division of chick embryo cells (Carney and Cunningham, 1977). Thus, it appeared that specific cell surface interactions of proteases determined their ability to generate mitogenic signals.

Thrombin proved to be even more potent than trypsin in stimulation of cell proliferation in chick embryo fibroblasts (Chen and Buchanan, 1975; Teng and Chen, 1975; Zetter *et al.*, 1976, 1977a). In addition, thrombin was the only protease which appeared to consistently stimulate mammalian fibroblasts in the absence of serum (Carney *et al.*, 1978). Similar to trypsin, thrombin appeared to selectively bind to chick cells and be internalized (Martin and Quigley, 1978; Zetter *et al.*, 1977a). However, immobilized thrombin appeared to stimulate cell proliferation by action at the cell surface without requirement for internalization (Carney and Cunningham, 1978a,c). Since thrombin was known to be the pivotal enzyme in blood clotting, this raised the possibility that thrombin might interact with specific receptors or substrates on the cell surface and that this initiation, unlike that generated by trypsin, might be physiologically related to wound healing and tissue repair.

In contrast to thrombin, prothrombin had no stimulatory activity, suggesting that the growth stimulatory effects of thrombin might be related to the activation of thrombin in blood or at the cell surface (Chen and Buchanan, 1975). Based on the amount of prothrombin in blood, initial estimates indicates that up to 30% of the growth factor activity of whole serum might be due to the effects of thrombin (Chen and Buchanan, 1975). Indeed, thrombin potentiates the response of cells to serum or purified growth factors (Cherington and Pardee, 1980; Zetter and Antoniades, 1979; Zetter *et al.*, 1977b). In addition, thrombin initiates cell proliferation in the absence of other growth factors (Carney *et al.*, 1978; Chen and Buchanan, 1975; Perdue *et al.*, 1981; Perez-Rodriguez *et al.*, 1981; Pohjanpelto, 1977, 1978; Van Obberghen-Schilling *et al.*, 1983). The efficacy of thrombin as a growth factor in the absence of other factors may relate to the multiple types of signals generated by thrombin (see Section 3).

1.2. The Search for Specific Thrombin Substrates or Receptors

Comparing the cell surface components cleaved by various proteases and the ability of these proteases to stimulate cell proliferation further indicated the specificity of thrombin interaction. Trypsin, thrombin, chymotrypsin, pronase, and a number of other proteases all cleaved cell surface proteins to various degrees. Fibronectin (then referred to as LETS) was one of the most prominent proteins which was cleaved by proteases (Hynes, 1974), leading to speculation that the cleavage of this protein released cells from contact inhibition. This did not appear to be the case. Chymotrypsin and pronase released most of the fibronectin from the surface of chick fibroblasts, but were not mitogenically active, whereas thrombin removed very little of the fibronectin (Teng and Chen, 1975; Zetter et al., 1976). Thus, the mitogenic effect of thrombin was not attributable to cleavage of fibronectin or other major identifiable surface proteins. Nevertheless, it was assumed that this stimulation involved proteolytic cleavage of a specific substrate because both trypsin and thrombin could initiate cell proliferation in different types of cells, and because proteolytically inhibited DIP-thrombin was not capable of stimulating cell proliferation (Carney et al., 1978; Perdue et al., 1981).

More recent studies utilizing specific labeling techniques have indicated that molecules of $M_r = 140,000$ and $55,000$ were cleaved by thrombin while others of $150,000$, $130,000$, and $45,000$ appeared to increase (Moss and Cunningham, 1981). Thrombin effects on responsive and nonresponsive (aged) chick embryo fibroblasts indicated that molecules of $M_r = 43,000$ and one of $150,000$ were removed from the surface of thrombin-responsive cells (Glenn and Cunningham, 1979), but the small molecule could be related to protease-nexin (described below) and the larger $150,000$ M_r band was removed from both responsive and nonresponsive cells. Although other cell types showed changes in molecules of similar sizes following thrombin treatment (Chen et al., 1976; Teng and Chen, 1975, 1976; Zetter et al., 1976), there does not appear to be a consistent pattern which would establish a causal relationship between a single proteolytic cleavage and mitogenic signaling.

A breakthrough in understanding the role of thrombin in stimulating cell proliferation came with the discovery of specific high-affinity receptors for thrombin on the surface of mouse (Carney and Cunningham, 1978b) and chick embryo fibroblasts (Martin and Quigley, 1978; Perdue et al., 1981). Binding of thrombin to these receptors appeared to correlate with the amount of thrombin required for mitogenesis (Carney and Cunningham, 1978b; Van Obberghen-Schilling et al., 1983). Furthermore, inhibition of thrombin binding inhibited its ability to stimulate cell proliferation

(Carney and Cunningham, 1978b). The mitogenic signals elicited by thrombin interaction with these receptors, however, appeared to involve both binding and proteolytic activity since DIP-inactivated thrombin could bind to these receptors, but could not stimulate cell proliferation (Carney *et al.*, 1986b). This has raised a number of questions about the role of this receptor in mitogenesis. (See review by Carney, 1987.) For example, it is not clear whether the receptor represents the substrate for thrombin and other proteases which stimulate cell proliferation, or whether this binding merely facilitates the ability of thrombin to interact with an adjacent substrate. In other systems, thrombin appears to act as a "bridging ligand" which binds to one molecule or receptor and then by the act of binding, its specific activity for cleaving a certain substrate is amplified manyfold (Fenton, 1988). Thus, thrombin might bind to its receptor and this binding may enhance its ability to cleave a yet unidentified cell surface substrate. In any event, the discovery of a specific receptor for thrombin on the surface of cells and its apparent relationship to stimulation of cell proliferation gave us a tool with which to begin to understand this process and the potential role of thrombin in physiological responses of cells to injury.

Binding studies also indicated that more than one receptor system may be involved in thrombin interaction with fibroblasts. Studies with mouse cells showed that after 2 h incubation at 23°C, 95% of the specifically bound thrombin could be removed by trypsin treatment, indicating that these receptors were located at the cell surface and each cell had approximately 200,000 binding sites with an apparent single affinity $K_d = 1.1$ nM (Carney and Cunningham, 1978b). These receptors appear to be clustered on the surface of cells prior to thrombin binding (Carney, 1983; Carney and Bergmann, 1982), and are not rapidly internalized via clathrin-coated pits (Bergmann and Carney, 1982; Carney and Bergmann, 1982). In contrast, chick fibroblasts showed specific internalization of thrombin and binding to approximately 3000 high-affinity sites ($K_d = 0.7$ nM) and up to 1,000,000 sites at a $K_d = 3$ nM (Perdue *et al.*, 1981). More recent studies showed that, indeed, these cells and others had two different mechanisms for interaction with thrombin. As described below, the high-affinity binding sites on mouse cells appear to represent the mitogenically linked thrombin receptor, while in chick and human cells at least part of the high-affinity binding and most of the rapid internalization represents binding of thrombin to protease-nexin (Baker *et al.*, 1980).

1.3. Protease-Nexin Negatively Regulates Thrombin Mitogenesis

Early studies showed that a large portion of the thrombin which bound to human fibroblasts could be recovered as a complex with apparent M_r = 68,000 (Baker *et al.*, 1979). A similar-sized complex was found when examining thrombin binding to mouse, chick, hamster lung, and human fibrosarcoma cells (Simmer *et al.*, 1979). This complex, which was stable to SDS, formed without cross-linking agents, but dissociated in the presence of 1 M hydroxylamine or when exposed to pH above 10 (Baker *et al.*, 1980). In addition, these studies showed that heparin increased the complex formation and that formation of the complex was dependent upon thrombin proteolytic activity since DIP-inactivated thrombin could not form complexes. Thus, many of the properties of this interaction were similar to the interaction between thrombin and antithrombin III (see review by Olson and Bjork, this volume). This molecule was smaller than antithrombin III (and not immunologically cross-reactive), formed complexes with urokinase and other proteases, and was synthesized by fibroblasts and released into the culture medium (Baker *et al.*, 1980). Because it could form complexes with a number of proteases, this new thrombin inhibitor was named protease-nexin (PN).

Although thrombin–PN complex binding added a new level of complexity to these studies, the PN complex binding could be easily distinguished from thrombin receptor binding. For example, DIP-inactivated thrombin binds to thrombin receptors with an affinity nearly equal to that of native α-thrombin, whereas PN–thrombin complex formation requires proteolytic cleavage of PN to form an aminoacyl complex (Baker *et al.*, 1980). In addition, since PN is secreted into the medium where it complexes with thrombin or other proteases, removal of the culture medium, rinsing the cells prior to incubation with thrombin, and incubation at 23°C instead of 37°C eliminated most thrombin–PN complex formation (Carney and Bergmann, 1982; Low *et al.*, 1982). Unlike thrombin receptors which are clustered on the surface of cells (Carney, 1983; Carney and Bergmann, 1982), PN appears to bind diffusely or localize with extracellular matrix molecules (Farrell *et al.*, 1988). In addition, thrombin–PN complexes are internalized and rapidly degraded (Low *et al.*, 1981), whereas receptor-bound thrombin appears to remain at the cell surface (Bergmann and Carney, 1982; Carney and Bergmann, 1982). Thus, the receptor systems involved in binding of thrombin and thrombin–PN complexes appear to be quite different.

The internalization and degradation of thrombin–PN complexes suggested that this represents a clearance mechanism for cells to regulate and perhaps negate the effects of extracellular proteases. Indeed, stimulation of DNA synthesis in mouse and IIC9 cells required up to tenfold higher concentrations of thrombin if cells were assayed in conditioned medium or medium to which PN was added, than that required when the medium was changed to fresh medium just prior to thrombin addition (Low *et al.*,

1982). Thus, PN has a negative effect on thrombin-stimulated cell proliferation. PN also appears to play a role in regulating extracellular levels of thrombin and other proteases including plasminogen activators in a number of different cell types including endothelial and epithelial cells (Eaton and Baker, 1983; Knauer *et al.*, 1983; Scott *et al.*, 1983). Recent studies have shown that much of the binding of thrombin to platelets is also mediated by PN (Gronke *et al.*, 1986, 1987). Thus, it will be important to consider potential thrombin–PN complex formation and binding in relationship to both past and future attempts to analyze thrombin interaction with a variety of cells.

2. CHARACTERIZATION OF THROMBIN BINDING SITES

2.1. Thrombin Domains Involved in Receptor Interaction

2.1.1. Binding of Thrombin and Thrombin Derivatives to Fibroblasts

Fibroblasts from mouse embryo, chick embryo, human skin, hamster lung, and several other types of cells all exhibit thrombin binding to high-affinity sites with K_d of approximately 0.7–2.4 nM (Carney, 1987). In contrast, other cells such as monocytes may have different receptors which interact with different domains on thrombin (Bar-Shavit and Wilner, 1986). Most fibroblastic cells examined appear to express 100,000 to 200,000 receptors on the surface of each cell. Thus, this appears to be an excellent cell type for structural and functional receptor characterization. Although the thrombin receptor on these cells has not yet been purified or sequenced, considerable information has been gained about these receptors and the thrombin domains with which they interact from studies with thrombin derivatives and synthetic peptides representing portions of the thrombin molecule.

Studies with various thrombin derivatives showed that DIP-inactivated thrombin bound to these receptors with approximately the same affinity as that of active thrombin molecules (Glenn *et al.*, 1980). Some other derivatives inactivated with larger groups, however, showed less ability to bind or compete for binding, suggesting that the domain of thrombin responsible for interaction with these receptors might be near or adjacent to the enzyme active site (Glenn *et al.*, 1980). As one examines more cells and more derivatives, the picture becomes more complex. Thrombin derivatives such as γ-thrombin (which retain full esterase activity, but cannot bind and cleave fibrinogen) show little ability to compete for binding sites on mammalian cells, suggesting that regions of thrombin distal to the proteolytic pocket are involved in receptor recognition (Glenn *et al.*, 1980). In contrast, in chick and human cells DIP-thrombin did not compete well for receptor binding and γ-thrombin showed much more competition than in other fibroblasts (Perdue *et al.*, 1981). As described above, part of the high-affinity binding to these cells appears to be due to the binding of thrombin–PN complexes (Baker *et al.*, 1980; Eaton and Baker, 1983; Simmer *et al.*, 1979). This would explain why DIP-thrombin did not compete with α-thrombin binding to these cells, while γ-thrombin which forms complexes with PN did compete (Glenn *et al.*, 1980; Perdue *et al.*, 1981). These observations indicate that useful information can be derived from derivative binding, but it is essential to know the nature of the binding interaction before using this information for structural predictions.

2.1.2. The Role of High-Affinity Thrombin Receptors in Mitogenesis

A number of investigators have questioned the role of thrombin receptors in mitogenesis (Low *et al.*, 1985; Van Obberghen-Schilling and Pouyssegur, 1985). As described above, thrombin initiation of cell proliferation required proteolytically active thrombin, yet DIP- and PMS-inactivated thrombin were able to bind to thrombin receptors on fibroblasts with the same apparent affinity as fully active thrombin. This suggested that if the mitogenic mechanism was a single enzymatic event, addition of DIP-thrombin should inhibit mitogenesis by blocking the interaction of active thrombin molecules with the receptor. Even with a 50-fold excess of DIP-thrombin, however, there was negligible inhibition or shifting of the thrombin dose–response curve (Low *et al.*, 1985; Van Obberghen-Schilling and Pouyssegur, 1985). This indicated that more than one site (or action) may be involved in mitogenesis or that the binding to high-affinity binding sites was not necessary for thrombin mitogenesis (Low *et al.*, 1985).

To further examine this problem, we carried out similar experiments with parallel binding studies. We found that there was a rapid exchange between receptor-bound DIP-thrombin and free α-thrombin (Carney *et al.*, 1986b). Thus, even if the receptors were saturated with a 50-fold excess of DIP-thrombin, at any one moment 2% of the receptors would be occupied by active thrombin which could cleave either the receptor or an adjacent substrate molecule. Further, if different signals were generated by thrombin receptor occupancy and proteolytic cleavage, then there might be negligible shifting of the dose–response curve for active thrombin in the presence of DIP-thrombin (Carney *et al.*, 1986b). This "double signal hypothesis" was verified by showing that addition of DIP-thrombin and γ-thrombin to ME cells even at times up to 6 h apart, stimulated DNA synthesis and cell proliferation to levels comparable to those of fully active α-thrombin (Carney et al., 1984, 1986b). Additional studies showed that DIP-inactivated thrombin could stimulate cells in the presence of low concentrations of phorbol esters which activated protein kinase C (Gordon and Carney, 1986). Moreover, a monoclonal antibody (TR-9) which recognizes thrombin receptors on fibroblasts and competes for thrombin binding to its receptor was shown to stimulate cell proliferation in the presence of either γ-thrombin, phorbol myristate acetate, or marginally active concentrations of α-thrombin (Frost et al., 1987). Thus, these studies demonstrated that occupancy of the receptor, even with a monoclonal antibody, was sufficient to generate part of the mitogenic signals necessary to stimulate cell proliferation.

If thrombin–receptor interaction were not required for thrombin mitogenesis, it should be possible to identify thrombin-responsive cells that lack thrombin receptors. CHEF IIC9 cells, selected for survival in prolonged culture and thrombin responsiveness, were reported to be up to 1000-fold more sensitive than mouse or other hamster fibroblasts to stimulation by thrombin and to be virtually devoid of high-affinity thrombin receptors (Low et al., 1985). Scatchard plots of [^{125}I]thrombin binding and binding of thrombin receptor antibody TR-9, however, showed that these cells had approximately the same number of thrombin receptors as normal hamster fibroblasts, but these receptors exhibited a much lower apparent affinity (Frost, 1987). These cells may thus have an altered thrombin receptor (perhaps with a more rapid off rate) that is more sensitive to activation by interaction with proteolytically active thrombin molecules. Additional studies are necessary to characterize the responsiveness of these cells, the difference in the receptor which alters its apparent affinity for thrombin, and the nature of signals normally stimulated by receptor occupancy to determine if some of these signals are constitutively activated in these cells. It is important, however, to recognize that these cells do have a normal number of thrombin receptors recognized by TR-9 monoclonal antibody to thrombin receptors. Thus, it is likely that these receptors still play a role in mitogenesis in IIC9 cells.

Further attempts to find thrombin-responsive cells which lack thrombin receptors have also been unsuccessful. Immunofluorescence studies of mouse cells revealed heterogeneity in number and density of thrombin receptors with approximately 37% of the cells having less than 60,000 receptors per cell; 48% having between 60,000 and 500,000; 13% having between 500,000 and 1 million; and 2% having more than 1 million receptors per cell (Carney, 1983). Although immunofluorescence and labeled nuclei studies have not been done on the same populations, only about 60–70% of the cells in these populations are mitogenically responsive to thrombin (Carney *et al.*, 1978). This may suggest that only cells with greater than 60,000 receptors per cell are mitogenically responsive to thrombin. In further studies, fractions of these mouse cells were separated on Percol gradients and examined for their receptor binding and mitogenic responsiveness. Consistent with the observations described above, populations of cells with low numbers of receptors appeared nonresponsive and maximal responsiveness occurred in populations with about 300,000 receptors per cell (Bernstein and Carney, 1991). Unexpectedly, some cells with very high numbers of receptors were also nonresponsive, raising the possibility that these receptors might generate both positive and negative signals (Thompson and Carney, 1984). That no cell populations responded to thrombin with fewer than approximately 50,000 receptors per cell, however, further indicates a role for these receptors in thrombin mitogenesis.

2.1.3. Use of Synthetic Thrombin Peptides to Identify Binding Domains

If both receptor binding and proteolytic activity are required for mitogenic activity in fibroblasts, it would appear that either thrombin binds to a single receptor and may cleave that receptor, or it binds and cleaves an adjacent protein. To better understand this potential relationship, we attempted to identify the portion of the thrombin molecule which interacts with the receptor by synthesizing a series of peptides and determining their ability to compete with thrombin for receptor binding and mitogenic signal production (Glenn *et al.*, 1988).

One of the synthetic thrombin peptides representing residues 508 to 530 of human prothrombin (residues 178–200 using chymotrypsin numbering with alignment at active serine 195), competed for up to 70% of the specific binding to mouse fibroblasts (Glenn *et al.*, 1988). Figure 1 shows a stylized model indicating the approximate location of this peptide relative to the catalytic site, the carbohydrate attachment site, and the regions of thrombin involved in fibrinogen interaction. This model was derived from previous models based on chymotrypsin (Fenton, 1988) and trypsin coordinates (Glenn *et al.*, 1988), with substitutions and insertion of amino acids unique to thrombin. This model predicts that the receptor-interacting peptide begins with a portion of the highly conserved region of thrombin (CR6) around the active site serine, extends up over the lip of the cleft into the outer portion of the fibrinogen binding grove (a highly variable region unique to thrombin), and curves back into the middle of the model structure (Fenton and Bing, 1986). Smaller peptides which include the RGDA portion of this sequence (residues 187–190) also inhibited throm-

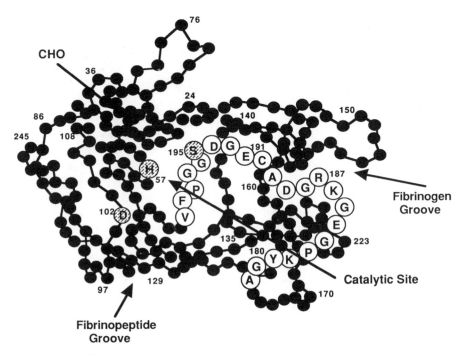

Figure 1. Simplified model of thrombin based on chymotrypsin coordinates with thrombin-specific substitutions to demonstrate the location of p508–530 (178–200), the thrombin receptor-binding peptide relative to the catalytic site (Ser-195, His-57, and Asp-102), the fibrinogen binding domain, and other regions of thrombin.

bin binding, suggesting that this region of the molecule might be at least a portion of the binding domain (Glenn *et al.*, 1988). Other proteases such as trypsin and chymotrypsin, which are nonmitogenic in these mouse fibroblasts, lack the variable region containing the RGDA sequence, but are similar or identical in the rest of the conserved protease-specific domain. In addition, other molecules containing portions of this sequence do not compete for thrombin binding. For example, RGDS sequences which are involved in fibronectin binding to cells do not compete nor do molecules such as urokinase which has a DS in this region rather than a DA (Glenn *et al.*, 1988). Thus, the interaction of this portion of the peptide with thrombin appears to be quite specific, suggesting that this region indeed represents at least a portion of the thrombin receptor binding domain.

One interesting aspect of this portion of the thrombin molecule interacting with thrombin receptors is that the proximity of this peptide to the proteolytic pocket would allow thrombin to bind and cleave this receptor to generate mitogenic signals. Recent x-ray structures for thrombin indicate that the variable region of thrombin which contains the RGDA sequence may be partially buried into the molecule as it exists when coupled to hirudin (Rydel *et al.*, 1990). The formation of the thrombin–hirudin complex, however, may alter the structure of thrombin, causing this portion of the molecule to close in. Previous studies have shown that thrombin–hirudin complexes are unable to bind to thrombin receptors (Low and Cunningham, 1982; Van Obberghen-Schilling *et al.*, 1982); thus, the conformation of the complex which appears to hide this domain may, in fact, contribute to its inability to interact with thrombin receptors. This may further indicate the importance of this region of the active thrombin molecule in interaction with thrombin receptors. If this variable region is also buried in native thrombin, it is possible that interaction of thrombin with its receptor alters the conformation of thrombin, exposing this region and allowing higher-affinity interaction with increased specificity or enzymatic activity toward the receptor or an interacting substrate. One might further speculate that thrombin derivatives, such as β-, γ-, or ζ-thrombin which are released from clots by various cell-secreted proteases, may have a more open structure in this region allowing them to interact enzymatically with thrombin receptors or cell-specific substrates while demonstrating only lower-affinity binding.

It should be noted that even at concentrations up to 4 μM, the synthetic receptor binding peptide would not compete for all of the thrombin bound to mouse cells (Glenn *et al.*, 1988). In other cell populations the relative amount of competition is sometimes as low as 30% of the total binding (Carney, unpublished). This may suggest that either the thrombin receptor exists in two states, or that there are more than one type of receptors on these cells. Because of this potential problem, it was important to determine if the receptor to which the peptide was binding was the same receptor that was involved in mitogenic signaling and whether binding of the peptide could activate receptor signals in these cells.

2.1.4. Activity of Thrombin Receptor Binding Peptides

The ability of synthetic peptide interaction with thrombin binding sites to generate mitogenic signals has been tested by addition of the peptides alone or in combination with thrombin to cultures of mouse embryo cells or NIL hamster fibroblasts. In these studies the p508 peptide (178–200) by itself, at concentrations up to 1 μM, had little if any effect (30 to 100% increases in [^{3}H]thymidine incorporation compared to 500–600% increases for maximally mitogenic concentrations of active thrombin), but in combination with either 4 nM thrombin or phorbol myristate acetate

(PMA) the peptide stimulated up to 600% increases in thymidine incorporation over peptide or PMA (Glenn *et al.*, 1988). Thus, this thrombin receptor binding peptide appears to activate receptor occupancy-dependent signals in a manner similar to the binding of DIP-inactivated thrombin or monoclonal antibody TR-9.

In these studies the effects of shorter peptides which inhibited thrombin binding were also examined. Intermediate-sized peptides containing 12 amino acids from p519 to 530 (189–200) were less effective than the 23-amino-acid peptide in enhancing the proliferative effects of low concentrations of α-thrombin, and the 4-amino-acid peptide RGDA was inhibitory (Glenn *et al.*, 1988). Concentrations of RGDA (0.5–2 µM) that inhibited greater than 50% of thrombin binding to these cells, also inhibited up to 85% of the maximal stimulation of thymidine incorporation stimulated by 10 nM thrombin (Glenn *et al.*, 1988). This suggests that the short peptide can interact with the receptor and block thrombin binding without itself generating a receptor occupancy-related signal. This may indicate that generation of receptor occupancy-related signals involves a conformational shift in the receptor which cannot be generated by the short peptide. More importantly, that this peptide acts as a competitive inhibitor for thrombin mitogenesis demonstrates that thrombin must interact with this receptor to generate its mitogenic effects. These inhibitory peptides are now being used to further demonstrate cellular events which require interaction of thrombin or other molecules with thrombin receptors.

2.2. Molecular Characterization of the Thrombin Receptor

2.2.1. Estimates of Thrombin Receptor Size Using Photoaffinity Labeling

Initial attempts to characterize the high-affinity receptor on fibroblasts employed the heterobifunctional photoactivatable cross-linking reagent *N*-(4-azido-2-nitrophenyl)-2-diaminoethane (Carney *et al.*, 1979). This cross-linker has a free amino group which couples to a carboxyl group on [125I]thrombin without loss of thrombin receptor binding or loss of proteolytic activity. This is an important consideration, since most cross-linkers bind to free amino groups which inactivate thrombin and inhibit high-affinity interaction with the receptor (see Carney, 1987). When derivative [125I]thrombin was incubated with mouse embryo cells under conditions identical to those used to demonstrate high-affinity binding and then photoactivated, approximately 60% of the specifically bound [125I]thrombin became cross-linked to a single molecule which migrated on reduced SDS gels with an apparent $M_r = 50,000$ (Carney *et al.*, 1979). Under parallel binding conditions, there is little if any PN–thrombin complex formation. Further, when PN–thrombin complexes are formed, the complex size indicates that the PN $M_r = 35,000$ (Baker *et al.*, 1979, 1980). Thus, it appears that this photoaffinity labeling may represent thrombin binding to its receptor, a subunit of the receptor, or a neighboring substrate, but not binding to PN (Carney *et al.*, 1979).

Later cross-linking studies using proteolytically inactivated thrombin photoaffinity probes (Moss *et al.*, 1983), or chemical cross-linking after incubation with PMS-thrombin (Van Obberghen-Schilling and Pouyssegur, 1985), demonstrated a major thrombin binding site with $M_r = 150,000$. In some cases, autoradiograms showed additional minor cross-linking to molecules of approximate $M_r = 400,000, 105,000, 50,000,$ and $38,000$ [after subtracting the size of thrombin itself (Moss *et al.*, 1983)]. The smallest complex, with $M_r = 38,000$, formed spontaneously, suggesting that this molecule was PN. The other molecules may represent additional binding sites, complexes of thrombin receptors, or portions of the intact receptor. The difference observed in major cross-linked species between studies with proteolytically inactive and active thrombin molecules (described above) may indicate: (1) that active thrombin and proteolytically inhibited thrombin primarily interact with different molecules; (2) that these forms of thrombin interact with the receptor in a slightly different manner so that cross-linking is to different subunits of the receptor; or (3) that active thrombin cleaves the receptor or an adjacent substrate and then cross-links to an $M_r = 50,000$ peptide fragment. The minor cross-linking of DIP-thrombin to molecules of $M_r = 105,000$ and $50,000$ may represent receptors cleaved by other proteases prior to thrombin interaction, the effects of a small number of active thrombin molecules in these preparations, or separate portions of a larger receptor complex.

2.2.2. Working Models for the Structure of Thrombin Receptors

The photoaffinity labeling studies described above identified two major thrombin receptors (or portions of the receptor) with apparent $M_r = 150,000$ and $50,000$. Several thrombin receptor models have been proposed to focus some of the data which have accumulated into a conceptual framework which might lead to experiments which will solve the structure of the receptor (see Fig. 2). The simplest model is that the receptor could be a single polypeptide chain of $M_r = 150,000$ (Fig. 2A,B). In this case, the cross-linking of receptor-bound thrombin to a $50,000$ M_r molecule may suggest that the receptor itself is cleaved by thrombin into fragments with

$M_r = 50,000$ and $100,000$ (see arrows in Fig. 2). Alternatively, the cross-linking of thrombin to an $M_r = 50,000$ species may suggest the presence of a separate $M_r = 50,000$ peptide that might be associated with the receptor (Fig. 2C). As suggested in Fig. 2C, if $M_r = 50,000$ subunits associate with either one or two $M_r = 150,000$ molecules, we might expect the receptor to exist as a higher-molecular-weight complex of $M_r = 200,000$, or perhaps as a heterodimer of approximate $M_r = 400,000$.

Another unknown feature of the receptor is whether the receptor traverses the membrane one time (Fig. 2A) similar to the EGF or platelet-derived growth factor receptors which activate tryosine kinases, or perhaps seven times (Fig. 2B) similar to a number of G-protein-linked receptors. Considerations of the types of mitogenic signals generated by thrombin may help predict these aspects of the receptor structure (see

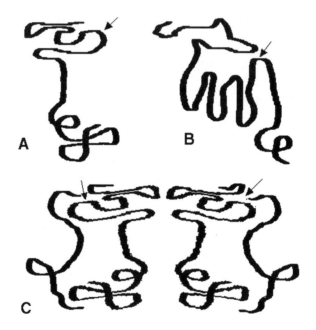

Figure 2. Proposed models for the subunit structure of the high-affinity thrombin receptor: A represents a single $M_r = 150,000$ subunit with potential cleavage site (arrows) which results in receptor fragments of $M_r = 100,000$ and $50,000$; B represents a single $M_r = 150,000$ receptor with seven transmembrane domains and a cleavage site which could also result in receptor fragments of $M_r = 1000,000$ and $50,000$; C represents a heterodimer structure with two dimers with subunits of $M_r = 150,000$ and $50,000$. The 150,000 subunit of this structure could also be cleaved to produce additional receptor fragments of approximately $M_r = 100,000$ and $50,000$.

Section 3), but these predictions will be highly speculative until the receptor is sequenced and compared to other growth factor receptors.

2.2.3. Attempts to Purify High-Affinity Thrombin Receptors

Attempts to purify the thrombin receptor have been hindered by the apparent inability of the thrombin to bind to this receptor once the receptor is solubilized in various detergents. Thus, simple thrombin affinity column separations such as those used to purify thrombomodulin (Esmon *et al.*, 1983) have been largely unsuccessful. Recent attempts to purify the receptor using a number of different approaches, however, have produced data which are consistent with initial affinity labeling data and the models proposed above.

Monoclonal antibody TR-9, which was shown to bind to thrombin receptors on fibroblasts and initiate mitogenic signals, appeared to bind to molecules with apparent $M_r = 150,000$ or $50,000$ in Western blots of SDS gels depending upon the preparation (Carney, 1987; Frost *et al.*, 1987). Furthermore, when this antibody was used as an affinity ligand to partially purify the receptor from an extract of octylglucoside-solubilized cells, the antibody pulled out a mixture of proteins which separated on reduced SDS gels into bands migrating with estimated $M_r = 150,000$, $105,000$, and $50,000$ (Frost *et al.*, 1987). Although were was no attempt in this study to quantitate the relative amounts of these different species, it was apparent from the gels that there was more $M_r = 50,000$ species than found in the other bands. This initial finding led to speculation that the $M_r = 50,000$ species might arise both from a cleavage of the $M_r = 150,000$ species and from a separate peptide subunit which might form a part of a higher-molecular-weight complex (Fig. 2C). Of special interest, in these studies it was possible to show that this antibody-affinity-purified material retained specific thrombin binding activity when adsorbed to plastic or nitrocellulose filters. Thus, these experiments demonstrated that this antibody did, in fact, recognize and partially purify the thrombin receptor.

High-performance liquid chromatography of CHAPS-solubilized membranes on TSK G4000 sizing columns indicated that the receptor in native form was found as a high-molecular-weight complex similar to that proposed in Fig. 2C. In these separations a major peak eluted with the excluded volume, followed by a number of smaller peaks. Analysis of these fractions using [^{125}I]thrombin binding to nitrocellulose membrane-adsorbed aliquots indicated the presence of thrombin binding proteins in fractions corresponding to $M_r > 400,000$ from mouse B11-C cells, but not in the same fractions obtained from FAZA cells which lack high-affinity thrombin binding (Frost, 1987; Frost and Carney, 1987). The same B11-C fractions were shown to bind monoclonal antibody TR-9, indicating that they indeed represented the thrombin receptor. These fractions had components of $M_r = 400,000$, 150,000, 105,000, and 55,000 on nonreduced SDS gels and components of $M_r = 155,000$, 105,000 and 50,000 on reduced gels (Frost, 1987). These molecular weight determinations are consistent with a complex heterodimer receptor structure (Fig. 2C), or with multiple binding sites. Attempts are currently under way to scale up the purification of the receptor to obtain sequence information and to more definitively define the nature of this receptor or receptors.

Recent attempts have been made to clone the functional thrombin receptor by injecting active mRNA fractions into frog oocytes and analyzing the oocytes for thrombin-stimulated ion transport (Van Obberghen-Schilling et al., 1990). In these studies, functional thrombin responsiveness could be generated in oocytes by injection of mRNA fractions corresponding to peptide chains with $M_r = 150,000$. The study did not, however, show increased binding of thrombin to the oocytes, or an increase in the synthesis of a molecule which might correspond to the receptor. Thus, it is as yet unclear whether the functional assay for response to thrombin measures insertion of new thrombin receptors or merely the coupling of signal-generating molecules to other molecules which might be cleaved or modified by thrombin.

2.2.4. Relationship between Binding Sites on Fibroblasts and Platelets

Since thrombin interacts with platelets causing platelet activation and aggregation, one might expect the thrombin binding sites on fibroblasts to be similar to those found on platelets. The interaction of thrombin with platelets has been recently reviewed (Berndt and Caen, 1984; Colman, 1990; and this volume). As in fibroblasts and other cells, at least part of the high-affinity thrombin binding to platelets is due to formation of covalent complexes with PN (Gronke et al., 1989; Knupp, 1989). This PN binding, however, can be separated from other high-affinity interactions by its dependence upon formation of aminoacyl complexes with active thrombin and its heparin sensitivity (see Section 1.3). Other high- and intermediate-affinity thrombin binding has been attributed to interaction with different platelet glycoproteins. Photoaffinity labeling studies indicate that α-thrombin complexes with molecules of $M_r = 400,000$, 200,000, 120,000, and 42,000 (Larsen and Simons, 1981). Interestingly, γ-thrombin cross-links to the molecule or molecules migrating at $M_r = 400,000$, but not to the other species (Jandrot et al., 1988). This has led to the speculation that the high-molecular-weight complex may be associated with moderate- or low-affinity thrombin interaction and enzymatic activation of phospholipase

C-generated signals, but that the high-affinity binding is primarily to GP-1b which has an approximate $M_r = 175,000$ to $200,000$. Monoclonal antibodies to GP-1b can inhibit thrombin-induced platelet aggregation (Yamamoto et al., 1985), and purified GP-1b or the glycocalicin portion of this molecule binds thrombin with both high and moderate affinities similar to that observed with intact platelets (Harmon and Jamieson, 1986). GP-1b contains two subunits, one of $M_r = 145,000$ and a smaller subunit of $M_r = 25,000$ (Berndt and Caen, 1984). Both subunits are glycosylated and are disulfide linked. In addition, at least the small 1b-β subunit may be linked to the cytoskeleton and may be phosphorylated by a cAMP-dependent kinase (Fox and Berndt, 1989; Fox, 1985). Although the sizes of these subunits do not correspond exactly to those reported above for thrombin receptors on fibroblasts, they may be close enough to suggest that the thrombin receptor on fibroblasts is related to these glycoproteins or has at least one subunit which may be similar. Indeed, a molecule immunologically related to GP-1b has been found on endothelial cells (Sprandio et al., 1988) and erythroleukemic cells (Adelman et al., 1985; Kieffer et al., 1986).

Because of the requirement for thrombin proteolytic activity in generating many of its activities on platelets, investigators have looked for a thrombin substrate which might be related to signal generation. GP-1b does not appear to be a substrate for thrombin, but it may be cleaved by chymotrypsin which may block signal generation (Tam et al., 1980). Thrombin does cleave GP-V ($M_r = 82,000$) to release a large fragment ($M_r = 69,500$) into the medium, but this cleavage does not appear to consistently correlate with platelet activation by thrombin (Detwiler and McGowan, 1985). Thus, as in fibroblasts there may be a number of different signals generated by the interaction of thrombin, making definitive identification of required proteolytic interactions difficult. Thrombin interaction with its receptor also appears to stimulate the activation of calpain, a cysteine protease, which cleaves aggregin, a glycoprotein of $M_r = 100,000$ (Puri et al., 1989). This cleavage also correlates with platelet activation by thrombin (Puri et al., 1989). Thus, the process of identifying a substrate molecule that is involved in platelet activation may be further complicated by action or activation of other proteases.

These comparisons suggest that the primary binding sites for thrombin on platelets may be different from those identified on fibroblasts and other cells. The possible parallel between GP-1b and the fibroblastic receptor, however, suggests the need for additional experiments to examine this question in more detail. Clearly as the thrombin receptor on fibroblasts is further purified and sequenced, it will be important to compare it to GP-1b and other platelet proteins.

3. GENERATION OF MITOGENIC SIGNALS BY RECEPTOR OCCUPANCY AND PROTEOLYTIC ACTIVITY

3.1. Thrombin Effects on Transmembrane Signals in Fibroblasts

3.1.1. Ion Transport and Cytoplasmic Alkalinization

Early studies showed that thrombin-stimulated amiloride-sensitive Na^+/H^+ antiport was a necessary event in thrombin-stimulated mitogenesis (Paris and Pouyssegur, 1983; Pouyssegur et al., 1982; Stiernberg et al., 1983). This stimulation appeared to be associated with phosphorylation of the S6 ribosomal protein (Pouyssegur et al., 1982), and with increasing intracellular pH (L'Allemain et al., 1984). In addition to the early increase, studies with amiloride added at times up to 8 h after thrombin suggested the need for a later event which required either continued activation of the Na^+/H^+ antiport, or a second activation (Stiernberg et al., 1983). Other types of transport including the $Na^+/H^+/Cl^-$ exchanger are also activated by thrombin, but their role in stimulation of cell proliferation is uncertain since blocking this transporter only inhibits thrombin mitogenesis by 15 to 25% (Paris and Pouyssegur, 1986a).

When various thrombin derivatives were tested to determine if receptor binding alone could stimulate amiloride-sensitive Na^+/H^+ antiport, it was discovered that DIP-inactivated thrombin was unable to stimulate antiport, whereas γ-thrombin stimulated transport almost as effectively as fully active α-thrombin (Stiernberg et al., 1984). These results indicated that stimulation of transport is associated with proteolytic activity of thrombin and not high-affinity receptor occupancy. Moreover, since γ-thrombin by itself is unable to stimulate cell proliferation, these studies also demonstrate that changes in transport, although they may be necessary, are not by themselves sufficient to stimulate mitogenesis without additional high-affinity occupancy-generated signals.

3.1.2. Thrombin Effects on Phosphoinositide Turnover, Ca^{2+} Mobilization, and Activation of Protein Kinase C

Because activation of Na^+/H^+ antiport appeared to be a required part of the thrombin mitogenic signal, it was important to determine how thrombin initiated this signal. Activators of protein kinase C were shown to stimulate Na^+/H^+ antiport, suggesting the involvement of phosphoinositide turnover in this process (Besterman and Cuatrecasas, 1984). Therefore, our laboratory and a number of others began to examine the effects of thrombin on this signal pathway (for review see Bar-Shavit and Wilner, 1986; Carney, 1987; Carney et al., 1986b; Pouyssegur et al., 1988).

Thrombin addition to NIL hamster fibroblasts caused an almost eightfold stimulation in phosphorylation of phosphoinositide bisphosphate (PIP_2), a nearly equal stimulation in release of inositol trisphosphate (IP_3), and mobilization of intracellular Ca^{2+} (Carney et al., 1985). Other cells show similar responses to thrombin (Paris and Pouyssegur, 1986b; Raben et al., 1987a,b). Neomycin, which selectively binds to PIP_2 to prevent its cleavage by phospholipase C, inhibited thrombin-stimulated cell proliferation (Carney et al., 1985). In addition, PI synthesis was not stimulated by proteolytic enzymes which were not mitogenic and nonresponsive cells showed little if any thrombin stimulation of PI synthesis (Raben et al., 1987b). Recent characterization of mutant lines of CCL39 cells selected for failure to respond to thrombin has further demonstrated that the failure of these cells to respond was linked to altered phospholipase C molecules which uncouple thrombin binding from its ability to activate PI turnover (Rath et al., 1989, 1990). Together, these studies indicate a direct involvement of PI turnover in stimulation of cell proliferation by thrombin.

Based on the finding that combinations of γ-thrombin and DIP-thrombin stimulated cell proliferation to levels approaching the level of stimulation by α-thrombin, we proposed, a "double signal hypothesis" which suggested that two different types of signals were generated by receptor occupancy and by the proteolytic activity of thrombin toward either the receptor or some adjacent substrate (Carney et al., 1986b). If stimulation of PI turnover is linked to activation of the Na^+/H^+ antiporter, one might predict that this activation would also be associated with proteolytic activity of thrombin. Indeed, several studies have now shown that proteolytically inactivated thrombin does not stimulate an early increase in PI turnover, but γ-thrombin, which retains esterase activity and the ability to cleave a number of substrate molecules, appears to stimulate phospholipase C and PI turnover to levels approaching that achieved with active α-thrombin (Carney, 1987). If proteolytic cleavage stimulates this signal pathway, two important questions are: What role is played by the thrombin receptor? Is the receptor the target of the proteolytic activity?

Other observations suggest that proteolytic cleavage of the receptor is involved in activation of phospholipase C. Thrombin increases arachidonate production in CCL39 cells as a result of phospholipase C or phospholipase A activation (Raben et al., 1987a). This activation can be elicited by γ-thrombin addition to these cells, but not by DIP-thrombin. Interestingly, the addition of chymotrypsin prior to thrombin blocks this stimulation (Raben et al., 1987a). It is not clear, however, whether this inhibition is due to chymotrypsin cleaving the receptor in such a way that it can no longer activate proper signals, or whether interaction of chymotrypsin

with cells generates another type of signal which prevents thrombin receptor-generated signals. Pretreatment of mouse and hamster cells with
chymotrypsin or γ-thrombin decreased the specific binding of [^{125}I]throm-
bin, suggesting that these proteases may directly modify the thrombin
receptor (Carney et al., 1986a). Further, in chick heart muscle cells and
astrocytoma cells, trypsin and thrombin both appear to act at the same site
to stimulate PI turnover (Jones et al., 1989). Addition of a 100-fold excess
of inactive PPACK-thrombin to these cells completely inhibited thrombin
stimulation of IP$_3$ release, and inhibited up to 80% of the stimulation
caused by addition of trypsin. Thus, binding of inactivated thrombin to the
receptor inhibits stimulation of PI turnover. These studies suggest that the
interaction of thrombin and perhaps trypsin (in chick cells) with thrombin
receptors is necessary for production of mitogenic signals related to activation of phospholipase C and that perhaps the receptor itself is cleaved as
a part of this action.

3.1.3. Desensitization of Phospholipase C-Coupled Responses: Are G-Proteins Involved?

Increased levels of PI turnover can be seen as early as 5 s after
α-thrombin addition to CCL39 cells, but the stimulated level then appears
to decrease to a lower steady-state level that persists throughout early
G$_0$ (L'Allemain et al., 1986). This suggests that continued occupancy of
thrombin receptors downregulates this signal. In so doing, thrombin may
activate other signals required to allow cells to proceed into a proliferative
cycle (Van Obberghen-Schilling et al., 1985). If PPACK is added to cultures
to inactivate thrombin at the time of thrombin addition or subsequently,
the stimulation of PI turnover and desensitization is attenuated (Van Obberghen-Schilling et al., 1985). Desensitization occurs with preincubation
of 4°C and without protein kinase C activation, but addition of hirudin
(which binds thrombin and prevents its rebinding to receptors) reverses
the desensitization with a half-life of approximately 1.5 h (L'Allemain et al.,
1986). These observations argue that desensitization may require both
receptor or substrate cleavage and an event coupled to receptor occupancy
which alters the association of thrombin receptors inside the cells with
G-proteins or other signal-generating molecules.

A number of studies have suggested that G-proteins are involved in
thrombin stimulation of phospholipase C (Magnaldo et al., 1987; Murayama and Ui, 1987; Paris and Pouyssegur, 1986b; Rebecchi and Rosen,
1987). Pertusus toxin inhibits thrombin activation of phospholipase C and
stimulation of cell proliferation, but has no effect on mitogenesis stimulated by other factors such as EGF which appear to function through

activation of tyrosine kinases (Chambard *et al.*, 1987; Paris and Pouyssegur, 1986b). G-protein-linked activation of phospholipase C also appears to be linked to activation of the Na^+/H^+ antiporter (Paris *et al.*, 1987).

If the thrombin receptor is similar to other G-protein-linked receptors, several predictions and speculations about receptor structure and signal modulation can be made. First, we might predict that the receptor or one of the active subunits of the receptor will have multiple membrane-spanning regions similar to other G-protein-linked receptors (see Fig. 2B). Second, the receptor is likely to exist in two states depending on the conformation of the receptor and its association with various G-proteins. Proteolytic activation of the thrombin receptor may be required to facilitate a conformational change which allows it to interact with G-proteins. The continued receptor occupancy by thrombin, antibodies, or peptides associated with mitogenesis may then be needed to prevent reversion of the receptor conformation. Once thrombin is removed, the receptor may revert to its previous state in which new G-proteins could be attracted by subsequent thrombin addition. This would coincide with a return of thrombin-sensitive activation of phospholipase C. If proteolysis of the receptor occurred during the initial occupation, however, reinitiation of the response may occur with subsequent addition of inactivated thrombin. This may explain why addition of DIP-thrombin can stimulate cell proliferation when added to cells up to 6 h after addition of γ-thrombin (Carney *et al.*, 1984).

3.1.4. Oncogenes and Other Signals Initiated by Thrombin

There is a great deal of interest in the possible role of cellular homologs to oncogenes in regulating proliferation of normal cells. α-Thrombin and insulin addition to CCL39 cells increased the amount of both c-myc and c-fos within 10 to 30 min (Blanchard *et al.*, 1985). These experiments further indicated that at least in the case of c-myc a large part of this increase appeared to be due to increased messenger RNA stability rather than stimulation of transcription. Continued occupancy of thrombin receptors with active thrombin was required to stimulate increased levels of c-myc, suggesting that this effect may also be related to an enzymatic event (Van Obberghen-Schilling *et al.*, 1985). PPACK addition to cultures at the same time or before thrombin addition, blocked thrombin-induced increase in c-myc. In addition, if PPACK was added 30 min after thrombin, it still blocked approximately 50% compared with continued treatment with α-thrombin. This suggested that full stimulation required prolonged treatment. Pertussis toxin inhibits c-myc accumulation, suggesting G-protein involvement (Paris *et al.*, 1987). Thus, we might expect that c-myc stimulation is coupled to other events stimulated by G-protein activation by thrombin. The relationship between these signals, however, appears to be more complex since c-myc levels increase even at nonpermissive pH_i where Na^+/H^+ antiport and phosphorylation of S6 protein appear to be blocked (Pouyssegur et al., 1985). Recent studies with rat aortic smooth muscle cells suggest that thrombin activates topoisomerase I through both a pertussis toxin-sensitive and -insensitive G-protein (Nambi et al., 1989). This may indicate that there are multiple receptors for thrombin on these cells or that more than one type of G-protein may associate with thrombin receptors.

3.2. Mitogenic Signals Generated by Thrombin or DIP-Thrombin Occupancy of Thrombin Receptors

The "double signal hypothesis" predicted that two different types of signals are required for thrombin to stimulate mitogenesis, one initiated by receptor occupancy, and the other by enzymatic interaction of thrombin with either the receptor or an adjacent substrate molecule. Most of the common mitogenic signals described above appear to require thrombin proteolytic activity, since most of these signals are not initiated by DIP- or PPACK-inhibited thrombin. This leads us again to question the role of receptor occupancy in the initiation sequence. Does receptor occupancy alone generate signals? Does it merely alter the conformation of the receptor so that thrombin can properly align with a cleavage site to initiate other signal events?

One signal that appears to be stimulated by receptor occupancy is a transient increase in levels of cyclic AMP (Gordon et al., 1986). Although early studies suggested that cAMP increases were inhibitory to cells, more recent experiments have indicated that increases may modulate or signal other mitogenic events (Rozengurt et al., 1983). When thrombin was added to quiescent NIL hamster fibroblasts, the level of intracellular cAMP increased 40 to 60% within 15 min (Gordon et al., 1986). The same or slightly greater stimulation was achieved with DIP-inactivated thrombin at concentrations of approximately 2 nM, whereas γ-thrombin which stimulated PI turnover in these cells had no apparent effect on activation of adenylate cyclase. This suggested that increased activity of adenylate cyclase was stimulated by receptor occupancy in a more classical sense for hormone receptor interaction without the requirement for proteolytic activity.

Attempts to bypass the requirement for receptor occupancy using cAMP analogs or inhibitors of phosphodiesterase resulted in inhibition rather than stimulation (Gordon *et al.*, 1986). More recent studies have shown that cAMP can inhibit PI turnover and stimulation of cell proliferation by thrombin and a number of growth factors in fibroblasts (Magnaldo *et al.*, 1989), and thrombin stimulation of oncogene expression in endothelial cells (Daniel *et al.*, 1987). This may suggest that there is cross talk between the signals generated by receptor occupancy and proteolytic activity either in the form of selective association of G-proteins, or perhaps in the phosphorylation of the receptor tail or receptor-interacting molecules by a cAMP-dependent kinase. Other examples of this type of cross talk and cross talk between tyrosine kinase receptors and G-protein-linked activities have been discussed in relation to these signals (Pouyssegur *et al.*, 1988).

4. THROMBIN INTERACTION WITH OTHER CELLS: ARE THE SAME THROMBIN DOMAINS AND RECEPTORS UNIVERSALLY INVOLVED IN CELL ACTIVATION?

Since it appears that thrombin plays a role in stimulation of cell proliferation and other cellular events which may contribute to tissue repair and other physiological processes, it is important to determine how many types of cells respond to thrombin, if these cells have thrombin receptors similar to those found on fibroblasts, and if the same thrombin domains and interactions are involved in generating these other cellular effects. In the confines of this review it is not possible to begin to cover all of the different cell types which are now being shown to respond to thrombin. Therefore, we will focus on cell types which may play a direct role in postclotting tissue damage and repair. If we can determine how thrombin exerts its effects on these cells, then we may be able to modulate some of these responses to overcome impairments to normal function, accelerate the time frame for tissue repair, or eliminate disease states where abnormal proliferation or cellular changes might lead to major health risks.

4.1. Effect of Thrombin on Neuronal Cells

A number of recent studies have shown that neuronal cells respond to thrombin either by generation of metabolic signals, mitogenic stimulation, or a dedifferentiation and retraction of neuritic processes. This area of research is becoming extremely important with potential involvement in cerebral hemorrhage, trauma to the central nervous system, and modulation of neuronal function and dysfunction.

4.1.1. Thrombin Stimulation of Cyclic GMP Formation

While studying the mechanism of receptor-mediated cyclic GMP formation in neuroblastoma cells, Snider and Richelson discovered that thrombin stimulated formation of cGMP (Snider and Richelson, 1983). This stimulation appeared to require thrombin proteolytic activity, yet other proteases including trypsin and γ-thrombin were approximately 50-fold less effective than α-thrombin (Snider *et al.*, 1984a). Interestingly, the response was half-maximal at approximately 1.5 nM which corresponds to the apparent K_d for thrombin interaction with its receptors on fibroblasts (see Section 2). This may suggest that these cells have thrombin receptors similar to those found on fibroblasts. Indeed, studies with [^{125}I]thrombin and immunofluorescence have shown specific thrombin binding to neuroblastoma cells (McKinney *et al.*, 1983a), homogenates from brain and spinal cord (McKinney *et al.*, 1983b), and neuronal cells in primary rat brain cultures (Means and Anderson, 1986). In addition, some of the signals generated by thrombin interaction with neuroblastoma cells are similar to those described above for fibroblasts (Snider *et al.*, 1984b), and thrombin appears to increase cGMP in fibroblasts (Bruhn and Pohl, 1981). Thus, it is possible that the effects in neuroblastoma cells are mediated by thrombin interaction with receptors that are closely related to those found on fibroblasts.

4.1.2. Effects of Thrombin and PN on Neurite Outgrowth

Neuronal cells are distinct from most types of cells in that in serum-free medium these cells stop proliferating and extend neuritic processes (Seeds *et al.*, 1970). This differentiation-related process can also be initiated in these cells by increasing cAMP (Seeds *et al.*, 1970), or addition of glial-conditioned medium containing glial-derived neurite-promoting factor (Monard *et al.*, 1983). One of the first indications that thrombin might be involved in regulating neuronal cell morphology and proliferative responses was the observation that hirudin stimulated neurite outgrowth in neuroblastoma cultures in the presence of serum (Monard *et al.*, 1983). This suggested that thrombin or similar proteases might be inhibiting neurite outgrowth. Additional experiments showed that, indeed, thrombin and urokinase were able to overcome the neurite-promoting effects of hirudin or glial-derived neurite-promoting factor (Cunningham and Gurwitz, 1989; Guenther *et al.*, 1985). A breakthrough in understanding of this process came with the sequencing of the glial-derived neurite-promoting factor and the discovery that this factor had 85% homology with the family of serpine protease inhibitors and appeared similar to PN (Sommer *et al.*,

1987). Sequencing of PN-1 demonstrated that the glial-derived neurite-promoting factor and PN were identical molecules (McGrogan *et al.*, 1988). These studies thus showed that if thrombin in serum was inhibited, neurite extension occurred. By inference, one could then predict that thrombin interaction with the neuronal cells was in some manner initiating proliferative signals which kept the cells from morphologically differentiating and forming neurites.

Experiments with direct addition of thrombin to neuroblastoma cells in serum-free medium confirmed the role of thrombin in this process. If thrombin was added to cultures of neuroblastoma cells at the time serum was removed, neurite outgrowth was inhibited (Gurwitz and Cunningham, 1988; Hawkins and Seeds, 1986). In these experiments, removal of this medium and replacement with fresh serum-free medium resulted in induction of neurite outgrowth, demonstrating that the cells retained their ability to form neurites. Interestingly, this effect of thrombin was maximal at thrombin concentrations of approximately 3 nM. DIP-inactivated thrombin was 500 to 1000 times less active, suggesting that this effect, like others stimulated by thrombin, required thrombin proteolytic activity. Other proteases (including urokinase, plasmin, and trypsin), however, were ineffective in blocking neurite outgrowth, indicating that this was not a general proteolytic effect, but rather a specific cellular interaction of thrombin (Gurwitz and Cunningham, 1988). It is not clear, however, if these effects are mediated through the thrombin receptor or specific cleavages of matrix components or their attachment sites (Hawkins and Seeds, 1986).

Another factor which has been studied in relation to neurite outgrowth has been named differentiation reversal factor (Grand *et al.*, 1989). Addition of this factor (DRF) to adenovirus-transformed retinoblasts which were cultured in serum-free medium containing cAMP, reverses differentiation, causing retraction of neurites and a mitogenic response in these cells. This factor has recently been shown to be identical to prothrombin (Grand *et al.*, 1989). Moreover, addition of thrombin (0.5 ng/ml) or prothrombin (20 ng/ml) to retinoblasts, caused half-maximal effects identical to those elicited by DRF. Similar to the effects on neuroblastoma cells described above, if thrombin proteolytic activity was inhibited, neither thrombin nor prothrombin was active (Grand *et al.*, 1989). This indicates that the proteolytic activity of thrombin or DRF is responsible for this activity, but the low concentration required for activity (0.5 ng/ml) suggests interaction of thrombin with sites at a K_d of 15 to 100 pM which would indicate interaction with specific high-affinity receptors. The activity of prothrombin may be related to the activation of prothrombin to thrombin on the surface of these cells.

Together, these results indicate that thrombin and thrombin-specific inhibitors play an important role in regulating the differentiation and proliferation of neuronal cells. These effects might be expected to play a critical role in early differentiation as the blood–brain barrier is established. This process, however, may also play a role in certain regenerative (and/or degenerative) processes following cerebrovascular or peripheral neuronal injury. Although mRNA for both PN (glial-derived neurite-promoting factor) and prothrombin have been found in brain (Cunningham and Gurwitz, 1989; Sommer *et al.*, 1987), additional studies need to be carried out to determine if changes in the amounts of these messengers correlate with neuronal development or the onset of disease states. The role of vascular development in these processes must also be considered in more detail since vascular patency in developing brain and adult tissues clearly plays an important role in determining the extent of thrombin interaction with neuronal cells.

4.2. Effect of Thrombin on Endothelial Cells

The effects of thrombin on vascular permeability in lung and brain, release of vasoactive molecules, presentation of thrombogenic and non-thrombogenic surfaces, and angiogenesis have become major areas of interest in recent years (for review see Garcia *et al.*, this volume; Malik, 1986; Shuman, 1986). For the purpose of this review, I will focus on the possible role of thrombin receptors in these effects, especially in relation to angiogenesis, and potential for modulation of endothelial cell functions.

4.2.1. What Endothelial Cells? What Function?

One problem associated with studying vascular endothelium is defining the types of cells one wishes to study. For example, a number of early studies looking at both the binding of [^{125}I]thrombin to endothelial cells (Lollar *et al.*, 1981) and the mitogenic activity of thrombin (Gospodarowicz *et al.*, 1978; Knauer and Cunningham, 1983; Zetter and Antoniades, 1979) utilized either bovine aortic endothelial cells or cells obtained from human umbilical vein. For the vascular system to function, one would predict that endothelial cells lining the lumens of large vessels would have different properties and responses than endothelial cells which form the capillaries and microvasculature that delivers nutrients, gases, fluid, and inflammatory cells into extravascular spaces in skin and different organs. Indeed, a number of studies have now shown that microvascular endothelial cells have different functions, different cell surface molecules, and different responses to thrombin than endothelial cells from large vessels (Zetter, 1988). For example, thrombin stimulates prostacyclin production in umbilical vein endothelial cells, but not in aortic endothelial cells (Goldsmith and Kisker, 1982; Hong, 1980), or microvascular endothelial cells (Charo et al., 1984). Thus, in small vessel endothelial cells and aortic endothelial cells where platelet activation may be required, there is no prostacyclin stimulation which would block platelet responses. To further complicate this issue, it appears that capillary endothelial cells from different organs have unique cell surface properties which relate to their specific functions and to the ability of certain cancer cells to metastasize to specific vascular beds (Belloni and Tressler, 1990). Thus, if one wants to determine whether thrombin plays a role in stimulating proliferation of endothelial cells or other responses of these cells that may relate to postclotting activities at the site of tissue injury, it is essential to look at the effects of thrombin on microvascular endothelial cells from specific organs, rather than those from large vessels.

4.2.2. Is Thrombin an Angiogenic Factor?

The role of thrombin in stimulating proliferation of endothelial cells has been somewhat controversial, in part due to differences in the types of endothelial cells used. Initial studies with bovine and human umbilical vein endothelial cells indicated that thrombin could enhance the stimulation of proliferation by EGF and PDGF in human umbilical vein endothelial cells, but not in bovine aortic cells (Gospodarowicz et al., 1978; Zetter and Antoniades, 1979). In ovine aorta cells, thrombin stimulated PI turnover and appeared to make cells competent to respond to insulin, but was not by itself mitogenic (Moscat et al., 1987). Other studies indicated that thrombin was not acting as a mitogenic factor or enhancing the effect of EGF in human umbilical vein endothelial cells (Knauer and Cunningham, 1983). In light of recent advances in culturing microvascular endothelial cells and the realization that these cells often behave differently than those from large vessels, we have reexamined this question using mouse and human capillary endothelial cells derived from different organs (Belloni et al., 1992). Thrombin stimulates rapid morphological change and increased proliferation of lung- and brain-derived microvascular endothelial cells, but has little effect on mouse endothelial cells derived from liver or on bovine aortic endothelial cells. Microvascular brain endothelial cells appeared to respond mitogenically to thrombin only when cultured on subendothelial cell matrix or in the presence of ECGF, suggesting that the brain-derived cells may require additional signals for responsiveness.

Binding of [^{125}I]thrombin to various endothelial cells demonstrated a correlation between mitogenic responsiveness to thrombin and the presence of high-affinity thrombin receptors. Microvascular endothelial cells from lung had the highest affinity (K_d = 0.9 nM) followed by those from liver and brain (Belloni *et al.*, 1992). BAE cells which did not respond to thrombin appeared to have only lower-affinity interactions with thrombin (K_d = 98.5 nM). To explore the possibility that these cells have the same receptor as that found on fibroblasts, cultures of these cells were fixed and incubated with the fibroblast thrombin receptor monoclonal antibody TR-9 (Frost *et al.*, 1987). Both lung- and brain-derived endothelial cells showed a dotlike immunofluorescent pattern nearly identical to that seen on fibroblasts (Belloni and Carney, unpublished). Moreover, if TR-9 was incubated with mouse organ-derived endothelial cells, up to 90% of the specific high-affinity binding of [^{125}I]thrombin was inhibited (Belloni *et al.*, 1992). Although additional studies are required to fully characterize these receptors and the response of microvascular endothelial cells to thrombin, current studies suggest that other antibodies made against the fibroblast thrombin receptor also bind to receptors on both the mouse microvascular endothelial cells and capillary endothelial cells viewed in histological sections of rat skin (McCroskey and Carney, unpublished). Thus, together these studies indicate that microvascular endothelial cells are mitogenically stimulated by thrombin and that this stimulation may occur through thrombin interaction with receptors similar to those on fibroblasts.

Corneal endothelial cells and human umbilical vein endothelial cells appear to have up to three different types of thrombin binding: a low-affinity binding which is rapid and reversible, a covalent binding of thrombin to a molecule of M_r = 30,000 to 35,000, a high-affinity binding which can be effected by DIP-thrombin (Awbrey *et al.*, 1979; Bauer *et al.*, 1983; Isaacs *et al.*, 1981; Lollar *et al.*, 1981; Savion *et al.*, 1981). The role of the lower-affinity binding is unclear; however, estimates of over 1 million binding sites have been made with a K_d of approximately 30 nM. Many of the effects of thrombin on umbilical vein endothelial cells including production of prostacyclin (Hong, 1980; Jaffe *et al.*, 1987; Weksler *et al.*, 1978), synthesis of tissue plasminogen activator (Levin *et al.*, 1986), and phosphorylation of vimentin (Bormann *et al.*, 1989), all require proteolytically active thrombin and in some cases this stimulation can be induced at least partially by lower-affinity interaction of trypsin with these cells. This may indicate that in these cells the receptor is already modified to allow lower-affinity interaction, or a prevalence of an active substrate subunit of the receptor complex. The higher-affinity covalent attachment may represent thrombin–PN complex formation (see Section 1.3), or complex formation with other plasminogen activator inhibitors. This covalent complex formation is not seen with DIP-inactivated thrombin, and can be formed in conditioned medium into which this binding protein has been secreted by the endothelial cells (Isaacs *et al.*, 1981). The presence of high-affinity binding sites on these cells to which DIP-inactivated thrombin will bind has been demonstrated, yet it is not clear if these receptors are functional in these cells and how they relate to the mitogenic receptors found in fibroblasts (Lollar and Owen, 1980). As better probes are made to study the mitogenically active thrombin receptor from fibroblasts, it will be much easier to determine if and how this receptor might be involved in proliferative responses of these cells.

The demonstration that thrombin stimulates proliferation of microvascular endothelial cells, indicated that thrombin might function as an angiogenesis factor in tissues following vascular injury. Indeed, in the chick chorioallantoic membrane angiogenesis assay, thrombin diffusing out of agar disks stimulates a reorientation of blood vessels toward the disk, and an increase in both the size and number of visible blood vessels and capillaries (Lehman and Carney, manuscript in preparation). In addition, in animal wound healing models, thrombin and synthetic thrombin receptor-activating peptides stimulate neovascularization in and beneath the site of tissue injury (see Section 5). Thus, thrombin appears to function as a potent angiogenic factor in stimulating proliferation of microvascular endothelial cells in culture and in experimental neovascularization of animal tissues.

The mechanism by which thrombin stimulates proliferation of endothelial cells remains to be determined. Several studies have shown that thrombin stimulates the synthesis and release of molecules which are similar to v-sis or PDGF (Daniel *et al.*, 1986; Harlan *et al.*, 1986), and that this stimulation can be inhibited by agents that increase levels of cAMP in these cells (Daniel *et al.*, 1987). Interestingly, in studies carried out with human umbilical vein endothelial cells, production of PDGF-like molecules appeared to require the proteolytic activity of thrombin (Harlan *et al.*, 1986), whereas in microvessel endothelial cells derived from renal cortex the response was triggered by DIP-inactivated thrombin (Daniel *et al.*, 1986). This may simply reflect the increased sensitivity of the microvascular cells due to the presence of thrombin mitogenic receptors, or a more fundamental deviation in the responsiveness of these cells from the requirement for both receptor binding and proteolytic cleavage of the receptor (Carney *et al.*, 1986b). We have recently shown that synthetic peptides representing the high-affinity binding domain of thrombin are angiogenic and effective in stimulating tissue repair (see Section 5). Since these peptides also do not have proteolytic activity, it will be important to establish how receptor occupancy-generated signals differ in these cells from the signals generated in fibroblasts.

4.2.3. Thrombin–Endothelial Effects Related to Inflammation

One of the early stages of wound healing and tissue repair is an inflammatory phase, in which leukocytes [first polymorphonuclear leukocytes (PMNs), then monocytes] invade the injured tissue. A number of studies have shown that thrombin alters the nonthrombogenic properties of endothelial cells and stimulates the adherence of platelets and PMNs (Bizios *et al.*, 1988; Prescott *et al.*, 1984; Zimmerman *et al.*, 1985, 1986). In human umbilical vein endothelial cells, this effect appears to be related to the production of platelet-activating factor (PAF) by the endothelial cells and not a direct effect on the PMNs (Prescott *et al.*, 1984). Other agents such as leukotrienes also stimulate production of PAF and neutrophil adhesion, but if one compares the time course of the effect elicited by thrombin with that of leukotrienes, it appears that the endothelial cells respond most rapidly (within the first few minutes) to thrombin (Prescott *et al.*, 1987). Of special interest for this review, this thrombin effect is half-maximal at thrombin concentrations of approximately 2 nM (Zimmerman *et al.*, 1985). Thus, this effect may also be mediated by thrombin interaction with high-affinity receptors on these cells. Studies with other proteases indicate that the effect is quite specific for thrombin, but the effect is lost if thrombin is inactivated with PMSF (Zimmerman *et al.*, 1985). Thus, the effect may be similar to that observed with mitogenic stimulation of fibroblasts where both interaction with specific thrombin receptors and proteolytic activity are required (Carney *et al.*, 1986b).

In endothelial cells derived from pulmonary arteries, thrombin also induces a similar increased adherence of PMNs, but this effect appears to be mediated over a longer time frame by production or release of adhesion molecules which interact with specific sites on the neutrophils rather than through synthesis of PAF (Bizios *et al.*, 1988); this type of adhesion may be mediated by molecules such as leukocyte adhesion molecule 1 (LAM-1) (Bevilacqua *et al.*, 1989). Monocytes and T cells also appear to have increased adherence to endothelial cells following thrombin treatment (Saegusa *et al.*, 1988), suggesting that thrombin may also play a role in accumulation of mononuclear cells into damaged tissue by increasing adherence and perhaps extravasation of these cells through the endothelium. These effects appear longer-lasting than those stimulated by PAF, peaking from 4 to 8 h after thrombin addition. In addition, this increase appears to require protein synthesis. Although the mechanism for these responses are not fully characterized, it appears that they play an important role in stimulating the accumulation of PMNs and mononuclear cells at the site of tissue injury where these cells may release additional factors to stimulate a cascade of inflammatory and wound healing responses. Inappropriate stimulation of these responses could lead to abnormal pathologies including cell damage, loss of vascular integrity, or atherogenesis.

4.3. Thrombin Effects on Cells of the Inflammatory Response

Although it is clear that thrombin can affect local inflammatory responses by increasing the adherence of PMNs and monocytes to microvessel endothelial cells, there is also evidence that thrombin can exert direct chemotactic and proliferative effects on these cells. Details of some of these effects have been reviewed elsewhere (Bar-Shavit and Wilner, 1986), and in this volume. However, we will consider some of these effects as they relate to possible interaction with thrombin receptors.

4.3.1. Thrombin Interaction with Neutrophils and Lymphocytes

One of the earliest responses to tissue injury following clotting is an inflammatory response characterized initially by an increase in the number of neutrophils. Some of this neutrophil sequestering might be explained by selective neutrophil adhesion to capillaries, but the response begins very shortly following injury and the neutrophils can be seen migrating through the capillary walls into the edematous wound fluid. This suggests that there is also an active migration of the cells toward the wound. Thrombin can act as a chemoattractant to sheep or human neutrophils stimulating both their directed movement and aggregation (Bizios et al., 1986). Interestingly, this effect of thrombin is observed with α-thrombin, γ-thrombin, and proteolytically inhibited thrombin (Bizios et al., 1986). Thus, it appears that this interaction may be different from the interaction of thrombin with fibroblasts, neuronal cells, and endothelial cells. Similar studies with lymphocytes also show aggregation effects that are independent of the proteolytic activity of thrombin (Bizios et al., 1985). This nonenzymatic activation of inflammatory cells may relate to the breakdown of thrombin to peptide fragments in the wound or a mechanism for activation of these cells while thrombin is still associated with fibrin in clots. Although 10 nM thrombin was required for chemotactic response through agarose, the amount of thrombin reaching and interacting with cells would initially be much lower. This would suggest that neutrophils may have specific high-affinity interactions with thrombin molecules. Indeed, studies with [^{125}I]thrombin binding to human neutrophils at 4°C indicate two different classes of thrombin receptors, one with a K_d of about 20 pM (40–100 sites per cell), and the second with a K_d of about 30 nM (6000 sites per cell) (Sonne, 1988). These results suggest that there might be a limited number of very high-affinity binding sites on these cells which could initiate chemotactic signals followed by activation or aggregation effects when the cells reached a location where the thrombin concentration was high enough to activate the lower-affinity receptors. It is too early to predict if these receptors are similar to those found on fibroblasts or if they are unique to inflammatory cells.

4.3.2. Thrombin Interaction with Monocytes

The effect of thrombin on monocytes has been reviewed (Bar-Shavit and Wilner, 1986; Bar-Shavit et al., this volume). Thrombin appears to be both a chemoattractant and a mitogen for monocytic cells (Bar-Shavit et al., 1983, 1986a). Thrombin interaction with human peripheral monocytes was shown to have two affinities, a high-affinity interaction ($K_d = 3.4$ nM) with approximately 30,000 sites and a low-affinity high-capacity binding estimated at 130 nM (Goodnough and Saito, 1982). Most other studies with monocytic cells have been carried out using a murine macrophage-like cell line, J774. In these cells thrombin binds to approximately 15,000 sites per cell with an affinity of approximately 7.5 nM (Bar-Shavit et al., 1983). Similar to thrombin binding to fibroblasts, DIP-inactivated thrombin binds to these cells, but unlike the effect of this binding on fibroblasts, DIP-thrombin appears to fully activate signals for chemotaxis and growth of these cells. This suggested that in these cells, as in neutrophils, proteolytic cleavage of the receptor or adjacent molecules is not required to generate signals necessary to complete the physiological response.

The interaction of thrombin with J774 cells appears to involve a different portion of the thrombin molecule than interaction of thrombin with fibroblasts. Studies with thrombin peptides have demonstrated that peptides near the site of thrombin carbohydrate attachment can compete with thrombin for binding to receptors on J774 cells and stimulate mitogenic and chemotactic signals (Bar-Shavit et al., 1986b). Interestingly, the fibroblast thrombin receptor binding domain peptide p508–530 (178–200) failed to stimulate these monocytic cells and the monocyte-activating peptide did not compete for thrombin binding to fibroblasts (Glenn et al., 1988). Thus, it appears that these cells have two different types of receptors which interact with different domains of the thrombin molecule.

5. TOWARD A STRATEGY FOR MODULATING WOUND HEALING *IN VIVO*

Initial studies demonstrating that thrombin, but not prothrombin, initiated cell proliferation suggested that this response might be important in postclotting effects of thrombin (Chen and Buchanan, 1975). Subsequently, we speculated that if thrombin were physiologically important in wound healing and other postclotting events, then there should be a receptor for thrombin on the surface of target cells which would be involved in these responses (Carney and Cunningham, 1978b). Indeed, as described above, over the last 12 years the number of cells which respond to thrombin and the number of responses which may relate to effects of signals generated by thrombin interaction with high-affinity receptors on various cells has increased exponentially. With the discovery of synthetic peptides that can enhance or inhibit mitogenic responses of thrombin (Glenn *et al.*, 1988), we can now move toward application of these peptides to better understand and manipulate various physiological responses to thrombin *in vivo*.

5.1. Potential Role of Thrombin in Cellular Aspects of Wound Healing

There are many different types of wounds (partial thickness, full dermal, incisions, biopsy excisions, burns, and deep tissue ulcers) which differ in the amount of tissue which has to be replaced or the type of cell layers needed for this replacement. In almost all cases, however, wound healing and tissue repair involve a series of common events following formation of blood clots which differ in different types of wounds in time sequence, duration, and magnitude needed to effect complete healing. These events include: (1) an initial inflammatory response, in which neutrophils, monocytes, and lymphocytes accumulate in the wound area; (2) a proliferative phase, in which fibroblasts, endothelial cells, and other cell types migrate into the wound, begin to proliferate, and secrete matrix materials, or form functional capillaries and epithelial cells proliferate and migrate as a layer over the surface of the open wound; and (3) a maturation phase, in which skin collagen and other matrix material is secreted and remodeled to restore tissue strength. There are a number of recent reviews on various aspects of wound healing (Barbul *et al.*, 1988; Goslen, 1988; Orgill and Demling, 1988). For our purpose, we will focus on cellular events which may be initiated directly or indirectly by thrombin.

As described above, recent studies have shown that a number of cells which are involved in wound healing can be stimulated by thrombin. Some of these interactions are depicted in Fig. 3. Once hemostasis is achieved, there is a vasodilation presumably caused by prostaglandins (Goslen, 1988). Subsequently, inflammatory cells begin to accumulate in the wound area. As noted previously, thrombin appears to increase adherence of neutrophils, monocytes, and lymphocytes to capillary endothelial cells

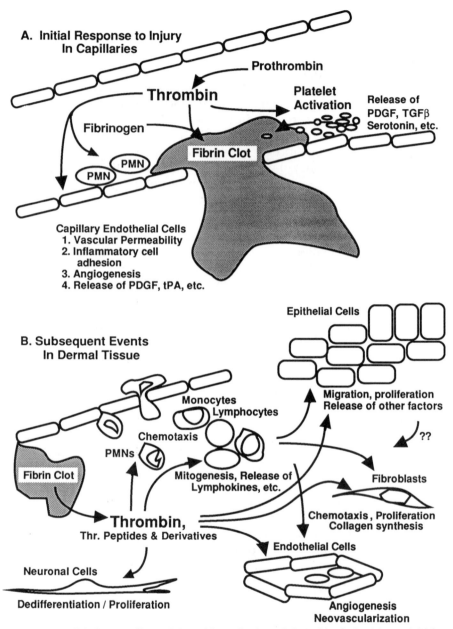

Figure 3. Models for (A) effects of thrombin and released thrombin derivatives on initial response to injury in capillaries, and (B) subsequent wound healing events which occur in dermal tissue. See text for details.

which may cause circulating inflammatory cells to accumulate in micro-vessels close to the site of tissue injury. Since thrombin increases vascular permeability, it may also increase the ease by which these inflammatory cells extravasate through vascular walls into the tissue. As noted, thrombin or portions of the thrombin molecule released from clots may also play a direct role in chemoattraction of these cells into wounds. A number of other growth and chemotactic factors including PDGF are released from endothelial cells following thrombin treatment. In addition, lymphokines, interleukins, and other factors secreted by the inflammatory cells may play a key role in recruitment of other cell types into the wound and stimulation of cell proliferation. Since fibroblasts and capillary endothelial cells re-spond mitogenically to thrombin and have high-affinity thrombin recep-tors, it appears likely that thrombin may also play a direct role in stimulat-ing proliferation of these cells as they migrate into the wound and contact active thrombin molecules or molecules which are partially degraded. Keratinocytes have also recently been shown to bind thrombin to high-affinity sites which are recognized in culture and in tissue slices with monoclonal antibodies to fibroblast thrombin receptors (McCaulley, McCroskey, and Carney, unpublished). Preliminary studies indicate that these cells do not respond mitogenically to thrombin; however, thrombin may stimulate these cells to secrete other factors which stimulate prolifera-tion or differentiation of cells in the granulating tissue beneath the epi-thelial layer.

5.2. Effects of Thrombin and Thrombin Receptor-Activating Peptides (TRAPs) on Wound Healing *in Vivo*

The discovery of synthetic peptides capable of interacting with throm-bin receptors on fibroblasts and other cells has given us the potential to modulate cellular effects that are mediated by receptor interaction (Glenn *et al.*, 1988). Studies using these peptides offer both a chance to better understand the role of thrombin receptors in various processes *in vivo*, and a means of stimulating or inhibiting various aspects of these responses. Although some of the inhibitory effects of these peptides may be extremely important, most of our initial studies have centered on the role of throm-bin in wound healing and the potential to use synthetic TRAPs to enhance the rate of healing in normal wounds and correct or overcome healing impairments which prevent healing of chronic wounds.

Initial studies with thrombin and TRAP were carried out using a full dermal punch biopsy assay in which 6-mm circular biopsies were removed from the backs of rats and the open wounds were treated with a single application of thrombin, TRAP, or saline control (Pernia *et al.*, 1990). Although control biopsies heal extremely fast in this model, TRAP was shown to accelerate the closure measured at 7 days by approximately 20%. In contract, thrombin appeared to have little if any effect. Other studies utilizing this model with hamsters have shown that neither thrombin, PDGF, insulin, nor EGF stimulated healing events enough to significantly affect the time required for complete closure of the wounds (Leitzel *et al.*, 1985). This suggested that the synthetic peptide may be better able to influence healing events than exogenously added thrombin either because thrombin may cause additional clotting (or negative effects), or because the peptides escape normal clearance mechanisms designed to regulate extracellular proteases.

   To more adequately assess the efficacy of TRAP and thrombin on wound healing, additional studies have been carried out using full dermal surgical incisions. In these studies, a 6-cm dorsal incision was cut through the dermis exposing the underlying muscle layer; the incisions were then closed with three interrupted sutures. Other laboratories have used a similar model with PDGF and TGF-β and shown significant enhancement of wound breaking strength over controls (Mustoe *et al.*, 1987; Pierce *et al.*, 1988). In these cases, growth factors were applied in a collagen gel matrix which retained the factor, but may also affect normal healing. In our studies a single application of TRAP in saline or PBS increased breaking strength from 30 to 70% when measured 7 days layer (Carney *et al.*, 1992; Pernia *et al.*, 1989). In these studies, thrombin also stimulated healing as measured by increased breaking strength, but only about 60% as well as TRAP. Consistent with our model for potential effects of thrombin in this process, histological examination of sections through the incision during the first 2 days following surgery shows increased numbers of inflammatory cells (unpublished). Sections cut 7 days after surgery show increased cellularity with fewer inflammatory cells, increased numbers of capillaries, and more mature type I collagen (Carney *et al.*, 1992). Thus, in thrombin- and TRAP-treated wounds one of the early effects may be to increase the inflammatory response which subsequently triggers additional signals necessary to ensure proper wound healing responses. The increased number of capillaries observed suggests that thrombin and TRAP might also stimulate healing through an enhanced neovascularization. Angiograms of rat backs following incisions showed, in fact, that TRAP and (to a lesser extent) thrombin increased the number and size of vessels which extend into and across the surgical incision (Carney *et al.*, 1992). These results suggest that thrombin and released thrombin peptides may play an important role in initiating the complex cascade of postclotting cellular signals required for neovascularization and proper wound healing. In addition, these peptides may offer new options for wound management following surgery or management of wounds in individuals with impaired healing.

Acknowledgments

The author thanks Dennis Cunningham, for his training and encouragement in early phases of this work; John Fenton, for his continuing support in gifts of highly purified thrombin (produced under grant HL-13160), and discussions which have helped shape the direction of many of the studies described from our laboratories and others around the country; and to my colleagues and students who have worked to make our efforts in this area both possible and enjoyable. Studies from the author's laboratory have been supported by the National Institute of Diabetes and Kidney Disease (DK-25807), The National Cancer Institute (RCDA, CA-00805), The Texas Coordinating Board Advanced Technology Program, and Monsanto Co.

6. REFERENCES

Adelman, E., Westin, E., Dalton, T., and Gorse, K., 1985, Production of a platelet glycoprotein Ib related protein by a human erythroleukemia cell line, *Thromb. Haemostas.* **54:**112a.

Awbrey, B. J., Hoak, J. C., and Owen, W. G., 1979, Binding of human thrombin to cultured human endothelial cells, *J. Biol. Chem.* **254:**4092–4095.

Baker, J. B., Simmer, R. L., Glenn, K. C., and Cunningham, D. D., 1979, Thrombin and epidermal growth factor become linked to cell surface receptors during mitogenic stimulation, *Nature* **278:**743–745.

Baker, J. B., Low, D. A., Simmer, R. L., and Cunningham, D. D., 1980, Protease-nexin: A cellular component that links thrombin and plasminogen activator and mediates their binding to cells, *Cell* **21:**37–45.

Barbul, A., Pines, E., Caldwell, M., and Hunt, T. K., 1988, *Growth Factors and Other Aspects of Wound Healing: Biological and Clinical Implications,* Liss, New York.

Bar-Shavit, R., and Wilner, G. D., 1986, Mediation of cellular events by thrombin, *Int. Rev. Exp. Pathol.* **29:**213–241.

Bar-Shavit, R., Kahn, A., Fenton, J. W., II, and Wilner, G. D., 1983, Receptor-mediated chemotactic response of a macrophage cell line (J774) to thrombin, *Lab. Invest.* **49:**702–707.

Bar-Shavit, R., Kahn, A. J., Mann, K. G., and Wilner, G. D., 1986a, Growth-promoting effects of esterolytically inactive thrombin on macrophages, *J. Cell Biochem.* **32:**261–272.

Bar-Shavit, R., Kahn, A. J., Mann, K. G., and Wilner, G. D., 1986b, Identification of a thrombin sequence with growth factor activity on macrophages, *Proc. Natl. Acad. Sci. USA* **83:**976–980.

Bauer, P. I., Machovich, R., Aranyi, P., Buki, K. G., Csonka, E., and Horvath, I., 1983, Mechanism of thrombin binding to endothelial cells, *Blood* **61:**368–372.

Belloni, P. N., and Tressler, R. J., 1990, Microvascular endothelial cell heterogeneity: Interactions with leukocytes and tumor cells, *Cancer Metastasis Rev.* **8:**353–389.

Belloni, P. N., Carney, D. H., and Nicolson, G. L., 1992, Organ-derived endothelial cells express differential responsiveness to thrombin and other growth factors, *Microvasc. Res.* **43:** 20–45.

Bergmann, J. S., and Carney, D. H., 1982, Receptor-bound thrombin is not internalized through coated pits in mouse embryo cells, *J. Cell Biochem.* **20:**247–258.

Berndt, M. C., and Caen, J. P., 1984, Platelet glycoproteins, in: *Progress in Hemostasis and Thrombosis*, Grune & Stratton, New York.

Bernstein, R. L., and Carney, D. H., 1992, Characterization of thrombin receptor number and mitogenic responsiveness of different subpopulations of mouse embryo cells to thrombin, submitted for publication.

Besterman, J. M., and Cuatrecasas, P., 1984, Phorbol esters rapidly stimulate amiloride-sensitive Na/H exchange in human leukemic cell lines, *J. Cell Biol.* **99:**340–343.

Bevilacqua, M. P., Stengelin, S., Gimbrone, M. A. J., and Seed, B., 1989, Endothelial leukocyte adhesion molecule 1: An inducible receptor for neutrophils related to complement regulatory proteins and lectins, *Science* **243:**1160–1165.

Bizios, R., Lai, L., Fenton, J. W., II, Sonder, S. A., and Malik, A. B., 1985, Thrombin-induced aggregation of lymphocytes: Non-enzymic induction by an hirudin-blocked thrombin exosite, *Thromb. Res.* **38:**425–431.

Bizios, R., Lai, L., Fenton, J. W., II, and Malik, A. B., 1986, Thrombin-induced chemotaxis and aggregation of neutrophils, *J. Cell. Physiol.* **128:**485–490.

Bizios, R., Lai, L. C., Cooper, J. A., Del Vecchio, P. J., and Malik, A. B., 1988, Thrombin-induced adherence of neutrophils to cultured endothelial monolayers: Increased endothelial adhesiveness, *J. Cell. Physiol.* **134:**275–280.

Blanchard, J. M., Piechaczyk, M., Dani, C., Chambard, J. C., Franchi, A., Pouyssegur, J., and Jeanteur, P., 1985, c-myc gene is transcribed at high rate in G_0-arrested fibroblasts and is post-transcriptionally regulated in response to growth factors, *Nature* **317:**443–445.

Blumberg, P. M., and Robbins, P. W., 1975, Effect of proteases on activation of resting chick embryo fibroblasts and on cell surface proteins, *Cell* **6:**137–147.

Bormann, B.-J., Huang, C.-K., Lam, G. F., and Jaffe, E. A., 1989, Thrombin-induced vimentin phosphorylation in cultured human umbilical vein endothelial cells, *J. Biol. Chem.* **261:**10471–10474.

Bruhn, H. D., and Pohl, J., 1981, Growth regulation of fibroblasts by thrombin, factor XIII and fibronectin, *Klin. Wochenschr.* **59:**145–146.

Burger, M. M., 1970, Proteolytic enzymes initiating cell division and escape from contact inhibition of growth, *Nature* **227:**170–171.

Carney, D. H., 1983, Immunofluorescent visualization of specifically bound thrombin reveals cellular heterogeneity in number and density of preclustered receptors, *J. Cell. Physiol.* **117:**297–307.

Carney, D. H., 1987, Characterization of the thrombin receptor and its involvement in initiation of cell proliferation, in: *Control of Animal Cell Proliferation* (A. L. Boynton and H. L. Leffert, eds.), Academic Press, New York, pp. 265–296.

Carney, D. H., and Bergmann, J. S., 1982, ^{125}I-Thrombin binds to clustered receptors on noncoated regions of mouse embryo cell surfaces, *J. Cell Biol.* **95:**697–703.

Carney, D. H., and Cunningham, D. D., 1977, Initiation of chick cell division by trypsin action at the cell surface, *Nature* **268:**602–606.

Carney, D. H., and Cunningham, D. D., 1978a, Cell surface action of thrombin is sufficient to initiate division of chick cells, *Cell* **14:**811–823.

Carney, D. H., and Cunningham, D. D., 1978b, Role of specific cell surface receptors in thrombin-stimulated cell division, *Cell* **15:**1341–1349.

Carney, D. H., and Cunningham, D. D., 1978c, Transmembrane action of thrombin initiates chick cell division, *J. Supramol. Struct.* **9:**337–350.

Carney, D. H., Glenn, K. C., and Cunningham, D. D., 1978, Conditions which affect initiation of animal cell division by trypsin and thrombin, *J. Cell. Physiol.* **95:**13–22.

Carney, D. H., Glenn, K. C., Cunningham, D. D., Das, M., Fox, C. F., and Fenton, J. W., 1979, Photoaffinity labeling of a single receptor for alpha-thrombin on mouse embryo cells, *J. Biol. Chem.* **254:**6244–6247.

Carney, D. H., Stiernberg, J., and Fenton, J. W., II, 1984, Initiation of proliferative events by human alpha-thrombin requires both receptor binding and enzymic activity, *J. Cell Biochem.* **26:**181–195.

Carney, D. H., Scott, D. L., Gordon, E. A., and LaBelle, E. F., 1985, Phosphoinositides in mitogenesis: Neomycin inhibits thrombin-stimulated phosphoinositide turnover and initiation of cell proliferation, *Cell* **42:**479–488.

Carney, D. H., Herbosa, G. J., Bergmann, J. S., and Gordon, E. A., 1986a, Involvement of high and low affinity thrombin receptor interactions in initiation of cell proliferation, *J. Cell Biol.* **103:**438a.

Carney, D. H., Herbosa, G. J., Stiernberg, J., Bergmann, J. S., Gordon, E. A., Scott, D., and Fenton, J. W., II, 1986b, Double-signal hypothesis for thrombin initiation of cell proliferation, *Semin. Thromb. Hemost.* **12:**231–240.

Carney, D. H., Mann, R., Redin, W. R., Pernia, S. D., Berry, D., Heggers, J. P., Hayward, P. G., Robson, M. C., Christie, J., Annable, C., Fenton, J. W., II, and Glenn, K. C., 1992, Thrombin and synthetic thrombin receptor-activating peptides enhance incisional wound healing and neovascularization, *J. Clin. Invest.*, submitted for publication.

Chambard, J. C., Paris, S., L'Allemain, G., and Pouyssegur, J., 1987, Two growth factor signalling pathways in fibroblasts distinguished by pertussis toxin, *Nature* **326:**800–803.

Charo, I. F., Shak, S., Karasek, M. A., Davison, P. M., and Goldstein, I. M., 1984, Prostaglandin I_2 is not a major metabolite of arachidonic acid in cultured endothelial cells from human foreskin microvessels, *J. Clin. Invest.* **74:**914–919.

Chen, L. B., and Buchanan, J. M., 1975, Mitogenic activity of blood components. I. Thrombin and prothrombin, *Proc. Natl. Acad. Sci. USA* **72:**131–135.

Chen, L. B., Teng, N. N. H., and Buchanan, J. M., 1976, Mitogenicity of thrombin and surface alterations on mouse splenocytes, *Exp. Cell Res.* **101:**41–46.

Cherington, P. V., and Pardee, A. B., 1980, Synergistic effects of epidermal growth factor and thrombin on the growth stimulation of diploid Chinese hamster fibroblasts, *J. Cell. Physiol.* **105:**25–32.

Colman, R. W., 1990, Platelet receptors, in: *Hematology Oncology Clinics of North America*, Vol. 4 (J. Rubin, ed.), Saunders, Philadelphia, pp. 27–42.

Cunningham, D. D., and Gurwitz, D., 1989, Proteolytic regulation of neurite outgrowth from neuroblastoma cells by thrombin and protease nexin-1, *J. Cell Biochem.* **39:**55–64.

Cunningham, D. D., and Ho, T.-S., 1975, Effects of added proteases on concanavalin A-specific agglutinability and proliferation of quiescent fibroblasts, in: *Proteases and Biological Control* (E. Reich, D. B. Rifkin, and E. Shaw, eds.), Cold Spring Harbor Press, Cold Spring Harbor, N.Y., pp. 795–806.

Daniel, T. O., Gibbs, V. C., Milfay, D. F., Garovoy, M. R., and Williams, L. T., 1986, Thrombin stimulates c-sis gene expression in microvascular endothelial cells, *J. Biol. Chem.* **261:**9579–9582.

Daniel, T. O., Gibbs, V. C., Milfay, D. F., and Williams, L. T., 1987, Agents that increase cAMP accumulation block endothelial c-sis induction by thrombin and transforming growth factor-beta, *J. Biol. Chem.* **262**:11893–11896.

Detwiler, T. C., and McGowan, E. B., 1985, Platelet receptors for thrombin, *Adv. Exp. Med. Biol.* **192**:15–28.

Eaton, D. L., and Baker, J. B., 1983, Evidence that a variety of cultured cells secrete protease nexin and produce a distinct cytoplasmic serine protease-binding factor, *J. Cell. Physiol.* **117**:175–182.

Esmon, N. L., Carroll, R. C., and Esmon, C. T., 1983, Thrombomodulin blocks the ability of thrombin to activate platelets, *J. Biol. Chem.* **258**:12238–12242.

Farrell, D. H., Wagner, S. L., Yuan, R. H., and Cunningham, D. D., 1988, Localization of protease nexin-1 on the fibroblast extracellular matrix, *J. Cell. Physiol.* **134**:179–188.

Fenton, J. W., II, 1988, Regulation of thrombin generation and functions, *Semin. Thromb. Hemostas.* **14**:234–240.

Fenton, J. W., II, and Bing, D. H., 1986, Thrombin active-site regions, *Semin. Thromb. Hemostas.* **12**:200–208.

Fox, J. E. B., 1985, Linkage of a membrane skeleton to integral membrane glycoproteins in human platelets, *J. Clin. Invest.* **76**:1673–1678.

Fox, J. E., and Berndt, M. C., 1989, Cyclic AMP-dependent phosphorylation of glycoprotein Ib inhibits collagen-induced polymerization of actin in platelets, *J. Biol. Chem.* **264**:9520–9526.

Frost, G. H., 1987, Characterization of the thrombin receptor and its involvement in stimulating cell proliferation, Ph.D., The University of Texas Graduate School of Biomedical Sciences, Galveston, Texas.

Frost, G. H., and Carney, D. H., 1987, HPLC purification of the thrombin receptor from hamster and mouse fibroblasts, *J. Cell Biol.* **105**:235a.

Frost, G. H., Thompson, W. C., and Carney, D. H., 1987, Monoclonal antibody to the thrombin receptor stimulates DNA synthesis in combination with gamma-thrombin or phorbol myristate acetate, *J. Cell Biol.* **105**:2551–2558.

Glenn, K. C., and Cunningham, D. D., 1979, Thrombin-stimulated cell division involves proteolysis of its cell surface receptor, *Nature* **278**:711–714.

Glenn, K. C., Carney, D. H., Fenton, J. W., II, and Cunningham, D. D., 1980, Thrombin active site regions required for fibroblast receptor binding and initiation of cell division, *J. Biol. Chem.* **255**:6609–6616.

Glenn, K. C., Frost, G. H., Bergmann, J. S., and Carney, D. H., 1988, Synthetic peptides bind to high-affinity thrombin receptors and modulate thrombin mitogenesis, *Peptide Res.* **1**:65–73.

Goldsmith, J. C., and Kisker, T., 1982, Thrombin–endothelial cell interactions: Critical importance of endothelial cell vessel of origin, *Thrombin. Res.* **25**:131–136.

Goodnough, L. T., and Saito, H., 1982, Specific binding of thrombin by human peripheral blood monocytes, *J. Lab. Clin. Med.* **9**:873–874.

Gordon, E. A., and Carney, D. H., 1986, Thrombin receptor occupancy initiates cell proliferation in the presence of phorbol myristic acetate, *Biochem. Biophys. Res. Commun.* **141**:650–656.

Gordon, E. A., Fenton, J. W., II, and Carney, D. H., 1986, Thrombin-receptor occupancy initiates a transient increase in cAMP levels in mitogenically responsive hamster (NIL) fibroblasts, *Ann. N.Y. Acad. Sci.* **485**:249–263.

Goslen, J. B., 1988, Wound healing for the dermatologic surgeon, *J. Dermatol. Surg. Oncol.* **14**:959–972.

Gospodarowicz, D., Brown, K. D., Birdwell, C. R., and Zetter, B. R., 1978, Control of proliferation of human vascular endothelial cells. Characterization of the response of human umbilical vein endothelial cells to fibroblast growth factor, epidermal growth factor and thrombin, *J. Cell Biol.* **77**:774–788.

Grand, R. J., Grabham, P. W., Gallimore, M. J., and Gallimore, P. H., 1989, Modulation of morphological differentiation of human neuroepithelial cells by serine proteases: Independence from blood coagulation, *EMBO J.* **8**:2209–2215.

Gronke, R. S., Curry, T. K., and Baker, J. B., 1986, Formation of protease nexin–thrombin complexes on the platelet surface, *J. Cell Biochem.* **32**:201–206.

Gronke, R. S., Bergman, B. L., and Baker, J. B., 1987, Thrombin interaction with platelets. Influence of a platelet protease nexin, *J. Biol. Chem.* **262**:3030–3036.

Gronke, R. S., Knauer, D. J., Veeraraghavan, S., and Baker, J. B., 1989, A form of protease nexin I is expressed on the platelet surface during platelet activation, *Blood* **73**:472–478.

Guenther, J., Nick, H., and Monard, D., 1985, A glia-derived neurite-promoting factor with protease inhibitory activity, *EMBO J.* **4**:1963–1966.

Gurwitz, D., and Cunningham, D. D., 1988, Thrombin modulates and reverses neuroblastoma neurite outgrowth, *Proc. Natl. Acad. Sci. USA* **85**:3440–3444.

Harlan, J. M., Thompson, P. J., Ross, R. R., and Bowen-Pope, D. F., 1986, Alpha-thrombin induces release of platelet-derived growth factor-like molecule(s) by cultured human endothelial cells, *J. Cell Biol.* **103**:1129–1133.

Harmon, J. T., and Jamieson, G. A., 1986, The glycocalicin portion of platelet glycoprotein Ib expresses both high and moderate affinity receitpr sites for thrombin. A soluble radioreceptor assay for the interaction of thrombin with platelets, *J. Biol. Chem.* **261**:13224–13229.

Hawkins, R. L., and Seeds, N. W., 1986, Effect of proteases and their inhibitors on neurite outgrowth from neonatal mouse sensory ganglia in culture, *Brain Res.* **398**:63–70.

Hodges, G. M., Livingston, D. C., and Franks, L. M., 1973, The localization of trypsin in cultured mammalian cells, *J. Cell Sci.* **12**:887–902.

Hong, S. H., 1980, Effect of bradykinin and thrombin on prostacyclin synthesis in endothelial cells from calf and pig aorta and human umbilical vein, *Thrombin. Res.* **18**:787–795.

Hovi, T., and Vaheri, A., 1975, Reversible release of chick embryo fibroblast cultures from density dependent inhibition of growth, *J. Cell. Physiol.* **8**:245–257.

Hynes, R. O., 1974, Role of surface alterations in cell transformation: The importance of proteases and surface proteins, *Cell* **1**:147–156.

Isaacs, J. D., Savion, N., Gospodarowicz, D., Fenton, J. W., and Shuman, M. A., 1981, Covalent binding of thrombin to specific sites on corneal endothelial cells, *Biochemistry* **20**:398–403.

Jaffe, E. A., Grulich, J., Weksler, B. B., Hampel, G., and Watanabe, K., 1987, Correlation between thrombin-induced prostacyclin production and inositol trisphosphate and cytosolic free calcium levels in cultured human endothelial cells, *J. Biol. Chem.* **262**:8557–8565.

Jandrot, P. M., Didry, D., Guillin, M. C., and Nurden, A. T., 1988, Cross-linking of alpha and gamma-thrombin to distinct binding sites on human platelets, *Eur. J. Biochem.* **174**:359–367.

Jones, L. G., McDonough, P. M., and Brown, J. H., 1989, Thrombin and trypsin act at the same site to stimulate phosphoinositide hydrolysis and calcium mobilization, *Mol. Pharmacol.* **36**:142–149.

Kieffer, N., Debili, N., Wicki, A., Titeux, M., Henri, A., Mishal, Z., Breton-Gorius, J., Vainchenker, W., and Clemetson, K., 1986, Expression of platelet glycoprotein Ib alpha in HEL cells, *J. Biol. Chem.* **261**:15854–15861.

Knauer, D. J., and Cunningham, D. D., 1983, A reevaluation of the response of human umbilical vein endothelial cells to certain growth factors, *J. Cell. Physiol.* **117:**397–406.

Knauer, D. J., Thompson, J. A., and Cunningham, D. D., 1983, Protease nexins: Cell-secreted proteins that mediate the binding, internalization, and degradation of regulatory serine proteases, *J. Cell. Physiol.* **117:**385–396.

Knupp, C. L., 1989, The interaction of thrombin with platelet protease nexin, *Thromb. Res.***56:**77–90.

L'Allemain, G., Paris, S., and Pouyssegur, J., 1984, Growth factor action and intracellular pH regulation in fibroblasts. Evidence for a major role of the Na^+/H^+ antiport, *J. Biol. Chem.* **259:**5809–5815.

L'Allemain, G., Paris, S., Magnaldo, I., and Pouyssegur, J., 1986, Alpha-thrombin-induced inositol phosphate formation in G_0-arrested and cycling hamster lung fibroblasts: Evidence for a protein kinase C-mediated desensitization response, *J. Cell. Physiol.***129:**167–174.

Larsen, N. E., and Simons, E. R., 1981, Preparation and application of a photoreactive thrombin analogue: Binding to human platelets, *Biochemistry* **20:**4141–4147.

Leitzel, K., Cano, C., Marks, J. G., and Lipton, A., 1985, Growth factors and wound healing in the hamster, *J. Dermatol. Surg.* **11:**617–622.

Levin, E. G., Stern, D. M., Nawroth, P. P., Marlar, R. A., Fair, D. S., Fenton, J. W., II, and Harker, L. A., 1986, Specificity of the thrombin-induced release of tissue plasminogen activator from cultured human endothelial cells, *Thromb. Haemostas.* **56:**115–119.

Lollar, P., and Owen, W. G., 1980, Evidence that the effects of thrombin on arachidonate metabolism in cultured human endothelial cells are not mediated by a high affinity receptor, *J. Biol. Chem.* **255:**8031–8034.

Lollar, P., Hoak, J. C., and Owen, W. G., 1981, Binding of thrombin to cultured human endothelial cells, *J. Biol. Chem.* **256:**10279–10283.

Low, D. A., and Cunningham, D. D., 1982, A novel method for measuring cell surface-bound thrombin, *J. Biol. Chem.* **257:**850–858.

Low, D. A., Baker, J. B., Koonce, W. C., and Cunningham, D. D., 1981, Released protease-nexin regulates cellular binding, internalization, and degradation of serine proteases, *Proc. Natl. Acad. Sci. USA* **78:**2340–2344.

Low, D. A., Scott, R. W., Baker, J. B., and Cunningham, D. D., 1982, Cells regulate their mitogenic response to thrombin through release of protease nexin, *Nature* **298:**476–478.

Low, D. A., Wiley, H. S., and Cunningham, D. D., 1985, Role of cell-surface thrombin-binding sites in mitogenesis, in: *Cancer Cells 3/Growth Factors and Transformation* (J. Faramisco, B. Ozanne, and C. Stiles, eds.) , Cold Spring Harbor Press, Cold Spring Harbor, N.Y., pp. 401–408.

McGrogan, M., Gohari, J., Li, M., Hsu, C., Scott, R., Simonsen, C., and Baker, J. B., 1988, Cloning and sequencing of protease-nexin, *Bio/Technology* **6:**172–175.

McKinney, M., Snider, R. M., Fenton, J. W., II, and Fichelson, E., 1983a, Thrombin binding and effects in intact murine neuroblastoma N1E-115 cells, *Fed. Proc.* **42:**883 (abstract).

McKinney, M., Snider, R. M., and Richelson, E., 1983b, Thrombin binding to human brain and spinal cord, *Mayo Clin. Proc.* **58:**829–831.

Magnaldo, I., Talwar, H., Anderson, W. B., and Pouyssegur, J., 1987, Evidence for a GTP-binding protein coupling thrombin receptor to PIP2-phospholipase C in membranes of hamster fibroblasts, *FEBS Lett.* **210:**6–10.

Magnaldo, I., Pouyssegur, J., and Paris, S., 1989, Cyclic AMP inhibits mitogen-induced DNA synthesis in hamster fibroblasts, regardless of the signalling pathway involved, *FEBS Lett.* **245**:65–69.

Malik, A. B., 1986, Thrombin-induced endothelial injury, *Semin. Thromb. Hemostas.* **12**:184–196.

Martin, B. M., and Quigley, J. P., 1978, Binding and internalization of ^{125}I-thrombin in chick embryo fibroblasts: Possible role in mitogenesis, *J. Cell. Physiol.* **96**:155–164.

Means, E. D., and Anderson, D. K., 1986, Thrombin interactions with central nervous system tissue and implications of these interactions, *Ann. N.Y. Acad. Sci.* **485**:314–322.

Monard, D., Niday, E., Limat, A., and Solomon, F., 1983, Inhibition of protease activity can lead to neurite extension in neuroblastoma cells, *Prog. Brain Res.* **58**:359–364.

Moscat, J., Moreno, F., and Garcia, B. P., 1987, Mitogenic activity and inositide metabolism in thrombin-stimulated pig aorta endothelial cells, *Biochem. Biophys. Res. Commun.* **145**:1302–1309.

Moss, M., and Cunningham, D. D., 1981, Cleavage of cell surface proteins by thrombin, *J. Supramol. Struct.* **15**:49–61.

Moss, M., Wiley, H. S., Fenton, J. W., II, and Cunningham, D. D., 1983, Photoaffinity labeling of specific alpha-thrombin binding sites on Chinese hamster lung cells, *J. Biol. Chem.* **258**:3996–4002.

Murayama, T., and Ui, M., 1987, Possible involvement of a GTP-binding protein, the substrate of islet-activating protein, in receptor-mediated signaling responsible for cell proliferation, *J. Biol. Chem.* **262**:12463–12467.

Mustoe, T. A., Pierce, G. F., Thomason, A., Gramates, P., Sporn, M. B., and Deuel, T. F., 1987, Accelerated healing of incisional wounds in rats induced by transforming growth factor-beta, *Science* **237**:1333–1336.

Nambi, P., Mattern, M., Bartus, J. O., Aiyar, N., and Crooke, S. T., 1989, Stimulation of intercellular topoisomerase I activity by vasopressin and thrombin: Differential regulation by pertussis toxin, *Biochem. J.* **262**:485–489.

Noonan, K. D., 1976, Role of serum in protease-induced stimulation of 3T3 cell division past the monolayer stage, *Nature* **259**:573–576.

Noonan, K. D., and Burger, M. M., 1973, Induction of 3T3 cell division at the monolayer stage, *Exp. Cell Res.* **80**:405–414.

Orgill, D., and Demling, R. H., 1988, Current concepts and approaches to wound healing, *Crit. Care Med.* **16**:899–908.

Paris, S., and Pouyssegur, J., 1983, Biochemical characterization of the amiloride-sensitive Na/H antiport in Chinese hamster lung fibroblasts, *J. Biol. Chem.* **258**:3503–3508.

Paris, S., and Pouyssegur, J., 1986a, Growth factors activate the bumetanide-sensitive $Na^+/K^+/Cl^-$ cotransport in hamster fibroblasts, *J. Biol. Chem.* **261**:6177–6183.

Paris, S., and Pouyssegur, J., 1986b, Pertussis toxin inhibits thrombin-induced activation of phosphoinositide hydrolysis and Na^+/H^+ exchange in hamster fibroblasts, *EMBO J.* **5**:55–60.

Paris, S., Chambard, J. C., and Pouyssegur, J., 1987, Coupling between phosphoinositide breakdown and early mitogenic events in fibroblasts. Studies with fluoraluminate, vanadate, and pertussis toxin, *J. Biol. Chem.* **262**:1977–1983.

Perdue, J. F., Lubenskyi, W., Kivity, E., Sonder, S., and Fenton, J. W., II, 1981, Protease mitogenic response of chick embryo fibroblasts and receptor binding/processing of human alpha-thrombin, *J. Biol. Chem.* **256**:2767–2776.

Perez-Rodriguez, R., Franchi, A., and Pouyssegur, J., 1981, Growth factor requirements of Chinese hamster lung fibroblasts in serum-free media: High mitogenic reaction of thrombin, *Cell. Biol. Int. Rep.* **5**:347–357.

Pernia, S. D., Redin, W. R., and Carney, D. H., 1989, Synthetic thrombin peptide enhances healing of full dermal wounds, *J. Cell. Biol.* **109**:22a.

Pernia, S. D., Berry, D. L., Redin, W. R., and Carney, D. H., 1990, A synthetic peptide representing the thrombin receptor-binding domain enhances wound closure *in vivo.*, *South. Assoc. Agric. Sci. Bull. Biochem. Biotechnol.* **3**:8–12.

Pierce, G. F., Mustoe, T. A., Senior, R. M., Reed, J., Griffin, G. L., Thomason, A., and Deuel, T. F., 1988, *In vivo* incisional wound healing augmented by platelet-derived growth factor and recombinant *c-sis* homodimeric proteins, *J. Exp. Med.* **167**:974–987.

Pohjanpelto, P., 1977, Proteases stimulate proliferation of human fibroblasts, *J. Cell. Physiol.***91**:387–392.

Pohjanpelto, P., 1978, Stimulation of DNA synthesis in human fibroblasts by thrombin, *J. Cell. Physiol.* **91**:387–392.

Pouyssegur, J., Chambard, J. C., Franchi, A., Paris, S., and Van Obberghen-Schilling, E., 1982, Growth factor activation of an amiloride-sensitive Na/H exchange system in quiescent fibroblasts: Coupling to ribosomal protein S6 phosphorylation, *Proc. Natl. Acad. Sci. USA* **79**:4492–4495.

Pouyssegur, J., Chambard, J. C., Franchi, A., L'Allemain, G., Paris, S., and Van Obberghen-Schilling, E., 1985, Growth-factor activation of the Na/H antiporter controls growth of fibroblasts by regulating intracellular pH, in: *Cancer Cells 3/Growth Factors and Transformation*, Cold Spring Harbor Press, Cold Spring Harbor, N.Y., pp. 409–415.

Pouyssegur, J., Chambard, J. C., L'Allemain, G., Magnaldo, I., and Seuwen, K., 1988, Transmembrane signalling pathways initiating cell growth in fibroblasts, *Philos. Trans. R. Soc. London B Ser.* **320**:427–436.

Prescott, S. M., Zimmerman, G. A., and McIntyre, T. M., 1984, Human endothelial cells in culture produce platelet-activating factor (1-alkyl-2-acetyl-sn-glycero-3-phosphocholine) when stimulated with thrombin, *Proc. Natl. Acad. Sci. USA* **81**:3534–3538.

Prescott, S. M., Zimmerman, G. A., and McIntyre, T. M., 1987, The production of platelet-activating factor by cultured human endothelial cells: Regulation and function, in: *Platelet-Activating Factor and Related Lipid Mediators* (F. Snyder, ed.), Plenum Press, New York, pp. 323–340.

Puri, R. N., Zhou, F. X., Colman, R. F., and Colman, R. W., 1989, Cleavage of a 100 kDa membrane protein (aggregin) during thrombin-induced platelet aggregation is mediated by the high affinity thrombin receptors, *Biochem. Biophys. Res. Commun.***162**:1017–1024.

Raben, D. M., Yasuda, K., and Cunningham, D. D., 1987a, Modulation of thrombin-stimulated lipid responses in cultured fibroblasts. Evidence for two coupling mechanisms, *Biochemistry* **26**:2759–2765.

Raben, D. M., Yasuda, K. M., and Cunningham, D. D., 1987b, Relationship of thrombin-stimulated arachidonic acid release and metabolism to mitogenesis and phosphatidylinositol synthesis, *J. Cell. Physiol.* **130**:466–473.

Rath, H. M., Doyle, G. A., and Silbert, D. F., 1989, Hamster fibroblasts defective in thrombin-induced mitogenesis. A selection for mutants in phosphatidylinositol metabolism and other functions, *J. Biol. Chem.* **264**:13387–13390.

Rath, H. M., Fee, J. A., Rhee, S. G., and Silbert, D. F., 1990, Characterization of phosphatidylinositol-specific phospholipase C defects associated with thrombin-induced mitogenesis, *J. Biol. Chem.* **265**:3080–3087.

Rebecchi, M. J., and Rosen, O. M., 1987, Stimulation of polyphosphoinositide hydrolysis by thrombin in membranes from human fibroblasts, *Biochem, J.* **245**:49–57.

Rozengurt, E., Stroobant, P., Waterfield, M. D., Deuel, T. F., and Keehan, M. G., 1983,

Platelet-derived growth factors elicit cyclic AMP accumulation in Swiss 3T3 cells: Role of prostaglandin production, *Cell* **34**:265–272.

Rydel, T. J., Ravichandran, K. G., Tulinsky, A., Bode, W., Huber, R., Roitsch, C., and Fenton, J. W., II, 1990, The structure of a complex of recombinant hirudin and human alpha thrombin, *Science* **249**:277–280.

Saegusa, Y., Cavender, D., and Ziff, M., 1988, Stimulation of mononuclear cell binding to human endothelial cell monolayers by thrombin, *J. Immunol.* **141**:4140–4145.

Savion, N., Isaacs, J. D., Gospodarowicz, D., and Shuman, M. A., 1981, Internalization and degradation of thrombin and up regulation of thrombin-binding sites in corneal endothelial cells, *J. Biol. Chem.* **256**:4514–4519.

Scott, R. W., Eaton, D. L., Duran, N., and Baker, J. B., 1983, Regulation of extracellular plasminogen activator by human fibroblasts: The role of protease nexin, *J. Biol. Chem.* **258**:4397–4403.

Seeds, N. W., Gilman, A. G., Amano, T., and Nirenberg, M. W., 1970, Regulation of axon formation by clonal lines of a neural tumor, *Proc. Natl. Acad. Sci. USA* **66**:160–167.

Sefton, B. M., and Rubin, H., 1970, Release from density dependent inhibition by proteolytic enzymes, *Nature* **227**:843–845.

Shuman, M. A., 1986, Thrombin–cellular interactions, *Ann. N.Y. Acad. Sci.* **485**:228–239.

Simmer, R. L., Baker, J. B., and Cunningham, D. D., 1979, Direct linkage of thrombin to its cell surface receptors in different cell types, *J. Supramol. Struct.* **12**: 245–247.

Snider, R. M., and Richelson, E., 1983, Thrombin stimulation of guanosine 3′,5′-monophosphate formation in murine neuroblastoma cells (clone N1E-115), *Science* **221**:566–568.

Snider, R. M., McKinney, M., Fenton, J. W., II, and Richelson, E., 1984a, Activation of cyclic nucleotide formation in murine neuroblastoma N1E-115 cells by modified human thrombins, *J. Biol. Chem.* **259**:9078–9081.

Snider, R. M., McKinney, M., Forray, C., and Richelson, E., 1984b, Neurotransmitter receptors mediate cyclic GMP formation by involvement of arachidonic acid and lipoxygenase, *Proc. Natl. Acad. Sci. USA* **81**:3905–3909.

Sommer, J., Gloor, S. M., Rovelli, G. F., Hofsteenge, J., Nick, H., Meier, R., and Monard, D., 1987, cDNA sequence coding for a rat glia-derived nexin and its homology to members of the serpine superfamily, *Biochemistry* **26**:6407–6410.

Sonne, O., 1988, The specific binding of thrombin to human polymorphonuclear leucocytes, *Scand. J. Clin. Lab. Invest.* **48**:831–838.

Sprandio, J. D., Shapiro, S., Thiagarajan, P., and McCord, S., 1988, Cultured human umbilical vein endothelial cells contain a membrane glycoprotein immunologically related to platelet glycoprotein Ib, *Blood* **71**:234–237.

Stiernberg, J., LaBelle, E. F., and Carney, D. H., 1983, Demonstration of a late amiloride-sensitive event as a necessary step in initiation of DNA synthesis by thrombin, *J. Cell. Physiol.* **117**:272–281.

Stiernberg, J., Carney, D. H., Fenton, J. W., II, and LaBelle, E. F., 1984, Initiation of DNA synthesis by human thrombin: Relationships between receptor binding, enzymic activity, and stimulation of $^{86}Rb^+$ influx, *J. Cell. Physiol.* **1209**:289–295.

Tam, S. W., Fenton, J. W., II, and Detwiler, T. C., 1980, Platelet thrombin receptors. Binding of alpha-thrombin is coupled to signal generation by a chymotrypsin-sensitive mechanism, *J. Biol. Chem.* **255**:6626–6632.

Teng, N. N. H., and Chen, L. B., 1975, The role of surface proteins in cell proliferation as studied with thrombin and other proteases, *Proc. Natl. Acad. Sci. USA* **72**:413–417.

Teng, N. N. H., and Chen, L. B., 1976, Thrombin-sensitive surface protein of cultured chick embryo cells, *Nature* **259**:578–580.

Thompson, W. C., and Carney, D. H., 1984, Mitogenic responsiveness of mouse embryo cell lines with high receptor number, *J. Cell Biol.* **99**:417a.

Vaheri, A., Ruoslahti, E., and Hovi, T., 1974, Cell surface and growth control of chick embryo fibroblasts in culture, in: *Control of Proliferation in Animal Cells* (B. Clarkson and R. Baserga, eds.), Cold Spring Harbor Press, Cold Spring Harbor, N.Y., pp. 305–312.

Van Obberghen-Schilling, E., and Pouyssegur, J., 1985, Affinity labeling of high-affinity alpha-thrombin binding sites on the surface of hamster fibroblasts, *Biochim. Biophys. Acta* **847**:335–343.

Van Obberghen-Schilling, E., Perez-Rodriquez, R., and Pouyssegur, J., 1982, Hirudin, a probe to analyze the growth-promoting activity of thrombin in fibroblasts: Reevaluation of the temporal action of competence factors, *Biochem. Biophys. Res. Commun.* **107**:359–369.

Van Obberghen-Schilling, E., Pérez-Rodriguez, R., Franchi, A., Chambard, J. C., and Pouyssegur, J., 1983, Analysis of growth factor "relaxation" in Chinese hamster lung fibroblasts required for tumoral expression, *J. Cell. Physiol.* **115**:123–130.

Van Obberghen-Schilling, E., Chambard, J. C., Paris, S., L'Allemain, G., and Pouyssegur, J., 1985, alpha-Thrombin-induced early mitogenic signalling events and G_0 to S-phase transition of fibroblasts require continual external stimulation, *EMBO J.* **4**:2927–2932.

Van Obberghen-Schilling, E., Chambard, J. C., Lory, P., Nargeot, J., and Pouyssegur, J., 1990, Functional expression of Ca-mobilizing alpha-thrombin receptors in mRNA-injected Xenopus oocytes, *FEBS Lett.* **262**:330–334.

Weksler, B. B., Ley, C. W., and Jaffe, E. A., 1978, Stimulation of endothelial cell prostacyclin production by thrombin, trypsin, and the ionophore A 23187, *J. Clin. Invest.* **62**:923–930.

Yamamoto, N., Kitgawa, H., and Tanoue, K., 1985, Monoclonal antibody to glycoprotein Ib inhibits both thrombin and ristocetin-induced platelet aggregation, *Thromb. Res.* **39**:751–759.

Zetter, B. R., 1988, Endothelial cell heterogeneity: Influence of vessel size, organ location, and species specificity, in: *Biology of Vascular Endothelial Cells* (U. Ryan, ed.), CRC Press, Boca Raton, Fla., pp. 63–80.

Zetter, B. R., and Antoniades, H. N., 1979, Stimulation of human vascular endothelial cell growth by platelet-derived growth factor and thrombin, *J. Supramol. Struct.* **11**:361–370.

Zetter, B. R., Chen, L. B., and Buchanan, J. M., 1976, Effects of protease treatment on growth, morphology, adhesion, and cell surface proteins of secondary chick embryo fibroblasts, *Cell* **7**:407–412.

Zetter, B. R., Chen, L. B., and Buchanan, J. M., 1977a, Binding and internalization of thrombin by normal and transformed chick cells. *Proc. Natl. Acad. Sci. USA* **74**:596–600.

Zetter, B. R., Sun, T.-T., Chen, L. B., and Buchanan, J. M., 1977b, Thrombin potentiates the mitogenic response of cultured fibroblasts to serum and other growth promoting agents, *J. Cell. Physiol.* **92**:233–240.

Zimmerman, G. A., McIntyre, T. M., and Prescott, S. M., 1985, Thrombin stimulates the adherence of neutrophils to human endothelial cells in vitro, *J. Clin. Invest.* **76**:2235–2246.

Zimmerman, G. A., McIntyre, T. M., and Prescott, S. M., 1986, Thrombin stimulates neutrophil adherence by an endothelial cell-dependent mechanism: Characterization of the response and relationship to platelet-activating factor synthesis, *Ann. N.Y. Acad. Sci.* **485**:349–368.

Chapter 11

REGULATION OF THROMBIN-INDUCED ENDOTHELIAL BARRIER DYSFUNCTION AND PROSTAGLANDIN SYNTHESIS

Joe G. N. Garcia, Judy L. Aschner, and Asrar B. Malik

1. INTRODUCTION

The development of tissue culture techniques which allowed for the successful isolation and maintenance of cultured endothelial cells nearly two decades ago has resulted in an explosion of information which has refuted the previously held perception of vascular endothelium as a passive, inert tissue. In recent years the high metabolic activity of endothelial cells has been documented and an exhaustive list of diverse endothelial cell func-

Joe G. N. Garcia • Departments of Medicine, Physiology, and Biophysics, Indiana University School of Medicine, Indianapolis, Indiana 46202. **Judy L. Aschner** • Departments of Physiology and Pediatrics, The Albany Medical College, Albany, New York 12208. **Asrar B. Malik** • Departments of Physiology and Cell Biology, The Albany Medical College, Albany, New York 12208.

Thrombin: Structure and Function, edited by Lawrence J. Berliner. Plenum Press, New York, 1992.

tions has been recognized as being modulated by interaction with the bioregulatory coagulant protein, thrombin.

As an organ system, the response of the endothelium to thrombin is diverse and complex. Thrombin stimulates endothelium to release substances which exert effects on vascular tone, vascular permeability, and airway smooth muscle. At the cellular level, thrombin binds to endothelial cells (Awbrey et al., 1979; Lollar and Owen, 1980a), potentiates endothelial cell proliferation (Gospodarowicz et al., 1978), rapidly stimulates the synthesis of platelet-activating factor (Camussi et al., 1983; Prescott et al., 1984), enhances expression of tissue factor on the cell surface (Galdal et al., 1985), increases transcription and release of platelet-derived growth factor (DiCorleto and Bowen-Pope, 1983; Harlan et al., 1986), the release of tissue plasminogen activator (Levine et al., 1982) and its inhibitor (Gelehrter and Sznycer-Laszuk, 1986), and the release of factor VIII and von Willebrand factor from vascular endothelium (Levine et al., 1982). At the biochemical level, thrombin is a potent activator of critical endothelial cell signal transducing mechanisms including phospholipase C-mediated phosphatidylinositol 4,5-bisphosphate (PIP_2) hydrolysis (Jaffe et al., 1987; Brock and Capasso, 1988; Halldorsson et al., 1988). For the induction of these effects, the proteolytic activity of this enzyme is a strict requirement.

Another important endothelial cell function altered by thrombin stimulation is the release and subsequent metabolism of arachidonate to vasoactive prostaglandins (Weksler et al., 1978). A second recently described, thrombin-mediated endothelial cell event is the ability of thrombin to induce profound alterations in cultured endothelial cell monolayer permeability secondary to alterations of cell–cell and cell–matrix interactions (Garcia et al., 1986). Despite the recognition that thrombin mediates these and other important endothelial cell events, information on the regulatory pathways involved in the transduction of the thrombin-induced endothelial cell responses remains limited. In this chapter, we will address thrombin-mediated alterations in lung fluid balance emphasizing the current concepts and information regarding the regulation of thrombin-induced endothelial prostaglandin synthesis and barrier dysfunction. Specific attention will be directed to the role of second messengers accomplishing these key endothelial cell events.

2. THROMBIN-MEDIATED ENDOTHELIAL CELL BARRIER DYSFUNCTION

The pulmonary microvascular endothelium functions as a selective permeability barrier between the blood and the pulmonary interstitial space. Encompassing an extensive surface area, these endothelial cells are a target for many inflammatory and thrombogenic mediators which can result in barrier disruption and increased permeability to plasma proteins; a characteristic feature of inflammatory lung injury seen in disease states such as the adult respiratory distress syndrome, pulmonary embolism, and neonatal bronchopulmonary dysplasia.

While the pulmonary vascular endothelial monolayer is structurally simple, the functional complexity of the endothelium is only recently beginning to be understood. The endothelial cells lining specific blood vessels likely differ with respect to their response to many inflammatory stimuli, including thrombin. For example, there are physiologic and biochemical differences between endothelial cells obtained from various species and between those derived from large vessels compared to those obtained from the microvasculature. These differences are being addressed in some of the current literature (Jaffe *et al.*, 1987; Ryan *et al.*, 1988; A. J. Carter *et al.*, 1989; Ryan, 1990). Although *in vitro* studies have most often utilized large-vessel endothelial cells obtained from either bovine and human pulmonary artery and aorta, or human umbilical vein because of the relative ease in harvesting and maintaining these cells in culture, it is the extensive microvascular bed of the lung which is responsible for processing vasoactive substances, and for producing and delivering endothelium-derived products into the circulation. Whenever possible, this review will focus on results obtained from studies of thrombin's interactions with pulmonary microvascular endothelial cells (Ryan *et al.*, 1982; Del Vecchio *et al.*, 1991), and relate these studies to *in vivo* observations.

2.1. Animal Studies of Thrombin-Mediated Lung Edema Formation

Intravenous infusion of α-thrombin has been used as a model of pulmonary intravascular coagulation and microembolism to study the mechanisms of lung vascular injury *in vivo*. Intravenous α-thrombin is a potent stimulus for increased pulmonary vascular permeability to proteins resulting in pulmonary edema (Johnson *et al.*, 1983a,b; Minnear *et al.*, 1983; Johnson and Malik, 1985; Lo *et al.*, 1985). Thrombin challenge is also associated with increased leukocyte influx into alveolar spaces (Garcia *et al.*, 1988). Thrombin mediates lung vascular injury both by its direct effects on vascular endothelium and indirectly by its effects on plasma constituents and bloodborne elements. Although the data point to both as important factors in the development of vascular barrier dysfunction, the relative contribution of each of these factors in the intact microvasculature is unclear.

2.1.1. Pulmonary Fluid Balance and Hemodynamic Changes

In awake and anesthetized sheep, the intravenous infusion of α-thrombin produces an increase in pulmonary lymph flow with an increase in the lymph-to-plasma protein concentration ratio indicating an increase in vascular permeability to proteins (Johnson *et al.*, 1983a,b; Johnson and Malik, 1985). Thrombin also increases pulmonary arterial pressure and pulmonary vascular resistance, indicating that the pulmonary edema that occurs with thrombin-induced intravascular coagulation results from both increased vascular permeability and increased capillary hydrostatic pressure (Lo *et al.*, 1985). In the isolated guinea pig lung perfused only with Ringers saline and albumin, α-thrombin, but not catalytically inactivated DIP-thrombin, causes a rapid dose-dependent increase in pulmonary arterial pressure, and capillary hydrostatic pressure as a result of postcapillary vasoconstriction. These findings support the argument that a portion of thrombin's effect on lung fluid balance and hemodynamics are direct and independent of the action of thrombin on blood components (Horgan *et al.*, 1987). Thrombin challenge is associated with an increase in the lung effluent thromboxane B_2 concentration which is a likely mediator of the vasoconstriction (Horgan *et al.*, 1987). Pulmonary edema is evident within 90 min after thrombin injection as demonstrated by a 60% increase over baseline lung weight (Horgan *et al.*, 1987).

2.1.2. Role of Fibrin

The mechanism by which the intravenous challenge of α-thrombin results in pulmonary intravascular coagulation involves, at least in part, the deposition of fibrin in the pulmonary vascular bed. Thrombin cleaves fibrinogen to fibrin, resulting in entrapment of fibrin clots in pulmonary microvessels. Thrombin-induced fibrin microemboli may serve as a meshwork for sequestration of leukocytes and platelets in the pulmonary microvessels (Fig. 1). Fibrin depletion by infusion of Ancrod, purified from Malayan pit viper venom, attenuates the thrombin-induced increase in pulmonary lymph flow ($\dot{Q}$lym) and transvascular protein clearance (Johnson and Malik, 1982a; Johnson *et al.*, 1983a). Pulmonary arterial pressure (Ppa) and pulmonary vascular resistance (PVR) do not increase significantly after α-thrombin infusion in defibrinogenated sheep. Similar results are seen after infusion of proteolytically active γ-thrombin (Garcia-Szabo *et al.*, 1984). These findings underscore the importance of fibrin deposition in mediating thrombin-induced increases in pulmonary transvascular fluid and protein exchange.

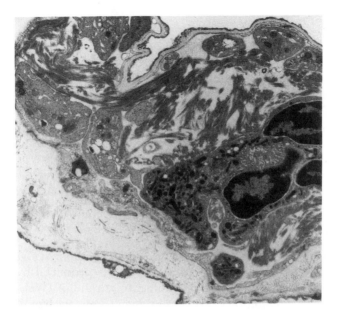

Figure 1. Sequestration of leukocytes and platelets in the pulmonary microcirculation following thrombin infusion. Pulmonary microvessel of a dog showed deposition of fibrin. Neutrophils and platelets are found associated with fibrin and degranulated platelets are found around the edges of the endothelium. The neutrophil is in the process of phagocytosing fibrin particles. The endothelium is swollen and edema fluid is present in the interstitial space. (Original magnification: × 9000.) Micrograph courtesy of Dr. F. L. Minnear.

2.1.3. Role of Plasminogen Activation

Pulmonary endothelial cells contain large amounts of tissue plasminogen activators which convert plasminogen to plasmin, the proteolytic enzyme responsible for fibrinolysis. Thrombin becomes sequestered in the clot during thrombus formation, and therefore may attain high local concentrations when released during clot lysis (Fenton *et al.*, 1977). Plasminogen activation appears to be necessary for leukostasis in the lung. Tranexamic acid-induced inhibition of fibrinolysis prevents the thrombin-induced systemic leukopenia and pulmonary neutrophil sequestration as measured by pulmonary uptake of ^{111}In oxide-labeled neutrophils (Johnson *et al.*, 1987). Recent data indicate that one possible mechanism by which plasmin can mediate vascular injury is by promotion of neutrophil adherence to endothelial cells (Lo *et al.*, 1989).

2.1.4. Role of Complement Activation

Generation of the complement-derived chemotactic and leukocyte-aggregating peptides, C3a and C5a, as well as deposition of fibrin microthrombi is associated with entrapment of platelets and neutrophils in the pulmonary microcirculation. Complement depletion using cobra venom factor partially protects sheep from thrombin-induced lung vascular injury (Johnson *et al.*, 1983b, 1986a, 1987). This may be the result of decreased neutrophil sequestration in the lungs. However, complement activation alone does not increase pulmonary permeability (Johnson *et al.*, 1986a), implicating the potential importance of other critical mediators, such as platelet-activating factor and leukotriene B_4, in lung vascular injury.

2.1.5. Role of Prostacyclin

Prostacyclin (PGI_2) is a potent systemic and pulmonary vasodilator and an inhibitor of platelet aggregation which is produced by the pulmonary vascular endothelium in response to thrombin (Weksler *et al.*, 1978). PGI_2 infusion attenuates the increase in pulmonary lymph flow and transvascular protein clearance after thrombin (Perlman *et al.*, 1986). Increases in PVR and Ppa are also inhibited by PGI_2 (Perlman *et al.*, 1986). These effects are related to the reduction in pulmonary capillary hydrostatic pressure rather than to a PGI_2-mediated reduction in lung vascular permeability.

Ibuprofen, a cyclooxygenase inhibitor and inhibitor of thromboxane and PGI_2 synthesis, prevents the thrombin-induced lung vascular injury (Johnson and Malik, 1985). Ibuprofen reduced pulmonary lymph flow, lymph protein clearance, and both Ppa and PVR by decreasing thrombin-induced neutrophil adherence to pulmonary vascular endothelium (Perlman *et al.*, 1986). Thus, the protective effect of ibuprofen may be due to a reduction in neutrophil sequestration in the lung rather than an effect on arachidonate metabolism.

2.1.6. Role of Neutrophils

Thrombin causes increased neutrophil adherence to the endothelium (Zimmerman *et al.*, 1985; Bizios *et al.*, 1988), as well as neutrophil chemotaxis and neutrophil aggregation. Neutrophils are a key effector cell mediating lung vascular injury after thrombin-induced pulmonary intravascular coagulation (Tahamont and Malik, 1983; Heath *et al.*, 1986; Garcia *et al.*, 1988). α-Thrombin infusion causes pulmonary leukostasis as assessed by ^{111}In oxide-labeled neutrophils and a marked influx of neutrophils into alveolar spaces (Garcia *et al.*, 1988). The severity of lung vascular injury may be related to the duration of neutrophil sequestration which is determined by the kinetics of fibrin deposition and fibrinolysis. Depletion of granulocytes in sheep, using hydroxyurea or antineutrophil serum, prevents the thrombin-induced increase in lung vascular permeability (Johnson and Malik, 1982b; Tahamont and Malik, 983; Heath *et al.*, 1986).

Recent studies indicate that neutrophil repletion causes a reexpression of vascular injury after thrombin challenge (Lo *et al.*, 1990). Neutrophil activation by α-thrombin may damage the endothelium by generation of free radicals and proteases. The effect of thrombin on Q̇lym and transvascular protein clearance is blunted in sheep infused with superoxide dismutase (Johnson *et al.*, 1986b). This is similar to the response in neutropenic sheep (Tahamont and Malik, 1983).

2.2. *In Vitro* Studies of Thrombin-Induced Endothelial Permeability

The development of cultured endothelial monolayer preparations has advanced the study of the functional characteristics of endothelial cells including their properties as a selective permeability barrier. An *in vitro* system has been developed which permits the direct measurement of endothelial permeability for a known surface area, eliminating oncotic and hydrostatic forces influencing transendothelial fluid and solute fluxes *in vivo* (Garcia *et al.*, 1986; Del Vecchio *et al.*, 1987). The advantage of this system over animal or organ models is its ability to assess the direct effects of thrombin on endothelial function without interference from thrombin's effect on other cell types and plasma constituents. The strength of this system lies in its potential to investigate cellular and biochemical mechanisms involved in the mediation of endothelial permeability changes. The obvious caveat is that by the very nature of its simplicity, extrapolation to intact endothelium *in vivo* must be made with caution.

2.2.1. *Direct Effect of α-Thrombin and Thrombin Derivatives on Endothelial Monolayer [^{125}I]-Albumin Flux*

α-Thrombin causes concentration-dependent and reversible increases in endothelial permeability as measured by the clearance rate of [^{125}I]albumin across monolayers of bovine pulmonary microvessel endothelial cells and pulmonary artery endothelial cells (Garcia *et al.*, 1986; Siflinger-Birnboim *et al.*, 1987; Aschner *et al.*, 1990) (Fig. 2). The effect of thrombin is rapid, occurring within 2 min (Fig. 3), and is reversible within 15 min after washing away the thrombin from the monolayer (Garcia *et al.*, 1986; Lum, *et al.*, 1992).

The native enzyme has several sites capable of mediating the increase in endothelial permeability including the active catalytic site, the anionic binding exosite, and the fibrinogen recognition site (Fenton *et al.*, 1977). Because of the complex ways in which thrombin interacts with endothelial cells, it has been difficult to assign cause-and-effect relationships to thrombin's cellular interactions and the observed biological responses, such as thrombin-mediated increases in endothelial permeability. In this regard, modified forms of thrombin have been useful tools in dissecting the complex interactions between thrombin and endothelial cells.

The interaction of α-thrombin with endothelial cell membranes simultaneously demonstrates the characteristics of an enzyme-catalyzed reaction and agonist–receptor equilibrium reaction. It has been shown that α-thrombin binds to cultured human (Awbrey *et al.*, 1979; Lollar *et al.*, 1980) and bovine (Isaacs *et al.*, 1981) endothelial cells in a rapid, reversible, and saturable manner behaving quantitatively like a classical ligand–receptor system. This high-affinity binding site is active-site independent, as

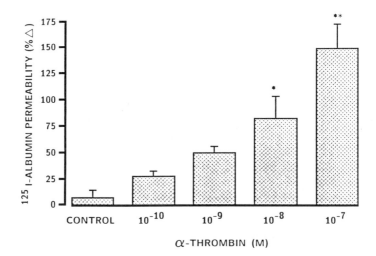

Figure 2. α-thrombin causes a concentration-dependent increase in endothelial monolayer permeability. After a 30-min baseline, [^{125}I]albumin clearance rate was obtained, control media or increasing concentrations of α-thrombin were added to endothelial monolayers, and a posttreatment clearance rate was recalculated over the subsequent 30 min. Results are expressed as % change over the baseline clearance rate, allowing each monolayer to serve as its own control. Values shown are the mean ± SEM of three separate experiments containing six to eight monolayer wells in each treament group per experiment. *$p < 0.05$ compared with the control group. **$p < 0.05$ compared with all other groups. (Data from Garcia *et al.*, 1986; Aschner *et al.*, 1990.)

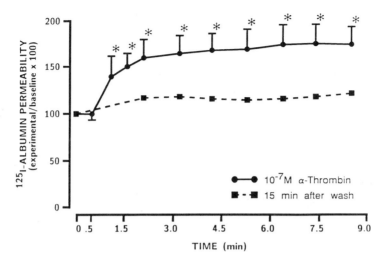

Figure 3. Rapidity of the α-thrombin-mediated increase in endothelial monolayer permeability. α-thrombin (10^{-7} M) increased the transendothelial [^{125}I]albumin clearance rate half-maximally as early as 1 min and maximally by 2 min after thrombin challenge. The thrombin response was reversed by washing the endothelial cell monolayer and reincubating with fresh DMEM for 15 min prior to measurement of [^{125}I]albumin flux. (H. Lum, unpublished data.)

catalytically inactive DIP-α-thrombin binds to the same site with an affinity similar to that of α-thrombin (Awbrey *et al.*, 1979; Lollar and Owen, 1980a). However, there is accumulating evidence that binding alone is an insufficient signal for activation of the transmembrane events that trigger the increase in permeability (Aschner *et al.*, 1990). DIP-α-thrombin is unable to increase permeability in concentrations up to 10 NIH units/ml (10^{-7} M) (Aschner *et al.*, 1990). α- or γ-thrombin which has been inactivated with D-phenylalanyl-prolyl-arginine chloromethyl ketone (PPACK) does not result in increased endothelial permeability (Fig. 4). Although this finding suggests that the catalytic site is needed for the permeability increase, it does not rule out the possibility that high-affinity binding and an active catalytic site are both needed to generate the appropriate transmembrane signals, as has been shown to be the case for fibroblast mitogenesis (Carney *et al.*, 1984, 1986).

In contrast to DIP-α-thrombin, γ-thrombin does not compete for high-affinity binding sites (Glenn and Cunningham, 1979; Alexander *et al.*, 1983; Carney *et al.*, 1986), but has an intact serine active site. γ-Thrombin consistently increased monolayer permeability, although on an equimolar basis, the response is less than with α-thrombin (Aschner *et al.*, 1990) (Fig.

4). This suggests that enzymatic activity alone can cause an increase in endothelial permeability, but it does not rule out a facilitative role for high-affinity binding.

DIP-α-thrombin, when added to endothelial monolayers to block high-affinity binding sites, blunts the effects of subsequently added α-thrombin, suggesting that high-affinity binding by α-thrombin facilitates contact between the active serine site and its substrate (Aschner *et al.*, 1990). This also suggests that the region of the thrombin molecule involved in high-affinity binding is close to the active serine site (Carney *et al.*, 1984). High-affinity binding may accelerate the active-site reaction by facilitating contact between the active site and its substrate or by conferring a conformational change on the receptor which enhances affinity of the enzyme for its substrate. γ-Thrombin may form the same enzyme–substrate complex, but without facilitation, due to its inability to bind to high-affinity sites.

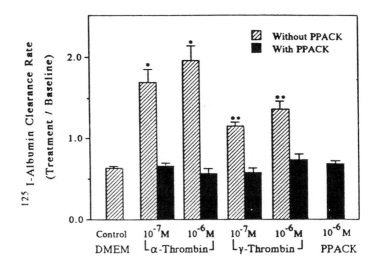

Figure 4. Incubation of α- or γ-thrombin with PPACK prevents the thrombin-induced increase in [^{125}I]albumin clearance rate. After a 30-min baseline, [^{125}I]albumin clearance rate was obtained, control media, α-thrombin, γ-thrombin (10^{-7} and 10^{-6} M) with and without PPACK, or PPACK alone was added to endothelial monolayers, and a posttreatment clearance rate was recalculated over the subsequent 30 min. The results are expressed at the ratio of the [^{125}I]albumin clearance rate before and after addition of the treatment, allowing each monolayer to serve as its own control. Values shown are the mean ± SEM of three separate experiments containing six to eight monolayer wells in each treament group per experiment. *$p < 0.05$ compared to the control DMEM treatment group and the respective α-thrombin treatment groups. (From Aschner *et al.*, 1990.)

2.2.2. Role of Actin Rearrangement in Thrombin-Induced Permeability Increase

The actin cytoskeleton plays a critical structural and mechanical role in maintaining the integrity of the endothelial cell monolayer. There is evidence that actin filaments may organize and stabilize junctional proteins which appear to be important regulators of permeability to ions and macromolecules (Phillips *et al.*, 1989). The cytoskeletal alterations that occur with thrombin challenge include the loss of peripheral actin bands and an increase in the number and thickness of stress fibers (Fig. 5). This is a reversible phenomenon implying that the response is not the result of endothelial injury and may be physiologically important. Treatment of bovine pulmonary artery endothelial cells with NBD-phallacidin, which is rapidly incorporated into cells and stabilizes actin filaments, prevents the thrombin-induced increase in permeability (Phillips *et al.*, 1989).

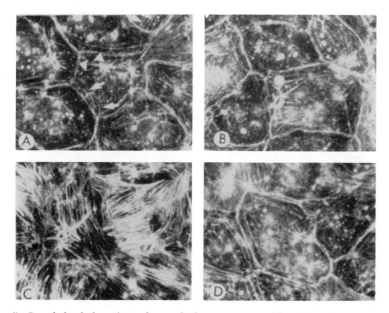

Figure 5. Cytoskeletal alterations observed after treatment with α-thrombin. Rhodamine-phalloidin staining of endothelial monolayers growing on filters. (A) Control; (B) NBD-phallacidin (0.3 μM) pretreatment for 3 h; (C) α-thrombin (10^{-7} M) treatment for 30 min; (D) pretreatment with 0.3 μM NBD-phallacidin for 3 h followed by α-thrombin (10^{-7} M) treatment for 30 min. After treatment, monolayers were fixed, permeabilized, stained with rhodamine phalloidin, and examined by fluorescence microscopy. Arrowheads indicate actin peripheral bands that delineate margins of cells; arrows point to individual cytoplasmic stress fibers. Magnification × 1240. (From Phillips *et al.*, 1989.)

3. THROMBIN-STIMULATED ENDOTHELIAL CELL PROSTAGLANDIN SYNTHESIS

The primary arachidonate metabolite produced by cultured endothelial cells is PGI_2. Additional prostaglandins such as PGE_2, $PGF_{2\alpha}$, thromboxane, and dihydroperoxyacids such as 5- and 15-HETES are produced in smaller quantities (Goldsmith and Needleman, 1982). The liberated arachidonate is sequentially converted to PGG_2 and PGH_2 by the cyclooxygenase enzyme and PGH_2 is subsequently metabolized to PGI_2 by prostacyclin synthetase (Marcus, 1978). PGI_2 is a potent vasodilator and inhibitor of platelet aggregation and is a key participant in maintaining vascular tone and vessel patency. Marked disruption of endothelial cell prostaglandin synthesis is observed in arteriosclerotic or diabetic vessels, implicating a pathogenetic role for PGI_2 synthesis in vascular disease (Weksler, 1984).

Thrombin is a potent stimulus for the release of free arachidonate and the generation of its biologically important metabolites such as PGI_2 (Weksler *et al.*, 1978) which are intimately involved in the control of vascular permeability, smooth muscle cell proliferation, and hemostasis. While the details of arachidonate metabolism leading to PGI_2 synthesis are well-recognized, the regulatory mechanisms controlling PGI_2 synthesis are incompletely understood. The following sections will address the current concepts regarding signalling mechanisms which appear to be involved in the regulation of arachidonic acid release and PGI_2 synthesis by thrombin.

3.1. Relationship of Thrombin Binding to Prostaglandin Synthesis

As mentioned above, the interaction between thrombin and its specific receptor on cultured human or bovine endothelial cells is rapid, reversible, and saturable (Awbrey *et al.*, 1979; Lollar *et al.*, 1980; Isaacs *et al.*, 1981). The receptors have been characterized as containing a limited number of distinct high-affinity states ($k_D \approx 0.1$ nM for human endothelium, 0.5 nM for bovine endothelium). In addition to these high-affinity sites, more numerous low-affinity sites with $k_D \approx 10$ nM and ≈ 50 nM have been identified on human and bovine endothelium, respectively (Awbrey *et al.*, 1979; Lollar *et al.*, 1980; Parkinson *et al.*, 1990). The difference in thrombin–receptor dissociation constants between the two species appears to be important in determining the potency of thrombin's effects on arachidonate metabolism with thrombin-stimulated bovine endothelium producing markedly reduced levels of prostaglandins than similarly challenged human endothelium. One high-affinity thrombin-binding site on the endothelial cell surface (and possibly the sole high-affinity site) is the surface glycoprotein, thrombomodulin (Esmon and Owen, 1981). Recent work has further identified that the glycosaminoglycan chain attachment of the thrombomodulin molecule appears to be critical for α-thrombin binding to thrombomodulin (Parkinson *et al.*, 1990). Substantial evidence, however, suggests that it is not the high-affinity binding site, i.e., thrombomodulin, but a low-affinity binding site that is the thrombin receptor which is functionally coupled to PGI_2 synthesis (Lollar and Owen, 1980b). As mentioned above, similar to other thrombin-mediated endothelial cell functions including enhanced permeability, proteolytic activity is required for thrombin-stimulated PGI_2 synthesis. Native α-thrombin stimulation of PGI_2 synthesis is rapid (Fig. 6) and near maximal at 10 nM. A 100-fold greater concentration of γ-thrombin (1 μM) produced only half the level of PGI_2 synthesis when compared with α-thrombin and proteolytically inactive DIP-α-thrombin, which does not possess clotting or proteolytic activity, failed to stimulate PGI_2 synthesis at any concentration (Fig. 6) (Garcia *et al.*, 1990). As DIP-thrombin is an excellent competitive inhibitor for

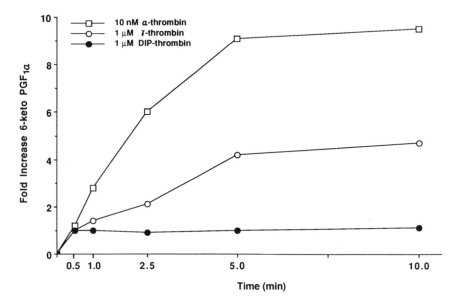

Figure 6. Effect of thrombin and modified thrombin forms on HUVEC PGI_2 synthesis. Confluent HUVEC monolayers were challenged with α-thrombin, γ-thrombin, and DIP-thrombin at the concentration shown and aliquots of cell-free supernatants removed at specified times. Rapid increases in 6-keto $PGF_{1\alpha}$ were observed with 10 nM α-thrombin, whereas proteolytically inactive DIP-thrombin (1 μM) did not increase levels of the PGI_2 metabolite. γ-thrombin produced significant levels of PGI_2 but required a 100-fold greater concentration when compared to α-thrombin. (From Garcia *et al.*, 1990.)

α-thrombin binding to thrombomodulin, thrombomodulin is unlikely to be the thrombin receptor which is functionally coupled to arachidonate metabolism and PGI_2 synthesis. As expected, repeated stimulation of human umbilical vein endothelial cells with α-thrombin results in desensitization to further PGI_2 production consistent with a receptor-mediated mechanism for PGI_2 synthesis (Halldorsson *et al.*, 1988).

3.2. Thrombin Activation of Arachidonate Releasing Pathways

The release of arachidonate by the hydrolysis of phosphatidylcholine to lysophosphatidylcholine in response to the activation of the membrane phospholipase A_2 is generally accepted as the rate-limiting step involved in thrombin-induced endothelial cell prostaglandin synthesis (Hong and Deykin, 1982; Hong *et al.*, 1985). Additional, albeit smaller, pools of arachidonate are derived from diacylglycerol lipase action on diacylglycerol. Both pathways are functional in cultured human umbilical vein endothelial cells with approximately two-thirds of the arachidonate released in thrombin-stimulated HUVEC being derived via the deacylating action of PLA_2 on membrane phospholipids. The relative importance of each pathway in determining arachidonate availability, however, depends on the specific cell type. As the acylhydrolase PLA_2 is exquisitely sensitive to fluctuations in cytosolic calcium, $Ca^{2+}{}_i$, studies aimed at delineating the signaling pathways involved in thrombin-induced PGI_2 synthesis have been directed toward the regulation of $Ca^{2+}{}_i$ release (Hallam *et al.*, 1988).

4. SECOND MESSENGERS INVOLVED IN THROMBIN-MEDIATED ENDOTHELIAL CELL EFFECTS

4.1. Regulation of Ca^{2+} in Cultured Endothelial Cells

Mediators that increase endothelial macromolecular permeability and stimulate prostaglandin synthesis also cause an increase in intracellular Ca^{2+} ($Ca_i{}^{2+}$) (Shasby *et al.*, 1985; Lum *et al.*, 1989; Halldorsson *et al.*, 1988; Ryan *et al.*, 1988; Garcia *et al.*, 1990, 1991). Using the photoprotein Ca^{2+} probe aequorin or fluorescent indicator probes such as Quin 2, Indo-1, and Fura-2, α-thrombin has been found to be a rapid and potent stimulus for increases in levels of cytosolic Ca^{2+} in cultured human umbilical vein and pulmonary artery endothelium (Garcia *et al.*, 1991; Brock *et al.*, 1988; Jaffe *et al.*, 1987) as well as in bovine pulmonary artery endothelium (Ryan *et al.*, 1988; Goligorsky *et al.*, 1989). Although the regulation of cytosolic Ca^{2+} concentrations following stimulation of thrombin is incompletely understood, thrombin has been demonstrated to increase polyphosphoinositide turnover in human and porcine endothelium (Hong and Deykin, 1982; Moscat et al., 1987; Jaffe et al., 1987; Brock and Capasso, 1988). The activation of phosphoinositol-specific membrane phospholipase C (PLC) results in the hydrolysis of phosphatidyl 4,5-bisphosphate and the generation of the short-lived but critical second messengers inositol trisphosphate (IP_3) and diacylglycerol (DAG). IP_3 is generally accepted as directly regulating $[Ca^{2+}_i]$ by mobilizing the release of Ca^{2+} from internal stores and indirectly by stimulating Ca^{2+} entry, possibly in concert with its phosphorylated metabolite inositol tetrakisphosphate (IP_4) (Berridge and Irvine, 1984). Consistent with PLC-mediated PIP_2 hydrolysis, α-thrombin stimulates rapid and transient formation of (1,4,5)-IP_3 in HUVEC (Jaffe et al., 1987; Brock and Capasso, 1988; Lampugnani et al., 1989; T. D. Carter et al., 1989) but not in bovine aortic endothelial cells (Jaffe et al., 1987) or in microvascular endothelial cells obtained from human omentum (A. J. Carter et al., 1989).Phosphoinositide turnover and Ca^{2+} mobilization show a similar dependence on thrombin dose (Lampugnani et al., 1989).

4.2. Role in Thrombin-Mediated PGI₂ Synthesis

As thrombin produces an immediate increase in inositol polyphosphates, this would appear to be consistent with the hypothesis that an early event in thrombin-mediated endothelial cell activation and resultant loss of barrier function and PGI_2 synthesis is the activation of phosphoinositide-specific PLC. Results from kinetic experiments from several laboratories suggest a sequential association between receptor-linked and Ca^{2+}-controlled hydrolysis of phosphoinositides and the ultimate generation of prostaglandins (Jaffe et al., 1987; Halldorsson et al., 1988). Phosphoinositide hydrolysis also precedes the onset of thrombin-mediated permeability changes (Lum et al., 1992). Similar temporal observations have been noted with bradykinin-challenged bovine endothelium (Lambert et al., 1986). Hong and Deykin (1982) also demonstrated that endothelial PGI_2 synthesis involved diacylglycerol formation. Thus, thrombin-induced PGI_2 synthesis involves the sequential action of two Ca^{2+}-dependent phospholipases, PLC and PLA_2. The mobilization of Ca^{2+} likely occurs as a result of thrombin effects on phosphoinositide metabolism by PLC and the transient generation of IP_3, events which precede the release of arachidonate acid by PLA_2 and the secretion of PGI_2.

Using Fura-2 to trace the kinetics of $[Ca^{2+}]_i$, a prolonged elevation of

$[Ca^{2+}]_i$ is seen in α-thrombin-stimulated cells. The initial component of the response is the result of rapid mobilization of Ca^{2+} from intracellular pools as it is insensitive to extracellular calcium, $[Ca^{2+}]_e$. This phase of the response is consistent with thrombin-induced PLC-mediated PIP_2 hydrolysis. The sustained increase in $[Ca^{2+}]_i$ is the result of a net increase in Ca^{2+} influx as it correlates well with the time course of $^{45}Ca^{2+}$ uptake and is abolished by a Ca^{2+}-free medium. As shown in Fig. 7 the increase in HUVEC cytosolic Ca^{2+} is derived primarily from endogenous Ca^{2+} stores as the removal of Ca^{2+}_e attenuates but does not abolish the early increase in Ca^{2+}_i. Similarly, the absence of Ca^{2+}_e availability does not significantly inhibit thrombin-induced PGI_2 synthesis (Garcia *et al.*, 1990). In contrast, the addition of Ca^{2+}_i chelators such as TMB-8 (Jaffe *et al.*, 1987) or BAPTA (Fig. 8) abolishes α-thrombin-induced prostaglandin synthesis.

4.3. Role of Ca^{2+} in Thrombin-Mediated Permeability Changes

Lum *et al.* (1989) have demonstrated that the thrombin-induced increase in endothelial permeability to albumin is dependent on both Ca^{2+} mobilization from intracellular stores and increased extracellular Ca^{2+} entry in endothelial cells. Removal of extracellular Ca^{2+} with EGTA or replacement of Ca^{2+} with lanthanum chloride, which competes for Ca^{2+} entry, diminishes the α-thrombin-induced increase in bovine endothelial cell permeability. Buffering cytosolic Ca^{2+} by preloading endothelial cells with Quin 2 produced a 50% reduction in the α-thrombin-induced permeability response.

Goligorsky *et al.* (1989) have reported that in bovine pulmonary artery endothelial cells, the sustained phase of Ca_i^{2+} transients was diminished by pretreatment with nordihydroguaiaretic acid, a lipoxygenase inhibitor, whereas indomethecin, a cyclooxygenase inhibitor, prolonged the sustained increase in Ca_i^{2+} after thrombin exposure (Fig. 9). This sustained increase in Ca_i^{2+} appears to be essential for endothelial cell retraction (Goligorsky *et al.*, 1989) which probably modulates the permeability re-

Figure 7. α-Thrombin-induced aequorin luminescence in human endothelium. HUVEC loaded with the photoprotein aequorin were embedded in agarose threads. Panel A shows the representative changes in aequorin luminescence induced by 10 nM α-thrombin in the presence or absence of 1.2 mM Ca^{2+}_e. Panel B plots the effect of pertussis toxin pretreatment on 20 mM NaF- and 10 nM α-thrombin-induced aequorin luminescence. Pertussis treatment alone (1 μg/ml × 2 h) did not alter basal luminescence. In additional experiments, HUVEC monolayers were scrape-loaded with either buffer or GDP βS (0.5 mM) prior to aequorin loading. Panel C shows 20 mM NaF-, 10 nM α-thrombin-, and 1 μM A23187-induced aequorin luminescence determined in the presence of 1.2 mM Ca^{2+} with or without prior loading with GDP βS. (From Garcia *et al.*, 1991, 1992.)

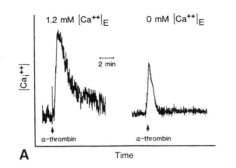

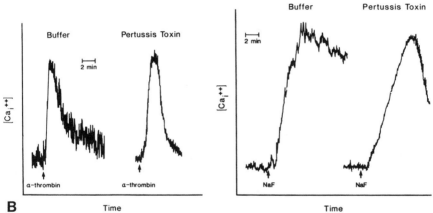

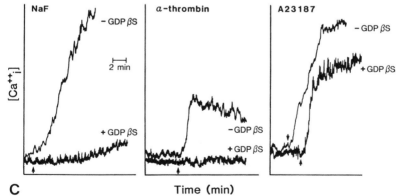

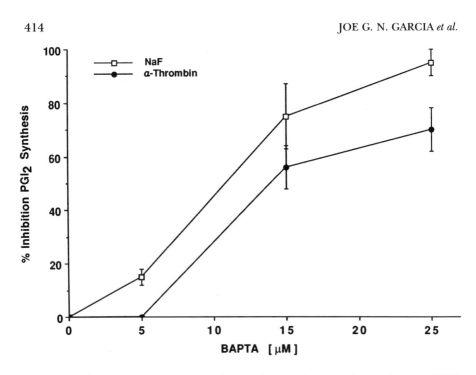

Figure 8. Effect of BAPTA-AM on agonist-stimulated PGI$_2$ synthesis. Confluent HUVEC monolayers in 24-well plates were pretreated with either buffer or increasing concentrations of BAPTA-AM for 30 min. The cells were then washed twice and cell-free supernatants harvested 15 min after agonists were added. Chelation of intracellular Ca^{2+} by BAPTA produces significant inhibition fo PGI$_2$ synthesis. (From Garcia *et al.*, 1991b.)

sponse (Garcia *et al.*, 1986). Using a microinjection technique with a concomitant recording of Fura-2 fluorescence, Goligorsky *et al.* (1989) have shown that microinjection of PIP$_2$-specific PLC, IP$_3$, or Ca^{2+} causes an immediate Ca$_i^{2+}$ increase but is not followed by the sustained elevation characteristic of thrombin. Only microinjection of phospholipase A$_2$ or co-injection of phospholipase A$_2$ with PIP$_2$-specific PLC results in the sustained Ca$_i^{2+}$ transients typically observed with thrombin. As in the case of PGI$_2$ synthesis these results are most consistent with a model of thrombin–endothelial cell interaction resulting in simultaneous activation of PLC and phospholipase A$_2$ (Fig. 10). This could occur by dual coupling of a thrombin receptor to GTP-binding proteins linked to PLC and phospholipase A$_2$ or by activation of two types of thrombin receptors with different coupling pathways. Consistent with this hypothesis, endothelial cell activation of PLC by thrombin would trigger phosphatidylinositol turnover and immediate Ca^{2+} mobilization, while phospholipase A$_2$ activation would lead

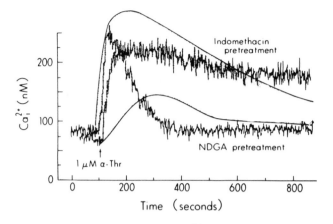

Figure 9. Modulation of the thrombin-induced Ca_i^{2+} transients by lipoxygenase inhibition and cyclooxygenase inhibition. The sustained phase of thrombin-induced Ca_i^{2+} transients is diminished by pretreatment with nordihydroguaiaretic acid (NDGA), a lipoxygenase inhibitor, and prolonged by pretreatment with indomethacin, a cyclooxygenase inhibitor. The area between the two solid lines represents the mean ± SD of the unconditioned thrombin response. (From Goligorsky *et al.*, 1989.)

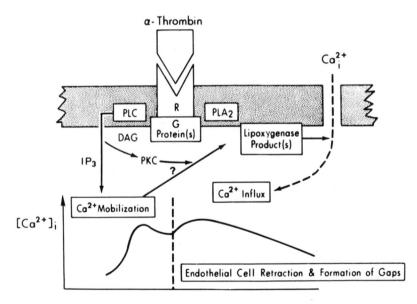

Figure 10. Hypothetical pathways of endothelial cell activation by thrombin. Interaction of thrombin with endothelial cell receptor(s) causes activation of PLC and phospholipase A_2 resulting in generation of second messengers which leads to endothelial cell retraction and alteration of barrier function as well as PGI_2 synthesis. (From Goligorsky *et al.*, 1989.)

to the formation of lipoxygenase products which would trigger the sustained phase of Ca^{2+} influx, which is a prerequisite for the thrombin-induced changes in endothelial cell topography (Goligorsky *et al.*, 1989).

Thrombin stimulation of inositol phosphate turnover in endothelial cells is dependent upon the presence of an active catalytic site. Both α- and γ-thrombin activate PLC resulting in (1,4,5)-IP_3 generation and activation of protein kinase C, whereas DIP- and PPACK-α-thrombin are inactive. DIP- and PPACK-α-thrombin do not increase $[Ca^{2+}]_i$ in HUVEC (Jaffe *et al.*, 1987; Brock and Capasso, 1988).

4.4. Role of Guanine Nucleotide Regulatory Proteins in PGI_2 Synthesis

In a variety of cell systems, stimulus/response coupling involves regulation by a family of heterotrimeric GTP-binding proteins (G-proteins). There is abundant evidence that a G-protein plays an essential transducing role in the coupling of receptors for Ca^{2+}-mobilizing hormones to PI-specific PLC in various cell types. Recent evidence suggests that guanine nucleotide-binding proteins (G-proteins) mediate receptor activation of PLC-stimulated polyphosphoinositide breakdown (Cockcroft and Gomperts, 1985; Brock *et al.*, 1988). The contribution of G-proteins in HUVEC membranes to the regulation of thrombin-induced PGI_2 synthesis has been recently explored (Garcia *et al.*, 1990; Magnusson *et al.*, 1989). The participation of G-proteins in the generation of PGI_2 is implicated by experiments wherein the addition of the known G-protein activators NaF and GTP γS to intact and permeabilized monolayers, respectively, results in time- and dose-dependent increase in the levels of 6-keto $PGF_{1\alpha}$ over control values (Figs. 11 and 12) (Garcia *et al.*, 1991, 1992). The involvement of G-proteins in regulating thrombin-induced Ca^{2+} mobilization and prostaglandin synthesis has been further demonstrated in studies using the G-protein inhibitor GDP βS. Pretreatment of HUVEC monolayers with the GDP βS (15–120 min exposure) does not significantly alter spontaneous PGI_2 synthesis (< 5% inhibition at all doses tested, 0.05–5 mM). However, there is marked dose-dependent attenuation of α-thrombin-stimulated PGI_2 responses in permeabilized cells pretreated with GDP βS for 1 h (Fig. 13) (Garcia *et al.*, 1991). To further address potential G-protein involvement in the regulation of α-thrombin-induced Ca^{2+} mobilization, HUVEC cells loaded with both aequorin and GDP βS were subsequently stimulated with either NaF, α-thrombin, or calcium ionophore A23187. As shown in Fig. 7C, α-thrombin- and NaF- but not A23187-mediated increases in Ca^{2+}_i are significantly reduced in GDP βS-treated cells (Garcia *et al.*, 1991). These results are consistent with the results of Brock and Capasso (1989), who found GTP γS to augment α-thrombin-induced inositol phosphate generation and indicate G-protein modulation of Ca^{2+}_i via an effect on PLC.

As these studies suggested that thrombin-induced Ca^{2+} mobilization and PGI_2 synthesis in human endothelium are regulated by GTP-binding proteins, attempts were made to further characterize the G-protein, G_p, which is functionally coupled to thrombin-stimulated PLC activity. Bacterial toxins, such as islet-activating protein from *Bordetella pertussis*, catalyze the ADP ribosylation of the α subunit of membrane-bound G proteins (Gilman, 1984, 1987) rendering some cell systems unresponsive to ligand stimulation and have thereby been useful tools to probe the involvement of G-proteins in cellular signal transduction. A 40-kDa pertussis toxin-sensitive, ADP ribosylation substrate has been identified in HUVEC membranes and determined to be immunologically related to other Gα subunits (Garcia *et al.*, 1990). In HUVEC monolayers, thrombin decreases the availability of a GTP-binding protein substrate susceptible to ADP ribosylation by pertussis toxin (Garcia *et al.*, 1990). Preincubation with pertussis toxin, however, fails to block either thrombin-induced PLC activity, IP_3

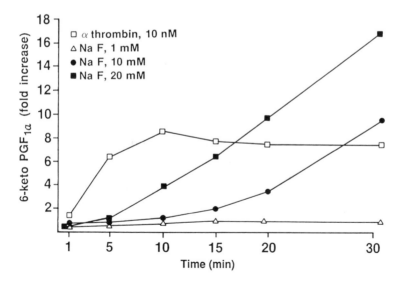

Figure 11. Effect of NaF on HUVEC production of 6-keto $PGF_{1\alpha}$. Varying concentrations of NaF were added to confluent HUVEC monolayers in 24-well plates and cell supernatants removed at specified intervals and assessed for levels of 6-keto $PGF_{1\alpha}$. NaF responses were compared to α-thrombin (10 nM). PGF_2 release is expressed as the fold increase in PGF_2 synthesis over buffer-treated monolayers ($n = 8$). (From Garcia *et al.*, 1990, 1991.)

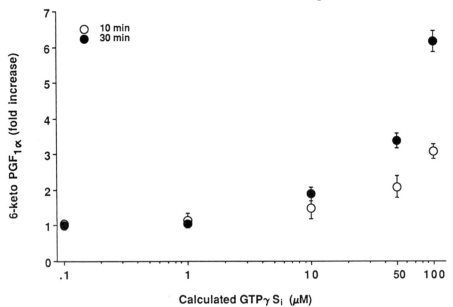

Figure 12. Effect of GTP γS on PGF$_2$ synthesis by HUVEC monolayers. Varying concentrations of GTP γS (the nonhydrolyzable GTP analog) were added to digitonin-permeabilized monolayers and cell supernatants removed at 10 or 30 min (n = 6 each concentration). The range of exogenous extracellular GTP γS levels for the depicted concentrations was 1 μM– mM. (From Garcia *et al.*, 1990.)

formation, Ca^{2+} mobilization (Fig. 7B) (Brock and Capasso, 1989; Garcia *et al.*, 1991), or PGI$_2$ synthesis (Fig. 14) (Garcia *et al.*, 1990, 1991). Similar pretreatment of bovine endothelial cell monolayers with pertussis toxin does not inhibit thrombin-induced increases in endothelial permeability (Aschner *et al.*, unpublished data). Nevertheless, thrombin and GTP γS act in a synergistic fashion to increase PIP$_2$ hydrolysis, as measured by IP$_3$ and diacylglycerol formation (Brock and Capasso, 1989). Thus, these studies suggest that G-proteins are involved in transducing signals from the thrombin receptor to its intracellular targets which result in prostaglandin synthesis. Since the response is pertussis toxin-insensitive, this further suggests that a pertussis toxin-insensitive G-protein may be a key intermediate in the signaling pathway linking thrombin receptor occupancy to PLC-mediated PIP$_2$ hydrolysis in HUVEC. Similar ADP ribosylation substrates for other bacterial toxins such as botulinum C and cholera toxin have also been identified. However, these bacterial toxin-sensitive substrates, like the pertussis toxin-sensitive G-protein, do not appear to be involved in thrombin-mediated PGI$_2$ synthesis in human endothelium

(Garcia *et al.*, unpublished data). Their role in endothelial cell signaling is currently under investigation.

The evidence to data for G-protein regulation of thrombin-mediated PLC activity does not exclude the possibility of a distinct G-protein directly regulating PLA$_2$ activity. Because PLA$_2$, like PLC, is a Ca^{2+}-requiring enzyme, and GTP-binding proteins appear to activate PLC by reducing the Ca^{2+} requirement of this enzyme (Smith *et al.*, 1986), G-proteins have also been suggested to couple receptors directly linked to PLA$_2$ (Okano *et al.*, 1987; Kajiyama *et al.*, 1989; Houslay *et al.*, 1986; Nakashima *et al.*, 1988) independently of PLC, inositol phosphate generation, and thus intracellular Ca^{2+} release. Jelsema and Axelrod (1987) have shown that the β-subunit dimer of G-proteins is capable of activating PLA$_2$ in retinal rod outer segments, supporting a link between the receptor and PLA$_2$ by GTP-binding proteins in specific cellular systems. As neither HUVEC PLC activity, PLA$_2$ activity, nor PGI$_2$ synthesis stimulated by either NaF or

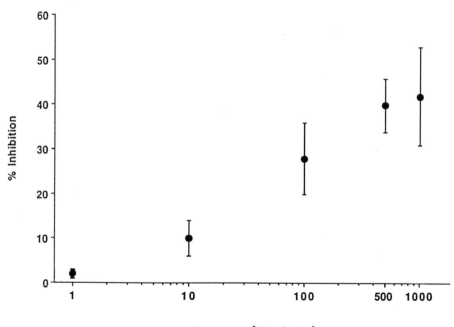

Figure 13. Effect of GDP βS on α-thrombin-induced PGI$_2$ synthesis. Increasing concentrations of GDP βS in M-199-digitonin (1 μg/ml) were added to HUVEC monolayers for 1 h prior to α-thrombin stimulation (10 nM). PGI$_2$ was assayed at 10 min (n = 5 each concentration). (From Garcia *et al.*, 1990.)

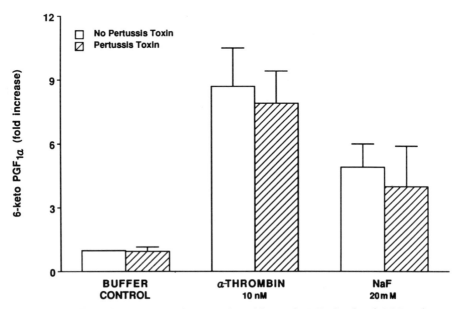

Figure 14. Effect of pertussis toxin on α-thrombin- and NaF-stimulated PGI_2 release. HUVEC monolayers were pretreated (120 min) with either pertussis toxin (cross-hatched bar) (1 μg/ml, no DTT activation) or buffer (open bar) and challenged with α-thrombin (10 nM) or NaF (20 mM). PGI_2 releases was assayed in cell supernatants removed at 10 min and is expressed as the fold increase in PGI_2 synthesis over buffer-treated monolayers ($n = 8$ each intervention). (From Garcia *et al.*, 1990, 1991.)

thrombin is inhibited by the G-protein probe, pertussis toxin (Garcia *et al.*, 1990, 1991), the existence of a distinct G-protein linked to PLA_2 (G_{PLA2}) in human endothelium remains speculative. Further work is clearly required to characterize the G-protein modulation of specific endothelial prostaglandin synthetic responses.

4.5. Role of Protein Kinase C in the Regulation of Thrombin-Mediated Prostaglandin Synthesis

The other major product of PLC-mediated PIP_2 hydrolysis is *sn*-1,2-diacylglycerol (DAG) which activates a Ca^{2+}- and phospholipid-sensitive protein kinase (PKC) (Kikkawa and Nishizuka, 1986). A number of investigators have been keenly interested in DAG, and its potential participation in the regulation of thrombin-induced PGI_2 synthesis as well as endothelial permeability by virtue of its ability to activate and translocate protein kinase C (PKC). Both α-thrombin and NaF activate PI-specific

PLC (Brock *et al.*, 1988; Garcia *et al.*, 1991) and result in the rapid generation of DAG with α-thrombin-stimulated DAG production occurring within 15 s (Brock *et al.*, 1988), whereas NaF-induced DAG formation is more delayed at > 2 min (Garcia *et al.*, 1991a). Recently, this increase in DAG has been confirmed to correlate with a translocation of PKC activity from the cytosolic compartment to the membrane compartment (Garcia *et al.*, 1992). Phorbol 12-myristate 13-acetate (PMA), α-thrombin, and NaF produce rapid, time-dependent translocation of PKC from the cytosolic compartment to the membrane compartment as assessed by phosphorylation of histone-1 (Lynch *et al.*, 1990; Garcia *et al.*, 1992). Evidence for a critical regulatory role of PKC in signal-transducing pathways involved in prostaglandin synthesis is available. Pretreatment with stimuli which activate PKC, such as PMA, mezerein, and *sn*-1,2-dioctanylglycerol, attenuates thrombin-induced increases in the concentration of intracellular cytosolic Ca^{2+}, as well as IP_3 formation (Brock and Capasso, 1988; Garcia *et al.*, 1992). The exact site of PKC effects on endothelial cell prostaglandin responses has not been delineated; however, as PMA inhibits both thrombin- and NaF-induced Ca^{2+} increases, the site of PKC regulation appears to be distal to the occupancy of the thrombin receptor (Brock and Capasso, 1988; Garcia *et al.*, 1992; Lynch *et al.*, 1990). These results strongly suggest a tight coupling between PLC and cytosolic Ca^{2+} and that this coupling is under regulation by PKC. An obvious possibility exists that PKC activation by endogenous DAG production during thrombin stimulation may represent an inhibitory feedback pathway for PLC activity, thus limiting phosphoinositide hydrolysis. The DAG formation after NaF and thrombin is sustained (greater than 10 min), implying an important physiologic role for PKC in regulating endothelial cell activation responses.

PKC appears to critically regulate prostaglandin synthesis, although the exact site of PKC effect is unidentified (Demolle and Boeynaems, 1988; T. D. Carter *et al.*, 1989; Garcia *et al.*, 1992). PKC has been implicated in endothelial cell desensitization involving PGI_2 synthesis via mechanisms which include receptor phosphorylation, receptor inactivation, phosphorylation of G-proteins, and/or activation of inositol phosphate kinase systems (Conolly *et al.*, 1986; King and Rittenhouse, 1989). Activation of PKC by PMA over a 60 min period does not increase prostaglandin synthesis; however, PKC activation produced by brief pretreatment with PMA (5 min) results in the inhibition of NaF-induced inositol phosphate increases and attenuation of both α-thrombin- and NaF- but not Ca^{2+} ionophore A23187-activated increases in Ca^{2+}_i detected by aequorin luminescence, again suggesting negative feedback inhibition of PI-specific PLC (Garcia *et al.*, 1990). Agonist-stimulated arachidonate release and PGI_2 synthesis in PMA-pretreated cultured human endothelial cells, however, are potentiated and the enhanced PGI$_2$ synthesis produced by A23187, NaF, and α-thrombin is dependent upon the dose of PMA (Fig. 15) (Garcia *et al.*, 1990). Treatment of HUVEC monolayers with the intracellular Ca^{2+} chelator, BAPTA-AM, does not inhibit PMA-induced PKC activation; however, BAPTA-mediated inhibition of agonist-stimulated PGI$_2$ synthesis is partially attenuated by prior PMA pretreatment. The agonist-stimulated production of PGI$_2$ induced by α-thrombin, NaF, and A23187 is further implicated as a PKC-dependent event by studies involving either staurosporine, a potent PKC inhibitor, or downregulation of PKC activity by prolonged (18 h) treatment with PMA (Garcia *et al.*, 1990; Lynch *et al.*, 1990). Staurosporine, at concentrations which inhibit PKC-induced phosphorylation of histone-1, augments α-thrombin- or NaF-induced production of inositol phosphates but markedly inhibits α-thrombin-, NaF-, and A23187-induced PGI$_2$ synthesis (Garcia *et al.*, 1990). The downregulation of PKC activity by prolonged PMA treatment produces similar inhibition of PGI$_2$ synthesis by these agonists (≈ 50% inhibition). Neither stauro-

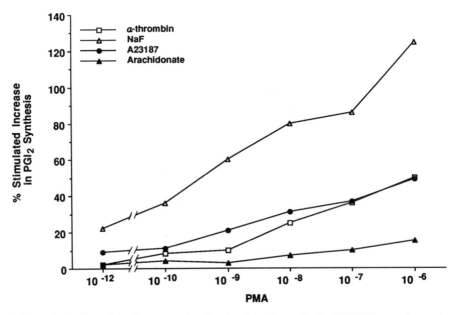

Figure 15. Effect of PMA on agonist-stimulated PGI$_2$ synthesis. HUVEC monolayers in 24-well plates were pretreated for 5 min with increasing concentrations of PMA prior to agonist challenge. Shown is the % increase in PGI$_2$ synthesis over buffer-pretreated, agonist-challenged wells. The concentrations of each agonist were as follows; α-thrombin 10 nM; NaF, 20 mM; A23187 1 μM; arachidonate, 10 μM. (From Garcia *et al.*, 1992.)

sporine nor PKC downregulation attenuates agonist-induced Ca^{2+} mobilization nor does it alter PGI_2 synthesis after exogenous arachidonate (T. D. Carter *et al.*, 1989; Garcia *et al.*, 1990c). These studies indicate that the integrated activities of phosphoinositol-specific PLC and PLA_2 which lead to arachidonate release and PGI_2 synthesis, are under complex regulation by factors which include both PKC activation and cytosolic Ca^{2+} concentrations. These results further suggest that PKC activation exerts dual effects on prostaglandin synthesis via negative regulation of G_p-coupled phosphoinositide-specific PLC, and positive feedback regulation of arachidonate release and PGI_2 synthesis possibly by lowering Ca^{2+} requirements for membrane PLA_2 enzymatic activity. PKC is thus a critical determinant in the regulation of human endothelial cell prostaglandin synthesis by receptor-mediated (α-thrombin) or G-protein-induced (NaF) cellular activation.

4.6. Role of Protein Kinase C Activation in Thrombin-Mediated Barrier Dysfunction

Increased $[Ca^{2+}]_i$ may increase permeability by activation of Ca^{2+}-sensitive regulatory proteins such as calmodulin and PKC or by direct interaction with cytoskeletal actin microfilaments and associated proteins, such as tubulin, vinculin, and myosin light chain (Werth *et al.*, 1983; Huang *et al.*, 1988; Stasek *et al.*, 1990; Wysolmerski and Lagunoff, 1990). PKC activation by α-thrombin may result in phosphorylation of cytoskeletal proteins (Stasek *et al.*, 1990), leading to changes in endothelial cell shape and altered cell–cell contact (Garcia *et al.*, 1986). The cytoskeletal proteins caldesmon, vinculin, vimentin, and myosin light chain are rapidly phosphorylated in response to PKC activation (Lynch *et al.*, 1990; Stasek *et al.*, 1990). This may result in the transient disruption of interendothelial junctional complexes, and thereby increase endothelial cell permeability. Lynch *et al.* (1990) have shown that PKC activation is correlated with increased [125I]albumin clearance rates in bovine pulmonary artery endothelial cells exposed to PMA, PLC, and α-thrombin (Fig. 16). Inhibition of endothelial PKC with H7 significantly reduces the permeability-increasing effects of α-thrombin (Lynch *et al.*, 1990). Endothelial PKC activation by thrombin results in a characteristic decrease in cytosolic PKC activity with a simultaneous increase in covalently bound membrane-associated PKC activity (Lynch *et al.*, 1990; Garcia *et al.*, 1992; Stasek *et al.*, 1990).

4.7. Thrombin's Enzymatic Activity

The strict requirement for an active catalytic site in the aforementioned thrombin-mediated cellular responses (PGI_2 synthesis and barrier

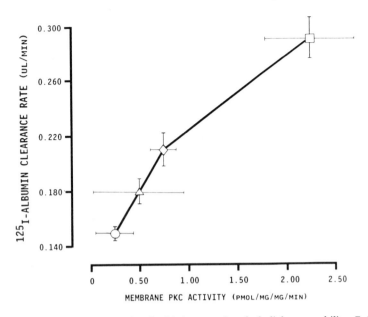

Figure 16. PKC activation is associated with increased endothelial permeability. Relationship of membrane PKC activity to transendothelial [^{125}I]albumin clearance rates after exposure of endothelial monolayers to (O) DMSO vehicle, ($\triangle$) 10^{-8} M, ($\Diamond$) 10^{-7} M, and ($\square$) 10^{-6}M PMA. Values represent the mean ± SEM of three experiments (PKC activity) or four experiments ([^{125}I]albumin clearance rates). (From Lynch *et al.*, 1990.)

dysfunction) allows for the possibility that thrombin-induced PLC activation is not a typical receptor-mediated event. If transmembrane signal generation is active-site independent, catalytic cleavage of a plasma membrane substrate may be involved. One possible substrate is the thrombin "receptor." This receptor may be the same site or distinct from the receptor that binds both DIP-α-thrombin and α-thrombin (thrombomodulin). Studies in fibroblasts (Glenn and Cunningham, 1979; Glenn *et al.*, 1980; Carney *et al.*, 1986) have shown that thrombin-stimulated cell division involves proteolysis of its cell surface receptor. Alternatively, there may be an active-site-dependent receptor for thrombin which is different from the receptor that binds both DIP-α-thrombin and active α-thrombin. However, this receptor must have low affinity or be few in number to explain the difficulty in distinguishing it from "nonspecific" binding seen in receptor assay systems.

The catalytic action of α-thrombin might also directly affect components of the guanine nucleotide-binding protein that interacts with PLC, perhaps by direct proteolysis of the α subunit of this GTP-binding protein.

Cleavage of the α subunit may correlate with the activation of PLC. Alternatively, thrombin may activate endogenous proteases causing a proteolytic modification of the GTP-binding protein that leads to PLC activation (Lapetina *et al.*, 1986).

Despite the strong association between the thrombin-induced increase in endothelial cell permeability and the PLC-mediated PIP_2 hydrolysis caused by thrombin, these events may be independent. Enzymatically active thrombin may increase endothelial monolayer permeability by degradation of the subcellular matrix directly or by activation of endothelial proteases. Thrombin may also alter membrane proteins involved in the formation of tight junctions which play a role in the barrier function of endothelial cells.

5. CONCLUSION

A great deal of progress has been made in understanding the transmembrane signals activated when endothelial cells are exposed to α-thrombin. A great deal more work is needed before the biochemical and cellular mechanisms involved in thrombin-mediated loss of barrier function and prostaglandin release are delineated so that pathophysiology of pulmonary edema following pulmonary intravascular coagulation can be fully understood.

Acknowledgments

Supported by NIH Research Grants HL45638, HL27016, HL02312, HL44746, awards from the Veteran's Administration Medical Research Council, and the American Heart Association, Indiana Affiliate. J.L.A. is a Fellow of the Parker B. Francis Foundation.

6. REFERENCES

Alexander, R. J., Fenton, J. W., II, and Detwiler, T. C., 1983, Thrombin–platelet interactions: An assessment of the roles of saturable and nonsaturable binding in platelet activation, *Arch. Biochem. Biophys.* **222**:266–275.

Aschner, J. L., Lennon, J. M., Fenton, J. W., II, Aschner, M., and Malik, A. B., 1990, Enzymatic activity is necessary for thrombin-mediated increase in endothelial permeability, *Am. J. Physiol.* **259**:L270–L275.

Awbrey, B. J., Hoak, J. C., and Owen, W. G., 1979, Binding of human thrombin to cultured human endothelial cells, *J. Biol. Chem.* **254**:4092–4095.

Berridge, M. J., and Irvine, R. F., 1984, Inositol trisphosphate, a novel second messenger in cellular signal transduction, *Nature* **308**:693–698.

Bizios, R., Lai, L. C., Cooper, J. A., Del Vecchio, P. J., and Malik, A. B., 1988, Thrombin-induced adherence of neutrophils to cultured endothelial monolayers. Increased endothelial adhesiveness, *J. Cell. Physiol.* **134**:275–280.

Brock, T. A., and Capasso, E. A., 1988, Thrombin and histamine activate phospholipase C in human endothelial cells via a phorbol ester-sensitive pathway, *J. Cell. Physiol.* **136**:54–62.

Brock, T. A., and Capasso, E. L., 1989, GTP γS increases thrombin-mediated inositol trisphosphate accumulation in permeabilized human endothelial cells, *Am. Rev. Respir. Dis.* **140**:1121–1125.

Brock, T. A., Dennis, P. A., Griendling, K. K., Diehl, T. S., and Davies, P. F., 1988, GTP γS loading of endothelial cells stimulates phospholipase C and uncouples ATP receptors, *Am. J. Physiol.* **225**:C667–C673.

Camussi, G., Aglietta, M., Malavasi, F., Tetta, C., Piacibello, W., Sanavio, F., and Bussolino, F., 1983, The release of platelet-activating factor from human endothelial cells in culture, *J. Immunol.* **131**:2397–2403.

Carney, D. H., Stiernberg, J., and Fenton, J. W., II, 1984, Initiation of proliferative events by human α-thrombin requires both receptor binding and enzymic activity, *J. Cell. Biochem.* **256**:181–195.

Carney, D. H., Herbosa, G. J., Stiernberg, J., Bergmann, J. S., Gordon, E. A., Scott, D., and Fenton, J. W., II, 1986, Double-signal hypothesis for thrombin initiation of cell proliferation, *Semin. Thromb. Hemostas.* **12**:231–240.

Carter, A. J., Eisert, W. G., and Muller, T. H., 1989a, Thrombin stimulates inositol phosphate accumulation and prostacyclin synthesis in human endothelial cells from umbilical vein but not from omentum, *Thromb. Haemostas.* **61**:122–136.

Carter, T. D., Hallam, T. J., and Pearson, J. D., 1989b, Protein kinase C activation alters the sensitivity of agonist-stimulated endothelial-cell prostacyclin production to intracellular Ca^{2+}, *Biochem. J.* **262**:431–437.

Cockcroft, S., and Gomperts, B. D., 1985, Role of guanine nucleotide regulatory binding proteins in the activation of polyphosphoinositide phosphodiesterase, *Nature* **314**:534–536.

Connolly, T. M., Lawing, W. J., Jr., and Majerus, P. W., 1986, Protein kinase C phosphorylates human platelet inositol trisphosphate 5′-phosphomonoesterase, increasing the phosphatase activity, *Cell* **46**:951–958.

Del Vecchio, P. J., Siflinger-Birnboim, A., Shepard, J. M., Bizios, R., Cooper, J. A., and Malik, A. B., 1987, Endothelial monolayer permeability to macromolecules, *Fed. Proc.* **46**:2511–2515.

Del Vecchio, P. J., Belloni, P. N., Holleran, L. A., Lum, H., Siflinger-Birnboim, A., and Malik, A. B., 1992, Culture and characterization of pulmonary microvascular endothelial cells, *In Vitro* (in press).

Demolle, D., and Boeynaems, J. M., 1988, Role of protein kinase C in the control of vascular prostacyclin: Study of phorbol esters effect in bovine aortic endothelium and smooth muscle, *Prostaglandins* **35**(2):243–257.

DiCorleto, P. E., and Bowen-Pope, D. F., 1983, Cultured endothelial cells produce a platelet-derived growth factor-like protein, *Proc. Natl. Acad. Sci. USA* **80**:1919–1923.

Esmon, C. T., and Owen, W. G., 1981, Identification of an endothelial cell cofactor for thrombin-catalyzed activation of protein C, *Proc. Natl. Acad. Sci. USA* **78**:2249–2252.

Fenton, J. W., II, Fasco, M. J., Stackrow, A. B., Aronson, D. L., Young, A. M., and Finlayson, 1977, Human thrombins. Production, evaluation and properties of thrombin, *J. Biol. Chem.* **252**:3587–3589.

Galdal, K. S., Lybert, T., Evensen, S. A., Nilsen, E., and Prydz, H., 1985, Thrombin induces thromboplastin synthesis in cultured vascular endothelial cells, *Thromb. Haemostas.* **54**:373–376.

Garcia, J. G. N., Birnboim, A. S., Bizios, R., Del Vecchio, P. J., Fenton, J. W., and Malik, A. B., 1986, Thrombin-induced increases in albumin clearance across cultured endothelial monolayers, *J. Cell. Physiol.* **128**:96–104.

Garcia, J. G. N., Perlman, M. B., Johnson, A., Ferro, T., Jubiz, W., and Malik, A. B., 1988, Inflammatory events following fibrin microembolization: Alteration in alveolar macrophage and neutrophil function, *Am. Rev. Respir. Dis.* **137**:630–635.

Garcia, J. G. N., Painter, R. G., Fenton, J. W., English, D., and Callahan, K. S., 1990, Thrombin-induced prostacyclin biosynthesis in human endothelium: Role of guanine nucleotide-regulatory proteins in stimulus coupling responses, *J. Cell. Physiol.* **142(1)**:186–193.

Garcia, J. G. N., Dominguez, J., and English, D., 1991, Sodium fluoride induces phosphoinositide hydrolysis, Ca^{2+} mobilization and prostacyclin synthesis in cultured human endothelium. Further evidence for regulation by a pertussis toxin-insensitive guanine nucleotide binding protein, *Am. J. Respir. Cell Mol. Biol.* **5**:113–124.

Garcia, J. G. N., Stasek, J., Natarajan, V., Patterson, C., and Dominguez, J., 1992, Role of protein kinase C in the regulation of phospholipase activities and prostaglandin synthesis in human endothelium, *Am. J. Respir. Cell Mol. Biol.* **6**:315–325.

Garcia-Szabo, R. R., Kern, D. F., Bizios, R., Fenton, J. W., II., Minnear, F. L., Lo, S. K., and Malik, A. B., 1984, Comparison of α- and γ-thrombin on lung fluid balance in anesthetized sheep, *J. Appl. Physiol.* **57**:1375–1383.

Gelehrter, T. D., and Sznycer-Laszuk, R., 1986, Thrombin induction of plasminogen activator-inhibitor in cultured human endothelial cells, *J. Clin. Invest.* **77**:165–169.

Gilman, A. G., 1984, G proteins and dual control of adenylate cyclase, *Cell* **36**:577–579.

Gilman, A. G., 1987, G proteins: Transducers of receptor-generated signals, *Annu. Rev. Biochem.* **56**:615–649.

Glenn, K. C., and Cunningham, D. D., 1979, Thrombin-stimulated cell division involves proteolysis of its cell surface receptor, *Nature* **278**:711–714.

Glenn, K. C., Carney, D. H., Fenton, J. W., II, and Cunningham, D. D., 1980, Thrombin active site regions required for fibroblast receptor binding and initiation of cell division, *J. Biol. Chem.* **255**:6609–6616.

Goldsmith, J. G., and Needleman, S. N., 1982, A comparative study of thromboxane and prostacyclin release from *ex-vivo* and cultured bovine vascular endothelium, *Prostaglandins* **24**:173–178.

Goligorsky, M. S., Menton, D. N., Laslo, A., and Lum, H., 1989, Nature of thrombin-induced sustained increase in cytosolic calcium concentration in cultured endothelial cells, *J. Biol. Chem.* **264**:16771–16775.

Gospodarowicz, D., Brown, C. D., Birdwell, C. R., and Zetter, B. R., 1978, Control of proliferation of human vascular endothelial cells, *J. Cell Biol.* **77**:774–788.

Hallam, T. J., Pearson, J. D., and Needham, L. A., 1988, Thrombin-stimulated elevation of human endothelial-cell cytoplasmic free calcium concentration causes prostacyclin production, *Biochem. J.* **S51**:243–249.

Halldorsson, H., Kjeld, M., and Thorgeirsson, G., 1988, Role of phosphoinositides in the regulation of endothelial prostacyclin production, *Arteriosclerosis* **8**:147–154.

Harlan, J. M., Thompson, P. J., Ross, R. R., and Bowen-Pope, D. F., 1986, α-Thrombin induced release of platelet-derived growth factor-like molecule(s) by cultured human endothelial cells, *J. Cell. Biol.* **103(3)**:1129–1133.

Heath, C. A., Lai, L., Bizios, R., and Malik, A. B., 1986, Pulmonary hemodynamic effects of antisheep serum-induced leukopenia, *J. Leukocyte Biol.* **39**:385–397.

Hong, S. L., and Deykin, D., 1982, Activation of phospholipases A$_2$ and C in pig aortic endothelial cells synthesizing prostacyclin, *J. Biol. Chem.* **257:**7151–7154.

Hong, S. L., McLaughlin, N. J., Tzeng, C., and Patton, G., 1985, Prostacyclin synthesis and deacylation of phospholipids in human endothelial cells: Comparison of thrombin, histamine and ionomycin, *Thromb. Res.* **38:**1–10.

Horgan, M. J., Fenton, J. W., II, and Malik, A. B., 1987, Alpha-thrombin-induced pulmonary vasoconstriction, *J. Appl. Physiol.* **63:**1993–2000.

Houslay, M. D., Bojanic, D., Gawler, D., O'Hagan, S., and Wilson, A., 1986, Thrombin, unlike vasopressin, appears to stimulate two distinct guanine nucleotide regulatory proteins in human platelets, *Biochem. J.* **238:**109–113.

Huang, C. H., Devanney, J. F., and Kennedy, S. P., 1988, Vimentin, a cytoskeletal substrate of protein kinase C, *Biochem. Biophys. Res. Commun.* **150:**1006–1011.

Isaacs, J., Savion, N., Gospodarowicz, D., and Shuman, M. A., 1981, Effect of cell density on thrombin binding to a specific site on bovine vascular endothelial cells, *J. Cell Biol.* **90:**670–674.

Jaffe, E. A., Grulich, J., Weksler, B. B., Hampel, G., and Watanabe, K., 1987, Correlation between thrombin-induced prostacyclin production and inositol trisphosphate and cytosolic free calcium levels in cultured human endothelial cells, *J. Biol. Chem.* **137:**8557–8565.

Jelsema, C. L., and Axelrod, J., 1987, Stimulation of phospholipase A$_2$ in bovine rod outer segment by the beta gamma subunits of transducing and its inhibition by the alpha subunit, *Proc. Natl. Acad. Sci. USA* **84:**3623–3627.

Johnson, A., and Malik, A. B., 1982a, Effect of defibrinogenation on lung fluid and protein exchange after glass bead embolization, *J. Appl. Physiol.* **53:**895–900.

Johnson, A., and Malik, A. B., 1982b, Pulmonary edema after glass bead microembolization. Protective effect of granulocytopenia, *J. Appl. Physiol.* **52:**155–161.

Johnson, A., and Malik, A. B., 1985, Pulmonary transvascular fluid and protein exchange after thrombin-induced microembolism, *Am. Rev. Respir. Dis.* **132:**70–76.

Johnson, A., Tahamont, M. V., and Malik, A. B., 1983a, Thrombin-induced lung vascular injury: Role of fibrinogen and fibrinolysis, *Am. Rev. Respir. Dis.* **128:**38–44.

Johnson, A., Blumenstock, F. A., and Malik, A. B., 1983b, Effect of complement depletion on lung fluid balance after thrombin, *J. Appl. Physiol.* **55:**1480–1485.

Johnson, A., Cooper, J. A., and Malik, A. B., 1986a, Effect of complement activation with cobra venom factor on pulmonary vascular permeability, *J. Appl. Physiol.* **61:**2202–2209.

Johnson, A., Perlman, M. B., Blumenstock, F. A., and Malik, A. B., 1986b, Superoxide dismutase prevents the thrombin-induced increase in lung vascular permeability: Role of superoxide in mediating the alteration in lung fluid balance, *Circ. Res.* **59:**405–415.

Johnson, A., Lo, S. K., and Malik, A. B., 1987, CVF-induced decomplementation: Effect on lung transvascular protein flux after thrombin, *J. Appl. Physiol.* **62:**863–869.

Kajiyama, Y., Murayama, T., and Nomura, Y., 1989, Pertussis toxin-sensitive GTP-binding proteins may regulate phospholipase A$_2$ in response to thrombin in rabbit platelets[1], *Arch. Biochem. Biophys.* **274:**200–208.

Kikkawa, U., and Nishizuka, Y., 1986, The role of protein kinase C in transmembrane signalling, *Am. Rev. Cell Biol.* **2:**149–178.

King, W. G., and Rittenhouse, S. E., 1989, Inhibition of protein kinase C by staurosporine promotes elevated accumulations of inositol trisphosphates and tetrakisphosphate in human platelets exposed to thrombin, *J. Biol. Chem.* **264:**6070–6074.

Lampugnani, M. G., Pedenovi, M., Dejana, E., Rotilio, D., Donati, M. B., Bussolino, F., Garbarino, G., Ghigo, D., and Bosia, A., 1989, Human α-thrombin induces phosphoinositide turnover and Ca^{2+} movements in cultured human umbilical vein endothelial cells, *Thromb. Res.* **54**:75–87.

Lapetina, E. G., Reep, B., and Chang, K., 1986, Treatment of human platelets with trypsin, thrombin, or collagen inhibits the pertussis toxin-induced ADP-ribosylation of a 41-kDa protein, *Proc. Natl. Acad. Sci. USA* **83**:5880–5883.

Levin, E. G., Marzec, U., Anderson, J., and Harker, L. A., 1984, Thrombin stimulates tissue plasminogen activator release from cultured human endothelial cells, *J. Clin. Invest.* **74**:1988–1995.

Levine, J. D., Harlan, J. M., Harker, L. A., Joseph, M. L., and Counts, R. B., 1982, Thrombin-mediated release of factor VIII antigen from human umbilical vein endothelial cells in culture, *Blood* **60**:531–534.

Lo, S. K., Perlman, M. B., Niehaus, G. D., and Malik, A. B., 1985, Thrombin-induced alterations in lung fluid balance in awake sheep, *J. Appl. Physiol.* **58**:1421–1427.

Lo, S. K., Ryan, T. J., Gilboa, N., Lai, L., and Malik, A. B., 1989, Role of catalytic and lysine-binding sites in plasmin-induced neutrophil adherence to endothelium, *J. Clin. Invest.* **84**:793–801.

Lo, S. K., Garcia-Szabo, R. R., and Malik, A. B., 1990, Leukocyte repletion reverses protective effect of neutropenia in thrombin-induced increase in lung vascular permeability, *Am. J. Physiol.* **259**:H149–H155.

Lollar, P., and Owen, W. G., 1980a, Clearance of thrombin from circulation in rabbits by high-affinity binding sites on endothelium, *J. Clin. Invest.* **66**:1222–1230.

Lollar, P., and Owen, W. G., 1980b, Evidence that the effects of thrombin on arachidonate metabolism in cultured human endothelial cells are not mediated by a high affinity receptor, *J. Biol. Chem.* **255**:8031–8034.

Lollar, P., Hoak, J. C., and Owen, W. G., 1980, Binding of thrombin to cultured human endothelial cells, *J. Biol. Chem.* **255**:10279–10283.

Lum, H., Del Vecchio, P. J., Schneider, A. S., Goligorsky, M. S., and Malik, A. B., 1989, Calcium dependence of the thrombin-induced increase in endothelial albumin permeability, *J. Appl. Physiol.* **66**:1471–1476.

Lum, H., Phillips, P. G., Aschner, J. L., Fletcher, P. W., and Malik, A. B., Relationship of cytosolic Ca^{2+} to the time course of thrombin-induced increase in endothelial permeability, *Amer. J. Physiol.* (in press).

Lynch, J. J., Ferro, T. J., Blumenstock, F. A., Brochenauer, A. M., and Malik, A. B., 1990, Increased endothelial albumin permeability mediated by protein kinase C activation, *J. Clin. Invest.* **85**:1991–1998.

Magnusson, M. K., Halldorsson, H., Kjeld, M., and Thorgeirsson, G., 1989, Endothelial inositol phosphate generation and prostacyclin production in response to G-protein activation by AlF_4, *Biochem. J.* **264**:703–711.

Marcus, A. J., 1978, The role of lipids in platelet function with particular reference to arachidonic acid pathways, *Lipid Res.* **19**:793–800.

Minnear, F. L., Martin, D., Taylor, A. E., and Malik, A. B., 1983, Large increase in pulmonary lymph flow after thrombin-induced intravascular coagulation, *Fed. Proc.* **42**:1274.

Moscat, J., Moreno, F., and Garcia-Barreno, P., 1987, Mitogenic activity and inositide metabolism in thrombin-stimulated pig aorta endothelial cells, *Biochem. Biophys. Res. Commun.* **145**:1302–1309.

Nakashima, S., Hattori, H., Shirato, L., Takenaka, A., and Nozawa, Y., 1988, Differential sensitivity of arachidonic acid release and 1,2-diacylglycerol formation to pertussis toxin, GDP βS and NaF in saponin-permeabilized human platelets: Possible evidence for distinct GTP-binding proteins involving phospholipase C and A_2 activation, *Biochem. Biophys. Res. Commun.* **148**:971–978.

Okano, Y., Yamada, K., Yano, K., and Nozawa, Y., 1987, Guanosine 5'-(gamma-thio) triphosphate stimulates arachidonic acid liberation in permeabilized rat peritoneal mast cells, *Biochem. Biophys. Res. Commun.* **145:**1267–1275.

Parkinson, J. E., Garcia, J. G. N., and Bang, N. U., 1990, Decreased thrombin affinity of cell-surface thrombomodulin following treatment of cultured endothelium with beta-Dxyloside, *Biochem. Biophys. Res. Commun.* **169:**177–183.

Perlman, M. B., Lo, S. K., and Malik, A. B., 1986, Effects of prostacyclin on pulmonary vascular response to thrombin in awake sheep, *J. Appl. Physiol.* **60:**546–553.

Phillips, P. G., Lun, H., Malik, A. B., and Tsan, M., 1989, Phallacidin prevents thrombin-induced increases in endothelial permeability to albumin, *Am. J. Physiol.* **257:**C562–C567.

Prescott, S. M., Zimmerman, G. A., and McIntyre, T. M., 1984, Human endothelial cells in culture produce platelet-activating factor (1-alkyl-2-acetyl-*sn*-glycero-3-phosphocholine) when stimulated with thrombin, *Proc. Natl. Acad. Sci. USA* **81:**3534–3538.

Ryan, U. S., 1990, Receptors on pulmonary endothelial cells, *Am. Rev. Respir. Dis.* **141:**S132–S136.

Ryan, U. S., White, L., Lopez, M., and Ryan, J. W., 1982, Use of microcarriers to isolate and culture pulmonary microvascular endothelium, *Tissue Cell* **14:**597–606.

Ryan, U. S., Avdonin, P. V., Posin, E. Y., Popov, E. G., Danilov, S. M., and Tkachuk, V. A., 1988, Influence of vasoactive agents on cytoplasmic free calcium in vascular endothelial cells, *J. Appl. Physiol.* **65:**2221–2227.

Shasby, D. M., Lind, S. E., Shasby, S. S., Goldsmith, J. C., and Hunninghake, G. W., 1985, Reversible oxidant-induced increases in albumin transfer across cultured endothelium: Alterations in cell shape and calcium homeostasis, *Blood* **65:**605–614.

Siflinger-Birnboim, A., Del Vecchio, P. J., Cooper, J. A., Blumenstock, F. A., Shepard, J. M., and Malik, A. B., 1987, Molecular sieving characteristics of the cultured endothelial monolayer, *J. Cell. Physiol.* **132:**111–117.

Smith, C. D., Cox, C. C., and Snyderman, R., 1986, Receptor-coupled activation of phosphoinositide-specific phospholipase C by an N protein, *Science* **232**(4746):97–100.

Stasek, J. S., Patterson, C. E., and Garcia, J. G. N., 1990, Cultured bovine pulmonary artery endothelium permeability is modulated by protein kinase C activation and cytoskeletal protein phosphorylation, *Clin. Res.* **38:**849 (Abstract).

Tahamont, M. V., and Malik, A. B., 1983, Granulocytes mediate the increase in pulmonary vascular permeability after thrombin embolism, *J. Appl. Physiol.* **54:**1489–1495.

Weksler, B. B., 1984, Prostaglandins and vascular function, *Circulation* **70**(III):63–71.

Weksler, B. B., Ley, C. W., and Jaffe, E. A., 1978, Stimulation of endothelial cell PGI$_2$ production by thrombin, trypsin and ionophore A23187, *J. Clin. Invest.* **62:**923–938.

Werth, D. K., Niedel, J. E., and Pastan, K., 1983, Vinculin, a cytoskeletal substrate of protein kinase C, *J. Biol. Chem.* **258:**11423–11426.

Wysolmerski, R. B., and Lagunoff, D., 1990, Involvement of myosin light-chain kinase in endothelial cell retraction, *Proc. Natl. Acad. Sci. USA* **87:**16–20.

Zimmerman, G. A., McIntyre, T. M., and Prescott, S. T., 1985, Thrombin stimulated the adherence of neutrophils to human endothelial cells *in vitro*, *J. Clin. Invest.* **76:**2235–2246.

INDEX

431